Introductory Combinatorics

Introductory Combinatorics

Third Edition

KENNETH P. BOGART
Dartmouth College

A Harcourt Science and Technology Company

San Diego   San Francisco   New York   Boston
London   Toronto   Sydney   Tokyo

This book is printed on acid-free paper.

Copyright © 2000, 1990, 1983 by Academic Press

Academic Press
A Harcourt Science and Technology Company
525 B Street, Suite 1900, San Diego, CA 92101-4495 USA
http:// www.academicpress.com

Academic Press
Harcourt Place, 32 Jamestown Road, London NW17BY
United Kingdom
http://www.hbuk.co.uk/ap/

Harcourt/Academic Press
200 Wheeler Road, Burlington, MA 01803 USA
http://www.harcourt-ap.com

Library of Congress Catalog Card Number: 99-68562

ISBN: 0-12-110830-9

PRINTED IN THE UNITED STATES OF AMERICA
99 00 01 02 03 IP 9 8 7 6 5 4 3 2 1

PREFACE

The subjects of discrete mathematics in general and combinatorial mathematics in particular have become increasingly important parts of the university mathematics curriculum. A clear reason for this can be found in the wealth of applications of combinatorial mathematics in computer science, operations research, statistics, and both the social and physical sciences. The rapid growth in the breadth and depth of the field of combinatorics in the last several decades, first in graph theory and designs and more recently in enumeration and ordered sets, has led to a recognition of combinatorics as a field with which the aspiring mathematician should become familiar.

The Scope of the Book

The purpose of this book is to present a broad comprehensive survey of modern combinatorics at an introductory level. The book begins with an introduction of concepts fundamental to all branches of combinatorics in the context of combinatorial enumeration. Chapter 2 is devoted to enumeration problems that involve counting the number of equivalence classes of an equivalence relation. Chapter 3 discusses somewhat less direct methods of enumeration, the principle of inclusion and exclusion and generating functions. The remainder of the book is devoted to a study of combinatorial structures. Chapters 4 and 5 provide a solid introduction to graph theory and algorithmic graph theory; with the exception of Section 3 of Chapter 4, most of the algorithmic graph theory is concentrated in Chapter 5. Chapter 6 is an introduction to experimental design, covering Latin squares, block designs, plane geometries, and enough coding theory to introduce the relationship between codes and designs. Chapter 7 introduces ordered sets, including interval orders, lattices, and Boolean algebras, from the points of view of various applications and concludes with Möbius inversion as an example of the interplay between combinatorial structures and the theory of enumeration. Chapter 8 is an extension of Chapters 2 and 3, beginning with a discussion of sufficient group theory to discuss counting orbits under the action of a group and concluding with an introduction to Pólya's theory of counting under the action of a group.

In choosing material for this book, I have aimed for core material of value to students in a wide variety of fields: mathematics, computer science, statistics, operations research, physical sciences, and behavioral sciences. A course from this book might be taught in departments of mathematics, computer science, operations research, or statistics, and so I have tried to write a book that can be taught by a mathematician or statistician, not necessarily with a discrete mathematics background, to a varied audience as well as by a computer scientist to a group of primarily computer science students.

A course in discrete mathematics could easily include material that is covered in courses in algebra, analysis of algorithms, data structures, digital circuit design, logic, probability, and the statistical design of experiments. Including a great deal of material from any of these areas would interfere with these well-established courses and lead away from combinatorics. Thus, I have only introduced ideas from these areas rather than devoted significant effort to them. Prior study from this book should enhance a student's performance in any of these courses. Since the book introduces applications to computer science, operations research, and statistics without going into detail in any of these subjects, I have been able to write it with the style and flavor of a traditional mathematics textbook. On the other hand, I have also tried to present the tantalizing algorithmic "aroma" that makes combinatorics so important to computer science and operations research.

Because the subjects we deal with in combinatorics are so concrete, and because the book is written in the style of a traditional mathematics textbook, this book should, if covered slowly and therefore selectively, be a good way for students to learn about proofs and become comfortable with mathematical abstraction. Thus it should enhance students' performance in more advanced mathematics courses, especially those that rely on ideas like mathematical induction and equivalence relations. In a department that wanted to develop a combinatorics course aimed at the transition between computationally oriented mathematics and abstract mathematics, this book could follow a rather computational linear algebra course and precede abstract courses in algebra and analysis.

Special Features of the Second Edition

In the second edition, new material was included throughout. The discussion of counting equivalence classes in Chapter 2 was reworked for clarity, completeness, and consistency of style. The treatment of generating functions in Chapter 3 was reworked, and although it still begins with Pólya's "picture writing," it begins with a more elementary version and leaves Pólya's approach to systems of simultaneous recurrence relations to a later section. A major revision in Chapter 4 was the inclusion of an entire section on the algorithmic theory of trees, including finding shortest paths and depth- and breadth-first search. This allowed the inclusion of material on orientation in the section on directed graphs and a treatment of backtracking algorithms as they arise naturally. The major revisions in Chapter 5 include an introduction to Edmonds' theory of minimum cardinality matchings in arbitrary graphs and a discussion of alternative network flow algorithms. Chapter 6 was made more comprehensive. It was revised to include a construction of 10-by-10 orthogonal Latin squares, both elementary and advanced techniques for building block designs up from smaller ones, affine plane geometries, and the coding theory mentioned earlier. Chapter 7 was revised to include an introduction to interval orders and an algorithm for finding minimum-sized chain decompositions of ordered sets. Chapter 8 was a new chapter on counting under the action of a group. New and interesting exercises and examples were added throughout and an appendix contains answers to odd-numbered exercises.

Special Features of the Third Edition

In this third edition the emphasis has been mainly on reworking material for clarity and correctness. The exercises and their answers were carefully revised. However, I have added optional material on Ramsey numbers and Catalan numbers in Chapter 1. As a result, students and instructors who would like to combine study of the fundamentals of combinatorics with a study of some of their nontrivial applications can do so. The instructor who would like to touch lightly on these topics or pass them by in the introductory part of the course may do so, but the instructor who would like to cover them more deeply will find exercises that challenge the student. (In the case of Ramsey numbers, the most challenging exercises come after Section 5 of Chapter 1, where Ramsey numbers motivate the (optional) use of double induction.) Chapter 8 has also been rewritten to develop Pólya Theory from the viewpoint of Pólya's idea of picture writing. This treatment reduces the reliance on group theory and is thus more elementary than the standard treatments. New exercises have also been added throughout the book. The appendix on matrix algebra has been removed; I am not aware of any courses that have used it in recent years.

In designing all editions of this book, I have tried to introduce a subject only if it has a major application or inherent mathematical interest and only if that application or interest can be explained in elementary terms. Virtually every section begins with a problem to be solved, develops tools to solve the problem, and then gives more examples of how the tools are used. At times, these motivating problems and examples are intentionally unsophisticated (e.g., distributing candy to children), so that the student does not carry the double burden of learning a sophisticated application and a new mathematical idea at the same time. As much as possible, definitions are introduced to name an idea that has come up in a motivating problem or example. Major terms being defined appear in bold italic type; somewhat lesser terms appear in italic type when they are defined. Theorems typically arise as answers to questions suggested by motivating problems or examples. Proofs are written as informal explanations. Unless there is a clear reference to a proof, skipping over it should not interfere with the flow of ideas, so the instructor—or reader—who chooses to deemphasize proofs should have no difficulty. On the other hand, proofs—especially those using induction—have been written carefully. Section 5 of Chapter 1 is a thorough treatment of mathematical induction so that students not familiar with it can learn it as part of the course. All proofs are designed to serve as models in courses intended in part to increase the student's mathematical sophistication. Proofs or topics that are either complex or unusually technical have been indicated with an asterisk; there is no harm in passing over such material. Exercises that depend on such material or are especially difficult are also marked with asterisks. The end of a proof or example is marked with a solid rectangle (■).

Courses That Can Use the Book

This book should be useful in a wide variety of courses. As the name suggests, the book is designed for introductory courses in combinatorics. As described later, it could also be used of a courses in discrete mathematics. I have been told of successful uses of the first edition in courses ranging from (honors) freshman level to beginning graduate level. At Dartmouth, we have used earlier versions of the book for an upper-division course. The prerequisites for this course are calculus through power series and matrix algebra, topics we teach in second-term calculus. Our audience has consisted of roughly equal numbers of students majoring in mathematics, computer science, and other subjects. Similarly, we have had roughly equal numbers of first-, second-, third-, and fourth-year students. The uses of matrix algebra in the book are concentrated in just a few places where they can be skipped over by the instructor whose students have not had matrix algebra. Although the book does make use of power series, it makes essentially no use of convergence and develops from scratch the algebraic properties of power series it uses. Thus I would have no hesitation using the book for students who have not been exposed to power series; however, for such students the algebraic manipulations of generating functions should be done rather slowly. Although virtually all the material in the book has been taught at one time or another, there is more material here than can be covered in one semester (unless the students have had a lower-division discrete mathematics course).

Typically we have covered a selection of material from at least six of the chapters. We often find it possible to use one of the chapters we have not covered as the basis for an independent project in which the student reads a chapter, chooses an appropriate selection of exercises, and works them out and turns them in. Chapters for which this has proved especially appropriate are Chapters 5 through 8. An instructor wishing to use Chapter 5 or 7 in this way may wish to cover Chapter 4 after Sections 1 and 2 of Chapter 2 in order to give students ample time to start their projects. Chapter 8 does use modular arithmetic, a topic developed in the first section of Chapter 6, and, because of the common algebraic emphasis, might be assigned in conjunction with the last section of Chapter 6. In this case, the first four sections of Chapter 6 could be covered any time after Section 2 of Chapter 2. Students who have already had a lower-division discrete mathematics course should be able to skip over most of Chapter 1 (a quick once-over to establish notation and review topics would be useful, and the optional material on Ramsey and Catalan numbers would be new to the student) and cover other sections quickly enough to finish much of the book in one term. Depending on the strength of the students' background, the instructor may have to choose between covering topics selectively and covering most of the book.

For some years this course substituted for a discrete mathematics course in the Dartmouth curriculum, and I have been told of other institutions where it is used in the same way. The result is a discrete mathematics course with a very combinatorial flavor, not a survey course or logic-oriented course that current discrete mathematics texts offer. There are two kinds of discrete mathematics courses that have been recommended by the committees of the Mathematical Association of America in recent years, an upper-division course and a lower-division course. It is straightforward to design a course covering the Committee on the Undergraduate Program in Mathematics (CUPM) outline for an upper-division discrete structures course from this book; in fact, most of the possible supplementary topics recommended by the CUPM are also covered in this book. The standard course in discrete structures in computer science typically covers the material in Chapter 1, the first two sections of Chapter 2, Chapter 3 (if a treatment of generating functions is desired), most of Chapter 4, and the first four sections of Chapter 7. Chapter 5 might be covered in such a course or might be left to a course in the design and analysis of algorithms. By covering Chapters 1–4 in one term and leaving Chapters 5–8 for a second term, it is possible to teach a two-term lower-division discrete mathematics course for highly prepared students from this book as well. The instructor would want to be selective in leaving out more sophisticated combinatorial topics throughout. Such a course could be profitably followed by courses in probability and linear algebra.

Dependences among the Sections

Although there is a natural progression in sophistication as one moves through the book, it has been designed with a minimum of dependence between chapters. Chapter 1 and the first two sections of Chapter 2 are fundamental to the rest of the book. Generally, each section in a chapter assumes familiarity with the preceding sections. However, in Chapter 4 the material on coloring and orientability is not used in subsequent sections and planarity occurs again only in connection with coloring, and in Chapter 7 the material on Boolean algebras is not used in subsequent sections. An outline of the other dependences of sections on earlier sections and prerequisite material is shown in Table P.1. Some exercises do not follow the pattern of the table; however, the statement of the exercise should make this clear.

Table P.1

Section	Depends directly on	Could benefit from
3.1		2.3
3.2–3.5	Power series	2.4, derivatives
3.4		Partial fractions
3.5	Exponential function	2.3
4.2		3.2–3.3
4.7	Matrix products	*Determinants
5.1	4.1–4.3	
5.2	4.3	
5.3–5.4	4.1–4.3, 4.5	
6.2	Matrix products, transposes	Determinants
6.4		Vector subspaces
6.5	Matrix products	Systems of equations
7.1	4.1–4.2, 4.5	4.3, 4.4
7.2		5.2–5.4
7.5–7.6	Matrix inverses; systems of equations	
8.1–8.3	6.1 (modular arithmetic)	
8.2		Matrix inverses, determinants

*Only asterisked subsection uses determinants.

Answers to Exercises

Answers and, in most complicated cases, complete solutions to virtually all the odd-numbered exercises are given in Appendix 1. In addition, there are selected solutions for a small number of even-numbered exercises that demonstrate a crucial point or are an essential part of a sequence of several exercises. At times when an even-numbered answer is given, an adjacent odd-numbered answer has been omitted for balance. Exercises that ask for computer programs do not have answers, although there may be a brief discussion of the appropriate algorithm to use in such an odd-numbered exercise. Most computer exercises could be used as the basis of a programming project.

Supplement

An instructor's manual with all answers and solutions is available from the publisher.

Acknowledgments

Many people have had an impact on this book as it has developed over the years. Several of my classes in combinatorics have cheerfully served as guinea pigs and offered valuable comments. Among the students in these classes, Drew Golfin deserves special thanks. Fred McMorris used an early version of the first edition in an experimental course at Bowling Green State University and Ed Scheinerman used a nearly final version of the second edition at Johns Hopkins. Their advice and experience have had a significant impact on the book. Robert Norman taught a course from the next-to-last draft of the first edition of the book at Dartmouth; Joe Bonin taught a course from the next-to-last draft of the second edition of the book at Dartmouth and read most of the book with great attention to detail. The enthusiastic advice and encouragement I have received from them and their students, as well as the many other faculty members and students who used the first edition of the book at Dartmouth, have also shaped the final product significantly. The reviewers of all three editions, some of them anonymous, have also influenced the book in many ways. Ronald D. Baker, Agnes Chan, Karen Collins, Daniel C. Coster, Jay Goldman, Eugene Lawler, Steven C. Locke, Steven Maurer, Anthony Ralston, Ed Scheinerman, David C. Sutherland, and Douglas B. West have each made special contributions. Several people, most notably Larry Langley, have helped me by checking most of the answers. Ernst Snapper gave me detailed comments about several sections of the second edition of the book, and Joe Bonin gave a detailed review of the entire second edition of the book. Their advice has had a major impact in shaping the third edition. Robin deGracia has given me an undergraduate's perspective on the exposition and suggested helpful changes throughout. Ruth Tucker Bogart's proof reading of the third edition at several different stages has helped me remove numerous bugs that might otherwise sting the reader.

Almost every author stands on the shoulders of those who have written before and in this I am no exception. All the books in the suggested reading lists are books that I have found valuable in my courses in combinatorics. While any list of names will undoubtedly have an important omission, I especially want to note that without what I have learned from the work of or personal associations with Martin Aigner, Claude Berge, Robert Dilworth, Jack Edmonds, Shimon Even, D. Ray Fulkerson, Curtis Greene, Jay Goldman, Marshall Hall, Jr., Frank Harary, Donald Knuth, Lásló Lovász, Robert Norman, Gian-Carlo Rota, Herbert Ryser, J. Laurie Snell, Robert Tarjan, and Herbert Wilf, this book would not be what it is and probably would not exist at all.

The preparation of the second edition was an adventure. It was a pleasure to see the additions and improvements to the text grow naturally from the foundation already there. It was also a pleasure to work with the editorial and production staff at Harcourt Brace Jovanovich as we learned how to use the Textures™ implementation of TEX and MacDraw™ in order to produce a high-quality book. Their patience, eagerness, and flexibility as we worked through the process were wondrous. In this third edition we have hopefully removed all the errors I inadvertently created by my mistakes in TEX in the second edition!

Finally, several members of the staff of Dartmouth's former Department of Mathematics and Computer Science made major contributions to the first two editions, from typing the first edition into computer files to converting these files to TEX format, adding the new material for the second edition, and preparing much of the art electronically. Without their help, the project might never have been completed. In thanks for their cheerful, helpful, and useful efforts, this book is dedicated to the staff of the Departments of Mathematics and Computer Science at Dartmouth College.

CONTENTS

2 *Equivalence Relations, Partitions, and Multisets*

3 *Algebraic Counting Techniques*

4 Graph Theory

5 *Matching and Optimization*

6 Combinatorial Designs

7 *Ordered Sets*

8 Enumeration under Group Action

1

An Introduction to Enumeration

Section 1   Elementary Counting Principles

What Is Combinatorics?

In combinatorial mathematics, we study how we may combine objects into arrangements. The objects might be computers; the arrangement might be a network connecting them. The objects might be headache relief tablets. They might be arranged into groups for testing on patients. Usually, the set of objects we wish to arrange is finite or the objects themselves are arranged among a finite number of sets. The problems studied are quite varied. Often, more than one kind of problem arises from a given application. One may ask if there are *any* arrangements at all satisfying certain conditions, or one may ask *how many* arrangements there are that satisfy these conditions. One may ask if all arrangements satisfying a given set of conditions must also satisfy certain additional conditions. Finally, one might ask for an efficient way to generate one arrangement or all arrangements satisfying certain conditions. Frequently, in combinatorial mathematics we attempt to represent an arrangement in concrete, but mathematical, terms in order to analyze it more easily.

The Sum Principle

We begin our study of combinatorial mathematics with elementary principles that show us how to count the number of arrangements that satisfy certain conditions or that lie in a certain set of arrangements built up from other sets. Suppose we have two sets A and B of objects (the objects might be arrangements of other objects). Their ***union***, $A \cup B$, is the set of objects in A or B or in both. Their ***intersection***, $A \cap B$, is the set of objects in both A and B. The union of a collection of sets is the set of objects *in at least one of the original sets*; the intersection of a collection is the set of objects *in every set of the collection*. Probably the most elementary "counting principle" deals with the union of a collection of sets no two of which have any elements in common. (In other words, the intersection of every pair of sets has no elements. We shall say the sets in such a collection are mutually **disjoint**.) The **sum principle** tells us that

> *the number of elements in the union of a finite collection of mutually disjoint sets is the sum of the numbers of elements in each of the sets.*

The illustrations in Figure 1.1 are called Venn diagrams. They show where the principle does and does not apply.

Figure 1.1   In each diagram the circles represent sets A and B chosen from some bigger set represented by the rectangle. Since the circles don't overlap in the first diagram, the sum principle applies. Since the circles representing A and B overlap nontrivially in the second diagram, the size of $A \cup B$ is not the sum of the sizes of A and B.

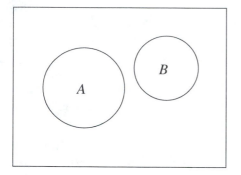 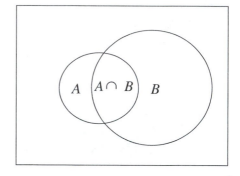

Example 1.1. The symbols that can be typed on a standard typewriter include the eight digits (2–9), 26 lowercase letters, 26 uppercase letters and 24 punctuation marks and special symbols. What is the total number of symbols available? It is obvious—and is a consequence of the sum principle—that the total number is $8 + 26 + 26 + 24 = 84$. ∎

By itself, the principle is so obvious as to be trivial. However, sometimes in the solution of a major problem, you can get started by asking yourself if the sum principle applies. If it does, divide the problem into cases, count the possibilities in each case, and add the results. Exercise 5 at the end of this section is an example where breaking a problem into cases is helpful.

At times, it will be useful to have a standardized notation for use with the sum principle. If we have two sets A and B, we use the symbol $A \cup B$ to stand for their union. For any set S, we use $|S|$ to stand for the size of S. Thus for two sets A and B, the sum principle says that if A and B are disjoint, then

$$|A \cup B| = |A| + |B| \,.$$

To deal with an arbitrary collection $A_1, A_2, \ldots, A_n$ of sets, we use either the notation

$$\bigcup_{i=1}^{n} A_i \quad \text{or} \quad A_1 \cup A_2 \cup \cdots \cup A_n$$

to stand for the union of the sets. Thus, for n sets $A_1, A_2, \ldots, A_n$ that are mutually disjoint, we can write either

$$|A_1 \cup A_2 \cup \cdots \cup A_n| = |A_1| + |A_2| + \cdots + |A_n|$$

or

$$\left| \bigcup_{i=1}^{n} A_i \right| = \sum_{i=1}^{n} |A_i| \,.$$

We read the second equation as "the size of the union from $i = 1$ to n of A_i is equal to the sum from $i = 1$ to n of the size of A_i."

Example 1.2. Suppose a club has 15 members and must choose a president and secretary–treasurer from among its members. In how many ways may it choose its officers?

We may regard a list of two names of club members as a choice first of a president and then of a secretary–treasurer. Since these lists are the arrangements we are interested in, we now ask how many such lists of two different club members there are. It may be obvious to some people that there are $15 \cdot 14 = 210$ such lists. (While this answer is correct, it is not obvious to everyone. A major reason why this book was written was so that one group of people would not have to arbitrarily accept what might be obvious to some other group. For this reason, will try to find ways to explain things from principles we understand.) This result also follows from an application of the sum principle. Suppose we divide the lists into piles (sets) such that each pile has all the lists with a given person as president and no other lists. Then we have 15 piles, one for each presidential candidate. Each pile has 14 lists because, once we choose a president, there are 14 people left for the second place on the list. Now, no two piles have any lists in common, so the number of elements in the union of the piles is the sum of the number of elements in each pile. Thus, we add 15 numbers, each equal to 14, to get a total of $15 \cdot 14 = 210$ lists in the union of all the piles. ∎

The Product Principle

The example suggests a second "counting principle," called the ***product principle***. One abstract statement of the product principle is that *a union of m disjoint sets each of size n has m·n elements.*

Clearly we can *prove* this principle by applying the sum principle as before. Another more concrete statement of the product principle suggested by our example says that

> *if the lists in a set of two-element lists have m possible first elements, and if for each first element, n different second elements are paired with it in lists in the set, then the set has m · n lists.*

Ordered Pairs

A list with two elements is normally called an ordered pair; to be more precise, an **ordered pair** consists of two objects, one called the first element of the pair and the other called the second. If the first element of the pair is a and the second element is b, we write (a, b) as the symbol for the ordered pair.

Example 1.3. A delicatessen offers a "simple sandwich" consisting of your choice of whole wheat, rye, or white bread filled with corned beef, roast beef, ham, or turkey. How many different simple sandwiches are possible?

Each sandwich may be represented as an ordered pair (bread, meat) with three choices for the first entry and four choices for the second. By the product principle, there are 12 sandwiches possible. ∎

With the same computation, we can prove another formula.

Theorem 1.1. If the set M has m elements and the set N has n elements, then there are mn ordered pairs whose first entry is in M and whose second entry is in N.

Proof. Similar to the computation of Example 2.1. ∎

Cartesian Product of Sets

The set of all ordered pairs whose first entry is in M and whose second entry is in N is sometimes called the **Cartesian product** of M and N and is denoted by $M \times N$. This notation is suggestive, since the size of $M \times N$ is mn. Once again we use $|S|$ to stand for the number of elements of a set S. In this notation, Theorem 2.1 says that $|M \times N| = |M| \cdot |N|$.

The General Form of the Product Principle

The concrete version of the product principle that we gave for lists takes more words than the abstract version, but it suggests a question we should ask: What if the lists have three or four elements? Does some kind of product principle apply?

Example 1.4. If the club in the preceding example has to choose a president, a secretary, and a treasurer, then the problem would ask us to count three-element lists. If the club needs a vice-president too, then the lists we would count are four-element lists. Our intuition suggests that the number of lists of four different people chosen from among the 15 club members is $15 \cdot 14 \cdot 13 \cdot 12$. ∎

This example suggests that a more general form of the product principle could deal with lists of length k. Then the principle we already know would be the special case for $k = 2$. The *general form of the product principle* can be stated as follows:

> *If the lists in a set S of lists of length k have the properties that*
> *(1) there are m_1 different first elements of lists in S;*
> *(2) for each way of specifying the first $i - 1$ entries of a list, there are m_i ways to specify the ith entry in the list;*
> *then S contains $m_1 \cdot m_2 \cdots m_k$ lists.*

(Later on we shall use the symbol $\prod_{i=1}^{k} m_i$ for the product from $i = 1$ to k of m_i.)

In Example 1.3, there are 15 different choices for president; and given a choice of president, there are 14 choices of vice-president; and once we have chosen a president and vice-president, there are 13 choices of secretary; and once we have chosen a president, vice-president, and secretary, there are 12 choices for treasurer. Then the product principle tells us that there are $15 \cdot 14 \cdot 13 \cdot 12$ ways to choose the four officers in the club. Just as we can represent possible selections of club officers or disk storage units as sets of lists, we can represent many practical or mathematical ideas as sets of lists, frequently as sets of ordered pairs. A list of n objects is sometimes called an n-tuple.

Lists with Distinct Elements

The product principle yields some useful formulas. Without these formulas, we would often need to appeal to the product principle to justify straight-forward computations. The symbol $n!$ is called n factorial and stands for the product of the first n positive integers; that is,

$$n! = n(n-1)\cdots 2 \cdot 1.$$

The symbol 0! is taken to mean the number 1.

> **Theorem 1.2.** The number of ways of listing all the elements of an n-element set (without repeating any elements) is $n!$.
>
> *Proof.* Theorem 1.2 is an immediate result of the product principle. ∎

A list of all the elements of a set (without repeats) is sometimes called a *permutation* of the set. A list of k distinct elements chosen from an n-element set S is called a **k-element permutation** chosen from S or sometimes a permutation of k elements chosen from S. If the member x of S is in the list, then we say x is a *member* of the permutation; if x occupies place j of the list then we say x is the jth member of the permutation. As in Example 1.3, we shall often need to know the number of k-element permutations of some set. The number of k-element permutations of an n-element set is denoted by the symbol $(n)_k$, called a **falling factorial** or *factorial power*. As we show in the next theorem, it is equal to $\frac{n!}{(n-k)!}$. This quantity is also sometimes denoted by $P(n, k)$ or $_nP_k$ (standing for "the number of permutations of n things taken k at a time").

> **Theorem 1.3.** The number of k-element permutations of an n-element set is
>
> $$n!/(n - k)! = (n)_k = n(n - 1) \cdots (n - k + 1) \quad \text{(for } k \leq n\text{)}.$$
>
> *Proof.* The proof is a natural application of the product principle. If $k = 1$, the number of lists in question is simply the number of elements in the set, which is $n = \frac{n!}{(n-1)!}$. For an arbitrary number k, there are n choices for the first element in the list, $n - 1$ choices for the second, $n - 2$ for the third, and so on down to $n - k + 1$ choices for the kth position. Thus, by the product principle, there are
>
> $$n(n - 1)(n - 2) \cdots (n - k + 1) = n!/(n - k)!$$
>
> lists. ∎

The preceding proof is somewhat distasteful because of the "and so on," especially since it comes at the point that requires the most thought, namely the reference to $n - k + 1$. By learning the principle of mathematical induction and becoming comfortable with it, we may replace such vague uses of "and so on" with clearer explanations. A later section of this chapter will be devoted to this principle; in that section we will derive the most general form of the product principle from the simpler version.

Lists with Repeats Allowed

Sometimes we will be interested in lists that are allowed to have repeat entries.

Example 1.5. The central processing unit of a computer must retrieve four files from disk storage units. Each file may be stored on any of the five different disk storage units connected to the central processing unit. In how many ways may the files be located on the disk storage units?

A description of where the files are located is a list that tells us which of the units has file 1, which has file 2, which has file 3, and which has file 4. Because there are five possibilities for each entry in the list, there are $5 \cdot 5 \cdot 5 \cdot 5 = 625$ possible arrangements of the files on the disk drives. ∎

> ***Theorem 1.4.*** The number of lists of length k, each of whose entries is chosen from a set with n elements (with repeats allowed), is n^k.

> *Proof.* For each position in the list, there are n choices, so the product principle tells us there are n^k lists. ∎

Stirling's Approximation for $n!$

A remarkable result due to James Stirling, appearing in his book *Methodus Differentialis*, published in 1730, is that for large values of n, $n!$ is approximately $\sqrt{2\pi n}(\frac{n}{e})^n$, where e is the base for the natural logarithm. More precisely, the ratio of $n!$ to this product approaches 1 as n increases (although the difference between $n!$ and this product increases as n increases). The approximation saves us some computation in getting estimates of $n!$ for large values of n since finding powers by successive squaring involves fewer computations than successive multiplications.

Example 1.6. To compute the approximation for 80! we would note that

$$\left(\frac{80}{e}\right)^{80} = \left(\frac{80}{e}\right)^{64} \cdot \left(\frac{80}{e}\right)^{16},$$

so we would compute the ratio $r = \frac{80}{e}$, find the second, fourth, eighth, sixteenth, thirty-second, and sixty-fourth powers of r in six multiplications, find r^{80} in one more multiplication, and multiply by the square root of 160π for a total of nine multiplications, including the multiplication by π. Also, we would have one square root extraction. ∎

More important, since $2^{n-1} < n! < n^{n-1}$ for $n > 2$, it is natural to ask whether, as a function of n, the growth rate of $n!$ is more like the exponential function 2^n or the function n^n. Stirling's approximation tells us that $n!$ is more like n^n in its growth rate than like 2^n. Stirling's formula also has important theoretical uses; it is used frequently, for example, in proofs in probability and statistics.

In the Exercises, we outline how to use elementary calculus to show that

$$\sqrt{2\pi n}\left(\frac{n}{e}\right)^n < n! < \sqrt{2\pi n}\left(\frac{n}{e}\right)^n\sqrt{1 + \frac{1}{2n}}.$$

This shows that the ratio of $n!$ to $\sqrt{2\pi n}(\frac{n}{e})^n$ does indeed approach 1 as n increases. More precise inequalities whose derivations use more advanced calculus techniques show that $n!$ is approximately $\sqrt{2\pi n}(\frac{n}{e})^n(1 + \frac{1}{12n})$. (In fact, it is just slightly larger than this approximation. See Volume 1 of Donald Knuth's *The Art of Computer Programming*, listed in the suggested reading for this chapter, for details.)

EXERCISES

1. A city council with seven members must elect a mayor and vice-mayor from among its members. In how many ways can the council choose these officers?

2. In how many ways can you distribute three distinct pieces of fruit among five children if no child can receive more than one? If any child can receive any number?

3. Imagine a large display of 10 different varieties of fruit in a grocery store.
   (a) In how many ways can five people choose one piece of fruit each?
   (b) In how many ways can five people choose one piece of fruit, each of a different type?
   (c) In how many ways can each person choose two different pieces of fruit each? (The same kind of fruit may be chosen by more than one person.)
   (d) In how many ways can each person choose two pieces of fruit if they need not be different?

4. A local restaurant offers a meat and cheese sandwich. You can choose one of three kinds of bread, one of four kinds of meat, and one of three kinds of cheese. How many sandwiches are possible?

5. Suppose the restaurant in Exercise 4 offers either a cheese sandwich or a meat sandwich or a meat and cheese sandwich. Now how many sandwiches are possible? What if we can also choose to have or not to have lettuce?

6. A 16-bit computer word is a list of 16 symbols, each a 0 or 1. How many such 16-bit computer words are there?

7. A hamburger shop offers you a "personalized hamburger." You can choose to have any of the following on your hamburger: lettuce, tomato, cheese, pickles, special sauce, catsup, mustard, onions. How many different kinds of personalized hamburgers can you choose? (Hint: One way to approach this is to use a list of 0's and 1's to describe the toppings on a sandwich.)

8. A coffee machine allows you to choose your coffee either plain or with single or double portions of sugar and/or cream. In how many ways can you choose your coffee?

9. What is the number of 10 digit numbers with no two successive digits equal? What is the number that have at least one pair of successive digits equal?

10. In order to hike along trails from Bing Mountain to South Mountain, you must either cross Joe's Peak or go through Narrow's Pass and cross Greene Mountain. There are two trails each from Bing Mountain to Joe's Peak and Narrow's Pass. There are three trails from Narrow's Pass to Greene Mountain, two trails from Greene Mountain to South Mountain, and two trails from Joe's Peak to South Mountain. In how many ways can you plan a hike from Bing Mountain to South Mountain?

11. An ice cream shop offers the following banana split. You get two scoops of ice cream, each having any of eight flavors. You get your choice of one of four toppings, whipped cream if you want it, and your choice of shredded coconut, chopped peanuts, or no nuts. All this sits on a split banana. (Do you think the order in which the scoops of ice cream sit on the banana matters?) How many different banana splits are possible?

12. A computer-aided instruction program asks a student each question on a list of 20 different questions. For each question, there are 10 choices for the answer. If the student answers a question correctly, the program says so and goes on to the next question. If the student gets the answer wrong, the program gives the student a second chance. If the answer is wrong the second time, the computer types the correct answer with a brief explanation; if the answer is right the second time, the computer says so and, in either case, goes on to the next question.
   (a) If we watch the process, observing only the computer's responses, how many different patterns could we conceivably see?
   (b) If we watch the process, observing both the computer's responses *and* the student's responses, how many outcomes could we conceivably see? (Note that this is considerably more complex than (a) unless you first divide the set of results for a given question into

those with a correct answer on the first try, those with a correct answer on the second try, and those with two incorrect answers, which is also a good strategy for part (a). This illustrates the earlier remark in the text that if you *search* for a way to apply the sum principle, you can make an apparently complex problem manageable.)

13. A professor has six test questions. Three of them form a unit and must be kept together in the order the professor has chosen. In how many ways can the professor arrange the test questions?

14. Answer the question of Exercise 13 if the three questions that are kept together can be arranged in any order.

15. Three reporters from the school newspaper and three reporters from the school radio station form the panel for a discussion. In how many ways can they be seated in a row behind a table if
(a) Each group must sit together?
(b) No two members of a group can sit together?

16. Use $Y - X$ to denote the set of elements of Y not in X. If every element of the set X is an element of the set Y, apply the sum principle to the sets X and $Y - X$ to explain why

$$|Y - X| = |Y| - |X|.$$

17. The intersection of two sets A and B, denoted by $A \cap B$, is the set of elements they have in common, and their union, $A \cup B$, is the set of all elements in one or the other set or in both of them. Explain why

$$|A \cup B| = |A| + |B| - |A \cap B|.$$

(Hint: You can find three disjoint sets whose union is $A \cup B$ by looking at part (b) of Figure 1.1. Problem 16 can be applied to compute the size of two of these sets.)

18. Stirling showed that $n!$ is approximately $\sqrt{2\pi n} n^n e^{-n}$. Using a computer or scientific calculator, determine the ratio of $n!$ to this approximation for values of n equal to multiples of 10 up to 50. For each Stirling approximation, multiply it by $1 + \frac{1}{12n}$ and find the ratio of the result to $n!$.

19. Approximately how many digits does 272! have?

20. In this exercise, we define a sequence c_n by the formula

$$c_n = n!/n^{n+1/2} e^{-n}.$$

In the next exercise, we show that $\lim_{n\to\infty} c_n$ exists; however, that method will not let us compute c_n. Here we show that the limit must be $\sqrt{2\pi}$.

(a) Use integration by parts to show that

$$m \int \sin^m t \, dt = -\sin^{m-1} t \cos t + (m-1) \int \sin^{m-2} t \, dt$$

and

$$\int_0^{\pi/2} \sin^m t \, dt = \frac{m-1}{m} \int_0^{\pi/2} \sin^{m-2} t \, dt.$$

(b) By applying the formula in part (a), first with $m = 2n$, then $2n-2$, etc., derive the formula

$$\int_0^{\pi/2} \sin^{2n}(t) \, dt = \frac{\prod_{i=1}^n (2i-1)}{\prod_{i=1}^n 2i} \frac{\pi}{2}.$$

(c) Derive the formula

$$\int_0^{\pi/2} \sin^{2n+1} t \, dt = \frac{\prod_{i=1}^n 2i}{\prod_{i=1}^n (2i+1)}.$$

(d) Show that $\prod_{i=1}^n (2i+1) = \frac{(2n+1)!}{2^n n!}$. Express $\prod_{i=1}^n 2i$ in terms of factorials in a similar way.

(e) From parts (b), (c), and (d) derive the inequalities

$$\frac{(2^n n!)^2}{(2n)!(2n+1)} \leq \frac{\pi}{2} \frac{(2n)!}{(2^n n!)^2}$$

and

$$\frac{\pi}{2} \frac{(2n)!}{(2^n n!)^2} \leq \frac{(2^n n!)^2}{(2n)!(2n)}.$$

(f) Substitute $cn^{n+1/2} e^{-n}$ for $n!$ in part (e) to show that the limit c of the sequence c_n defined at the beginning of the problem satisfies the inequalities

$$\sqrt{2\pi} \leq c \leq \sqrt{2\pi \left(1 + \frac{1}{2n}\right)}$$

and explain why Stirling's approximation follows.

21. In this exercise we illustrate why it is natural to *expect* to find a constant c such that $n!$ is approximately $cn^{n+1/2} e^{-n}$. One standard technique for

analyzing products of numbers is to take logarithms in order to convert the products to sums. We see that

$$\log(n!) = \log\left(\prod_{i=1}^{n} i\right) = \sum_{i=1}^{n} \log(i).$$

Sums of this form frequently arise in the study of integrals; if we were to experiment with Riemann sums, trapezoid rule sums, midpoint rule sums, etc., for the integral of the logarithm, we would see that each of these involves a similar sum. Thus to approximate the sum in the best way possible, we want to take the variation of the sum that gives the best approximation to the integral and somehow figure out how to replace the sum by the integral. Since we will be doing calculus, the logarithm that will make our work easiest is the natural logarithm, the logarithm to the base $e = 2.71828\ldots$.

(a) Show that the approximation obtained to $\int_1^n \log(x)\,dx$ by the trapezoid rule using trapezoids whose vertical sides are the vertical lines through $x = 1, 2, 3, \ldots, n$ is $\sum_{i=1}^{n} \log(i) - \frac{1}{2}\log(n)$.

(b) Show that the approximation to the integral $\int_{3/2}^{n+1/2} \log(x)\,dx$ that we obtain by adding areas of trapezoids whose tops are above the curve and tangent to the graph of $y = \log(x)$ at the integer points $(i, \log(i))$ and whose vertical sides are the vertical lines through $x = \frac{3}{2}, \frac{5}{2}, \frac{7}{2}, \ldots, \frac{2n+1}{2}$ is $\sum_{i=1}^{n} \log(i)$. (This uses the "tangent rule" for approximation of integrals.)

(c) Explain why

$$\sum_{i=1}^{n} \log(i) \leq \int_1^n \log(x)\,dx + \frac{1}{2}\log(n) \leq \frac{1}{2}\log\left(\frac{3}{2}\right) + \sum_{i=1}^{n} \log(i).$$

(Would the inequality remain true if $\leq$ were replaced by $<$? The point to this question is that it is perfectably acceptable to say $a \leq b$ even when you know that the more precise statement $a < b$ is true.)

(d) Show that if d_n (for difference number n) is defined by

$$d_n = \int_1^n \log(x)\,dx + \frac{1}{2}\log(n) - \sum_{i=1}^{n} \log(i),$$

then d_n is an increasing sequence bounded above by $\frac{1}{2}\log(\frac{3}{2})$, so that d_n has a limit d.

(e) Use integration by parts to show that

$$d_n = \left(n + \frac{1}{2}\right)\log(n) - n + 1 - \sum_{i=1}^{n}\log(i) .$$

(f) Show that if d is the limit in part (d), then

$$e^{d-1} = \lim_{n\to\infty}\frac{n^{n+1/2}e^{-n}}{n!} .$$

(g) What numerical bounds does this place on the limit e^{d-1} of part (f)? How do these compare numerically with the value $\sqrt{2\pi}$ computed in the previous exercise?

Section 2 Functions and the Pigeonhole Principle

Functions

The idea of a function is central in all mathematics. Typically, in discussions for mathematical beginners, you will see the phrase "f is a function from A to B" defined by the statement, "f is a rule which associates with each element x of A a unique element $f(x)$ of B." For most purposes, this definition is entirely adequate. However, if we ask ourselves "How do we tell whether two functions are really different?" we see a small problem.

The rule $f(x) = x + 2$ clearly describes a function from $\{-1, 0, 1, 2\}$ to $\{1, 2, 3, 4\}$. The rule given by $g(x) = x^4 - 2x^3 - x^2 + 3x + 2$ is also a rule for a function from $\{-1, 0, 1, 2\}$ to $\{1, 2, 3, 4\}$. In fact, $f(x) = g(x)$ for each x in the domain set $\{-1, 0, 1, 2\}$. Thus, we have two apparently different rules that describe the same function rather than two different functions. If we agreed upon a standardized way of stating the relationship that the rule specifies, then we could say that two functions are the same if they describe the same relationship. One way to describe the relationship that f specifies is to write down the set $\{(-1, f(-1)), (0, f(0)), (1, f(1)), (2, f(2))\}$ of all ordered pairs whose first entry is a domain element and whose second entry is the related range element which the function rule assigns to the first entry. For f we get

$$f = \{(-1, 1), (0, 2), (1, 3), (2, 4)\}$$

and for g we get

$$g = \{(-1, 1), (0, 2), (1, 3), (2, 4)\}.$$

In other words, by using ordered pairs we have found an unambiguous way to describe functions. We will call the set of ordered pairs associated with some function the *relation* (short for relationship) of the function. A way to visualize this relation, called a "directed graph" or "digraph" of the relation, is shown in Figure 2.1.

Figure 2.1 A way of visualizing the relation of the function given by $f(x) = x + 2$.

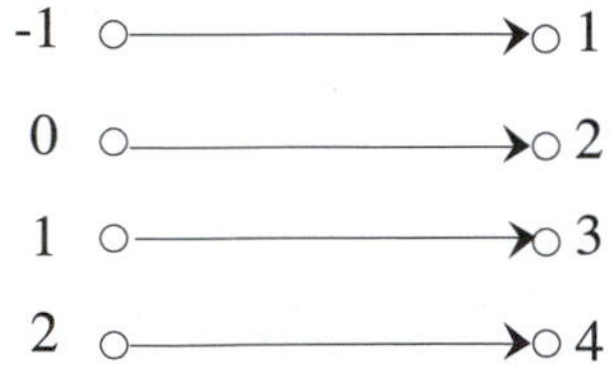

The idea of using a set of ordered pairs to describe a relationship is useful in many situations, so we will try to make this idea more specific.

Relations

We define a **relation** *from* a set A *to* a set B to be a set of ordered pairs whose first entries are in A and whose second entries are in B. A is called the *domain* of the relation and B is called the *range*. If (a, b) is an ordered pair in a relation, we say that the relation *associates* b with a.

Example 2.1. The set $S = \{(1,2),\ (1,3),\ (1,4),\ (2,3),\ (2,4),\ (3,4)\}$ is a familiar relation. What is the usual description of the relation and what are its domain and range?

An ordered pair (a, b) is in S if and only if $a < b$. Thus, we would ordinarily say S is the "less than" relation. There is more than one answer to the question about domain and range. For example, we could agree to let the domain A and range B both be the set $\{1, 2, 3, 4\}$. However, we could use $\{1, 2, 3\}$ for A and $\{2, 3, 4\}$ for B without changing our informal description that S is the "less than" relation. Figure 2.2 shows how to visualize this relation. ∎

Figure 2.2   A directed graph for visualizing the relation of Example 2.1.

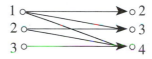

Definition of Function

Now we *define* a **function** from A to B to be a relation from A to B that associates with each element a of A a unique element b of B. We frequently use a letter such as f to stand for a function and write $f(x)$ for the element f associates with x. We say f *maps* x to $f(x)$. (Can you find reasons why the relation of Example 2.1 is not a function?) The definition of a function we just gave has an intuitive advantage. If we use just our intuitive understanding that the word "relation" (or "relationship") means about the same as the word "rule," then our new definition of function is the same as the old definition we gave using the word "rule." If, on the other hand, we use the definition of a relation as a set of ordered pairs, we have a very precise and unambiguous definition of a function. With this definition, it makes perfectly good sense to ask how many functions there are from a finite set A to a finite set B.

The Number of Functions

Theorem 2.1. If A has k elements and B has n elements, then there are n^k functions from A to B.

Proof. List the elements of A in some order, say $a_1, a_2, \ldots, a_k$. Then each function f from A to B gives a list of k elements chosen from B, namely $f(a_1), f(a_2), \ldots, f(a_k)$. Also each list of k elements chosen from the set B, say $b_1, b_2, \ldots, b_k$, gives a function, namely the one whose relation is

$$\{(a_1, b_1), \ (a_2, b_2), \ldots, \ (a_k, b_k)\}.$$

Clearly, different lists give different functions, so the number of functions from A to B is the same as the number of lists of k elements chosen from B (with repeats allowed). By Theorem 1.3, there are n^k such lists; thus there are n^k functions from A to B. ∎

A standard notation for the set of all functions from A to B is B^A; in this notation Theorem 2.1 may be expressed as

$$|B^A| = |B|^{|A|} .$$

Example 2.2. You have five different chairs to paint and four different colors of paint. In how many ways can you paint the chairs?

Assigning a color of paint to each chair gives a function from the set of chairs to the set of paints. Different functions yield different paint jobs. Thus, by Theorem 2.1, there are $4^5 = 1024$ possible paint jobs. ∎

In the proof of Theorem 2.1, we used a natural correspondence between functions and lists. This might be slightly disconcerting since the concept of function has been defined precisely and the concept of list has not. In fact, we can use the concept of a function to define the concept of a list precisely. We define a *k-element list* or *k-tuple* chosen from a set S to be a function from the set $K = \{1, 2, \ldots, k\}$ to the set S. We say that x is the jth *member* of the list if x is the element assigned to j. In this context, the product principle for lists becomes a principle that we accept, without proof, for certain sets of functions as we already did for sets of lists. However, with this more precise definition of a list, we will actually be able, later on, to derive the product principle for lists of length n from the product principle for lists of length two.

One-to-One Functions

A function f is called a ***one-to-one function*** or ***injection*** if it associates *different* values in the range with different values in the domain. A common example of a one-to-one function is given by $f(x) = x^3$ on the set of real numbers. On the other hand, the function with the rule $f(x) = x^2$ is not a one-to-one function on the domain of real numbers because, for example, $f(-1)$ and $f(1)$ are the same real number. The function illustrated in Figure 2.1 is a one-to-one function.

Notice that in these terms a k-element permutation chosen from a set B is a one-to-one function from the set $K = \{1, 2, \ldots, k\}$ to the set B. Figure 2.1 may help you visualize the concept of a one-to-one function. Note that each arrow points to a different number. None of the paint jobs in Example 2.2 corresponds to a one-to-one function because with five chairs and four colors of paint, two chairs will have to have the same color. This is one example of the *pigeonhole principle*. Informally, the principle says that if more than n pigeons are to be placed in n pigeonholes, one hole will end up with more than one pigeon. Another way to say the **pigeonhole principle** is that there are no one-to-one functions from a k-element set to an n-element set if $k > n$.

On the other hand, if $k \leq n$ there is a straightforward formula using the falling factorial function we introduced in Section 1 for the number of one-to-one functions from a k-element set to a n-element set.

Theorem 2.2. If $k \leq n$, there are $(n)_k = \frac{n!}{(n-k)!}$ one-to-one functions from a k-element set A to an n-element set B.

Proof. As in Theorem 2.1, we may associate lists of k elements chosen from B with functions from A to B. To say that some function is one-to-one is to say that the associated list is a list of distinct elements of B. By Theorem 1.2, the number of lists is $\frac{n!}{(n-k)!} = (n)_k$. ∎

Example 2.3. If five chairs are to be painted different colors and seven colors are available, how many ways are there to paint the chairs?

Since the paint jobs correspond to one-to-one functions, there are

$$7!/(7-5)! = 7!/2! = 7 \cdot 6 \cdot 5 \cdot 4 \cdot 3 = 2520$$

possible paint jobs. ∎

Example 2.4. If seven chairs are to be painted and five colors are available, then at least two of the chairs must be painted the same color by the pigeonhole principle. ∎

Applications of the pigeonhole principle can be much more subtle than the last example.

Example 2.5. The powers of two are all even and the powers of five all end in five. Show that for any other prime p, among the first six powers of p, one of these numbers must end in one.

If a prime is not two, then no power of it is even, so a power can only end in one of the five odd digits. Then the function given by

$$f(i) = \text{the last digit of } p^i$$

from $\{1, 2, 3, 4, 5, 6\}$ to the set of digits $\{1, 3, 5, 7, 9\}$ cannot be one-to-one. Therefore p^i and p^j have the same last digit for some i and j between 1 and 6 with $i < j$. Thus $p^j - p^i$ ends in a zero, so it is a multiple of 10. Thus $p^i(p^{j-i} - 1)$ is a multiple of 10, and since neither 2 nor 5 is a factor of p, $p^{j-i} - 1$ must be a multiple of 10 (and therefore end in a zero), so adding 1 to each side tells us that p^{j-i} ends in a 1. ∎

Notice that the empty set of ordered pairs satisfies the conditions that make it a function—in fact, a one-to-one function—from the empty set to a set B. It will be easier to be consistent in our later terminology and formulas if we agree to consider the empty set of ordered pairs to be a function from the empty set to any set B, a one-to-one function from the empty set to any set B, and a zero-element permutation of any set B.

Onto Functions and Bijections

A function f from A **onto** B is defined as a function from A to B such that each member of B is associated with at least one member of A. A function from A onto B is also called a **surjection** from A to B. In Figure 2.3 we show a directed graph of a function from $\{-2, -1, 0, 1, 2\}$ *onto* the set $\{0, 1, 4\}$ and a directed graph of a function from $\{-2, -1, 0, 1, 2\}$ which is *not onto* the range set $\{0, 1, 2, 3, 4\}$. Both functions have the relation given by $f(x) = x^2$.

Figure 2.3   A way of visualizing a function that is onto its range and a function that is not.

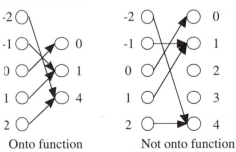

Onto function   Not onto function

As part (b) Exercise 14 (where functions assign candy to children) illustrates, the computation of the number of functions from A onto B is more involved than the kinds of computations we have made so far. A one-to-one function from a set A onto a set B is called a **bijection**. The function illustrated in Figure 2.1 is a bijection. Bijections play a fundamental role in combinatorial mathematics. A bijection from a set onto itself is usually called a **permutation** of that set. In Section 1, we defined a permutation of a set as a listing of that set. Given a set listed in one way, say (a, b, c, d), a second list such as (b, c, a, d) determines a one-to-one function from the set onto itself, in this case $\{(a, b), (b, c), (c, a), (d, d)\}$. Thus, the two uses of the word "permutation" are closely related. Typically the first definition is used when there is some intuitively understood and naturally preferred way to list the set (as with alphabetical order for $\{a, b, c, d\}$) and so the list really does correspond to a function. Since both uses of the word are common, it is necessary to determine from the context whether a particular use of the word "permutation" refers to a listing of the set or a one-to-one function from a set onto itself.

The concept of a bijection underlies the concept of size. The intuitive idea that we should capture in a definition of size is that we count a set by choosing one element and saying "one," choosing a different element and saying "two," and continuing in this way until we have chosen every element of our set. As we count a set in this way we are setting up a bijection between the set and a set of consecutive integers beginning with 1. Thus we define a set A to have **size** n if there is a bijection from the set $N = \{1, 2, \ldots, n\}$ onto the set A. We refer to this bijection as a *counting function*.

We have already used the relationship between bijections and size in earlier work. In the proof of Theorem 2.1, we described (without explicitly saying so) a bijection between the set of functions in question and a corresponding set of lists. Then on the basis of that bijection we claimed that the two sets had the same size. This is the essence of the **bijection principle:**

> *Two sets have the same size if and only if there is a bijection between them.*

Not only is the bijection principle clear on intuitive grounds, it also follows quickly from our definition of size. For this purpose, we introduce the idea of composition of functions. To see what the idea is, suppose two sets A and B each have size k and we wish to find a bijection between them. There are bijections f_1 from A onto K and f_2 from K onto B. We now define a new function f from A to B by $f(x) = f_2(f_1(x))$; intuitively it is clear that f is a bijection from A to B. The function f is called the *composition* of f_2 with f_1. In general, if $f : A \to B$ and $g : B \to C$, then the **composite function** $g \circ f$ is defined by $g \circ f(x) = g(f(x))$.

Theorem 2.3. Suppose f is a function from A to B and g is a function from B to C. Then $g \circ f$ is a function from A to C and
    (1) If f and g are one-to-one, then so is $g \circ f$.
    (2) If f and g are onto, then so is $g \circ f$.
    (3) If f and g are bijections, then so is $g \circ f$.

Proof. By definition, $g \circ f$ is a relation from A to C. To see that $g \circ f$ is a function, observe that for each x, there is one and only one value $f(x)$ related to it, and since $f(x)$ is in B, the domain of g, there is one and only one $g(f(x)) = g \circ f(x)$ related to it. Thus, $g \circ f$ is a function. Suppose for the proof of 1 that f and g are one-to-one. Then, since g is one-to-one, for each $y = g \circ f(x)$ there is one and only one value z in B such that $y = g(z)$; this z must be $f(x)$ since $f(x)$ is such a value. But since f is one-to-one, x is the only value that f relates to z, so x must be the only value that $g \circ f$ relates to y. Thus $g \circ f$ is one-to-one. Part 2 is dealt with in a similar fashion to part 1 by applying the definition of onto, and part 3 follows from 1 and 2. ∎

The relation of having the same size is symmetric; that is, if A has the same size as B, then B has the same size as A. Thus, for the bijection principle to be valid, if there is a bijection from A to B, then there must be a bijection from B to A as well.

Theorem 2.4. If f is a bijection from A to B, then the relation f^{-1}, defined by $f^{-1} = \{(y, x) | (x, y) \in f\}$, is a bijection from B to A.

Proof. This follows from Exercises 15 and 16. It is also straightforward to prove by verifying that the relation f^{-1} defined is a function, that it is one-to-one, and that it is onto. ∎

The function f^{-1} defined in the theorem is called the ***inverse*** of f. Inverses of functions are discussed in detail in the Exercises.

Corollary 2.5. Two sets A and B have the same size if and only if there is a bijection between them.

Proof. If A and B have the same size n, then there is a bijection f from A to $N = \{1, 2, \ldots, n\}$ and a bijection h from B to N. By Theorem 2.3, there is a bijection g from N to B; by Theorem 2.4, $g \circ f$ is a bijection from A to B.

    Now if there is a bijection f from A to B and B has size n, then $g \circ f$ is a bijection from A to N so that A has size n also. ∎

The Extended Pigeonhole Principle

If we are painting 13 chairs and have only three colors of paint, then at least one color must be used on five or more chairs. While it is certainly clear intuitively, there is also a natural explanation using counting principles. If each color were used on four or fewer chairs, then the set of painted chairs would be a union of three disjoint sets (one for each color), each of size at most four. Then by the product principle, at most 12 chairs would be painted.

As with the pigeonhole principle, this may be expressed in the language of functions. Our function that assigns colors to chairs is a function from the 13-element set of chairs to the three-element set of colors; our conclusion is that there is some color that the function assigns to at least five different chairs. Thus the function is "constant" on this set of chairs. Although this use of the word constant seems strange at first (constancy seems to imply a discussion of numbers), to say a function is constant on a set simply means its value is the same for each element of the set. The preceding analysis using the product principle may be used to prove the **extended pigeonhole principle** that follows.

> **Theorem 2.6.** If f is a function from a set of $mk + 1$ elements to a set of m elements, then there is a set of at least $k + 1$ elements of the domain on which f is constant.
>
> *Proof.* Exercise 26. ∎

Ramsey Numbers

In a group of n people, either they all know each other or else there are two who don't know each other. Since this is obvious, it is perhaps surprising that it takes some thought to show, as Exercise 27 requests, that in a group of six people, there are either three mutual acquaintances or three mutual strangers. How many people would we need in a group to ensure that there are either four mutual acquaintances or three mutual strangers? Experimentation is a good way to work on such a question, so let's experiment with twice as many people as we said would guarantee either three mutual acquaintances or three mutual strangers.

Example 2.6. Show that among 12 people, there are either four mutual acquaintances or three mutual strangers.

Let's single out one person, say Tom. Now there are 11 other people, so by the extended pigeonhole principle, either six of them are acquaintances of Tom or else six of them are strangers to Tom. (Here m was 2, k was 5, giving us our set of $2 \cdot 5 + 1$ people, our function went from these 11 people to the set {acquaintance, stranger} giving us either $5 + 1$ acquaintances, or $5 + 1$ strangers.) We will examine these two cases separately.

Case 1. If Tom has six acquaintances, then, by Exercise 27, either three of these people are mutual acquaintances, giving us, with Tom included, four mutual acquaintances, or three of these people are mutual strangers, giving us the number of mutual strangers we asked for.

Case 2. If six people are strangers to Tom, then either two of these six people are strangers to each other, giving us, with Tom included, the three mutual strangers we asked for, or all six people know each other, giving us even more than the four mutual acquaintances we asked for.

Thus in both cases we have either four mutual acquaintances or three mutual strangers. ∎

The analysis in case 2 of the example suggests that in that case we need only have four people who are strangers to Tom. Thus it is natural to continue our experimentation by asking if 10 people will work in place of 12.

Example 2.7. Show that among 10 people, there are either four mutual acquaintances or three mutual strangers.

As before, if we choose one person, Tom, then by the sum principle, among the remaining nine people, there are either six who are acquaintances of Tom or four who are strangers to Tom. (Just how is the sum principle being used here?) We divide the problem into two cases, case 1 being identical to the previous exercise.

In case 2, there are four people who are strangers to Tom. Either two of the four people are strangers, giving us, with Tom, three mutual strangers, or all four are acquaintances, giving us four mutual acquaintances. ∎

The number of people we need to have in order to have either m mutual acquaintances or n mutual strangers is called the **Ramsey number** $R(m, n)$. We have now seen that $R(m, 2) = m$, and $R(4, 3) \leq 10$. Exercise 27 is equivalent to showing that $R(3, 3) \leq 6$. In the constext of Exercise 27, Exercise 28 is equivalent to showing that $R(3, 3) = 6$. It is natural to expect from the smooth way Example 2.6 worked out that $R(4, 3) = 10$ as well; however $R(4, 3) = 9$ as Exercises 29–31 show. By interchanging the roles of acquaintance and stranger, we get $R(2, n) = n$, and $R(3, 4) = 9$. Remarkably few Ramsey numbers are known; as of this writing $R(3, n)$ is known for $n \leq 9$, $R(4, 4) = 18$, and $R(4, 5) = 25$.

Although describing Ramsey numbers in terms of acquaintances and strangers is intuitively appealing, it also has the potential for ambiguity. (Jan knows who Jill is, but not the reverse, etc.) In the exercises of Section 6 of Chapter 4 we will explore a precise way to describe Ramsey numbers using the ideas of graphs. This approach has the advantage of appealing to our geometric intuition.

*Using Functions to Describe Ramsey Numbers

For those who have an algebraic intuition we can also use the idea of a function in our definition. To see how, suppose we define a function on a set of pairs of distinct people by $f(P_1, P_2) = A$ if P_1 and P_2 are acquainted and $f(P_1, P_2) = S$ if they are strangers. Our function will be *symmetric* in that $f(P_1, P_2) = f(P_2, P_1)$. With this in mind we give a formal definition of Ramsey numbers as follows.

The **Ramsey number** $R(m, n)$ is the smallest number k such that if f is a symmetric function from the pairs of distinct elements of a k-element set K to the two element set $\{A, S\}$, then there is either an m-element set M of elements of K such that $f(x, y) = A$ for every two elements x and y in M or else an n-element set N of elements of K such that $f(x, y) = S$ for every two elements x and y in N.

EXERCISES

1. Write down all the sets of ordered pairs that correspond to possible functions from $\{0, 1\}$ to $\{1, 2, 3\}$.

2. Write down all the sets of ordered pairs that correspond to possible functions from $\{1, 2, 3\}$ to $\{0, 1\}$.

3. Write down all the sets of ordered pairs that correspond to possible relations from the set $\{0, 1\}$ to the set $\{a, b\}$.

4. Write down the set of ordered pairs that corresponds to the "greater than" relation from the set $\{1, 2, 3, 4\}$ to the set $\{1, 2, 3, 4\}$.

5. Let $f(x) = x^2 - 1$. Write down the set of ordered pairs that is the relation f if the domain of f is $\{1, 2, 3, 4, 5\}$.

6. Do Exercise 5 with domain $\{-3, -2, -1, 0, 1, 2, 3\}$.

7. Draw pictures like those in Figures 2.1 and 2.2 to represent the functions of the last two problems. Which of these functions is one-to-one?

8. Which of the following relations are functions? If not, why not?
   (a) $\{(a, 1), (b, 2), (c, 1)\}$
   (b) $\{(1, a), (2, b), (3, b), (1, c)\}$
   (c) $\{(-1, 1), (0, 0), (1, 1), (2, 4), (-2, 4)\}$
   (d) $\{(1, -1), (0, 0), (1, 1), (4, 2), (4, -2)\}$
   (e) $\{(0, 0), (1, 1), (4, 2)\}$

9. On the set of integers from 1 to 12 inclusive, there is a relation containing (a, b) if a is a factor of b (but not b itself). Write down the set of ordered pairs corresponding to this relation.

10. How many functions from a four-element set to a five-element set are not one-to-one? How many functions from a five-element set to a four-element set are not one-to-one?

11. A computer must assign each of four jobs to one of 10 different slave computers.

(a) In how many ways can it do so?

(b) In how many ways can it make the assignments if no slave is to get more than one job?

12. A group organizing a faculty–student tennis match must match five faculty volunteers with five of the 12 students who volunteered to be in the match. In how many ways can they do this?

13. The chairperson of a department has to assign advisors to three seniors from among 10 faculty members in the department. How many different assignments are there?   How many give no faculty member three advisees? How many give no faculty member two or three advisees?

14. (a) In how many ways can you pass out 10 different pieces of candy to three children?

(b) What if each child must get at least one piece? (Hint: To answer this question, ask yourself in how many distributions does one child get candy? In how many distributions do exactly two children get candy?)

15. The inverse of a relation R, denoted by R^{-1}, is the set of all ordered pairs (y, x) such that (x, y) is in R. Show that the inverse relation of a function f is a function (whose domain is some subset of the range of f) if and only f is one-to-one. Show that the inverse relation of a function f is a function whose domain is the range of f if and only if f is a bijection.

16. The inverse of a function is defined in the previous exercise. Show that the inverse of a bijection is a bijection. How does this prove Theorem 2.4?

17. (a) There is a function from the set of subsets of $N' = \{1, 2, \ldots, n-1\}$ to the subsets of $N = \{1, 2, \ldots n\}$ given by $f(S) = S \cup \{n\}$. Is this function one-to-one? Is this function onto?

(b) There is a function from the set of subsets of $N = \{1, 2, \ldots n\}$ to the subsets of $N' = \{1, 2, \ldots, n-1\}$ given by $f(S) = S \cap N'$. Is this function one-to-one? Is this function onto?

18. (a) Consider the function from the set of lists of k elements chosen from $\{1, 2, \ldots, n\}$ to the set of lists of $k+1$ elements chosen from $\{1, 2, \ldots, n\}$ for which $f(L)$ consists of the k elements of L followed by n. Is this function one-to-one? Is it onto?

(b) Consider the function from the set of lists of k elements chosen from $\{1, 2, \ldots, n\}$ to the set of lists of $k+1$ elements chosen from $\{1, 2, \ldots, n\}$ for which $f(L)$ consists of the k elements of L followed by the first element of L. Is this function one-to-one? Is it onto?

    (c) Consider the function from the set of lists of $k+1$ elements chosen from $\{1, 2, \ldots, n\}$ to the set of lists of k elements chosen from $\{1, 2, \ldots, n\}$ for which $f(L)$ is the list obtained from L by deleting its last element. Is this function one-to-one? Is it onto?

19. Give a proof of Theorem 2.4 without appealing to Exercises 15 and 16.

20. Show that if S and T are of the same size, then a function $f\colon S \to T$ is onto if and only if it is one-to-one. Discuss what this means about testing a function between two sets of the same size to see if it is a bijection.

21. Prove part 2 of Theorem 2.3.

22. Show that if A has the same size as B and B has the same size as C, then A has the same size as C.

23. (a) Explain why we may conclude in Example 2.5 that, among the first five powers of p, one of these numbers must end in one.
    (b) May we make the same conclusion as in part (a) about the first four powers of p? Why or why not?
    (c) May we make the same conclusion as in part (a) about the first three powers of p? Why or why not?

24. Show that if a prime is not 2 or 5, then one of its powers must have 01 as its last two digits. How many powers do you need to guarantee this? Explain why, for each prime different from 2 or 5 and for each n, some power of the prime ends with a string of n zeros and a one.

25. Show that for each positive integer n there is a number divisible by n whose only decimal digits are zero and one.

26. (a) Prove the extended form of the pigeonhole principle.
    (b) Show that for any set of seven integers at least three of the positive differences between them have the same last digit.

27. In a group of six people, there is either a set of three people all of whom know each other or a group of three people none of whom know each other. Why?

28. Show that it is possible to choose acquaintances among the five people Al, Bo, Ed, Jo, Mo so that no three people are mutually acquainted and no three people are mutual strangers.

29. Show (assuming as we have that being acquainted is a symmetric relationship) that among an odd number of people, someone is acquainted with an even number of people and is thus a stranger to an even number of people.

30. By the previous exercise, in a set of nine people, at least one of them, say Pam, is acquainted with an even number of people and is a stranger

to an even number of people. Show that either Pam and two other people are mutual strangers, some group of four people are acquainted, or three people with whom Pam is acquainted are mutual strangers. What does this tell you about $R(4,3)$?

31. Define a symmetric function f from the pairs of distinct elements of $\{1,2,3,4,5,6,7,8\}$ to $\{A,S\}$ so that there is neither a four-element subset of $\{1,2,3,4,5,6,7,8\}$ so that $f(x,y) = A$ for all elements x and y of that subset nor a three-element subset such that $f(x,y) = S$ for all elements x and y of that subset. What does this say about $R(4,3)$?

*32. Using the pigeonhole principle, show that in a list of $n^2 + 1$ distinct numbers, there are either $n + 1$ numbers (not necessarily consecutive) in increasing order or $n+1$ numbers in decreasing order. (For example, in the list 1, 5, 3, 4, 2, we have both the increasing list 1, 3, 4 and the decreasing lists 5, 4, 2 and 5, 3, 2.)

Section 3   Subsets

The Number of Subsets of a Set

We say a set A is a *subset* of a set B (written $A \subseteq B$) if each element of A is an element of B. A typical way to visualize the subset relation is shown in Figure 3.1.

Figure 3.1

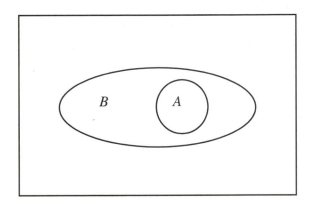

As we saw in Section 2, it is convenient to imagine a set called the *empty set*, denoted by $\emptyset$, which has no elements whatsoever. We say that $\emptyset \subseteq A$ for every set A. To help you visualize the empty set, suppose that the law requires the energy committee of the U.S. Senate to have a subcommittee on the extraction of oil from oil shale, but the head of the energy committee refuses to appoint anyone to the subcommittee. Then the set of members of the subcommittee is the empty set.

Theorem 3.1. The number of subsets of an n-element set is 2^n.

Proof. Let the n-element set be given by $N = \{1, 2, \ldots, n\}$. To decide on a subset A, for each element of N we decide whether or not it is in A. This gives a list of n choices, each choice being yes or no. By the product principle, there are 2^n such lists. ∎

Subsets and functions are intimately related. For each subset A of $N = \{1, 2, \ldots, n\}$, we define the function f_A by

$$f_A(x) = \begin{cases} 1 & \text{if } x \in A \\ 0 & \text{if } x \notin A. \end{cases}$$

Think of 1 as meaning "yes" and 0 as meaning "no."

The function f_A is called the **characteristic function** of A. Note that two different subsets A and B will have different characteristic functions f_A

and f_B. Thus, each subset A of N determines a unique function from N into the set $\{0, 1\}$.

Now suppose f is a function from N to $\{0, 1\}$. Let A be the set of all x in N such that $f(x) = 1$. Then $f = f_A$. Thus, every function from N to $\{0, 1\}$ is an f_A for some subset A of N. This one-to-one correspondence between subsets and functions gives us a second proof that the number of subsets of N is 2^n, because 2^n is the number of functions from N to $\{0, 1\}$. Notice how this proof uses the bijection principle by establishing a one-to-one correspondence between subsets and functions. We will frequently use the bijection principle without explicitly referring to it. Look back, for example, at the proof of Theorem 3.1, and you will see a one-to-one correspondence between sets and lists of yeses and noes.

Example 3.1. A psychology professor asks for volunteers from a nine-person class to participate in a perception experiment. How many different groups could the professor get?

Since any subset of the class (including the empty set!) could be the set of volunteers, there are $2^9 = 512$ possible experimental groups. ∎

Frequently we will be interested in subsets of a set of a certain size.

Example 3.2. The professor in Example 3.1, realizing that there could be no volunteers, decides instead to choose three students and assign the experiment to them as a special project. How many different experimental groups can the professor choose?

First, the number of distinct lists of three different names that can be chosen from the names of all the nine students is $9!/6! = 9 \cdot 8 \cdot 7 = 504$. However, each three-element subset occurs in $3! = 6$ different lists. Thus if s is the number of sets of three names that can be chosen, $6 \cdot s$ is, by the product principle, the number of lists. Therefore,

$$6 \cdot s = 504$$

so that $s = 504/6 = 84$ is the number of ways of choosing three names from among all nine possible names. ∎

Binomial Coefficients

We use the symbol $\binom{n}{k}$ or $C(n; k)$ to stand for the number of k-element subsets of an n-element set. It is standard to read either symbol as "n choose k." (Another common notation that we will not use is $_nC_k$.) The usual name for these quantities is *binomial coefficient*; the reason for the name will become apparent later. $C(n; k)$ is also called "the number of combinations of n things taken k at a time," but we shall not use this terminology. Note that since a set with n elements has no subsets of negative size or of size larger than n, $\binom{n}{k}$ is 0 when $n \geq 0$ but $k < 0$ or $k > n$. In a later chapter, we will define the symbol for negative values of n.

k-Element Subsets

Theorem 3.2. If $0 \leq k \leq n$, then the number of k-element subsets of an n-element set is given by the formula

$$\binom{n}{k} = \frac{n!}{k!\,(n-k)!}.$$

Proof. The set of all lists of k elements chosen from the n-element set contains $\frac{n!}{(n-k)!}$ lists. However, each k-element subset can be listed in $k!$ ways. The number of ways first to choose a k-element subset and then to list the elements of that subset is, by the product principle,

$$\binom{n}{k} \cdot k!.$$

However, each of these lists is a different list of k elements chosen from among all the n elements, and each list of k distinct elements arises from choosing a subset and listing it in a certain order. Thus

$$\binom{n}{k} \cdot k! = \frac{n!}{(n-k)!}$$

which yields

$$\binom{n}{k} = \frac{n!}{k!(n-k)!}. \qquad \blacksquare$$

Labelings with Two Labels

There is another interpretation of $\binom{n}{k}$ that leads to a useful generalization analogous to our generalization of the product principle from ordered pairs to lists. Suppose we have a set of n elements and two kinds of labels: k labels of one kind and $(n-k)$ labels of the other kind. In how many ways can we assign these labels to the elements of the set so that each element is labeled?

Example 3.3. We have seven different chairs and we are supposed to paint two of them blue and five of them red. In how many different ways can we paint the chairs?

Once we select the two chairs to paint blue, we have determined the color of the rest of the chairs also, so the number of different paint jobs is the number of ways of choosing two objects out of seven, namely $C(7; 2) = 7!/(5! \cdot 2!) = 21$. Of course, we might have chosen the red chairs first, but we should end up with the same number of paint jobs anyhow. In fact, our formula says $C(7; 5) = 7!/(2! \cdot 5!) = 21$, just as it must. ∎

Two theorems are immediately suggested by Example 3.3; their proofs consist of picking out the ideas displayed in the example and are left as exercises.

Theorem 3.3. The number of ways to label n objects with k labels of one kind and $(n - k)$ labels of a second kind is $\binom{n}{k}$.

Proof. Exercise 15. ∎

A labeling of a set N with the labels 0 and 1 (or the labels "out" and "in") assigns the number 1 to k elements of N and the number 0 to the other $n - k$ elements of N. Thus, the labeling gives us a function f from N to $\{0, 1\}$. This is the same as the "characteristic function" description given for subsets following the proof of Theorem 3.1. In Section 4 we will generalize binomial coefficients to count labelings using more than two labels, much as we generalized the product principle from pairs to lists.

The other fact Example 3.3 suggests is that the number of labelings with k labels of type 1 and $n - k$ of type 2 must be the same as the number of labelings with k labels of type 2 and $n - k$ of type 1.

Theorem 3.4. $\binom{n}{k} = \binom{n}{n-k}$.

Proof. Exercise 16. ∎

Pascal's Triangle

To obtain the number of 15-element subsets of a 25-element set, our formula would have us compute

$$\frac{25!}{15! \cdot 10!} = \frac{25 \cdot 24 \cdots 16}{10 \cdot 9 \cdot 8 \cdot 7 \cdots 1}.$$

The numerator in this fraction will have 10 factors whose average value is approximately 20, so it would be "close" in size to

$$20^{10} = 2^{10} \cdot 10^{10} = 1024 \cdot 10^{10} \doteq 10^{13}.$$

(The symbol "$\doteq$" stands for "approximately equal.") The calculation of such a large product involves considerable effort, with considerable chance for error. On the other hand, a rough estimate of $C(25;15)$ can be obtained by replacing each term in both the numerator *and* the denominator by an approximate average value. We get

$$C(25;15) \doteq \frac{20^{10}}{5^{10}} = 4^{10} = 2^{20} = (1024)^2 \doteq 10^6.$$

Thus, the number $C(25;15)$, although large, is not itself so large as to be difficult to calculate with. This suggests that we should search for a method of computing the numbers $C(n;k)$ by building them up from smaller numbers.

An array of numbers called **Pascal's triangle** gives us such a method for computing binomial coefficients. Table 3.1 was created by listing all nonzero values of $\binom{0}{k}$ in one row, next listing all nonzero values of $\binom{1}{k}$ in the next row, then all values of $\binom{2}{k}$ and so on, so each row starts right below where the last one started. Thus in row n and column k, we find the number $\binom{n}{k}$.

Table 3.1

Row number	Column number						
	0	1	2	3	4	5	6
0	1						
1	1	1					
2	1	2	1				
3	1	3	3	1			
4	1	4	6	4	1		
5	1	5	10	10	5	1	
6	1	6	15	20	15	6	1

Table 3.2, usually called Pascal's triangle, is formed from Table 3.1 by removing the row and column labels and arranging the rows so that their centers line up instead of their leftmost entries. Table 3.2 has two important features.

Table 3.2

```
            1
          1   1
        1   2   1
      1   3   3   1
    1   4   6   4   1
  1   5  10  10   5   1
1   6  15  20  15   6   1
```

First, the left and right sides of the table consist entirely of 1's. This corresponds to the fact that $C(n;0) = C(n;n) = 1$. Not quite so obvious is the fact that each entry not on either the left or right border is the sum of the two elements above and to the left and right of it. This suggests an important theorem about binomial coefficients. To understand what the theorem says, we first note that the corresponding observation about Table 3.1 is that each entry is the sum of a pair of entries in the row directly above it, with the first entry of the pair in the same column and the second in the preceding column. The kth entry in row n is thus apparently the sum of entry $k - 1$ and entry k in the row above, which is a list of the binomial coefficients $C(n - 1; h)$. This suggests the relationship, which we call the *Pascal relation*, in Theorem 3.5.

Theorem 3.5. $\binom{n}{k} = \binom{n-1}{k-1} + \binom{n-1}{k}$ whenever $0 < k < n$.

Proof. The formula says that the number of k-element subsets of an n-element set is the sum of the number of $(k - 1)$-element subsets *and* the number of k-element subsets of an $(n-1)$-element set. This is what we now prove. Divide the k-element subsets of $N = \{1, 2, \ldots, n\}$ into two collections of sets; collection 1 consists of those subsets containing n, and collection 2 consists of those subsets not containing n. By the sum principle, the number of k-element subsets is the sum of the sizes of collection 1 and collection 2.

We will show that the size of collection 1 is $\binom{n-1}{k-1}$. To do so, let $N' = \{1, 2, \ldots, n - 1\}$. We will describe a bijection between the sets in collection 1 and the $(k - 1)$-element subsets of N'. We let each set S in collection 1 correspond to the set $S' = S - \{n\}$ consisting of the elements in S other than N. This is a $k - 1$-element subset of N'. This correspondence is one-to-one because if we remove n from two different k-element sets containing n, we get two different sets. Now suppose we have an $n - 1$-element subset T' of N'. We must show that it corresponds to some set in N. However let T be the set $T' \cup \{n\}$. This is a k element set containing n. But when we remove n from it, we clearly get T'. Therefore our correspondence is onto. Thus by the bijection principle, the number of sets in collection 1 is the same as the number of $(k - 1)$-element subsets of N'. Since this number is $\binom{n-1}{k-1}$, we have shown that the size of collection 1 is $\binom{n-1}{k-1}$.

Each set in collection 2 is a k-element subset of N', and each k-element subset of N' is in collection 2. Thus, by the bijection principle, the number of sets in collection 2 is the number of k-element subsets of N', which is an $(n - 1)$-element set. Therefore $C(n; k) = C(n - 1; k - 1) + C(n - 1; k)$. ∎

Example 3.4. Find $C(8,3)$.
From Table 3.2, $C(6;1) = 6, C(6;2) = 15$, and $C(6;3) = 20$. Thus $C(7;2) = 6 + 15 = 21, C(7;3) = 15 + 20 = 35$ and therefore $C(8;3) = 21 + 35 = 56$. ∎

In our statement of the Pascal relation, we assumed $0 < k < n$. In fact, since the number of n-element subsets of an $(n-1)$-element set is zero,

$$C(n;n) = C(n-1;n-1) + C(n-1,n).$$

Similarly, a set of size n has no subsets of size -1, so that

$$C(n;0) = C(n-1;0) + C(n-1;-1).$$

It may seem silly to define $C(n;k)$ when k is negative or when $k > n$, but as these examples show, having such definitions sometimes makes it easier to state results without having to give restrictions that keep us from dealing with annoying special cases. Thus we have defined $C(n,k)$ to be 0 for $n \geq 0$ but $k < 0$ or $k > n$, saying, in effect, that an n-element set has no subsets of negative size and no subsets of size larger than n. As we shall see later, there is another natural way to define $C(n,k)$ when n is negative.

How Fast Does the Number of Subsets Grow?

We have observed that the total number of subsets of an n-element set is 2^n. An exponential function of n, such as this one, grows very quickly. To see the practical implications of this, suppose we had five pieces of fruit and were trying to find a fruit selection whose weight was as close as possible to being an integer. We would try each of the 2^5 subsets on our scale and pick the one that best fit our criterion. With 20 pieces of fruit, the same problem would require that we examine $2^{20} = (1024)^2 > 1,000,000$ sets of fruit to find the one we wanted. On the other hand, if we wanted a selection of three pieces of fruit, we would have to examine $\binom{5}{3} = 10$ sets of fruit in the first case and $\binom{20}{3} = 1140$ sets of fruit in the second. Although large, this number is not unfeasibly large. We see that the total number of subsets of a set grows quite dramatically as the size increases while the number of three-element subsets grows less dramatically. Why is this? Since $\binom{n}{3} = \frac{n(n-1)(n-2)}{3!}$, this second number is a third-degree polynomial function of n rather than an exponential function of n. Although polynomial functions can grow quickly, their growth rates are dwarfed by those of exponential functions. We can see that the number of subsets of size i is always a polynomial of degree i, assuming we have fixed i. Unfortunately, if i can increase as n increases, then the number of subsets of size i might no longer be a polynomial.

If, in our fruit selection example, the subset we wanted could contain half the fruit, then the number of subsets of the n pieces of fruit we examine would be $\binom{n}{n/2}$, assuming n is even. To estimate how big this number is, we can use Stirling's approximation for n factorial. Doing so gives us

$$\binom{n}{n/2} = \frac{n!}{(n/2)!(n/2)!}$$

$$\approx \frac{n^n e^{-n}\sqrt{2\pi n}}{(1/2)^{n/2}n^{n/2}e^{-n/2}\sqrt{\pi n}\,(1/2)^{n/2}n^{n/2}e^{-n/2}\sqrt{\pi n}} = 2^n \frac{\sqrt{2}}{\sqrt{\pi n}}.$$

Thus, the number of subsets of size $n/2$ grows exponentially with n; in fact, it is the fraction $\sqrt{\frac{2}{\pi n}}$ of the total number of subsets of an n-element set.

Recursion and Iteration

The formula for computing the numbers $C(i;k)$ given in Theorem 3.5 is called a *recursion formula*, because the numbers $C(j;k)$ with j smaller than i *recur* (which we may somewhat carelessly think of as short for reoccur) in the computation of $C(i;k)$. The regularity of the computation lets us develop an *iterative method* or *iterative algorithm* for computing row i of the Pascal triangle. First write down $C(0;0) = C(1;0) = C(0;1) = 1$. This gives us rows 0 and 1. Then apply Theorem 3.5 and the rules $C(n;0) = 1$, $C(n;n) = 1$ to get row 2. Repeat (or *iterate*) this process until you reach row i. We shall use the phrase *iterative method* to describe a computation of a function that repeats a process again and again for increasing values of one or more parameters. A *recursive method*, on the other hand, is one that computes a function by starting with the parameter values for which the function is desired, attempting to evaluate the function in terms of the same function with smaller parameter values, setting the main problem aside while dealing with these problems with smaller parameter values in like fashion, and then using these newly computed values in the main computation.

Example 3.5. Given the recursion formula $n! = n \cdot (n-1)!$ for $n > 0$ and the value $0! = 1$, show an iterative and a recursive computation of $3!$.

To compute $3!$ iteratively, we compute

$$1! = 1 \cdot 0! = 1 \cdot 1 = 1$$
$$2! = 2 \cdot 1! = 2 \cdot 1 = 2$$
$$3! = 3 \cdot 2! = 3 \cdot 2 = 6 \ .$$

To compute $3!$ recursively, we compute

$$3! = 3 \cdot 2! = 3 \cdot (2 \cdot 1!) = 3 \cdot (2 \cdot (1 \cdot 0!))$$
$$= 3 \cdot (2 \cdot (1 \cdot 1)) = 3 \cdot (2 \cdot 1) = 3 \cdot 2 = 6 \ . \qquad ∎$$

Although the difference between iteration and recursion is small in the example, it is significant when applied to a computation with the Pascal relation, as seen in Exercises 33 and 34.

EXERCISES

1. Suppose you have a final exam with eight questions. The instructions state "Choose six of the eight questions and answer them in the order given." In how many ways can you choose the questions you will answer in order to complete the exam?

2. Write out all subsets of the set $\{a, b, c\}$.

3. How many two-element subsets does the four-element set $\{A, B, C, D\}$ have? List them.

4. A candy store stocks 10 different kinds of chocolate candy. In how many ways can someone select a bag of six pieces of candy, each piece of a different kind?

5. A test has two sections of five questions each. The instructions say to answer three questions in each section. In how many ways can a student choose the questions to answer in order to complete the exam?

6. Write down row 7 of Pascal's triangle (this means the row corresponding to subsets of a seven-element set).

7. Some rows of Pascal's triangle consist of even numbers with the exception of the two entries at the ends. Show that row 8 has this property. Do any rows beyond row 8 (that is, with $n > 8$) have this property? Explain why not or give an example.

8. In how many ways can a three-person executive committee and a four-person administrative committee be chosen from a 20-member club if the committees can overlap? What if the membership of the committees cannot overlap?

9. In how many ways can a 20-person club select a president, a vice-president, a secretary, a treasurer, and a three-person social committee if the officers cannot be on the committee? What if the officers can be on the committee? Suppose that one of the three people on the social committee is to be designated as the chair. Then what is the total number of selections in each of the two previous cases?

10. A test has three sections of four questions each. According to the instructions, the student must answer two questions from each of any two sections and one question from the other section. In how many ways can a student choose questions to answer?

11. A gift basket consists of two of four different kinds of cheeses, two of five different kinds of tinned meats, five of eight different kinds of fancy fruits, two of three different kinds of crackers, and three of six different kinds of cookies. In how many ways can the gift basket be completed?

12. In how many ways can a two-, a three-, and a four-element set containing all the elements of a nine-element set be chosen?

13. In how many ways can the committees of Exercise 8 be chosen if they can have at most one member in common?

14. A local ice cream shop offers the following banana split. You get three scoops of ice cream, each having any of eight flavors. You get your choice of one of four toppings, whipped cream if you want it, and your choice of shredded coconut, chopped peanuts, or no nuts. All this sits on a split banana. (Assume that the order in which the scoops of ice cream sit on the banana does not matter.) How many different banana splits are possible? (Compare this exercise with Exercise 11 of Section 1.)

15. Prove Theorem 3.3.

16. Prove Theorem 3.4.

17. (a) Give two proofs, only one of which uses the formula

$$(n)_k = n!/(n-k)!,$$

that the number of k-element permutations of an n-element set satisfies the recurrence

$$(n)_k = (n-1)_k + k(n-1)_{k-1}.$$

To get a proof that does not use the formula, think about what the expression on the left-hand side means and how that relates to the meanings of the expressions on the right-hand side.

   (b) Use this recurrence to write down the first five rows of a table of values of $(n)_k$.

18. Find (and prove the correctness of) a formula similar to the Pascal formula for the number $f(n,k)$ of functions from the n-element set of the first n integers onto a k-element set by considering whether or not another element has the same image as n.

19. For each of the following questions indicate whether it can be answered by using the formula for the number of k-tuples, the number of k-element permutations, or the number of k-element subsets of an n-element set or none of these.

(a) In how many ways can k distinct pieces of fruit be passed out to n children, if any child can get any number of pieces of fruit?

(b) In how many ways can n distinct pieces of fruit be passed out to k children, if any child can get any number of pieces of fruit?

(c) In how many ways can k distinct pieces of fruit be passed out to n children, if no child can get more than one piece of fruit?

(d) In how many ways can n distinct pieces of fruit be passed out to k children, if no child can get more than one piece of fruit?

(e) In how many ways can k identical apples be passed out to n children, if any child can get any number of pieces of fruit?

(f) In how many ways can n identical apples be passed out to k children, if any child can get any number of pieces of fruit?

(g) In how many ways can k identical apples be passed out to n children, if no child can get more than one piece of fruit?

(h) In how many ways can n identical apples be passed out to k children, if no child can get more than one piece of fruit?

20. (a) How many paths are there along horizontal or vertical lines through the integer points on the axes, always moving to the right or up, from the point $(1,2)$ to the point $(9,7)$?

(b) Find an appropriate general statement of what you discovered in part (a).

21. Prove the formula $n\binom{n-1}{k-1} = k\binom{n}{k}$ without using the formula for $\binom{n}{j}$. (Hint: Think about choosing a committee and a chair for that committee. The two sides of the equation represent two ways to do this.)

22. Prove the formula $\binom{n}{m}\binom{n-m}{k} = \binom{n}{k}\binom{n-k}{m}$ without using the formula for $\binom{n}{j}$. Hint: Consider pairs of subsets.

23. Prove the formula $\binom{n}{m}\binom{m}{k} = \binom{n}{k}\binom{n-k}{m-k}$ without using the formula for $\binom{n}{j}$ or the formula of the preceding exercise.

24. Prove the formula

$$\sum_{j=0}^{k} \binom{m}{j}\binom{n}{k-j} = \binom{m+n}{k}$$

without using the formula for a binomial coefficient.

*25. Prove the formula

$$\sum_{j=0}^{m} \binom{m}{j}\binom{n}{k+j} = \binom{m+n}{m+k}$$

without using the formula for a binomial coefficient.

26. If the version of Pascal's triangle in Table 3.1 were to have rows -4, -3, -2, and -1, what would row -4 have to be in order for the recurrence relation for binomial coefficients to give us the correct result for row 0? (A 1 in the 0,0 position with zeros to the right of it.) Assume that $\binom{n}{0} = 1$ for negative n as well as positive n. (What happens to the number of answers to the question if we don't make this assumption?)

*27. Prove the formula

$$\binom{0}{m} + \binom{1}{m} + \binom{2}{m} + \cdots + \binom{n}{m} = \binom{n+1}{m+1}$$

without using the formula for a binomial coefficient.

28. What value of k makes $\binom{n}{k}$ a maximum for a given value of n? Prove this by describing precisely the conditions under which $\binom{n}{m}$ is larger than or smaller than $\binom{n}{m+1}$.

29. Use Stirling's formula and a calculator to check the approximation we gave to $\binom{25}{15}$ by using average values. Did we have the correct number of digits (to the left of the decimal place)?

30. Use Stirling's formula to see whether the method of replacing a factorial in the numerator and denominator with a power of its average term gives a reasonable approximation to the value of $\binom{100}{50}$. In particular, estimate the difference in the number of digits (to the left of the decimal place) in the two approximations.

31. Write a brief paragraph outlining which values of $C(j; k)$ you *must* compute if you want to find $C(m; n)$ by the iterative method of Pascal's triangle. (For example, to compute $C(5; 2)$, you need not compute $C(4; 3)$ or $C(4; 4)$. The values of $C(j; k)$ you *do* have to compute include $C(4; 1)$, $C(4; 2)$, $C(3; 0)$, $C(3; 1)$, $C(3; 2)$ and some others. Try to make a general statement rather than just giving examples.)

32. Write a computer program that computes binomial coefficients by the iterative method of Pascal's triangle. Use your program to compute $C(30; 5)$ and $C(30; 15)$.

33. If you are familiar with a computer language that permits the use of recursive algorithms, attempt Exercise 32 by means of recursion. Compare the amounts of computer resources used by the two methods.

34. If you are familiar with the concept of recursion in computer languages or computer algorithms, discuss the advantages or disadvantages of using the Pascal relation as the basis of a recursive program (or algorithm) for computing the first n rows of Pascal's triangle versus

using the Pascal relation as the basis of an iterative program (or algorithm) for computing the first n rows of Pascal's triangle.

*35. On the basis of Exercise 7, make a conjecture about rows of even or odd numbers in Pascal's triangle and prove it.

Section 4 Using Binomial Coefficients

The Binomial Theorem

Our analysis of subsets and labelings makes it easy to prove a fundamental theorem of elementary algebra, the **binomial theorem**. It is because of their use in expanding the power of a binomial that the binomial coefficients get their name.

Theorem 4.1. For any integer $n \geq 0$,

$$(x+y)^n = \sum_{i=0}^{n} C(n; i) x^i y^{n-i}$$

This formula may also be written as

$$(x+y)^n = \sum_{i=0}^{n} \binom{n}{i} x^i y^{n-i} = \sum_{j=0}^{n} \binom{n}{j} x^{n-j} y^j.$$

Proof. $(x+y)^n$ is a product of n factors, each equal to $x+y$. We multiply them together by repeatedly choosing one of x or y from each factor, multiplying the choices together, and then adding the results from all possible sequences of choices. For example, by choosing x from all the terms, we get x^n in the product. By choosing x in one term and y in the rest, we get the product xy^{n-1}; this product occurs n times, once for each possible term from which we select x. Selecting x from i terms and y from the other $n-i$ terms amounts to labeling i of the terms with an x and $n-i$ of the terms with a y. Thus, there are $C(n; i)$ ways in which the product $x^i y^{n-i}$ occurs in the result. Adding the terms $C(n; i) x^i y^{n-i}$ that result gives

$$\sum_{i=0}^{n} C(n; i) x^i y^{n-i}.$$

We have now proved the equality in the first line of the statement of the Theorem. The first equality in the last line of the statement of the theorem is just a translation of notation; the last equality that appears in the statement of the theorem follows from the substitution of j for $n-i$ and the use of the fact that $\binom{n}{j} = \binom{n}{n-j}$. ∎

Note that the formula we get for $(x+y)^2$ from the first equality in Theorem 4.1 is $(x+y)^2 = y^2 + 2xy + x^2$. Ordinarily, we would write down

40

$(x+y)^2 = x^2+2xy+y^2$. The last equality in Theorem 4.1 gives the expansion for $(x+y)^2$ in the order we are used to. Our first two examples review how the binomial theorem is used in high school algebra. The importance of the binomial theorem in combinatorics is illustrated by Examples 4.3 and 4.4.

Example 4.1.

$$
\begin{aligned}
(2x+3)^4 &= \sum_{k=0}^{4} C(4;k)(2x)^k 3^{4-k} \\
&= C(4;0)(2x)^0 3^4 + C(4;1)(2x)^1 3^3 + C(4;2)(2x)^2 3^2 \\
&\qquad + C(4;3)(2x)^3 3^1 + C(4;4)(2x)^4 3^0 \\
&= 81 + 4 \cdot 2 \cdot 27x + 6 \cdot 4 \cdot 9x^2 + 4 \cdot 8 \cdot 3x^3 + 1 \cdot 16 \cdot x^4 \\
&= 81 + 216x + 216x^2 + 96x^3 + 16x^4.
\end{aligned}
$$
∎

Example 4.2.

$$
\begin{aligned}
(x-y)^6 &= \sum_{k=0}^{6} C(6;k)x^k(-y)^{6-k} \\
&= \sum_{k=0}^{6} C(6;k)(-1)^{6-k}x^k y^{6-k} \\
&= y^6 - 6xy^5 + 15x^2y^4 - 20x^3y^3 + 15x^4y^2 - 6x^5y + x^6,
\end{aligned}
$$

using the last row of the Pascal triangle, Table 3.2, rather than computing each number $C(6;k)$ individually. ∎

Example 4.3. Surprising formulas sometimes come from applying the binomial theorem to simple sums. For example, let's see what happens when we expand $(2-1)^{10}$.

$$
\begin{aligned}
(2-1)^{10} &= \sum_{k=0}^{10} C(10;k)2^k(-1)^{10-k} \\
1^{10} &= \sum_{k=0}^{10} C(10;k)2^k(-1)^{10}(-1)^{-k} \\
&= \sum_{k=0}^{10} C(10;k)2^k(-1)^{10}(-1)^{k}
\end{aligned}
$$

This gives us

$$1 = (-1)^{10} \sum_{k=0}^{10} C(10;k)(-2)^k$$

$$1 = (-1)^{10} \left(\binom{10}{0} 2^0 - \binom{10}{1} 2^1 + \binom{10}{2} 2^2 - \cdots + \binom{10}{10} 2^{10} \right)$$

$$1 = 1 - 2 + \binom{10}{2} 2^2 - \cdots + 2^{10},$$

giving us a rather strange-looking way to write the number 1. ∎

As you might guess from the preceding example, the binomial theorem is a good tool to try for discovering and proving formulas involving a sum of binomial coefficients.

Example 4.4. What formula for binomial coefficients corresponds to the factorization

$$(x+1)^m (x+1)^n = (x+1)^{m+n} \,?$$

Since

$$(x+1)^m (x+1)^n = \sum_{i=0}^{m} \binom{m}{i} x^i \sum_{j=0}^{n} \binom{n}{j} x^j$$

$$= \sum_{i=0}^{m} \sum_{j=0}^{n} \binom{m}{i} x^i \binom{n}{j} x^j$$

$$= \sum_{i=0}^{m} \sum_{j=0}^{n} \binom{m}{i} \binom{n}{j} x^i x^j$$

$$= \sum_{k=0}^{m+n} \left(\sum_{i=0}^{k} \binom{m}{i} \binom{n}{k-i} \right) x^k,$$

and since the coefficient of x^k must also be $\binom{m+n}{k}$, we get

$$\binom{m+n}{k} = \sum_{i=0}^{k} \binom{m}{i} \binom{n}{k-i}. \qquad\qquad 4.1$$

Note that in the last equality in the sequence, when $k > m$ or $k > n$ the number i is permitted to be larger than m and $k-i$ is permitted to be larger than n. Thus the corresponding values of $\binom{m}{i}$ or $\binom{n}{k-i}$ are 0. Formula 4.1 is sometimes called *Vandermonde's formula*. ∎

The derivation we gave for Vandermonde's formula is purely algebraic. The formula itself says something about k-element subsets of an $(m + n)$-element set. By analyzing the formula, we can get a much more intuitive proof that uses the bijection principle. A proof that uses a bijection to establish an equality is called a ***bijective proof***.

Example 4.5. Give a bijective proof of Vandermonde's formula (Formula 4.1).

The term on the left counts the number of k-element subsets of a union of two disjoint sets M and N. Each product on the right counts the number of ways to choose an ordered pair consisting of an i-element subset of M and a $(k - i)$-element subset of N. Thus the sum on the right-hand side counts the number of pairs of a subset of M and a (disjoint) subset of N whose union has k elements. The function

$$f(K) = (K \cap M, K \cap N)$$

is a bijection between the set of k-element subsets of $M \cup N$ and the set of pairs of disjoint subsets of total size k of M and N, respectively. Thus, by the bijection principle, the left- and right-hand sides are equal. ∎

A more natural way to describe the bijective proof is to avoid explicit mention of the bijection. This lets us say "The left-hand side of Vandermonde's formula counts the number of k-element subsets of a union of two disjoint sets M and N. The right-hand side computes this number of subsets as the number of ways to find subsets of M and N whose union has k elements. Thus the left-hand and right-hand sides are equal." Since this less explicit form is easier to read, we shall give most of our bijective proofs in this way.

Bijective proofs almost always add to our understanding of the result being proved. For this reason finding bijective proofs of results that have been proved in other ways is a central theme in modern combinatorics.

Multinomial Coefficients

We have seen the value of binomial coefficients in expanding a power of a binomial. It is natural to ask whether a similar family of coefficients might prove helpful when we wish to expand a power of a "trinomial" $(x + y + z)$ or general "multinomial" $(x_1 + x_2 + \cdots + x_m)$. In the same vein, we might ask whether labelings with more than two types of labels might also be counted with numbers like binomial coefficients. In fact, these two questions have related answers.

Example 4.6. In expanding $(x_1 + x_2 + x_3)^7$, we think of writing down seven $(x_1 + x_2 + x_3)$ terms in a row and then adding up $x_1^i x_2^j x_3^k$ for all ways of selecting x_1 from i of the terms, x_2 from j of the terms, and x_3 from k of the terms. Note that $i + j + k$ will have to be 7 in each case. The coefficient of $x_1^4 x_2^1 x_3^2$ is, for example, the number of ways of labeling four terms with the label "pick x_1," one term with "pick x_2," and two terms with "pick x_3." ∎

Suppose we have n objects, k_1 labels of type 1, k_2 labels of type 2, $\ldots$, and k_m labels of type m, and suppose that $k_1 + k_2 + \cdots + k_m = n$. In how many ways can we assign the labels to the objects? Another way to phrase this question is as follows. We are given a set N with n elements and a second set $M = \{1, 2, \ldots, m\}$. How many functions f are there from N to M that map k_i elements of N to element i of M? We denote the number of labelings by $C(n; k_1, k_2, \ldots, k_m)$ or $\binom{n}{k_1, k_2, \ldots, k_m}$; we refer to these numbers as **multinomial coefficients**.

Theorem 4.2. $\displaystyle \binom{n}{k_1, k_2, \ldots, k_m} = \frac{n!}{k_1! k_2! \cdots k_m!}.$

Proof. If f is a function from an n-element set N to $\{1, 2, \ldots, m\} = M$, we use the notation $f^{-1}(\{i\})$ to denote the *set* of elements of N to which f assigns i. In terms of a labeling, $f^{-1}(\{i\})$ stands for the set of elements with a label of type i. We may specify a function f by specifying

$$f^{-1}(\{1\}), \; f^{-1}(\{2\}), \; \ldots, \; f^{-1}(\{m\}),$$

(that is, by specifying which elements of N are related to i, or labeled with label i, for each i). This specification gives a list of m sets, the first of which must have size k_1, the second size k_2, and so on. Of course the function given by each such list is the type we desire, and so the number of functions in question is the same as the number of possible lists. We have $\binom{n}{k_1}$ choices for $f^{-1}(\{1\})$, and once we have chosen $f^{-1}(\{1\})$, we have $n - k_1$ elements left for the remaining choices. Once we have chosen the first i sets in the list, we have $n - k_1 - k_2 - \cdots - k_i$ elements left for our remaining choices; in particular, the number of choices for $f^{-1}(\{i+1\})$ is $C(n - k_1 - k_2 - \cdots - k_i; k_{i+1})$. Note that since $k_m = n - k_1 - \cdots - k_{m-1}$, the last binomial coefficient is $C(k_m; k_m)$. Thus, by the product principle, the total number of lists is given by the formula

$$\binom{n}{k_1}\binom{n - k_1}{k_2}\binom{n - k_1 - k_2}{k_3} \cdots \binom{n - k_1 - k_2 - \cdots - k_{m-1}}{k_m}$$

$$= \frac{n!}{k_1!(n - k_1)!} \cdot \frac{(n - k_1)!}{k_2!(n - k_1 - k_2)!} \cdots \frac{(n - k_1 - k_2 - \cdots - k_{m-1})!}{k_m! \cdot 0!}$$

$$= \frac{n!}{k_1! k_2! \cdots k_m!}.$$ ∎

Example 4.7. We must paint nine different chairs with green, red, and blue paint. We have enough blue for two chairs, enough red for three chairs, and enough green for four chairs. In how many different ways can we paint all the chairs?

The number of paint jobs is

$$C(9; 2, 3, 4) = \frac{9!}{2! \cdot 3! \cdot 4!} = \frac{9 \cdot 8 \cdot 7 \cdot 6 \cdot 5}{2! \cdot 3!} = 9 \cdot 4 \cdot 7 \cdot 5 = 1260.$$ ∎

The Multinomial Theorem

The ***multinomial theorem*** tells us how to expand a power of a multinomial.

> **Theorem 4.3.** $(x_1 + x_2 + \cdots + x_m)^n$ is the sum of all possible terms of the form
>
> $$C(n; k_1, k_2, \ldots, k_m) x_1^{k_1} x_2^{k_2} \cdots x_m^{k_m}$$
>
> using nonnegative integers k_i such that $k_1 + k_2 + \cdots + k_m = n$. We write this sum in the form
>
> $$\sum_{\substack{(k_1, k_2, \ldots, k_m): \\ k_1 + k_2 + \cdots + k_m = n}} C(n; k_1, k_2, \ldots, k_m) x_1^{k_1} x_2^{k_2} \cdots x_m^{k_m} = (x_1 + x_2 + \cdots + x_m)^n.$$

Proof. A direct analogy with the proof of the binomial theorem. ∎

We read the sum sign that appears in the theorem above as "the sum over all lists $k_1, k_2, \ldots, k_m$ such that $k_1 + k_2 + \cdots + k_m = n$."

Example 4.8. What is the coefficient of $x^6 y z^2$ in $(x + y + z + w)^9$? What is the coefficient of $x^6 y z^2$ in $(x + y + z + w)^{10}$?

The coefficient of the monomial $x^6 y z^2$ in $(x + y + z + w)^9$ is, by Theorem 4.3, $C(9; 6, 1, 2, 0)$, and

$$C(9; 6, 1, 2, 0) = \frac{9!}{6! \cdot 1! \cdot 2! \cdot 0!} = \frac{9 \cdot 8 \cdot 7}{2} = 252.$$

The coefficient of $x^6 y z^2$ in $(x + y + z + w)^{10}$ is *zero* because $x^6 y z^2$ is *not* one of the terms shown in Theorem 4.3. For this reason, we write $C(10; 6, 1, 2, 0) = 0$, and, more generally, $C(n, k_1, k_2, \ldots, k_m) = 0$ when the k_i's do not add to n. ∎

Multinomial Coefficients from Binomial Coefficients

It would be helpful to have a method like that of Pascal's triangle for computing multinomial coefficients. As we see in Exercise 31, there is a formula analogous to the Pascal formula. Since it does not involve just two variables n and k, it does not give us a two-dimensional table, so it is slightly less convenient to use by hand. (However, this limitation is irrelevant to the computation of multinomial coefficients in a computer language that allows multidimensional arrays.) However, if we have a Pascal triangle written down, we can use it to compute multinomial coefficients as well by applying the following theorem, which was proved as part of the proof of Theorem 4.2.

Theorem 4.4.

$$C(n; k_1, k_2, \ldots, k_m) = C(n; k_1) \cdot C(n-k_1; k_2) \cdots C(n-k_1-k_2-\cdots-k_{m-1}; k_m).$$

Proof. See the proof of Theorem 4.2 ∎

In a different notation, we can write

$$C(n; k_1, k_2, \ldots, k_m) = \prod_{i=1}^{m} C\left(n - \sum_{j=1}^{i-1} k_j; k_i\right).$$

We read the symbol $\prod_{i=1}^{m}$ as "the product from i equals 1 to m of ...."

Example 4.9. $\binom{7}{3\ 2\ 2} = \binom{7}{3}\binom{4}{2}\binom{2}{2} = 35 \cdot 6 \cdot 1 = 210.$ ∎

Lattice Paths

Example 4.10. A person wants to go from one street corner in a city to another one eight blocks east and seven blocks north. Thus the person must walk a total of 15 blocks (at a minimum). If the city streets are laid out as a square grid, in how many different ways may the person choose the 15-block walk?

For each of the 15 blocks, the person walks either north or east. Thus a walk can be described as a sequence of 15 N's and E's containing exactly 7 N's. The number of ways to choose the seven positions for the N's is $\binom{15}{7}$. ∎

There is a nice geometric representation of the preceding example. A *lattice path* in the (coordinate) plane is a path made up of horizontal and vertical line segments, each of which has integer coordinates for its endpoints. We construct a lattice path by starting at (0,0) and moving one unit parallel to the x-axis for each block the person walks east and one unit parallel to the

Figure 4.1

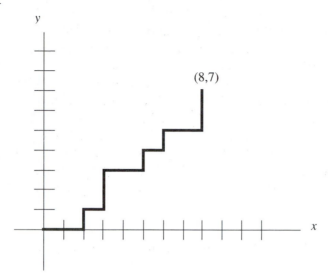

y-axis for each block the person walks north. Thus, in our example, we were asking for the number of lattice paths of length 15 from $(0,0)$ to $(7,8)$. The path corresponding to the sequence $EENENNEENENENEENN$ is shown in Figure 4.1.

With the same series of computations we used in Example 4.10, we may prove the following theorem about lattice paths. (See Exercise 20 of the previous section.)

> **Theorem 4.5.** The number of lattice paths of shortest length from the point (i_1, j_1) to the point (i_2, j_2) in the plane with $i_1 \leq i_2$ and $j_1 \leq j_2$ is
> $$\binom{i_2 + j_2 - i_1 - j_1}{i_2 - i_1}.$$

Proof. Essentially given in Example 4.10. ∎

Visualizing problems involving a sequence of choices with two possibilities for each choice as lattice path problems sometimes helps us find solutions.

As an example, suppose a charitable group has organized a benefit concert with a donation of $10.00 per person, payable at the door. On the evening of the concert, the organizers realize they have made no arrangements to have change on hand. Suppose i of the people who arrive have the exact change and j do not; suppose further that each person who does not have exact change has a $20.00 bill. (Assume also that people arrive one at a time.) If it is always the case that more people with exact change have arrived so far than those without, it will be possible to make change for everyone as they arrive. In order to compute the probability that things work out, we would need to divide the total number of patterns in which people can arrive into the number of patterns in which at least as many exact-change as non-exact-change people have arrived at each stage. The denominator is $(i + j)!$. What is the numerator?

Using E to stand for "exact change" and N to stand for "not exact change," we first compute the number of sequences of $i + j$ terms, i of which are E's and j of which are N's, such that at any place in the sequence, the number of E's so far is at least the number of N's so far. Then we multiply this number by $i! j!$ (to count the number of ways to assign the people who do have exact change to their places and the people who don't have exact change to theirs).

As with our paths in the city, a sequence of E's and N's may be thought of as a path in the plane in which we use x-coordinates to keep track of the number of E's and y-coordinates to keep track of the number of N's. To say that at each stage of the sequence we have at least as many E's so far as we have N's is to say that the path never goes below the line $y = x$. Thus we want to compute the number of paths from $(0,0)$ to (i, j) that never go below the line $y = x$.

Asking for the number of paths from $(0,0)$ to (i, j) that never *cross* the line $y = x$ is the same as asking for the number of paths from $(0,0)$ to (i, j) that never *touch* the line $y = x - 1$. Shifting all graphs up one unit shows that the number of paths from $(0,0)$ to (i, j) that never cross the line $y = x$ is the same as the number of paths from $(0, 1)$ to $(i, j + 1)$ that never *touch* the line $y = x$. We will compute this number by first noting that the total number of paths from $(0, 1)$ to $(i, j + 1)$ is

$$\binom{i + j + 1 - 1}{i} = \binom{i + j}{i}, \tag{4.2}$$

and then subtracting the number of paths from $(0, 1)$ to $(i, j + 1)$ that *do* touch the line $y = x$. We compute this second number with a clever bijection. Consider a path such as that shown in Figure 4.2(a) or 4.2(b) that *does* touch the line $y = x$.

Figure 4.2(c) and 4.2(d) show what happens if we take the segment of the path from $(0, 1)$ to the first touch on the line $y = x$ and reflect *this segment* around the line $y = x$. (Because reflecting the point (a, b) through the line $y = x$ gives the point (b, a), horizontal parts of the path reflect to vertical parts of the new path, and vice versa.)

Figure 4.2

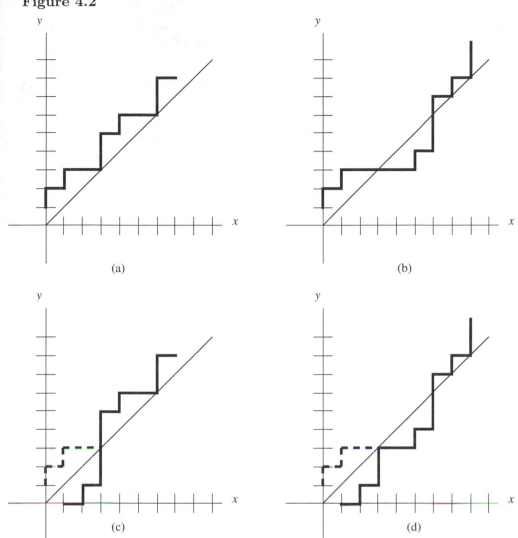

(a)

(b)

(c)

(d)

Thus each path from $(0,1)$ to $(i, j+1)$ that touches $y = x$ gives us a path from $(1,0)$ to $(i, j+1)$. Furthermore, given a path from $(1,0)$ to $(i, j+1)$, it must cross the line $y = x$ *somewhere* because $(1,0)$ is below the line and $(i, j+1)$ is above the line. Thus there will be a first point where it touches the line $y = x$. The segment of the path from $(1,0)$ to this first touching point will be below the line $y = x$. Reflecting the segment of the path from $(1,0)$ to the first touch gives us a new path that starts from $(0,1)$, touches the line $y = x$, and goes to $(i, j+1)$. Since reflecting a segment twice takes it to its original position, we have just described a bijection between paths from $(0, 1)$ to $(i, j + 1)$ that touch the line $y = x$ and paths from $(1, 0)$ to $i, j + 1)$. The number of paths of this second kind is

$$\binom{i - 1 + j + 1}{i - 1} = \binom{i + j}{i - 1}. \tag{4.3}$$

Subtracting the number in Equation (4.3) from the number in Equation (4.2) gives us

$$\binom{i + j}{i} - \binom{i + j}{i - 1} = \frac{(i + j)!}{i!j!} - \frac{(i + j)!}{(i - 1)!(j + 1)!} = \frac{(i + j)!}{(i - 1)!j!}(\frac{1}{i} - \frac{1}{j + 1})$$

$$\frac{(i + j)!}{(i - 1)!j!} \cdot \frac{j + 1 - i}{i(j + 1)}$$

$$= \binom{i + j}{i} \frac{j - i + 1}{j + 1}$$

for the number of paths from $(0, 1)$ to $(i, j + 1)$ that do not touch the line $y = x$. From our initial remarks, this is also the number of paths from $(0, 0)$ to (i, j) that do not cross the line $y = x$, and this is the solution to the problem.

In the case that $i = j = n$, we get

$$\frac{1}{n + 1}\binom{2n}{n}$$

for the number of paths from $(0, 0)$ to (n, n) that do not cross the line $y = x$. This number is called the **Catalan number** C_n, and arises in a surprisingly large number of combinatorial situations.

EXERCISES

1. Write out the expansion of $(x + y)^4$.
2. Write out the expansion of $(x + 2)^5$.
3. Write out the expansion of $(2x - 3)^4$.
4. Using Pascal's triangle, write out the expansion of $(x + 2)^6$.
5. Using Pascal's triangle, write out the expansion of $(2x - y)^6$.
6. What is the coefficient of x^5 in $(6x + \frac{1}{2})^7$?
7. What is the coefficient of $\sqrt{2}$ in $(1 + \sqrt{2})^7$? (Rewrite $(1 + \sqrt{2})^7$ in the form $a + b\sqrt{2}$ to answer this question.)
8. What is the coefficient of x^4 in $(4x - \frac{1}{2})^9$?

9. The number 2.01 is $2 + 0.01$. To compute $(2.01)^6$ in a straightforward way would require six multiplications. Use the binomial theorem and Pascal's triangle to compute $(2.01)^6$. (Note that while you still perform many operations of arithmetic, they are additions and easy multiplications.) How close would you have come to the correct answer if you had stopped after you had the sum of only the largest three terms?

10. Repeat the procedure of Exercise 9 with $(1.99)^6$. Why is the sum of the largest three terms (in absolute value) closer to the correct answer than the corresponding sum was to the answer in Exercise 9?

11. (a) Is $(1.001)^{1000}$ greater than 1.1?
    (b) Is $(1.0001)^{10,000}$ greater than 2?

12. (a) Compute $(.999)^{50}$ to five decimal places without using a calculator or computer except to check your work.
    (b) Estimate how many more terms you would need to include from the binomial expansion to get 10-place accuracy and 15-place accuracy.

13. Write out the expansion of $(x + y + z)^3$.

14. Write out the expansion of $(1 + \sqrt{2} + \sqrt{3})^3$ and collect like terms.

15. What is the coefficient of $x^2 y^3$ in $(x + 2y + 3)^7$?

16. In how many ways can nine different pieces of candy be given to Sam, Mary, and Pat so that Sam gets two pieces, Mary gets three pieces, and Pat gets four pieces?

17. In how many ways can nine different pieces of candy be given to three children so that each child gets three pieces?

18. In how many ways can nine different pieces of candy be given to three children so that one child gets two pieces, a second child gets three pieces, and the third child gets four pieces?

19. (a) In how many ways can a central processing unit assign nine jobs to three slave computers so that each computer gets three jobs?
    (b) In how many ways can the CPU assign the jobs so that no slave gets fewer than two jobs?

20. An *ordered partition* of a set S into k parts is a list $S_1, S_2, \ldots, S_k$ of disjoint nonempty subsets of S such that $S = S_1 \cup S_2 \cup \cdots \cup S_k$. How many ordered partitions of an n-element set into k parts have the property that, for each i, the ith set, S_i, has size s_i? (If some of the numbers s_i are 0, the list is sometimes called a *composition* of S, rather than a *partition*.)

21. What is $\sum_{i=1}^{10} \binom{10}{i} \cdot 3^{10-i}$?

22. Write down the formula for the expansion by the binomial theorem of $(x - 1)^n$. What can you conclude about the alternating sum $\binom{n}{0} - \binom{n}{1} + \binom{n}{2} - \cdots \pm \binom{n}{n}$ if $n > 0$?

23. By taking derivatives, show that
    (a)
$$\binom{n}{1} + 2\binom{n}{2} + 3\binom{n}{3} + \cdots = n2^{n-1}.$$
    (b)
$$\binom{n}{1} - 2\binom{n}{2} + 3\binom{n}{3} - 4\binom{n}{4} + \cdots = 0.$$

24. Write down the formula you get by computing the coefficient of $x^n y^n$ on both sides of the equation $(x + y)^n (x + y)^n = (x + y)^{2n}$. Apply the binomial theorem individually to both factors on the left side of the equation and apply it to the right side of the equation as well.

25. Derive from the binomial theorem the identity
$$\sum_{i=0}^{n} \binom{m}{i}\binom{n}{k+i} = \binom{m+n}{m+k}.$$

26. Show that
$$\sum_{i=0}^{n} \binom{n}{i}^2 = \binom{2n}{n}.$$

27. Show that
$$\left(\sum_{i=0}^{n} \binom{n}{i}\right)^2 = \sum_{j=0}^{2n} \binom{2n}{j}.$$

*28. Derive the identity
$$\sum_{k=0}^{n} \frac{(2n)!}{k!^2 (n-k)!^2} = \binom{2n}{n}^2.$$

29. It is clear that the total number of even-sized subsets of a set with an odd number of elements is the total number of odd-sized subsets because of the symmetry of the binomial coefficients. However, for any set, even one of even size, this statement is true. Use the binomial theorem to explain why.

30. (a) Show that
$$\binom{k}{m}\binom{n}{k} = \binom{n}{m}\binom{n-m}{n-k}.$$

(b) Show that
$$\sum_{k=m}^{n} \binom{k}{m}\binom{n}{k} = \binom{n}{m}2^{n-m}.$$

31. Show that
$$C(n; k_1, k_2, \ldots, k_m) = \sum_{i=1}^{m} C(n-1; k_1, k_2, \ldots, k_i - 1, \ldots, k_m).$$

32. A person wants to walk from a certain point in a city to a point seven blocks north and eight blocks east. In how many ways may the person go from one point to the other and walk exactly 17 blocks? In how many ways may the person go from one point to the other and walk exactly 16 blocks?

33. In how many ways may we choose a sequence a_i of $n + 1$ positive ones and n negative ones so that the "partial sums"
$$\sum_{i=1}^{m} a_i$$

are positive for each m from 1 to n?

34. A sequence of n left parentheses and n right parentheses is said to be *balanced* if as we count from left to right, we always count at least as many left parentheses as right parentheses. Thus (()) and ()(()()) are balanced while)(and (()))(() are not. How many balanced sequences of n left and n right parentheses are there?

35. Two candidates are running for office. The winner receives n votes and the loser receives m votes. As the votes are counted, the counters announce who is ahead at each stage. In what fraction of the counting sequences is the winner ahead at each stage? (Notice that we are asking for the probability that the winner is always ahead at each stage; in this context this problem is called the 'ballot problem.")

Section 5 Mathematical Induction

The Principle of Induction

A number of proofs we have given have been made hard to read because the phrase "and so on" appears in the middle of the proof. As proofs get more and more complex, this lack of precision becomes more and more confusing. In such situations, a reasonably simple solution is to appeal to a principle called the ***principle of mathematical induction***. There are two closely related versions of the principle. The second version is often referred to as the principle of *strong mathematical induction*.

> **Version 1.** Suppose the set S contains the integer i. Suppose also that for each integer j such that j is in S, the integer $j + 1$ is in S. Then S contains every integer greater than or equal to i.

> **Version 2.** Suppose the set S contains the integer i. Suppose also that for each integer j such that all integers between i and j inclusive are in S, the integer $j + 1$ is in S. Then S contains every integer greater than or equal to i.

Once you understand the wording of the principles you may react to them as people often do the to sum principle: "How can anything so obvious possibly be useful?" After all, they just capture the simple idea that if the number i is in a set and you add 1 to it again and again, you will eventually see any larger integer. Making this simple idea precise as a principle does give us a powerful new proof technique. (For the student interested in the foundations of mathematics, the principle is an axiom that we use to capture this idea, just as in geometry we use an axiom to capture the idea that there is one and only one line betweeen two points.)

Frequently you will find versions of the principle based on the idea that some statement involving an integer i is true for certain values of the integer. These versions may seem more difficult to understand because they involve the idea of a statement about an integer i. They are, however, entirely equivalent to the two versions given here, and the two versions given here are also equivalent to each other. For your own understanding, you should learn to use the principle as stated here. Then you should have no difficulty adapting to some other version. For the remainder of this section, we shall give examples of the use of induction in proving theorems.

Proving That Formulas Work

Figure 5.1 suggests a formula for the sum of consecutive odd integers. We can use the pictures to give a convincing geometric proof that the sum of n

Figure 5.1

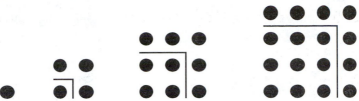

consecutive odd integers is n^2. What do we do, however, when we are sure a formula is true but we *don't* see a clever trick like Figure 5.1 to prove it?

Example 5.1. Prove that $1 + 3 + 5 + \cdots + 2n - 1 = n^2$ for all $n \geq 1$.

Let S be the set of all integers n such that $1 + 3 + 5 + \cdots + 2n - 1 = n^2$. The integer 1 is in S since $1 = 1^2$.

Suppose the integer k is in S. Then $1 + 3 + \cdots + 2k - 1 = k^2$ and

$$(1 + 3 + 5 + \cdots + 2k - 1) + 2(k + 1) - 1$$
$$= \underbrace{1 + 3 + 5 + \cdots + 2k - 1}_{k^2} + 2k + 1$$
$$+2k + 1 = (k + 1)^2.$$

Thus, the integer $k + 1$ is in S. Then by the principle of mathematical induction, S contains all positive integers and so our formula is proved. ∎

Notice that a proof by induction has three primary steps, illustrated by the three paragraphs of Example 5.1.

- Step 1, called a **base step**, establishes that an appropriately chosen number is in S as a starting point.
- Step 2, called the **inductive step**, shows how, by accepting that the integer n is in S, we can derive as a conclusion that $n + 1$ is in S. (Alternatively, we might show that by assuming the integer $n - 1$ is in S, we can derive as a consequence that the integer n is in S. The important thing is that assuming one integer is in S allows us to derive that the next one is also in S.) You can recognize the beginning of an inductive step since the natural starting place is what is called the **inductive hypothesis**, a statement which makes the assumption that a certain variable stands for a member of S (i.e., makes the theorem we wish to prove true.) At this stage it may be important to restrict the values of n considered to be at least the starting value used in step 1. (At any rate, these are the only relevant values, so no harm is done in making the restriction even if you do not use it.)

- Step 3 concludes that the set contains all integers beyond the starting point and interprets the meaning of this fact in the context of the result being proved.

Informal Induction Proofs

There is no need for proofs by induction to be as formal as they are in this section. The formalism is intended to illustrate exactly what is happening in an inductive proof. We will slowly dispense with the formalism in the text as you becomes more familiar with inductive proofs.

Example 5.2. Do Example 5.1 with no formal reference to a set S.

We wish to show that $1 + 3 + 5 + \cdots + 2n - 1 = n^2$ for all $n \geq 1$. Clearly this is the case when $n = 1$. Now suppose it is the case when $n = k - 1$, so that

$$1 + 3 + \cdots + 2(k - 1) - 1 = (k - 1)^2.$$

Then, by addition,

$$\begin{aligned}
1 + 3 + \cdots + 2(k - 1) - 1 + 2k - 1 &= (k - 1)^2 + 2k - 1 \\
&= k^2 - 2k + 1 + 2k - 1 \\
&= k^2.
\end{aligned}$$

Thus, by the principle of mathematical induction, the formula holds for all n. (We have illustrated, by the way, that it doesn't matter whether we derive the case $n = k + 1$ from the case $n = k$ or the case $n = k$ from the case $n = k - 1$.) ∎

Inductive Definition

In the last two examples, the three dots between the plus signs stand for "and so on up to." In part because of the three dots and in part because of the long strings of symbols, the algebra may have been difficult to follow. The purpose of summation notation is to overcome this sort of difficulty. Recall that

$$\sum_{i=1}^{n} a_i$$

stands for the sum of the numbers a_i for each i between 1 and n. Although this verbal description is fine for understanding the summation concept, it does not give us any insight into how to do algebra involving summation signs. Notice that the facts

$$\sum_{i=1}^{1} a_i = a_1 \qquad \text{and} \qquad \sum_{i=1}^{n} a_i = \left(\sum_{i=1}^{n-1} a_i \right) + a_n \qquad \text{for } n > 1$$

follow intuitively from our definition of summation notation. In fact, these two algebraic properties of summation completely define summation notation.

Example 5.3. Show that the two properties

$$\sum_{i=1}^{1} a_i = a_1 \qquad \text{and} \qquad \sum_{i=1}^{n} a_i = \left(\sum_{i=1}^{n-1} a_i \right) + a_n \qquad \text{for } n > 1$$

define summation notation

We let S be the set of integers k such that $\sum_{i=1}^{k} a_i$ is defined. Then 1 is in S because of the first property. But whenever $n - 1$ is in S, n is in S by the second property. Therefore by the principle of mathematical induction, S contains all positive integers, so that $\sum_{i=1}^{k} a_i$ is defined by the two properties for every positive integer k. ∎

This description of summation notation is an example of what is called an ***inductive*** or ***recursive*** definition. Inductive definitions are ideal for use in inductive proofs.

Example 5.4. State the formula of Examples 5.1 and 5.2 in summation notation and use the inductive definition of summation notation to aid in writing the proof.

The formula we are to prove may be written as

$$\sum_{i=1}^{n} (2i - 1) = n^2.$$

Now the formula is true when $n = 1$ since

$$\sum_{i=1}^{1} (2i - 1) = 2 \cdot 1 - 1 = 1 = 1^2.$$

Now suppose the formula holds when $n = k - 1$, so that

$$\sum_{i=1}^{k-1} (2i - 1) = (k - 1)^2.$$

Adding $2k - 1$ to both sides gives

$$\left(\sum_{i=1}^{k-1} (2i - 1) \right) + 2k - 1 = (k - 1)^2 + 2k - 1 = k^2 - 2k + 1 + 2k - 1 = k^2.$$

But then the inductive definition of summation notation applied to the expression before the first equal sign gives

$$\sum_{i=1}^{k}(2i-1) = k^2.$$

Thus, by the principle of mathematical induction, the formula holds for all positive integers n. ∎

The General Sum Principle

Up until now, the product principle has been a principle we have accepted without proof. Although that seems reasonable on intuitive grounds, we have now reached the point where we can *derive* it from the sum principle for two sets. In much the same way, we can derive the general sum principle from the sum principle for two sets. Just as we can never eliminate all assumptions in geometry, we cannot eliminate all assumptions in any other area of mathematics. However, suppose we can derive as a theorem a principle that has previously been an assumption. This makes us sure that it applies to all conceivable circumstances and not merely to the circumstances that fit our intuition. Here we shall derive the general form of the sum principle from the sum principle for two sets. In the exercises, we suggest how to derive the "union of disjoint sets" form of the product principle from the sum principle for two sets. The computations we used to derive the "lists of length 2" form of the product principle from the "union of disjoint sets" form of the principle can be used to prove the "lists of length 2" form of the principle as a theorem. In the exercises, we also suggest how to derive the product principle for lists of arbitrary length and how to derive the sum principle for two sets from the definition of size.

Example 5.5. The simplest version of the sum principle states that if two sets A and B have no elements in common, then the size of $A \cup B$ is the sum of the sizes of A and B; in symbols, $|A \cup B| = |A| + |B|$. Use this principle and the principle of mathematical induction to prove that if $A_1, A_2, \ldots, A_n$ is a list of sets such that for every distinct i and j, $A_i \cap A_j = \emptyset$ (that is, A_i and A_j have no elements in common), then

$$\left| \bigcup_{i=1}^{n} A_i \right| = \sum_{i=1}^{n} |A_i|.$$

(The notation $\bigcup_{i=1}^{n} A_i$ or $\cup_{i=1}^{n} A_i$ stands for the union of all the sets in the list and is read as "the union from i equals 1 to n of A_i." Notice that we are implicitly using an inductive definition of this notation in what follows.)

Let S be the set of all n such that for n mutually disjoint sets $A_1, A_2, \ldots, A_n$,

$$\left| \bigcup_{i=1}^{n} A_i \right| = \sum_{i=1}^{n} |A_i|.$$

Then 2 is in S by the sum principle (and 1 is in S because the statement then reads $|A_1| = |A_1|$).

Now suppose k is in S, and let $A_1, A_2, \ldots, A_k, A_{k+1}$ be a list of mutually disjoint sets. Then

$$\bigcup_{i=1}^{k+1} A_i = \left(\bigcup_{i=1}^{k} A_i \right) \cup A_{k+1}.$$

However, since A_{k+1} has no elements in common with any A_i for $i \leq k$, it has no elements in common with $\bigcup_{i=1}^{k} A_i$. Thus, by the sum principle for pairs of sets,

$$\left| \bigcup_{i=1}^{k+1} A_i \right| = \left| \bigcup_{i=1}^{k} A_i \right| + |A_{k+1}|.$$

However, k is in S, so that

$$\left| \bigcup_{i=1}^{k} A_i \right| = \sum_{i=1}^{k} |A_i|.$$

Thus

$$\left| \bigcup_{i=1}^{k+1} A_i \right| = \left| \bigcup_{i=1}^{k} A_i \right| + |A_{k+1}| = \left(\sum_{i=1}^{k} |A_i| \right) + |A_{k+1}| = \sum_{i=1}^{k+1} |A_i|.$$

Thus, $k + 1$ is in S, so by version 1 of the induction principle, S contains all positive integers; this proves the formula. ∎

An Application to Computing

Example 5.6. Show that, given a list of 2^n numbers in increasing order, it is possible to determine whether a particular number k is in the list by comparing k with at most $n + 1$ numbers of the list.

Let $S = \{n \mid$ with at most $n+1$ comparisons it is possible to tell if a number k is in a list of 2^n numbers in increasing order$\}$.

Then 0 is in S, for 2^0 is 1 and in a list of one number, only one comparison is required to determine whether k is in the list.

Now suppose we have a list of 2^{m+1} numbers in increasing order and let m be in S. Given the number k, we can use one comparison to compare it with the number a in position 2^m of the list.

If $k = a$, we have determined that k is in the list.

If $k < a$, we can determine whether k is between positions 1 and 2^m, using at most $m + 1$ comparisons. (This is possible since m is in S.)

If $k > a$, we can determine whether k is between position $2^m + 1$ and position 2^{m+1} of the list, using at most $m+1$ comparisons. (This is possible since m is in S.)

Thus, we can find whether k is in the list in at most a total of $1 + (m + 1)$ comparisons, so $m + 1$ is in S. Therefore, by the principle of mathematical induction, all nonnegative integers are in S. This proves that we can determine whether k is in a list of 2^n numbers by using a procedure that makes at most $n + 1$ comparisons. ∎

The process described in the preceding example is called *binary search*. The example shows how to find an entry in a list of one million things (in increasing order) by making just 21 comparisons. In essence, our proof demonstrated that the process of binary search will, in fact, determine whether something is an element of the list as well as verifying the upper limit on the number of comparisons needed. The description we have given for binary search is called a *recursive* description. A recursive description of how to solve a problem involving a number n begins by telling us how to solve the problem for some base value, such as $n = 1$; then it tells us how to use solutions of the problem involving numbers smaller than n to solve the problem for the current value of n. A recursive description of an algorithm is thus an inductive definition of a process to be carried out. Because of the similarity to the structure of an inductive proof, it should be no surprise that mathematical induction is an important tool in the analysis of recursive algorithms.

Proving That a Recurrence Works

Example 5.7. Show that if a function $D(n, k)$, which is defined for pairs n, k with $0 \leq k \leq n$, has the properties that

(1) $D(n, 0) = D(n, n) = 1$ and

(2) $D(n, k) = D(n - 1, k - 1) + D(n - 1, k)$ for all k with $0 < k < n$,

then $D(n, k) = n!/(n - k)!k!$

Let S be the set of all n such that $D(n, k) = \frac{n!}{(n-k)!k!}$ for all k with $0 \leq k \leq n$. Since $D(0, 0) = 1$ and $\frac{0!}{(0-0)!0!} = 1$, we may conclude that 0 is in S.

Suppose $n - 1$ is in S. Then for each k with $0 < k < n$

$$D(n, k) = D(n - 1, k - 1) + D(n - 1, k)$$
$$= \frac{(n - 1)!}{(k - 1)!(n - k)!} + \frac{(n - 1)!}{k!(n - k - 1)!}$$
$$= \frac{k(n - 1)!}{k!(n - k)!} + \frac{(n - 1)!(n - k)}{k!(n - k)!}$$
$$= \frac{n(n - 1)!}{k!(n - k)!} = \frac{n!}{k!(n - k)!}.$$

(The reason we wrote the equations for $0 < k < n$ rather than $0 \le k \le n$ was the two multiplications of the numerator and denominator of a fraction by k or $n - k$ in the next to last line of the equations.) Also $D(n, 0) = D(n, n) = 1 = \frac{n!}{(n - 0)!0!}$, so for all k with $0 \le k \le n$, we have the equation $D(n, k) = \frac{n!}{(n - k)!k!}$. Thus, n is in S. By the principle of mathematical induction, S contains all nonnegative integers. Thus, $D(n, k) = \frac{n!}{k!(n - k)!}$ for all n. ∎

Note that this example shows that the binomial coefficients must be completely determined by the rules we used to generate Pascal's triangle.

A Sample of the Strong Form of Mathematical Induction

Example 5.8. A prime number is a positive integer greater than 1 that has no positive factor other than itself and 1. Show that every positive number greater than 1 is a product of prime numbers or is prime itself.

Let S be the set of positive integers that are prime or products of prime numbers. Then 2 is in S because 2 is prime. Suppose now that all numbers between 2 and k are in S. We shall show that $k + 1$ is in S. If $k + 1$ has no factors other than 1 or $k + 1$, it is prime and thus is in S. If $k + 1$ has a positive factor m different from itself and 1, then $k + 1 = mj$ for some positive integer j. Since $m > 1$, j cannot be greater than or equal to $k + 1$. Thus $j < k + 1$. Also $j > 1$, for $m \ne k + 1$. Then both m and j either are prime or are products of prime numbers. Thus $mj = k + 1$ is a product of prime numbers or powers of prime numbers, so $k + 1$ is in S. Thus, by the principle of strong mathematical induction, S contains all integers greater than 1, so every integer greater than 1 is a prime number or a product of prime numbers. ∎

This example shows how useful the second (strong) form of the principle of mathematical induction can be. How, though, do we know which form to use? Generally, you do not decide on a proof by induction on the basis of some magical insight that induction is required. Instead, one of the following circumstances may lead you to it. You may first work out examples for quite a few values of the parameter (the k we've used in Examples 5.1–5.5 and 5.8 and also the m and n of Examples 5.6 and 5.7), and you see how one example seems to follow naturally from the immediately previous one—or from several of the previous ones (not necessarily immediately preceding). This would lead you to try induction. Perhaps, instead, you see how to divide the problem into several cases, each one with the parameter value being one smaller—or else into several cases each with a smaller parameter value (which is not necessarily just one smaller). Again you would try induction. *It is on the basis of whether you are reducing the problem with parameter value k to a problem with parameter value $k - 1$ or merely to **some** smaller parameter value that you decide which form is more useful.*

The two forms of the induction principle are equivalent mathematically. In other words, given one version of the principle, we could derive the other as a theorem. Proving this equivalence is important to the student interested in the foundations of mathematics; however, since the proof is more a matter of logic than combinatorics, we will not include it here.

Double Induction

There are times when a problem breaks into cases that involve two numbers m and n, and we can derive the result for one case from results with a smaller m or a smaller n or both a smaller m and n. Although it is possible to deal with such a situation by one of the forms of induction we have already discussed, it is usually easier to apply the following *double induction principle*.

> **Mathematical Induction, Version 3.** Suppose the set S contains the ordered pair (i, j) of integers. Suppose also that whenever S contains the ordered pairs $(r - 1, s)$ and $(r, s - 1)$, then it also contains the ordered pair (r, s). Then S contains all pairs of integers (m, n) with $m \geq i$ and $n \geq j$.

There is a similar strong version of double induction.

*Ramsey Numbers

When we gave our definition of Ramsey numbers, we said that the Ramsey number $R(m, n)$ is the smallest integer k so that for any set K of size k, if f is a symmetric function from pairs of distinct elements of K to $\{A, S\}$, then there is an m-element set on whose pairs f has constant value A or an n-element set on whose pairs f has the constant value S. We ignored the possibility that there may be no such integer k because there might not be any set K for which all symmetric functions of pairs have the desired property. Thus we don't really know that Ramsey numbers $R(m, n)$ exist except for the values of m and n for which we have explicitly computed them. The next theorem shows that Ramsey numbers do exist; it is a special case of a theorem called *Ramsey's theorem*.

Theorem 5.1. If K is a set of size $k = \binom{m+n-2}{m-1}$ (with $m, n \geq 2$) and f is a symmetric function from the pairs of distinct elements of K to $\{A, S\}$, then K has either a subset of size m on whose pairs f has the constant value A or a subset of size n on whose pairs f has the constant value S.

Proof. We will use double induction on m and n. When $m = n = 2$, if we have a set K of size 2 with a symmetric function defined on its pairs, then either $f(x, y) = A$ or $f(x, y) = S$ for the two distinct elements x and y of K and thus either f has the constant value A on the pairs of our two-element set K or f has the constant value S on the pairs of our two-element set K.

Thus we assume that the theorem holds when $m = r - 1$ and $n = s$ and when $m = r$ and $n = s - 1$. We now consider a set K of size $\binom{r+s-2}{r-1}$ and a function f from the pairs of distinct elements of this set to $\{A, S\}$. From the Pascal relation, we get

$$\binom{r + s - 2}{r - 1} = \binom{r + s - 3}{r - 2} = \binom{r + s - 3}{r - 1}.$$

Consider a particular element x of K. There are $\binom{r-s-2}{r-1} - 1$ other elements of K, and for each of them $f(x, y)$ is either A or S. Thus by the sum principle, there must be either $\binom{r+s-3}{r-2}$ elements y such that $f(x, y) = A$ or $\binom{r+s-3}{r-1}$ elements y such that $f(x, y) = S$. This gives us two cases.

In case 1, we have $\binom{r+s-3}{r-2} = \binom{r-1+s-2}{r-1-1}$ elements y such that $f(x, y) = A$, and so by our inductive hypothesis (that the theorem holds when $m = r - 1$ and $n = s$), one of two things must happen. Either there is a set of s elements $\{y_1, y_2, \ldots y_s\}$ on whose pairs f has constant value S, or there is a set of $r - 1$ elements $\{y_1, y_2, \ldots y_{r-1}\}$ on whose pairs f has the constant value A. In this second event, $f(x, y_i) = A$ for all these elements y_i, so

$$\{x, y_1, y_2, \ldots y_r\}$$

is a set of size r on whose pairs f has the constant value A. Thus in case 1, either we have a set of size s on whose pairs f has the constant value S or we have a set of size r on whose pairs f has constant value A. In case 2 an analogous argument yields the same result. Therefore the theorem holds when $m = r$ and $n = s$, and so by the principle of mathematical induction, it is true for all integers $m, n \geq 2$. ∎

EXERCISES

1. Use induction to prove that

$$\sum_{i=1}^{n} i = 1 + 2 + \cdots + n = \frac{n(n+1)}{2}.$$

2. Illustrate the result of Exercise 1 for $n = 3$ geometrically by drawing dots on x and y axes at the points $(0,0)$, $(1,0)$, $(0,1)$, $(2,0)$, $(1,1)$, and $(0,2)$. Now how many dots must you add to get a triangle with four dots on a side? Given a triangle with $n-1$ dots on a side, how many would you have to add to get a triangle with n dots on a side? Use the total number of dots in such a triangular array to explain the formula.

3. Exercise 1 may be thought of as finding

$$\binom{1}{1} + \binom{2}{1} + \binom{3}{1} + \cdots + \binom{n}{1}.$$

In this light we see that the formula says that

$$\sum_{i=1}^{n} \binom{i}{1} = \binom{n+1}{2}.$$

Guess and prove by induction a formula for

$$\sum_{i=2}^{n} \binom{i}{2}.$$

4. The formula in Exercise 1 can be interpreted as saying that if we choose two elements from $\{0, 1, 2, \ldots, n\}$, the larger one is either 1 or 2 or $\ldots$ or n. (If, for example, the larger one is i, then the smaller one may have any value from 0 to $i-1$. This accounts for the i in the sum $1 + 2 + \cdots + n$, for it is the number of pairs in which the larger element is i.) Give a similar combinatorial interpretation of the formula you found in the previous exercise.

5. Guess and prove a formula for $\sum_{i=k}^{n} \binom{i}{k}$.

6. Guess and prove a formula for the sum of the first n cubes.

7. Express n^3 as

$$a\binom{n}{3} + b\binom{n}{2} + c\binom{n}{1} + d\binom{n}{0}$$

for some numbers a, b, c, and d. Use the formula you found in Exercise 5 to obtain a formula for the sum of the first n cubes. This shows how we could discover the formula we were told to guess for problem 6.

8. Use induction to prove that

$$\sum_{i=1}^{n} i^2 = \frac{n(n+1)(2n+1)}{6}.$$

9. Prove that 6 is a factor of $n^3 + 5n$ for all positive integers n.

10. Use induction to prove that $(1+x)^n \geq 1+nx$ if $x \geq 0$ and n is a positive integer.

11. Prove that $n^3 > 2n^2$ if $n \geq 3$.

12. Prove that $2^n > n^2$ for $n > 5$.

13. Prove that for each integer $n \geq 2$, the number of distinct prime factors of n is less than $\log_2 n$. (Note, 1 is not a prime.)

14. Prove by induction that a set with n elements has 2^n subsets. To get started, look at the following questions for several values of n. How many subsets of $\{1, 2, \ldots, n\}$ contain n? How many don't contain n?

15. Use induction to reprove Theorem 1.2.

16. Use induction on n to prove there are m^n functions from an n-element set to an m-element set.

17. The simplest version of the product principle states that if n disjoint sets each have m elements, then their union has mn elements. Prove this using induction and the simplest form of the sum principle. (Example 5.5 is a good model because the problem stated here can be regarded as a special case of the problem stated in Example 5.5.)

18. Prove by induction on m that the union of a set of size n and a disjoint set of size m has size $n + m$. (The point is that you may assume you have a counting function from $\{1, 2, \ldots, m + n\}$ to the set $N \cup M$ and then when you add a new element to M, you need only define your new counting function by extending the old one so that it takes $n+m+1$ to the new element; finally, you show that this new function is a bijection.)

19. Using induction and the simplest form of the product principle stated in Exercise 17, prove the product principle for lists of length n. (Hint: Divide the set of lists up into disjoint sets, say one set for each possible first element of a list.)

20. Suppose $a \neq 0$ and let g be a function from the nonnegative integers to the positive integers with the properties that $g(1) = a$ and $g(m + n) = g(m)g(n)$. Prove that $g(0) = 1$. Prove that $g(n) = a^n$ for $n \geq 0$.

21. Show that it is possible to sort a list of 2^n numbers in no special order into a list in increasing order by making no more than $n \cdot 2^n$ comparisons. (Hint: Use a method like that of Example 5.6 and observe that two lists of j numbers each in increasing order can be merged into a single list in increasing order in at most $2j$ steps.)

22. Show that if $m \neq n$, there is no bijection from the set $\{1, 2, \ldots, n\}$ to the set $\{1, 2, \ldots, m\}$. This is why we can, without contradiction, count the number of elements of a set by placing it in a one-to-one correspondence with a set whose size we already know.

23. Prove that

$$\sum_{j=0}^{n} \binom{j}{k} = \binom{n+1}{k+1}.$$

24. Prove the second (strong) form of the mathematical induction principle from the first. (You will need to consider a set that has j in it if and only if all numbers from i through j are in the set S of the strong induction principle.)

25. The two-set distributive law states that $A \cap (B \cup C) = (A \cap B) \cup (A \cap C)$. Use this and mathematical induction to prove the distributive law

$$A \cap \bigcup_{i=1}^{n} B_i = \bigcup_{i=1}^{n} (A \cap B_i).$$

26. Two special cases of the rules of exponents are

$$a^0 = 1 \qquad a^{m+1} = a^m \cdot a.$$

(a) Prove that for each nonnegative integer n, the value of a^n is determined by these rules. Explain what you have done in terms of recursive definition.

(b) Prove that $a^{n+k} = a^n a^k$.

(c) Prove that $a^{mn} = (a^m)^n$.

27. A sequence a_i is called a *geometric progression* if there is a constant r such that $a_{i+1} = ra_i$ for all $i \geq 0$, so that $a_i = a_0 r^i$ (if the sequence starts at $i = 0$). Prove that if a_i is a geometric progression, then

$$\sum_{i=0}^{n} a_i = a_0 \frac{1 - r^{n+1}}{1 - r}.$$

28. A sequence a_i is called an *arithmetic progression* if there is a constant c such that $a_{i+1} = a_i + c$. Prove that the sum of terms m through n (inclusive) of an arithmetic progression is the average of the first and last terms times the number of terms.

29. Show that each integer greater than seven is a sum of a nonnegative multiple of three and a nonnegative multiple of five.

30. Use mathematical induction and the product rule to prove the formula for the derivative of x^n for $n > 0$.

31. Find the error in the following inductive "proof" that all positive integers n are equal. Let S be the set of all n such that n equals all integers between 1 and n. Then 1 is in S. Now suppose all integers up to and including k are in S. Then $k = k - 1$, so adding 1 to both sides gives $k + 1 = k$. Therefore, by the principle of mathematical induction, S contains all positive integers, and so all positive integers are equal.

32. Find the error in the following "proof." Let S be the set of integers n such that all the elements of an n-element set are equal. Then 1 is in S. Now assume $n - 1$ is in S and let N be an n-element set. Then the first $n - 1$ elements of N are equal, and the last $n - 1$ elements of N are equal. Thus, all the elements of N must be equal to the value of the common elements among the first $n - 1$ elements and the last $n - 1$ elements. Therefore, all elements of N are equal. Therefore, S contains all positive integers, so that for any integer n, all elements of a set of size n are equal.

33. (a) Show that the Ramsey numbers $R(m, n)$ satisfy the inequality

$$R(m, n) \leq R(m - 1, n) + R(m, n - 1)$$

(b) What does this tell us about the largest possible value of $R(5, 3)$?

(c) Consider 13 people arranged in a circle so that each person is acquainted with only the first and fifth person to the left and the first and fifth person to the right. Show that there is no subset of three mutual acquaintances and no subset of five mutual strangers.

(d) What do the two previous parts imply about the Ramsey number $R(5,3)$?

34. (a) What upper bound on the value of $R(4,4)$ is implied by the inequality of part (a) of the previous exercise?

(b) Consider 17 people arranged in a circle so that each person is acquainted with the first, second, fourth, and eighth people to the right and the first, second, fourth, and eighth people to the left. Show that there is no subset of four mutual acquaintances and no subset of four mutual strangers.

(c) What do the two previous parts imply about the Ramsey number $R(4,4)$?

Suggested Reading

Berge, C. 1971. *Principles of Combinatorics*, ch. 1. New York: Academic Press.

Feller, W. 1968. *An Introduction to Probability Theory and Its Applications*, ch. 3, 3d ed. New York: Wiley.

Knuth, D. E. 1973. *The Art of Computer Programming, Fundamental Algorithms*, vol. 1, 2d ed. Reading, MA: Addison Wesley.

Ryser, H. J. 1963. *Combinatorial Mathematics*, Carus Mathematical Monographs 14, ch. 1. Washington, DC: Mathematical Association of America.

2

Equivalence Relations, Partitions, and Multisets

Section 1   Equivalence Relations

The Idea of Equivalence

Suppose we are writing out seating charts to arrange four people clockwise around a circular table for a game of cards. We are going to assume that no place at the table is special. Thus if there is a "dealer" in the game, this person is decided in some way (perhaps by drawing the high card) after the people sit down at the table.

If we obtain one arrangement from a previous arrangement by shifting each person's place one seat to the right, the relative positions of all four players will remain the same. Thus, the two arrangements are equivalent for the purpose of playing a game of cards. Because of the possibility of equivalence, the number of truly distinct seating patterns is not the same as the total number of seating charts. Without a more precise understanding of what we mean by equivalence, it is hard even to say what we mean by "the number of truly distinct seating patterns." It will often happen that two arrangements of objects that are technically different—such as two different lists of places at a table—are so closely related that they are equivalent for certain purposes. Thus, under one set of conditions, we might see two objects as being equivalent, but under another set of conditions, we might see the same two objects as being inequivalent. However, there are certain properties that we intuitively associate with the phrase *is equivalent to*. For example, any object should be equivalent to itself. Also, if object A is equivalent to object B, then object B should be equivalent to object A. Further, if object A is equivalent to object B, and object B is equivalent to object C, then objects A and C should be equivalent. Although these are not the only properties we could list, there is good evidence (which we shall see later) that other properties we might wish to include can be derived from these three. We take these properties as the defining properties of the idea of equivalence.

Equivalence Relations

We gave a precise definition of the word *relation* in our examination of the definition of a function; now we use this precise concept to define equivalence precisely. Recall that a relation R on a set X is a set of ordered pairs of elements of X. We shall say that a relation on a set X is an **equivalence relation** *on* X if

(1) (x, x) is in R for all x in X. (**reflexive** law)

(2) If (x, y) is in R, then (y, x) is in R for all x and y in X. (**symmetric** law)

(3) If (x, y) and (y, z) are in R, then (x, z) is in R for all x, y, and z in X. (**transitive** law)

On any set X, the relation of "equality" is an equivalence relation. Less obvious examples of equivalence relations arise in many different ways.

Example 1.1. Let X be the set of college students attending some college. Let R be the relation "has the same grade point average as." Show that R is an equivalence relation on X.

(1) Any person has the same GPA as himself or herself.

(2) If person A has the same GPA as person B, then person B has the same GPA as person A.

(3) If person A has the same GPA as B and B has the same GPA as C, then A has the same GPA as C.

Thus, R is an equivalence relation. ∎

Notice that we verified the defining conditions of an equivalence relation without mentioning ordered pairs. When we have an intuitive understanding of a relation, informal statements of the reflexive, symmetric, and transitive laws such as we used here are easiest to work with. In this case, the translation of the intuitive statements into precise ones is so straightforward that we may omit it.

Circular Arrangements

Example 1.2. Let X be the set of all possible lists of the four names

$$\{\text{Bill, Chuck, Maria, Sarah}\}.$$

Regard the lists as seating charts for four people around a circular table having four places marked 1, 2, 3, 4. Define a relation R on X by letting

(K, L) be in R if the list K may be obtained from the list L by shifting each person the same number of places to the right or the left. For example, (using initials) B, C, M, S is related by R to S, B, C, M. In terms of ordered pairs, we would say that $(\langle B, C, M, S \rangle, \langle S, B, C, M \rangle)$ is in R. (The angle brackets around the lists have no special meaning; they were chosen just as a convenient form of "punctuation.") Show that R is an equivalence relation.

(1) For each list L, $(L, L) \in R$ because L may be obtained from L by shifting each person no places (or four places!).

(2) If K may be obtained from L by shifting everyone n places to the right, then L may be obtained from K by moving everyone n places to the left.

(3) We will say a left shift through n places is a right shift through $-n$ places. A right shift through m places followed by a right shift through n places results in a right shift through $m + n$ places. (Why is this true even if m or n is negative?) Thus, if $(K, L) \in R$ and $(L, M) \in R$, then $(K, M) \in R$.

Therefore, R is an equivalence relation. ∎

In the preceding example, suppose we start with the list $\langle B, C, M, S \rangle$ and ask which other lists are equivalent to it. By means of right shifts through 1, 2, and 3 places, respectively, we get $\langle S, B, C, M \rangle$, $\langle M, S, B, C \rangle$, and $\langle C, M, S, B \rangle$. By means of left shifts through 1, 2, and 3 places, respectively, we get $\langle C, M, S, B \rangle$, $\langle M, S, B, C \rangle$, and $\langle S, B, C, M \rangle$. Thus, the set of *all* lists equivalent to $\langle B, C, M, S \rangle$ is

$$S_1 = \{ \langle B, C, M, S \rangle, \ \langle S, B, C, M \rangle, \ \langle M, S, B, C \rangle, \ \langle C, M, S, B \rangle \}.$$

We could begin with another list and write the set S_2 of all lists equivalent to it. Continuing, we would get *sets of lists* $S_1, S_2, \ldots, S_k$ such that each list is in one and only one set, with all the lists in any one set being equivalent. Further, if two lists are in different sets, they will not be equivalent. Thus, each *set of lists* corresponds to a seating pattern, and different sets of lists correspond to different patterns.

The question "How many distinct seating patterns do we have?" may be reworded as "How many sets of lists do we have?" The answer is the number k of sets $S_1, S_2, \ldots, S_k$. It is clear that we have four lists in each set and that each list of initials appears in one and only one set. Thus, the union of all of the four-element sets $S_1, S_2, \ldots, S_k$ is the set of *all* lists of the four names and so has 4! elements. On the other hand, by the product principle, the union of k disjoint sets, each of size 4, has $4 \cdot k$ elements. Therefore

$$4! = k \cdot 4$$

so $k = 3! = 6$. Therefore, there are six distinct seating arrangements.
This kind of analysis leads us to a general principle.

Equivalence Classes

Theorem 1.1. Let R be an equivalence relation on a set X. Then
there is a collection $\mathbf{C}$ of nonempty subsets of X such that
(1) Each element x of X is in some set C_x of $\mathbf{C}$.
(2) Any two different sets in $\mathbf{C}$ are disjoint.
(3) For each set C_y in $\mathbf{C}$, all elements of C_y are equivalent relative
    to R.
(4) If two elements x and y are equivalent relative to R, then they lie
    in the same set of $\mathbf{C}$ (in fact $C_x = C_y$).

Proof. For each x in X, define C_x by

$$C_x = \{y \mid (x, y) \in R\}.$$

Let $\mathbf{C}$ be the collection of sets C_x. Note that C_x is not empty (it contains
the element x). Condition 1 of the theorem holds because x is in C_x. To
check condition 2, let x and y be in X. If $C_x \cap C_y$ is not empty, then C_x
and C_y have an element z in common. Using this common element, we
will be able to conclude that $C_x = C_y$. To see why, note that $(x, z) \in R$
since $z \in C_x$, and $(z, y) \in R$ since $z \in C_y$. Then $(x, y) \in R$ by the
transitive law. Thus, $x \in C_y$ and $y \in C_x$. In order to conclude that
$C_x = C_y$, note that if $w \in C_x$, then (w, x) and (x, y) are in R, so that
(w, y) is in R. Therefore, $w \in C_y$. That is, every element of C_x is in C_y.
Similarly, every element of C_y is in C_x, so $C_x = C_y$. In other words, for
any x and y, either $C_x = C_y$ or C_x and C_y are disjoint.
    Condition 3 follows immediately from the definition of C_x and the
symmetric and transitive laws. Condition 4 follows from condition 2,
because if x and y are equivalent then both x and y are in $C_x \cap C_y$
(so then C_x and C_y cannot be different because they are not disjoint).
Thus, the collection $\mathbf{C}$ given by

$$\mathbf{C} = \{C_x \mid x \in X\}$$

is the desired collection. (Note that this notation *does not mean* that if
C_x and C_y are equal, they appear twice. This notation means that the
collection consists of all the distinct sets among the C_x's, each included
once.) ∎

    The sets C_x in the collection $\mathbf{C}$ are called ***equivalence classes***. A
collection $\mathbf{C}$ of nonempty, mutually disjoint sets whose union is X is called

a *partition* of X, or, if there is a chance for confusion, a *set-partition* of X. Thus, given an equivalence relation, we have a partition **C** of X into equivalence classes. It is straightforward to show that given a partition **P** of X, there is an equivalence relation whose equivalence classes are exactly the sets of **P**.

> **Theorem 1.2.** If **P** is a partition of X, then there is one and only one equivalence relation whose equivalence classes are the classes of **P**.
>
> *Proof.* Exercise 7. ∎

At first, the idea of equivalence classes may appear quite abstract. Think, however, of any situation in which you would term some pairs of objects in a set as *equivalent* and others as *not equivalent*. In such a situation, you could divide up the objects into mutually equivalent objects. The groups of mutually equivalent objects are the equivalence classes. Further, given a way of dividing up a set of objects into groups for some purpose, we may think of the objects in a group as equivalent to each other for that purpose, just as we did in Example 1.2. Thus, any use of the word *equivalence* should divide up a set into equivalence classes, and any way of dividing a set up into equivalence classes should be associated with a natural use of the word *equivalence*. This explains our earlier remark that all properties we would associate with equivalence can be derived from the reflexive, symmetric, and transitive properties.

Counting Equivalence Classes

In our example of seating arrangements, all the equivalence classes had the same size. This made it easy to compute the number of equivalence classes. In the exercises, there are examples of useful equivalence relations whose equivalence classes *do not* have the same size. On the other hand, many practical problems give rise to equivalence relations whose equivalence classes all have the same size. In this case we can apply the following theorem, which we will call the **equivalence principle**.

> **Theorem 1.3.** If R is an equivalence relation on a set X with n elements and each equivalence class has m elements, then R has n/m equivalence classes.
>
> *Proof.* This is a direct consequence of the product principle. ∎

Example 1.3. A company has built new corporate headquarters in which five different, but compatible, computer file servers are to be installed for its five divisions. There are five floors linked by a communication cable with a connector on each floor; the cable makes a loop through all five floors and then returns to the first floor. Assume that the file servers have yet to be assigned to the floors. Then the communication patterns around the loop will be determined by which file server goes on which floor. If communications can go in only one direction along the loop (for example, in the order 1, 2, 3, 4, 5 and back to 1), in how many ways can the file servers be arranged along the loop? (Two ways are different if they differ in communications patterns.) How many arrangements are there if communications can go in both directions along the loop?

The first question is just like the question about seating people to play cards. There are 5! assignments of file servers, but shifting each file server the same number of places to the right or left around the loop gives an equivalent assignment. In any other change, some machine would send its messages to a different machine than before. Each assignment belongs to an equivalence class consisting of five different assignments, so there are

$$\frac{5!}{5} = 4! = 24$$

equivalence classes of assignments. Thus, 24 different communications patterns are possible.

On the other hand, if two-way communications are possible, an arrangement with A on floor 1, B on floor 2, C on floor 3, D on floor 4, and E on floor 5 has the same communications pattern as the reversed arrangement $EDCBA$ (because each computer can communicate with exactly the same machines as before). Thus, taking all shifts and reversals, we discover that the arrangement $ABCDE$ generates the equivalence class

$$\{ABCDE, BCDEA, CDEAB, DEABC, EABCD,$$
$$EDCBA, AEDCB, BAEDC, CBAED, DCBAE\}.$$

Now if we shift or reverse any arrangement in this set, we get another arrangement in the set. Thus, the set is indeed the equivalence class generated by $ABCDE$. We have 10 arrangements per equivalence class and $5!/10 = 120/10 = 12$ different communications patterns possible. ∎

The Inverse Image Relation

There is another reason why equivalence relations will be important to us. Given a function f from a set A to a set B, we can define a relation $R(f)$ by saying $(x, y) \in R(f)$ if $f(x) = f(y)$. That is, x is equivalent to y relative to $R(f)$ if $f(x) = f(y)$. It is straightforward to check that $R(f)$ is indeed an equivalence relation. For reasons that we shall see later on, it is natural to call this the *inverse image relation* of f.

Example 1.4. Let f be the function that assigns to each student in a college the postal (zip) code of that person's home town. Describe the inverse image relation $R(f)$ and its equivalence classes.

Two students are related by $R(f)$ if and only if they have the same postal code. Each equivalence class consists of all students with a given postal code. ∎

Example 1.5. Let $f((x,y)) = y - x$ be a function defined on the plane with its usual real number coordinate system. (Why are there double parentheses in the formula?) Describe $R(f)$ and its equivalence classes.

The inverse image relation $R(f)$ is an equivalence relation on ordered pairs (x_1, y_1), (x_2, y_2), etc. of real numbers in which (x_1, y_1) is equivalent to (x_2, y_2) if $y_1 - x_1 = y_2 - x_2$. Thus for each real number b (think of b as the common value of $y_1 - x_1$ and $y_2 - x_2$), there is an equivalence class of all ordered pairs (x, y) such that $y - x = b$ or $y = x + b$. Therefore, each equivalence class is a straight line with slope 1. In particular, when b is zero, the equivalence class we get is the straight line through the origin given by $y = x$. ∎

An example that helps us visualize how inverse image relations could arise in problems of enumeration follows.

Example 1.6. We have ample supplies of oranges, pears, apples, nectarines, and tangerines. In how many ways can we choose one piece of fruit each for six children and give them to the children if Sam, Jo, and Pat all get the same kind of fruit, Bill and Mary get a second kind of fruit, and Sue gets a third kind?

We are asking for the number of functions with inverse image partition

$$\{\{\text{Sam}, \text{Jo}, \text{Pat}\}, \ \{\text{Bill}, \text{Mary}\}, \ \{\text{Sue}\}\}.$$

A choice of fruit determines one fruit for our first set S_1, a second (different) fruit for our second set S_2, and a third fruit for our third set S_3. Thus, it gives a one-to-one function from $\{S_1, S_2, S_3\}$ to $\{O, P, A, N, T\}$. Since there are, in the falling factorial notation of Section 2 of Chapter 1, $(5)_3 = 60$ such one-to-one functions, there are 60 such distributions of fruit. ∎

By applying the same techniques as in the example, we can prove the following theorem.

Theorem 1.4. Suppose B is an n-element set. The number of functions from a set A to B whose inverse image partition is a given partition of A into m equivalence classes (parts) is the number of injections from an m-element set to an n-element set, $(n)_m = n!/(n-m)!$.

Recall that multinomial coefficients (introduced in Example 4.6 of Chapter 1) also gave us information about functions from a set A to a set B in terms of inverse images. However, in the case of multinomial coefficients, not only did we know the number m of equivalence classes, but we also knew that the elements of class 1 were to be labeled with a certain label x_1 from the set B, the elements of class 2 were to be labeled with x_2, and so on. When we asked how many labelings there were with k_1 elements labeled (mapped to) x_1, k_2 elements labeled (mapped to) x_2, and so on, we saw that the answer was

$$C(k; k_1, k_2, \ldots, k_m) = \frac{k!}{k_1! k_2! \cdots k_m!}$$

where $k = k_1 + k_2 + \cdots + k_m$ is the size of A.

In much the same way that we viewed arrangements of people or computers as equivalence classes of lists, we can find a very pretty proof of the formula for multinomial coefficients by regarding labelings with k_i objects receiving label x_i as equivalence classes of lists.

Theorem 1.5. The number of ways to label k objects with labels x_1 through x_m so that label x_i is used k_i times is

$$\frac{k!}{k_1! k_2! \cdots k_m!}.$$

Proof. Given a list of all the elements of an k-element set K, label the first k_1 elements with label x_1, the next k_2 elements with label x_2, ..., and so on. Given a labeling, we can create a list (possibly in many ways) by first listing the elements with label x_1, then the elements with label x_2, ..., and so on. Define two lists to be equivalent if they correspond in this way to the same labeling. The number of ways to list the first k_1 elements is $k_1!$, the next k_2 elements $k_2!$, and so on, so by the product principle, the number of (mutually equivalent) lists per labeling is $k_1! k_2! \cdots k_m!$. This number is the same for each equivalence class of lists that corresponds to a labeling giving k_i elements label x_i. Thus, the number of elements per equivalence class is $k_1! k_2! \cdots k_m!$, so the number of equivalence classes and thus the number of labelings is

$$\frac{k!}{k_1! k_2! \cdots k_m!}.$$

∎

The Number of Partitions with Specified Class Sizes

At times what is important in a given problem is not the actual labeling used but simply the way in which a set is being partitioned.

Example 1.7. A class of 30 college students is to be divided into two discussion sections with 10 students each and two discussion sections with 5 students each to test the effect of size on the value of discussion sections. All the discussions will be led by the same person. In how many ways can the class be divided into these four groups?

Because the same person will lead each section, there is no way to distinguish among (in other words, to label) the sections, so we are asking for the number of partitions of a 30-element set into two classes of size 10 plus two classes of size 5. One way to proceed would be to label the 30 elements with labels 1 through 4 corresponding to discussion sections 1–4. Thus, 10 elements get label 1, 10 elements get label 2, 5 elements get label 3, and 5 elements get label 4. For our purposes, two labelings would be equivalent if we got one from the other by interchanging the people with label 1 and label 2, or by interchanging those with label 3 and label 4, or both. Thus, each labeling is equivalent to three other labelings. We now have an equivalence relation on the set of labelings with four labelings in each equivalence class. All the labelings in an equivalence class are the same for our purposes; the number of divisions of the class is the *number* of equivalence classes. The total number of labelings is $C(30; 10, 10, 5, 5)$, and so by Theorem 1.3 the number of classes is $C(30; 10, 10, 5, 5)/4$. ∎

The partitions we are considering for the students are said to have "type vector" $(0, 0, 0, 0, 2, 0, 0, 0, 0, 2)$, meaning that the partition has two parts of size 5, two parts of size 10, and 0 parts of size 1, 2, and so on. In general, we say a partition has **type vector** $(j_1, j_2, \ldots, j_n)$ if the partition has j_i classes of size i for $i = 1$ to n. Thus, if $j_i = 0$, we have no classes of size i; if $j_i = 1$, we have one class of size i, and so on. (As before, we drop trailing zeros, so the length of a type vector is the size of the largest part.)

Theorem 1.6. The number of partitions of a set with n elements into j_1 classes of size 1, j_2 classes of size 2, up to j_n classes of size n is

$$\frac{n!}{\prod_{i=1}^{n}(i!)^{j_i} \cdot j_i!}.$$

Proof. We can form a partition with the type vector $(j_1, j_2, \ldots, j_n)$ by listing all the n elements and taking the first j_1 elements to be classes of size 1, then taking the next $2j_2$ elements two at a time to be classes of size 2 listed one after another, then taking the next $3j_3$ elements to be classes of size 3 listed one after another, and so on. We say that two lists are equivalent if they yield the same partition of type $(j_1, j_2, \ldots, j_n)$. Since this is an equivalence relation, the number of partitions of type $(j_1, j_2, \ldots, j_n)$ is the number of equivalence classes. One way in which two lists may be equivalent is if one is obtained from the other by any rearrangement of the j_1 classes of size 1. There are $j_1!$ ways to do so. Similarly, listing classes of size i in a different order gives an equivalent list. There are $j_i!$ such arrangements. (Note that if j_i is zero, $j_i!$ is one.) Also, a given class of size i may be listed in $i!$ different ways. Since there are j_i classes of size i, there are, by the multiplication principle, $(i!)^{j_i}$ ways to arrange the elements within each class of size i. Since there are an additional $j_i!$ ways to arrange the classes themselves, by the multiplication principle there are $(i!)^{j_i} \cdot j_i!$ ways to arrange the elements of classes of size i among the appropriate positions in the list. By one more application of the multiplication principle, there are

$$(1!)^{j_1} j_1! (2!)^{j_2} j_2! \cdots (n!)^{j_n} j_n! = \prod_{i=1}^{n} (i!)^{j_i} j_i!$$

ways to arrange all the n elements in a list that represents a given partition of type $(j_1, j_2, \ldots, j_n)$. Thus each equivalence class has this number of lists. The total number of lists is $n!$. Therefore, by Theorem 1.3, the number of equivalence classes is

$$\frac{n!}{\prod_{i=1}^{n}(i!)^{j_1} \cdot j_i!}. \qquad \blacksquare$$

It is possible to give a proof based on labelings as in the example that precedes the theorem. However, the notation of type vectors differs from the notation we used to describe labelings. In the exercises, we shall reformulate the notation of labelings so that you can relate the two ideas for yourself. Note, however, that $\frac{30!}{(5!)^2 \cdot 2! \cdot (10!)^2 \cdot 2!}$ is the same as $C(30; 10, 10, 5, 5)/4$, so our example and our theorem agree.

EXERCISES

1. Write down the relation determined by "x is related to y if $|x - y| \leq 1$" for the set of integers from 1 to 5. Is this relation an equivalence relation?

2. Is the relation "is a brother of" an equivalence relation on the set of all males? On the set of all people? Answer the same questions for "is a brother of or is." (The point is that a male *is not* his own brother but *is* himself.) Answer the same questions for "is a stepbrother of or is."

3. Two people are siblings if they are brother or sister. Answer the questions of Exercise 2 for "is a sibling of." Answer the questions for "is a sibling of or is."

4. Two people are cousins if a parent of one is a sibling of a parent of the other. Is the relation "is a cousin of or is" an equivalence relation?

5. Write down the set of ordered pairs that is the relation "is a factor of" for the set of integers between 1 and 12. (Note, $2 = 2 \cdot 1$, so 2 is a factor of 2.) Is this relation an equivalence relation?

6. For this exercise, write an ordered pair of positive integers as $a \mid b$. Is the relation given by "$a \mid b$ is related to $c \mid d$ if $ad = bc$" an equivalence relation? Can you give an everyday mathematical description of this relation?

7. Let $\mathbf{P} = \{C_1, C_2, \ldots, C_k\}$ be a partition of the set S. Suppose the relation R is defined by $(x, y) \in R$ if x and y are both in the same set of $\mathbf{P}$, i.e., the same C_i. Show that R is an equivalence relation. What are the equivalence classes of R? What part of Theorem 1.2 have you just proved? Finish the proof of Theorem 1.2.

8. In how many ways can k people be arranged around a circular table?

9. In bridge, four people play and the person who sits opposite you is your partner.
   (a) If we consider two arrangements to be equivalent if and only if each person has the same partner in both arrangements, then in how many ways can we arrange four people around a table for a game of bridge?
   (b) Answer the question of part (a) if two arrangements are equivalent if and only if each person has the same partner in both and the same person to the right in both.

10. In how many ways can five men and five women be arranged around a circular table so that no two neighbors are of the same sex?

In Exercises 11–16, two arrangements of beads are equivalent as necklaces if one can be obtained from the other by picking it up, moving it around, perhaps flipping it, and putting it back down.

11. In how many ways can six distinct beads be arranged on a string as a necklace?

12. Suppose we are given two red beads (R) and two black beads (B). Define two lists of two R's and two B's as equivalent if the same necklace is obtained by stringing the red and black beads in the orders given in the two lists. Show that this is an equivalence relation on the set of lists of two R's and two B's. Write down the equivalence classes and note that they have different sizes. How many necklaces are possible?

13. In how many ways can k distinct beads be arranged on a necklace?

14. In how many ways can k distinct red beads and two distinct black beads be arranged on a necklace?

15. We have k distinct red beads and k distinct black beads. In how many different ways can the beads be arranged around the edge of a circular table so that they alternate in color? In how many ways can they be strung on a necklace so that they alternate in color?

16. Answer the two questions of Exercise 15 assuming that all the beads of one color are together in a row (in place of the assumption that the beads alternate in color).

17. Six computers are to be arranged on a network as if they are at the vertices of a hexagon. Each computer is to be connected for two-way communications to all but the computer that is at the "opposite" vertex. How many inequivalent (for communications purposes) arrangements are there? (Assume the computers can be distinguished from one another.)

18. The substance dichlorobenzene is sometimes visualized as having a core consisting of a ring of six carbon atoms, each connected to the two adjacent to it. The ring has hydrogen atoms attached to four of the carbon atoms (one hydrogen for one carbon) and two chlorine atoms attached to the other two carbon atoms (again, one each). Using Cl for chlorine and H for hydrogen, we may represent such an arrangement as a ring of four H's and two Cl's, or an equivalence class of lists of four H's and two Cl's. Write down the possible equivalence classes and, assuming that any geometric motion that does not break any connections will give an equivalent list, determine how many forms of dichlorobenzene are theoretically possible. (Those with some experience in chemistry will note that the model we have chosen here assumes that the six

carbon bonds are all equivalent rather than alternating between single and double bonds. It is interesting from the point of view of chemistry, but not part of this problem, to answer the question for this second model.)

19. What are the maximum possible size and minimum possible size of an equivalence relation on an n-element set?

20. What is the inverse image relation $R(f)$ associated with $f(x) = x^2$ for all real numbers x and what are its equivalence classes? (Note: You should *not* have to deal with ordered pairs as we did in Example 1.5. Why?)

21. How many equivalence classes will $R(f)$ have for a function f from an n-element set to a k-element set if
    (a) f is one-to-one?
    (b) f is onto?

22. How many labelings are there for n-element sets with j_1 of the labels used once, j_2 of the labels used twice, ..., and j_n of the labels used n times?

23. Derive Theorem 1.6 from the result of Exercise 22.

24. In how many ways can 40 students be grouped into four groups of 5 each for a discussion section led by a graduate student and two groups of 10 each for a section led by a professor? For the purposes of this exercise, assume that the groups are not yet assigned a time slot or a discussion leader, so that the only differences among the groups are in their sizes.

25. How many partitions of a 15-element set have two classes of size 4, three classes of size 2, and one class of size 1? How many have one class each of sizes 1, 2, 3, 4, and 5?

26. In how many ways can 100 distinct beads be used to make three necklaces with 20 beads and four necklaces with 10 beads?

27. Suppose we have chosen a fixed positive integer n and we define two integers to be related to each other if their difference is a multiple of the fixed number n. (Note that 0 is a multiple of n since $0 = 0 \cdot n$.) Show that this is an equivalence relation. In the case $n = 2$, what are the equivalence classes?

28. Remember that we have defined a relation to be a special kind of set.
    (a) Is the intersection of two equivalence relations on the same set an equivalence relation? (Give reasons.)
    (b) Is the union of two equivalence relations on the same set an equivalence relation? (Give reasons.)

29. The ***transitive closure*** of a relation is defined to be the intersection of all the transitive relations that contain it as a subset.
    (a) Show that the transitive closure of a relation is in fact transitive.
    (b) Show that if a transitive relation T contains R as a subset, then it contains the transitive closure of R as a subset.
    (c) Describe the circumstances under which the transitive closure of a symmetric relation will be an equivalence relation.

Section 2 Distributions and Multisets

The Idea of a Distribution

Most of the formulas we derived in Chapter 1 can be regarded as solutions to problems of distributing objects to recipients. For example, a function f from a domain D to a range R may be thought of as telling us for each object x in D, to distribute x to the recipient $f(x)$ in R. Of course, if $y = f(x)$ for several different objects x, then all these objects are distributed to y. Similarly, if y is not $f(x)$ for any object x, then y receives nothing. In this way, we view a function as distributing objects in D to recipients in R. Thus, the formula n^k for the number of functions from a k-element set to an n-element set is also the formula for the number of distributions of k distinct objects to n distinct recipients.

Example 2.1. In how many ways may we distribute three different jobs to five workers?

Since there are $5^3 = 125$ functions from a three-element set to a five-element set, there are 125 possible ways to distribute the jobs. (You could also argue from scratch that there are five workers who could be given the first job, five who could be given the second, and five who could be given the third and then use the product principle, but this would simply repeat the argument we used to count functions.) ∎

Perhaps we think it is unfair in the last example to be able to distribute all three jobs to one worker and leave the remainder without any jobs. It might be more fair to ask for a distribution that gives at most one job to each worker. Since one-to-one functions assign different objects to different recipients, $(n)_k$, the first k terms of $n!$, is the formula for the number of distributions of k distinct objects to n recipients so that no recipient gets more than one.

Example 2.2. There are $5 \cdot 4 \cdot 3 = 60$ ways to assign three different jobs to five workers so that each gets at most one job. ∎

Similarly, the number of onto functions, for which we have no formula as yet, is the number of distributions such that each recipient gets at least one. The number of one-to-one and onto functions, or bijections, is the number of ways to distribute n objects to n recipients so that each gets exactly one.

To view the formula for the number of k-element subsets as the solution of a similar distribution problem, we can use the labeling interpretation of the binomial coefficients. Recall that choosing a k-element subset of an n-element set S amounts to giving the label "in" to k elements of S and the label "out" to the remaining elements. Once the "in" labels have been distributed, there is no choice for where the other labels go, so the number $\binom{n}{k}$ of k-element subsets is the number of ways to distribute the k identical "in" labels to the n elements of the set. Thus, binomial coefficients represent solutions to problems of distributing identical objects, the labels that say "in."

Example 2.3. There are $\binom{5}{3} = 10$ ways to choose three workers from among five who will work together to do one job. Similarly, there are $\binom{5}{3} = 10$ ways to distribute three identical jobs (such as packing boxes of 36 assorted dishes at a factory making dishes) to five workers. ∎

To unify and extend our counting techniques, we begin a systematic study of distributions of objects to recipients. We may allow the objects to be identical (as in distributing identical marbles to children) or distinct (as in assigning different jobs to workers). We may allow the recipients to be identical (as putting fruit into paper bags) or distinct (as children presumably would be).

We will also consider restrictions on how the recipients receive the objects. The most obvious possible restrictions we might consider are that no recipient receives more than one object and that each recipient receives at least one object. There is another restriction on the receipt of objects that also arises in practice. Imagine arranging books for a display on the shelves of a bookcase or arranging cheeses for a cheese tasting. We would say two arrangements of books on the shelves are different when the books are in different orders, even if each shelf has exactly the same set of books. Similarly, the order in which people receive cheese to taste affects their perception of the flavor of the cheese. Thus, when we are distributing distinct objects to recipients and the recipients may receive more than one object, we distinguish between distributions, called *ordered distributions* (defined more precisely later), in which the order in which a recipient receives objects matters and distributions in which the order does not matter. Distributions in which the order doesn't matter are simply *functions* whose domain is the set being distributed and whose range is the set of recipients.

How many kinds of distributions are there? Let us set ourselves up to use the product principle to count them. We have two choices of whether or not the domain and two choices of whether or not the range of the distribution are distinct objects. We have two conditions on whether or not each person gets at most one object, two more on whether or not each person gets at least one object, and two more on whether or not the order of receipt matters. By the product principle, we get $2 \cdot 2 \cdot 2 \cdot 2 \cdot 2 = 32$ different kinds of distribution problems.

Some of these problems are trivial or silly—such as distributing objects to recipients who may receive at most one and saying order matters. To avoid the silly problems, we divide any situation with distinct objects into three cases: the case in which the order in which the objects are received matters, the case in which each recipient receives at most one object, and the case in which neither restriction is enforced. Now we can use the product principle again. For each of these three possibilities, we have either identical or distinct recipients. Finally, we could have either the additional restriction that each recipient must receive at least one object or we could have no additional restriction so that some recipients may receive no objects. This gives us $3 \cdot 2 \cdot 2 = 12$ cases that appear in the first six rows of Table 2.1, our first of several versions of a table describing the solution of various kinds of distribution problems.

Table 2.1   First Table of Distribution Problems

Domain (size k)	Range (size n)	May receive 0	Each receives ≥ 1
Distinct	Distinct	Functions n^k	Surjections [?]
Distinct	Distinct (Each receives ≤ 1)	Injections $(n)_k$	Bijections $\begin{cases} n! \text{ if } n = k \\ 0 \text{ otherwise} \end{cases}$
Distinct	Distinct (Order received matters)	Ordered distributions [?]	?
Distinct	Identical	?	?
Distinct	Identical (Each receives ≤ 1)	1 if $k \leq n$ 0 otherwise	1 if $k = n$ 0 otherwise
Distinct	Identical (Order received matters)	?	?
Identical	Distinct	?	?
Identical	Distinct (Each receives ≤ 1)	Subsets $\binom{n}{k}$	1 if $n = k$ 0 otherwise
Identical	Identical	?	?
Identical	Identical (Each receives ≤ 1)	1 if $k \leq n$ 0 otherwise	1 if $k = n$ 0 otherwise

Each possible entry of Table 2.1 to the right of the vertical line represents a problem. We have filled in the information we have about the problems we have discussed so far. For example, the entry in row one and the first column to the right of the vertical line (which we shall call column one in the following discussion) corresponds to the fact that functions are distributions of distinct objects to distinct recipients. The entry in row one and column two should be the number of distributions of k distinct objects to n distinct recipients in which each recipient receives at least one object. As we noted earlier, this is the number of functions from the domain *onto* the range, or the number of surjections from the domain to the range. We note this fact in the table, but since we have no method for computing this number, we include a question mark as well. The entry in the second row and first column should be the number of distributions of k distinct objects to n distinct recipients. We have noted before that this is the number of one-to-one functions or injections from the domain to the range and that this number is $(n)_k$, and so that is what is filled in in the table. Similarly, the entry in row two and column two is a familiar number, the number of bijections from the domain to the range, which is $n!$ if the domain and range have the same size and is 0 otherwise. In row three and column one, we have named the number that is desired, but have no method to compute it, and in row three and column two, we have neither a name nor a formula.

Some of the entries in the table are trivial in the sense that once we analyze their situation, we see that either there is no distribution satisfying the conditions or all the distributions are equivalent. For example, in row five and column one we see that if the recipients are identical (for example grocery bags) and each is to receive at most one object, then the problem is impossible if there are more objects than recipients; however, if there are not too many objects, then all ways of giving each of the objects to a different one of the identical recipients are equivalent. On the other hand, if each recipient is to receive exactly one object, then there must be exactly as many objects as recipients in order for the number of distributions to be nonzero, and in this one case, all ways of distributing the objects are equivalent (because it doesn't matter which of the identical bags we put which object in; we wouldn't be able to notice the difference if we did it differently). In rows four and six we have only question marks because the problems the entries correspond to have not yet been discussed.

The horizontal double line near the middle of the table divides the problems into those with distinct objects and those with identical objects. When the objects to be distributed are identical, a restriction on the order in which objects are received is silly, so we distinguish only the case in which each recipient receives at least one object and the case without this restriction. Again, we have either identical or distinct recipients and either the restriction that each recipient must get at least one object or the possibility that some recipients can get none. This gives us the $2 \cdot 2 \cdot 2 = 8$

cases that appear in the last four rows of Table 2.1. Of these problems, the only nontrivial one we have solved so far is that in row eight and column one, the number of ways to distribute k identical objects to n distinct recipients. As we remarked before, this is the same as the number of k-element subsets of an n-element set, explaining the binomial coefficient in that place. Our goal in this chapter is to learn as much as we can about the other entries of the table. We will present more complete versions of this table at the end of this section and the next; rather than waiting for these summary tables, you may find it useful to note as we go which question marks we are working on.

Ordered Distributions

In order to decide which problems to take up next in filling in the table, let us analyze the techniques we have used to obtain the (nontrivial) answers we have so far. Computing the number of functions or injections (one-to-one functions) was a straightforward application of the product principle. If you analyze our computation of the number of k-element subsets of a set, you will see that we were considering two k-element permutations of the n-element set N to be equivalent if they permute (have as their image) the same subset of N. When we divided $(n)_k$ by $k!$, we were dividing the number of k-element permutations by the number of permutations per equivalence class to compute the number of equivalence classes. Thus, our previous experience suggests that a reasonable approach to filling in the rest of the table might be first to look for problems that we can solve with the product principle, then to look for ways to apply the equivalence principle to these problems to count the number of equivalence classes.

We have used the product principle so far to solve problems involving distinct objects and distinct recipients. This suggests that we try to find other problems with distinct objects and distinct recipients. Row three provides us with two such problems; since the function problem was easier in the case where we could distribute any number of objects to any recipient (that is, in column one), let us try to compute the number of *ordered distributions* of k objects to n recipients.

For this purpose it will be useful to have a precise definition of an ordered distribution. Since we are thinking of distributing the objects in a certain order to a recipient, we are assigning a permutation of some of the elements of the domain to each recipient. (Of course, some can get the empty permutation.) Thus, we define an ***ordered distribution*** of the objects in a set K to the recipients in a set N to be a function that assigns a permutation of some of the elements of K to each element x of N in such a way that each element of K is in one and only one of these permutations. To visualize a counting technique for ordered distributions of k distinct objects to n distinct recipients, we imagine placing books on the shelves of a bookcase. For the first book, there are n places to put it (one at the beginning of each shelf.) Now for the second book, there are all the previously available places except for the one we just used; it has been converted to *two* places, one before and one after the book just placed. Thus, there are $n + 1$ places to place book 2. Placing book 2 converts another place into two places, so there are $n + 2$ places for book 3. We continue this way, increasing the number of places by 1 when we place a book, so when we are about to place book k, we have $n + k - 1$ places to put it. Thus, by the product principle, we have

$$\prod_{j=0}^{k-1} n + j = \prod_{i=1}^{k} n + i - 1$$

ways to put the k books onto the shelves. Notice that this number is

$$(n + k - 1)!/(n - 1)!.$$

By converting the discussion of books and shelves to objects and recipients, we could prove the following.

Theorem 2.1. The number of ordered distributions of k distinct objects to n distinct recipients is

$$\prod_{i=1}^{k} n + i - 1 = (n + k - 1)!/(n - 1)! = (n + k - 1)_k.$$

Proof. Similar to the preceding remarks about bookshelves. ∎

Example 2.4. The number of ways to arrange six books on three shelves is $(3 + 6 - 1)_6 = 8 \cdot 7 \cdot 6 \cdot 5 \cdot 4 \cdot 3 = 20{,}160$. ∎

By thinking in terms of choosing the books that must be the first book on each shelf and then distributing the remaining books, you can prove the following theorem, which tells us how to fill in the second column of row three.

Theorem 2.2. The number of ordered distributions of k objects to n recipients so that each recipient receives at least one object is

$$(k)_n (k - 1)_{k-n} = \frac{k!(k - 1)!}{(k - n)!(n - 1)!} = k!\binom{k - 1}{n - 1}.$$

Proof. Exercise 23. ∎

Distributing Identical Objects to Distinct Recipients

We have now solved the distribution problems which analogy suggested might fall to the product principle. Our next task is to examine what new problems we should be able to solve by applying the equivalence principle. Since the problem of distributing identical objects to distinct recipients so that no one gets more than one, the subset problem, was solved by dividing the number of permutations by the number of permutations per equivalence class, a natural next problem to try is the problem of distributing k identical objects to n recipients without any restrictions. This is a good problem to do next for practical reasons also. Although, for the sake of intuition, we will begin by speaking of distributing candy or fruit to children, the problem of distributing identical objects to distinct recipients has a wide variety of applications. Feller's book on probability (see the suggested reading list at the end of the chapter), for example, gives 16 possible applications.

When we solved the subset problem in Chapter 1 we did not have the language of equivalence relations and equivalence classes. Thus, to make reasoning by analogy easier, let us reformulate the solution in terms of equivalence classes. When we computed the number of k-element subsets of an n-element set, say N, we were examining k-element permutations of the n-element set. These are one-to-one functions from $K = \{1, 2, \ldots, k\}$ into the n-element set. When we say the elements of the domain K are identical, we mean that for our purposes they are equivalent. We want to consider two permutations f and g as equivalent if, after we make the assignment and then regard the objects as identical, the objects assigned by f are assigned in exactly the same way as the identical objects assigned by g.

What matters is not which element of the domain is assigned to a range element but whether a domain element is assigned at all. Thus, any of the $k!$ ways to rearrange the assignment that f makes of domain elements to range elements results in an equivalent permutation g. Further, if g is equivalent to f, then the only difference between them may be in which domain element is assigned to a specific range element. Thus, a g equivalent to f must be obtained from f by rearranging which domain elements are assigned to which range elements. Therefore, f is equivalent to exactly $k!$ permutations, so the equivalence class containing f has size $k!$. Since there was nothing special that we assumed about f, each equivalence class has size $k!$. Therefore, the number of equivalence classes is the total number of k-element permutations, namely $(n)_k$, divided by the size of an equivalence class, namely $k!$, giving us the binomial coefficient formula

$$\binom{n}{k} = (n)_k/k!.$$

Our analogy with subsets suggests that we try to regard counting distributions of identical objects to distinct recipients as counting equivalence classes of distributions of distinct objects to distinct recipients. We have solved two problems involving such distributions, the problem of functions and the problem of ordered distributions. To apply the correspondence principle, we must find a way to regard the distributions of identical objects as *equal-sized* equivalence classes of functions or ordered distributions. Since the formula for ordered distributions is quite similar to the formula for permutations, let us see how making the domain objects equivalent to each other influences an ordered distribution. Suppose we have an ordered distribution of a set K to N. If we exchange two elements of K and they were distributed to different members of N, then we have converted one ordered distribution d to another one, d'. However, these two distributions are equivalent when the objects are identical. Suppose, instead, that we interchange two elements of K that were assigned to the same member of N. Then they are now assigned in a different order, so we have again converted one ordered distribution d to another one, d'. Once again, these two distributions are equivalent when the domain objects are equivalent. In this way, we see that every way of rearranging—that is, permuting—the domain objects converts an ordered distribution to another one, giving us $k!$ ordered distributions equivalent to the original one. Further, an ordered distribution equivalent to the original one will specify a permutation (rearrangement) of the domain objects, so the number of distributions per equivalence class is $k!$. Dividing the number $(n + k - 1)_k$ of ordered distributions by $k!$ gives us the number of distributions of k identical objects to n distinct recipients. We have proved the following theorem.

Theorem 2.3. The number of distributions of k identical objects to n distinct recipients is

$$(n + k - 1)_k/k! = \binom{n + k - 1}{k} = \frac{(n + k - 1)!}{(n - 1)!k!}.$$

Proof. We have already given the proof that the left-hand side of the equations is the desired number of distributions; the equalities follow from the definition of binomial coefficients. ∎

Notice that the fraction on the right-hand side of the theorem has a product of two factorials as its denominator. This suggests that we may be able to interpret the problem as a problem of counting equivalence classes in which the size of an equivalence class is the denominator, $(n-1)!k!$, and the set on which we are defining the equivalence relation has size equal to the numerator, or $(n+k-1)!$. The most natural kind of a set of this size would be a set of lists of length $n+k-1$. In order to understand the formula better, we shall examine such an interpretation. Rather than talk in abstract terms, let us work with an example.

Example 2.5. In how many ways can k identical candy bars be distributed among n children?

We can number the candy bars (in our imagination) with the numbers 1 through k. Imagine the n children sitting in a row, and imagine that the candy is distributed by placing the pieces of candy intended for some child directly in front of this child. Then we have k objects divided into n piles (some of the piles might be empty). To make sure the piles don't get mixed up, we can put $n-1$ dividers between the piles. (Why do we need $n-1$?) We regard all the objects as equivalent and note that the $n-1$ dividers can be used in any order. Thus, we have k pieces of candy $c_1, c_2, \ldots, c_k$ and $n-1$ dividers $d_1, d_2, \ldots, d_{n-1}$, and we determine a division by making a list of these $n+k-1$ objects. A rearrangement of the pieces of candy will not give a different division, nor will a rearrangement of the dividers. Thus, two lists of c_i's and d_j's are equivalent if one can be obtained from the other by rearranging the c_i's among themselves or by rearranging the d_j's among themselves (or both). By the product principle, each list is equivalent to $k! \cdot (n-1)!$ other lists. Since this equivalence is an equivalence relation on the set X of all $(n+k-1)!$ lists, Theorem 1.3, the equivalence principle, tells us there are

$$\frac{(n+k-1)!}{k!(n-1)!}$$

equivalence classes of lists. ∎

Now suppose we want to know the number of distributions such that each child gets a piece of candy. (This is the number desired in row seven and column two of our table of distribution problems.) We just give each child one piece of candy and then distribute the remaining $k-n$ pieces to the n children. This gives, once we work out the details, the following corollary to Theorem 2.3.

Corollary 2.4. The number of distributions of k identical objects to n distinct recipients so that each recipient gets at least one is $\binom{k-1}{k-n}$.

Proof. See Exercises 18 and 20. ∎

Theorem 2.3 and Corollary 2.4 give us both entries in row seven of our table of distribution problems.

Before discussing an important application of these computations, we note one other interpretation of the formula in Theorem 2.3. The binomial coefficient $\binom{n+k-1}{k}$ is the number of k-element subsets of a set of the appropriate size; is there a natural interpretation of the problem in these terms? In Example 2.2 we saw how to make a list of dividers and recipients to solve the problem; this list has $n + k - 1$ entries. By choosing the k places where the candy goes or the $n - 1$ places where the dividers go, we completely determine the distribution. Since there are $\binom{n+k-1}{k}$ ways to choose these places, this is the number of distributions.

We now have seeen three different solutions of the one problem of computing the number of ways to distribute k identical objects to n distinct recipients. Each solution contributes a different aspect to our understanding and possibly one makes more sense to you than the others. This illustrates an important principle about how to learn mathematics. To the extent you can, solve every problem in as many different ways as you can, and you will strengthen your understanding of the subject. In fact, you will build mental links between concepts that will help you discover new ideas much more easily.

Ordered Compositions

From a distribution of the candy to the children, we can define a function f from the set of children to the nonnegative integers by letting $f(x)$ be the number of pieces of candy given to child x. We let the symbols $x_1, x_2, \ldots, x_n$ stand for the names of the children. Then a distribution of k identical pieces of candy to the n children corresponds to a function f such that

$$\sum_{i=1}^{n} f(x_i) = k.$$

Another way to state the solution of the distribution problem involves such a function and is our next theorem.

> **Theorem 2.5.** The number of functions f from a set $\{x_1, x_2, \ldots, x_n\}$ to the nonnegative integers such that
>
> $$\sum_{i=1}^{n} f(x_i) = k$$
>
> is
>
> $$C(n + k - 1; k).$$

Proof. The proof consists of setting up a bijection between equivalence classes of ordered distributions and the functions described. ∎

In number theory, the type of function given in Theorem 2.5 would be called an *ordered composition* (or simply a composition) of k with n parts, because it gives a list (in a specific order) of n nonnegative numbers that add up to k.

Multisets

Our final application of the concept of a distribution of identical objects tells us the number of ways to make possibly repeated selections of objects out of a set. For example, the set of letters used in the word "roof" is the set $\{f, o, r\}$ (in alphabetical order). The full selection of letters used, however, is $\{f, o, o, r\}$. Because this notation suggests a set with multiple occurrences of elements allowed, we call such a selection a "multiset." Thus, the multiset of letters of the word "feet" is $\{e, e, f, t\}$ while the set of letters used is $\{e, f, t\}$. If we throw a pair of dice 15 times, the *multiset* of resulting sums has exactly 15 entries. However, since the sum must be a number between 2 and 12, the *set* of results has no more than 11 entries and could even have just 1 entry. Sometimes a multiset is called a "combination with repetitions," but we will not use this terminology.

To make the idea of a multiset precise, we define a k-element **multiset** *chosen from a set* S as an ordered pair $(S, m) = M$ where m is a function from S to the nonnegative integers such that the sum of the values of $m(x)$ for all x in S is k. The number $m(x)$ is called the **multiplicity** of x in M. The idea is that $m(x)$ tells us how often x appears in our multiset. Such a formal definition of a multiset is important primarily because it shows us how to determine precisely when an object is or is not a multiset. We can take advantage of the precise definition to determine exactly how many multisets of size k we can choose from a set S with n elements. It is clear that we have one multiset for each function m. Thus by Theorem 2.5, the number of multisets of size k chosen from an n-element set is $C(n + k - 1; k)$. This is the number of distributions of k identical objects to n distinct recipients. Notice the similarity to the distribution interpretation of subsets. Multisets of k elements chosen from a set of size n are sometimes called "combinations (with repetitions) of n things taken k at a time."

Example 2.6. How many fruit baskets with 10 pieces of fruit can be constructed using any number of oranges, apples, pears, and bananas?

We are asking for the number of 10-element multisets chosen from a 4-element set, so our answer is

$$C(10 + 4 - 1; 10) = \binom{13}{10} = 13 \cdot 22 = 286.$$ ∎

Example 2.7. What is the multiplicity of each letter in the word *excellent*?

Since x, c, n, and t each occur once, $m(x) = m(c) = m(n) = m(t) = 1$. Since l appears twice, $m(l) = 2$, and since e appears three times, $m(e) = 3$. Note that m is a multiplicity function for a nine-element multiset chosen from a six-element set. If, instead, we want to regard the multiset as a multiset chosen from the entire alphabet, then we simply define the multiplicity of each other alphabet letter to be 0. In that case we have a multiset chosen from a 26-element set. ∎

If the function m happens to take on only the values zero and one, then the multiset it describes can in fact be regarded as a subset of S. In fact, in this case m is the function, called the characteristic function, that we introduced to represent subsets in our second proof that an n-element set has 2^n subsets.

Broken Permutations of a Set

An ordered distribution of K to N assigns nonoverlapping permutations of subsets of K to the distinct elements of N. If the recipients are identical, then an assignment of permutations to recipients may be thought of as simply the set of permutations being assigned. Since all the elements in K must be distributed, we have a set of (nonempty) permutations of subsets of K such that each element of K is in exactly one of the permutations. One natural name for such a collection is a "permutation-partition" of K, in analogy to a set-partition of K, which is a set of (nonempty) *sets* such that each element of K is in one and only one of them. We might also think of the individual permutations as the result of breaking up a permutation of K into parts and so to avoid using the word partition in too many different ways, we define a **broken permutation** of K with n parts to be a set of n nonempty permutations (of subsets of K) such that each member of K is in one and only one of them.

Example 2.8. One broken permutation of $\{1, 2, 3, 4, 5\}$ is

$$\{(2, 4, 5), \ (3, 1)\}.$$ ∎

The number $L(k, n)$ of broken permutations is thus the number of ordered distributions of k distinct objects to n identical recipients so that each recipient receives at least one object. We call the numbers $L(k,n)$ the **Lah numbers**. The total number of ordered distributions of k distinct objects to n identical recipients is then the number of broken partitions with n or fewer parts, $L(k,1) + L(k,2) + \cdots + L(k,n)$. There does not seem to be a simpler formula than this sum, but the kinds of methods we used earlier give a formula for $L(k,n)$ similar to the other formulas we have found for distribution problems.

Theorem 2.6. The number of broken permutations of a k-element set with n parts is given by

$$L(k,n) = \binom{k}{n}(k-1)_{k-n}.$$

Proof. Exercise 21. ∎

Note that if $n > k$, the formula tells us that $L(k,n) = 0$, which is what we expect since we cannot break a permutation into more parts than it has elements. The formula also gives us $L(0,0) = 1$; this is consistent with having one broken permutation of the empty set, namely the empty set of nonempty permutations of the empty set. Rather than using this somewhat arcane (though completely appropriate and correct) interpretation of the Lah number $L(0,0)$, we may simply agree that $L(0,0) = 1$, $L(0,n) = 0$ for $n > 1$, and $L(k,n)$ is defined in terms of broken permutations only when $k > 0$.

Example 2.9. In how many ways can we divide 10 books into six stacks of books?

We assume that the order of the books in a stack matters, but since we have not said anything about where the stacks are placed, we view the recipients of the stacks as identical. Thus, the number of ways to divide the 10 books into six stacks is

$$L(10,6) = \binom{10}{6}(10-1)_{10-6} = 210 \cdot 9 \cdot 8 \cdot 7 \cdot 6 = 635,040. \qquad ∎$$

The other entries of our table of distribution problems will take still more effort to fill in—in fact, for some of them we will not get *formulas* at all. We show our progress so far in the second version of the table of distribution problems, Table 2.2.

Table 2.2   Second Table of Distribution Problems

Domain (size k)	Range (size n)	May receive 0	Each receives ≥ 1
Distinct	Distinct	Functions $\quad n^k$	Surjections [?]
Distinct	Distinct	Injections $\quad (n)_k$	Bijections $\quad \begin{cases} n! \text{ if } n = k \\ 0 \text{ otherwise} \end{cases}$
(Each receives ≤ 1)			
Distinct	Distinct	Ordered distributions	$k!\binom{k-1}{n-1} = k!\binom{k-1}{k-n}$
Order received matters)		$(n+k-1)_k$	
Distinct	Identical	?	?
Distinct	Identical	1 if $k \leq n$	1 if $k = n$
(Each receives ≤ 1)		0 otherwise	0 otherwise
Distinct	Identical	$\sum_{i=0}^{n} L(k, i)$	Broken permutations
(Order received matters)			$L(k, n) = \binom{k}{n}(k-1)_{k-n}$
Identical	Distinct	Multisets $\quad \binom{n+k-1}{k}$	$\binom{k-1}{n-1} = \binom{k-1}{k-n}$
Identical	Distinct	Subsets $\quad \binom{n}{k}$	1 if $n = k$
(Each receives ≤ 1)			0 otherwise
Identical	Identical	?	?
Identical	Identical	1 if $k \leq n$	1 if $k = n$
(Each receives ≤ 1)		0 otherwise	0 otherwise

EXERCISES

1. In how many ways can 10 identical candy bars be distributed to four children?

2. In how many ways can 10 identical candy bars be distributed to four children if each child must get at least one candy bar? (Hint: If you first give each child one piece, how many pieces can you distribute at will?)

3. In how many ways can k identical pieces of candy be distributed among n children if each child is to get at least j pieces?

4. In how many ways can you arrange r distinct books on s shelves if each shelf must have at least two books?

5. Write down the multiplicity function for the letters in the multiset of letters of the phrase *ulterior motive*.

6. Give a sensible definition of a *multisubset* of a *multiset*, and explain why you think it is sensible.

7. Three dice are thrown. The three tops give a multiset of numbers chosen from $\{1, 2, 3, 4, 5, 6\}$. How many different multisets of three numbers could be facing up?

8. A store sells eight different kinds of candy. In how many ways can you choose a bag of 15 pieces?

9. A fast food restaurant sells four different breakfasts. How many orders for breakfast for a family of six are possible? (Ignore any question of which family member is to receive which breakfast.)

10. Two committees of four people each are chosen from a 10-member board of directors. The people on the two committees form an eight-element multiset. How many such multisets are possible?

11. Explain what formula and what concept should be applied to solve each part of Exercise 19 of Section 3 of Chapter 1.

12. The neighborhood betterment committee has been given r trees to distribute to s families living along one side of a street.
    (a) In how many ways can they distribute all of them if the trees are distinct, there are more families than trees, and each family can get at most one?
    (b) In how many ways can they distribute all of them if the trees are distinct, any family can get any number, and a family may plant its trees where it chooses?
    (c) In how many ways can they distribute all the trees if the trees are identical, there are no more trees than families, and any family receives at most one?
    (d) In how many ways can they distribute them if the trees are distinct, there are more trees than families, and each family receives at most one (so there could be some leftover trees)?
    (e) In how many ways can they distribute all the trees if they are identical and anyone may receive any number of trees?
    (f) In how many ways can all the trees be distributed and planted if the trees are distinct, any family can get any number, and a family must plant its trees in an evenly spaced row along the road?
    (g) Answer the question in part (f) assuming that every family must get a tree.
    (h) Answer the question in part (e) assuming that each family must get at least one tree.

13. In how many ways can n identical chemistry books, r identical mathematics books, s identical physics books, and t identical astronomy books be arranged on three bookshelves? (Assume there is no limit on the number of books per shelf.)

14. (a) (De Moivre) Show that the number of ways of obtaining the positive integer k as a sum of a list of n nonnegative integers is $C(k+n-1; k)$.
    (b) What is the result if you require that the integers be positive?

15. If k identical dice and n identical coins are thrown, how many results can be distinguished?

16. In game one we throw 6 dice and win if we get a one at least once. In game two we throw 12 dice and win if we get at least two ones. What fraction of the outcomes leads to a win in each game?

17. Explain the entries in row 10 and the entries in column 2 of rows 5 and 8 in the table of distribution problems.

18. Write out the details of the proof of Corollary 2.4.

19. Write out the proof of Theorem 2.5 by describing the bijection explicitly and showing that it is a bijection.

20. Prove Corollary 2.4 by counting equivalence classes of appropriate kinds of ordered distributions of k distinct objects to n distinct recipients rather than by using the argument of first passing out n items.

21. Prove Theorem 2.6 by counting equivalence classes of ordered distributions of an appropriate kind.

22. The formula in Theorem 2.6 is a product of two terms. Find a proof in which we first select an appropriate subset and then do something else, arriving at the formula by the product principle.

23. Give the proof of Theorem 2.2 suggested in the text.

24. The formula in Theorem 2.2 is a product. Give a proof that explains the product.

25. Consider the relation on functions from $K = \{1, 2, 3, 4\}$ to $\{a, b, c\}$ given by f is related to g if and only if there is a bijection h from K to K (i.e., a permutation of K) such that $f(x) = g(h(x))$ for every x in K.
    (a) Show that this is an equivalence relation.
    (b) To what kind of distribution do the equivalence classes correspond?
    (c) Determine all possible sizes of equivalence classes and show a typical equivalence class. To do this part of the problem you will have to write down some functions explicitly. Remember that a function defined on K can be thought of as a list of four things.
    (d) What does part (c) tell us about the possibility of using the equivalence principle to count the distributions described in (b)?

26. Discuss, with appropriate proofs or examples, the possibility of computing the number of surjections from the second entry in row two or three of the table of distribution problems by use of the equivalence principle.

Section 3   Partitions and Stirling Numbers

Partitions of an _m_-Element Set into _n_ Classes

We have computed the number of partitions of an m-element set into n_1 classes of size 1, n_2 classes of size 2, and so on. We now take up the computation of the number of partitions of an m-element set into n classes. In terms of distributions, we are asking for the number of distributions of m distinct objects among n identical recipients. Notice that our usual notation for distributions would have us distributing k objects. Being consistent at this point would make it hard to see some useful analogies, so we are changing notation.

Although we are not yet in a position to give an exact formula for the number of partitions of an m-element set into n classes, we can say a good deal about the number of such partitions. We use $S(m, n)$ to stand for the number of partitions of an m-element set into n classes or parts, and we call the numbers $S(m, n)$ **_Stirling numbers of the second kind_**. (Stirling studied two important families of coefficients; our discussion of Stirling numbers of the first kind comes more naturally later.) It is implicit in this definition that $S(m, n) = 0$ unless m and n are integers and $1 \leq n \leq m$. However, in the case $m = n = 0$, we define $S(0, 0) = 1$, in effect saying that the empty set has a single partition (which we call the _empty partition_) into no parts. These numbers have some interesting analogies with the numbers $C(m; n)$, one of which is Stirling's triangle of the second kind.

Stirling's Triangle of the Second Kind

> **_Theorem 3.1._** $S(m, n) = S(m - 1, n - 1) + nS(m - 1, n)$.
>
> _Proof._ Let $M = \{1, 2, \ldots, m\}$ and $M' = \{1, 2, \ldots, m - 1\}$. Then a partition of M may have one class consisting of m alone; the number of such partitions is the number of partitions of M' with $n - 1$ parts. A partition of M may have m in a class containing some elements of M' as well. Deleting m gives a partition of M' with n parts, and since m may have been in any of the n parts, n different partitions of M yield the same partition of M'. Since m either is or isn't in a class by itself, this proves the theorem. ∎

Example 3.1. It is customary to write Stirling's triangle in the form of a right triangle as in the portion on and below the main diagonal in Table 3.1. Note that $S(3, 2) = S(2, 1) + 2S(2, 2) = 1 + 2 \cdot 1 = 3$ says that $S(3, 2)$ is the sum of twice the number above it plus the number immediately to the left of the number directly above it. The remainder of the table is filled in similarly. ∎

Table 3.1   Stirling's triangle.

$m\backslash n$	0	1	2	3	4
0	1				
1	0	1			
2	0	1	1		
3	0	1	3	1	
4	0	1	7	6	1

There is another recursive relationship that relates the value of $S(m,n)$ to the numbers $S(j, n-1)$ for all smaller values of j. Since $S(j, n-1)$ is zero when j is smaller than $n-1$, there are cases where this relation can be useful because of the small number of terms in the sum. Note that the relation given in Theorem 3.2 says that the entry in row m and column n of Table 3.1 is a weighted sum of the part of column $n-1$ that lies above row m.

Theorem 3.2. For $m > 0$, $S(m,n) = \displaystyle\sum_{j=0}^{m-1} \binom{m-1}{j} S(j, n-1)$.

Proof. If $\mathbf{P}$ is a partition of $M = \{1, 2, \ldots, m\}$ and if we eliminate from $\mathbf{P}$ the class containing m, we get a partition of a *subset* $J \subseteq \{1, 2, \ldots, m-1\}$. The resulting partition has $n-1$ parts. Every partition of every subset J of $\{1, 2, \ldots, m-1\}$ into $n-1$ parts arises exactly once from this kind of construction, so the number of partitions of M into n parts is the number of partitions of subsets of $\{1, 2, \ldots, m-1\}$ into $n-1$ parts. The formula totals the number of such partitions of a subset J for all possible sizes of J. ∎

Example 3.2. As you can see by examining Table 3.1,

$$S(4,3) = \sum_{j=0}^{3} \binom{3}{j} S(j, 2) = 1 \cdot 0 + 3 \cdot 0 + 3 \cdot 1 + 1 \cdot 3 = 6.$$

The Inverse Image Partition of a Function

The numbers $S(m,n)$ are closely related to functions. Suppose that we have a function f from $M = \{1, 2, \ldots, m\}$ onto $N = \{1, 2, \ldots, n\}$. We use the notation

$$f^{-1}(i) = \{x \mid f(x) = i\};$$

that is, $f^{-1}(i)$ is the subset of M consisting of elements that f maps onto i. The set $f^{-1}(i)$ is called the ***inverse image*** of i. Thus, we have a partition, called the *inverse image partition*,

$$\mathbf{P} = \{f^{-1}(1), f^{-1}(2), \ldots, f^{-1}(n)\}$$

of M into n classes. This is the equivalence class partition of the inverse image relation $R(f)$ (see Section 1 of this chapter) associated with f. In addition, we have a function g defined on $\mathbf{P}$ by $g(f^{-1}(i)) = i$; note that g is a one-to-one function from $\mathbf{P}$ to N.

Onto Functions and Stirling Numbers

On the other hand, given a partition

$$\mathbf{Q} = \{C_1, C_2, \ldots, C_n\}$$

of M into n classes and a one-to-one function g from $\mathbf{Q}$ to N, we can define a function f from M onto N by

$$f(x) = i \ \ \text{if } x \text{ is in } C_i.$$

Thus, the number $F(M, N)$ of functions from M onto N is, by the product principle, $S(m, n)n!$, the product of the number of partitions and the number of one-to-one functions. This proves the following theorem.

> **Theorem 3.3.** The number of functions from an m-element set onto a n-element set is $S(m, n)n!$.

Stirling Numbers of the First Kind

Stirling regarded these numbers not as representing the number of partitions but rather as the coefficients of a certain polynomial. To see why, we first examine the Stirling numbers of the first kind. These arise from studying polynomials and at first glance appear to have nothing to do with partitions. Suppose m and n are nonnegative integers and let $m \leq n$. Recall that the number $\frac{n!}{(n-m)!}$ of one-to-one functions from an m-element set to a n-element set is denoted by $(n)_m$. As we saw in Section 2 of Chapter 1,

$$(n)_m = n(n-1) \cdots (n-m+1),$$

which is a polynomial of degree m in n. Because it is a polynomial, there are numbers $s(n, j)$ such that it can be written in the form

$$(n)_m = \sum_{j=0}^{m} s(m, j)n^j.$$

We use this equation to define Stirling numbers of the first kind.   In other words, the numbers $s(m,j)$ are the coefficients of n^j that arise when we expand the product $(n)_m$.   The numbers $s(m,j)$ are called **Stirling numbers of the first kind**.

For any real or even complex number x, we can define

$$(x)_m = x(x-1)\cdots(x-m+1).$$

(We define $(x)_0 = 1$.) Then $(x)_m$ is a polynomial in x of degree m. We call $(x)_m$ a **factorial power** of x of degree m. Although, technically speaking, we defined $s(m,j)$ in terms of the coefficients of a polynomial with an integral variable, it is not surprising that changing n to x does not change the coefficients $s(m,j)$.

Theorem 3.4. For any x,

$$(x)_m = \sum_{j=0}^{m} s(m,j)x^j.$$

Proof. A polynomial in x of degree m is completely determined by its value at any $m+1$ distinct values of x. However, for each integer n larger than m, the substitution $x = n$ into both sides of the equation yields the equations we used to define the Stirling numbers. This means that the left- and right-hand sides of the equation are equal for infinitely many values of x and thus for $m+1$ values of x. Therefore, the left- and right-hand sides are equal as polynomials in x. ∎

Example 3.3. Write $(x)_3$ as an ordinary polynomial.

$$(x)_3 = x(x-1)(x-2) = x(x^2 - 3x + 2)$$

$$= x^3 - 3x^2 + 2x.$$

Note that we can tell from this example that

$$s(3,0) = 0, \quad s(3,1) = 2, \quad s(3,2) = -3, \quad \text{and } s(3,3) = 1. \qquad \blacksquare$$

Stirling Numbers of the Second Kind as Polynomial Coefficients

The numbers $S(m,n)$ (which we defined as the number of partitions of an m-element set into n classes) are called Stirling numbers of the second kind because Stirling discovered they also relate the factorial powers of n to ordinary powers of n.

Theorem 3.5. $n^m = \sum_{j=0}^{n} S(m,j)(n)_j.$

Proof. We begin with three observations. From Theorem 3.3, we know that for a particular set J, there are $S(m,j)j!$ functions from the n-element set onto J. We also know that every function from M to N maps onto *some* subset J of N. Third, we know there are $\binom{n}{j}$ different subsets J of size j that a function can map M onto. Using the sum and product principle, we put these three ideas together to get

$$n^m = \sum_{j=0}^{n} \binom{n}{j} S(m,j)j!$$

$$= \sum_{j=0}^{n} \frac{n!}{(n-j)!} S(m,j)$$

$$= \sum_{j=0}^{n} S(m,j)(n)_j.$$

Note: Our definition that $S(0,0) = 1$ makes this theorem true for $m = 0$ as well. ∎

Replacing n by x in this theorem (as in Theorem 3.4) presents certain technical difficulties because x would have to be the upper index of the sum. We finesse this difficulty by observing that for an integer n, $(n)_j = 0$ if $j > n$, and $S(m,j) = 0$ if $j > m$, so that

$$\sum_{j=0}^{n} S(m,j)(n)_j = \sum_{j=0}^{m} S(m,j)(n)_j.$$

(It is instructive to ask why we need both observations made before the equation.) Then, just as we proved Theorem 3.4, we may prove the following theorem.

Theorem 3.6. $x^m = \sum_{j=0}^{m} S(m,j)(x)_j.$

Example 3.4. Write x^3 as a combination of falling factorials.

Applying Theorem 3.6, we get

$$x^3 = \sum_{j=0}^{3} S(3,j)(x)_j$$
$$= S(3,0)(x)_0 + S(3,1)(x)_1 + S(3,2)(x)_2 + S(3,3)(x)_3$$
$$= 0 + 1(x)_1 + 3(x)_2 + 1(x)_3$$
$$= x + 3(x)(x-1) + 1(x)(x-1)(x-2).$$

To verify the use of the theorem, we multiply out the polynomials and add:

$$x + 3(x^2 - x) + 1 \cdot x(x^2 - 3x + 2) = x + 3x^2 - 3x + x^3 - 3x^2 + 2x = x^3. \;\blacksquare$$

If you have had linear algebra, you may recognize that the Stirling numbers are "change of basis" coefficients. That is, they can be used to switch from one basis for the vector space of polynomials of degree n or less to another basis.

Stirling's Triangle of the First Kind

The numbers $s(m,n)$ also satisfy a recurrence relation that can be used to compute them in a triangular array. Note that $s(m,0) = 0$ and $s(m,m) = 1$. Because of the following theorem, we can compute a triangular array containing all the values $s(m,n)$, the Stirling numbers of the first kind.

Theorem 3.7. $s(m,n) = s(m-1,n-1) - (m-1)s(m-1,n).$

Proof. Note that

$$\sum_{j=0}^{m} s(m,j)x^j = (x)_m = (x)_{m-1}(x - m + 1)$$

$$= \sum_{i=0}^{m-1} s(m-1,i)x^i(x - m + 1)$$

$$= \sum_{i=0}^{n-1} s(m-1,i)x^{i+1} - (m-1)s(m-1,i)x^i.$$

For a given j other than 0 or m, two terms on the right contain x^j, namely

$$s(m-1,j-1)x^j \text{ and } -(m-1)s(m-1,j)x^j.$$

Equating coefficients of x^j proves the theorem. $\;\blacksquare$

It is interesting to note that since some of the values of $s(m, n)$ are negative, we cannot interpret $s(m, n)$ as representing the size of some set of objects as we can interpret $S(m, n)$. (See, however, Exercise 21 where their absolute values are interpreted.)

The Total Number of Partitions of a Set

The total number B_m of partitions of an m-element set is a sum of Stirling numbers of the second kind, namely

$$B_m = \sum_{n=0}^{m} S(m, n).$$

The numbers B_m are called the **Bell numbers** after E. T. Bell who studied how the value of B_m increases as m increases. Much of the information known about the Bell numbers can be derived from properties of the Stirling numbers.

For example, from the second recursion formula for $S(m, n)$, we get the following theorem.

Theorem 3.8. For $m > 0$, $B_m = \sum_{j=0}^{m-1} \binom{m-1}{j} B_j$.

Proof. Since

$$B_m = \sum_{n=0}^{m} S(m, n),$$

$$B_m = \sum_{n=0}^{m} \sum_{j=0}^{m-1} \binom{m-1}{j} S(j, n-1)$$

$$= \sum_{j=0}^{m-1} \sum_{n=0}^{m} \binom{m-1}{j} S(j, n-1)$$

$$= \sum_{j=0}^{m-1} \binom{m-1}{j} \sum_{n=0}^{m} S(j, n-1)$$

$$= \sum_{j=0}^{m-1} \binom{m-1}{j} \sum_{i=0}^{m-1} S(j, i),$$

since $S(j, -1) = 0$. However,

$$\sum_{i=0}^{m-1} S(j, i) = \sum_{i=0}^{j} S(j, i) = B_j$$

because $S(j, i)$ is zero if $i > j$, and we know that $j \leq m-1$. Substituting B_j for the second sum yields the formula. ∎

Theorem 3.8 may be interpreted as follows. In a partition of $M = \{1, 2, \ldots, m\}$, one class must contain m; the remaining classes must be a partition of some j-element subset of the $(m - 1)$-element set $M' = \{1, 2, \ldots, m - 1\}$. This may be used as the basis of a more elementary (and more conceptual) proof (see Exercise 20) of Theorem 3.8. Thus, although the computations we gave as a proof might well have led us to discover the result, a combinatorial mathematician who has discovered the result would prefer the second technique to explain why it is true.

EXERCISES

1. Write out rows 0 through 8 of Stirling's triangle for Stirling numbers of the second kind.

2. Write down the values of the Bell numbers B_m for n between 1 and 8. Observe the ratio of each value to the preceding one. Can you make any intuitive statement about how fast B_m grows as a function of m?

3. Write out rows 0 through 8 of the triangular array of Stirling numbers of the first kind.

4. Express $(x)_4$ as an ordinary polynomial.

5. Express x^4 in terms of falling factorial polynomials.

6. Express the polynomial $3x^4 + 2x^2 + 1$ in terms of factorial polynomials $(x)_4$, $(x)_3$, etc.

7. Write down the partitions of $\{1, 2, 3, 4\}$ into two parts. Did you get $S(4, 2)$ partitions?

8. What are the possible type vectors of partitions of $\{1, 2, 3, 4, 5, 6\}$ into three parts? How many partitions do you have for each type vector? Use this information to compute $S(6, 3)$.

9. A computer has eight jobs to divide among five different slave computers. Assuming each slave gets at least one job, in how many ways can this be done? Answer the same questions assuming the five slave computers are indistinguishable to the master computer.

10. In how many ways can nine distinct pieces of candy be placed into three identical paper bags so that there is some candy in each bag? What if we don't require that each bag must have some candy?

11. How many functions are there from an eight-element set onto a four-element set?

12. Express $\binom{m}{5}$ as a polynomial in m.

13. Show that $S(m, m-1)$ is $\binom{m}{2}$.

14. Show that $|s(m, m-1)|$ is $\binom{m}{2}$.

15. Show that, for $m > 0$, $S(m, 2) = 2^{m-1} - 1$.

16. The *rising factorial polynomial* $((n))^m$ is defined as

$$((n))^m = n(n+1)\cdots(n+m-1),$$

and is thus a polynomial of degree m in n, so

$$((n))^m = \sum_{j=0}^{m} b(m, j)n^j.$$

(The double parentheses are required to avoid ambiguity in case we want to use an expression like $k+1$ in place of n.) Find a recurrence relation that is analogous to that for $s(m, n)$ and use it to write out the first five rows of a triangular array of b-values. How do the numbers $b(m, n)$ relate to the numbers $s(m, n)$?

17. Show that the number $L(m, n)$ of broken permutations of an m-element set with n parts satisfies a recurrence similar to that in Theorem 3.1.

18. Show that the number $L(m, n)$ (of broken permutations) relates falling factorial powers of x to rising factorial powers

$$((x))^j = x(x+1)\cdots(x+j-1)$$

in much the same way that Stirling numbers of the second kind relate falling factorial powers of x to ordinary powers.

*19. The delta function $\delta(j, m)$ is 0 if $j \neq m$ and is 1 if $j = m$. Show that if r is the larger of j and m

$$\sum_{n=0}^{r} s(m, n)S(n, j) = \delta(m, j)$$

and

$$\sum_{n=0}^{r} S(m, n)s(n, j) = \delta(m, j).$$

Interpret this as a theorem about a product of matrices.

20. Use the interpretation of Theorem 3.8 given after the proof of the theorem to find another (shorter) proof of the theorem analogous to the proof of Theorem 3.2.

*21. Recall that a one-to-one function from the set $M = \{1, 2, \ldots, m\}$ onto M is called a *permutation* of M; it is also called a *permutation on m letters*. If $f : M \to M$ is a permutation, then the sequence

$$a, f(a), f(f(a)), f(f(f(a))), \ldots$$

must repeat because it has at most m different values. Suppose we write the nonrepeating beginning of the sequence as

$$(a, f(a), f^2(a), f^3(a), \ldots, f^{i-1}(a)),$$

where $f^i(a)$ is the first repeated element.
(a) Show that $f^i(a) = a$.
(b) We call the i-tuple in parentheses a *cycle* of the permutation f. Two cycles are equal if they correspond to the same function defined on the same subset of M. Show that each element of M is in one and only one cycle of f.
(c) Let $c(m, n)$ be the number of permutations of m letters with exactly n cycles. Show that $c(m, n) = c(m - 1, n - 1) + (m - 1)c(m - 1, n)$.
(d) How does $c(m, n)$ relate to the absolute values of the Stirling numbers of the first kind?

*22. Show that the Stirling numbers of the second kind satisfy the recurrence

$$S(m, n) = \sum_{i=1}^{m} S(m - i, n - 1)\binom{n - 1}{i - 1}.$$

23. Show that $S(m, n) = \sum_{j=0}^{m-n} (-1)^j((n+1))^j S(m+1, n+j+1)$, in which $((n + 1))^j$ means the jth rising factorial power of $n + 1$.

24. The *signed Lah numbers* $L'(m, j)$ are used to convert between falling factorial polynomials and rising factorial polynomials by the rule

$$(x)_m = \sum_{j=0}^{m} L'(m, j)((x))^j .$$

We let $L'(0, 0) = 1$ and, for $m > 0$, $L'(m, 0) = 0$. (These choices are consistent with the convention that $(x)_0$ and $((x))^0$ are both 1.) Show that

$$L'(m, n) = (-1)^{m-n} \frac{m!}{n!}\binom{m - 1}{n - 1}.$$

How do the Lah numbers relate to the signed Lah numbers?

25. Show that the signed Lah numbers defined in the previous exercise satisfy the recurrence

$$L'(m, n) = L'(m - 1, n - 1) - (m + n - 1)L'(m - 1, n).$$

26. Find how the sum $\sum_{j=0}^{m}(-1)^j s(m, j)S(n, j)$ is related to the Lah numbers.

Section 4   Partitions of Integers

Distributing Identical Objects to Identical Recipients

We have studied distributions of both identical and distinct objects to distinct recipients and we have studied partitions of sets of distinct objects, that is, distributions of distinct objects to identical recipients. We have not yet discussed the problem of distributing identical objects to identical recipients. For example, if we have 10 identical pieces of candy and three identical bags, in how many ways can we place some candy in each of the three bags and use up all the candy? We are really asking in how many ways we can specify three numbers that add to 10. The order in which we specify the numbers does not matter because the bags are identical, so our specification is not a list. We can repeat a number in our specification, so our specification is not a set of numbers but a multiset. We formalize the idea in the definition of a partition of an integer.

A ***partition*** of a positive integer m is a multiset of positive numbers that add up to m. A partition of an integer is thus a different concept from a partition of a set. To distinguish between partitions of integers and partitions of sets, it is common to use the phrase *set partition* to refer to a partition of a set and the phrase *integer partition* to refer a partition of an integer when the kind of partition being discussed is unclear from context.

Type Vector of a Partition and Decreasing Lists

There are two different ways to describe such a multiset of integers that prove useful in working with partitions. First, we can specify the multiset by specifying the multiplicity of each element in it. A list of nonnegative numbers $(j_1, j_2, \ldots, j_m)$ such that

$$\sum_{i=1}^{m} i j_i = m$$

tells us the multiplicity of each integer i in the multiset. Our multiset has j_1 ones, j_2 twos, and so on up to j_m summands of m. Of course, j_m could only be zero or one. (Note that m is *not* the number of elements in the multiset; it is the sum of the elements of the multiset.) Such a list could also be a type vector of a partition of an m-element set, specifying the number of classes of size i for each i. We call such a list a ***type representation*** of an integer partition. In effect, when we study partitions of integers, we are studying type vectors of partitions of sets. (When there is no chance of confusion between set partitions and partitions of an integer, it is common to use the phrase *type vector* in place of type representation.)

Our second description of partitions uses the idea of a "decreasing list" of integers. A decreasing list is a list $(i_1, i_2, \ldots, i_n)$ with $i_1 \geq i_2 \geq \cdots \geq i_n$. A decreasing list whose entries add up to m describes a unique partition of m because a multiset of integers can be listed in decreasing order in exactly one way. A ***decreasing list representation*** of a partition of m is a decreasing list whose entries add up to m.

Example 4.1. The multiset $\{1, 1, 3, 3, 4\}$ is a partition of 12. Its type vector is $(2, 0, 2, 1, 0, 0, 0, 0, 0, 0, 0, 0)$. It is customary to delete trailing zeros and write the type vector as $(2, 0, 2, 1)$ or $(2, 0, 2, 1, 0, \ldots)$. Its decreasing list representation is 4, 3, 3, 1, 1. ∎

The Number of Partitions of m into n Parts

By the number of *parts* of a partition, we mean the size of the multiset. This size can be computed as either the sum of the multiplicities or the length of the decreasing list in our two representations. We use $P(m, n)$ to denote the number of partitions of the integer m into n parts. Note that $P(m, n) = 0$ if $n > m$ or $n < 0$. Our next theorem allows us to compute $P(m, n)$ recursively.

Theorem 4.1. $\sum_{i=1}^{n} P(m, i) = P(m + n, n)$.

Proof. The left-hand side of the equation is the number of partitions of m into n or fewer parts. Suppose we are given a partition P of m into i parts with type vector $(j_1, j_2, \ldots, j_m)$. From this partition, we can construct a partition of $m + n$ with type vector $(n - i, j_1, j_2, \ldots, j_m)$ obtained by adding 1 to each of the i nonzero parts of P and then adding $n - i$ parts of size 1. Since

$$n - i + \sum_{h=1}^{m} j_h(h + 1) = n - i + m + i = n + m$$

and

$$n - i + \sum_{h=1}^{m} j_h = n - i + i = n,$$

the partition constructed is a partition of $m + n$ into n parts. Also, given a partition of $m + n$ into n parts, we can construct a partition of m into n or fewer parts by deleting all parts of size 1 and subtracting 1 from the remaining parts. Since these two constructions reverse each other, the number of partitions of $m + n$ into n parts is the number of partitions of m into n or fewer parts. ∎

Example 4.2. The numbers of partitions of 3 into 1, 2, and 3 parts are 1, 1, and 1. What is the number of partitions of 6 into 3 parts?

By Theorem 4.1,

$$P(3+3,3) = P(3,1) + P(3,2) + P(3,3) = 1 + 1 + 1 = 3. \qquad \blacksquare$$

By letting $m+n$ be k (and thus letting $m = k-n$), we obtain a corollary that simplifies the recursive computation.

Corollary 4.2. For $n < k$, $P(k,n) = \sum_{i=1}^{n} P(k-n,i)$.

Note that $P(m,1) = P(m,m) = 1$. Thus, if we write a table of values of $P(m,n)$ with the rows corresponding to various values of m, then we have 1's as the first and last nonzero entries in each row. Further, to get the nth entry in row m, we go up n rows and add up the first n entries of that row. Since the number of partitions of m into n parts is 0 if $n > m$, we have zeros in the upper right-hand portion of our table. We don't have a zero row or column because we have only defined partitions of positive integers. Table 4.1 shows these properties.

Table 4.1

$m \backslash n$	1	2	3	4	5	6	7
1	1	0	0	0	0	0	0
2	1	1	0	0	0	0	0
3	1	1	1	0	0	0	0
4	1	2	1	1	0	0	0
5	1	2	2	1	1	0	0
6	1	3	3	2	1	1	0
7	1	3	4	3	2	1	1

Ferrers Diagrams

We can think of a partition of m as dividing m identical objects into n parts. This suggests a convenient geometric visualization like the example shown in Figure 4.1. We start with m identical squares, write our partition in decreasing list form as $(m_1, m_2, \ldots, m_n)$, and then place m_1 squares in a row, each touching the previous one. Next, we make a row of m_2 squares with the first square of row 2 below and touching the first square of row 1. In general, row i has m_i squares and begins below the first square of row $i-1$. The geometric figure that results is called a **Ferrers diagram** of the partition. The decreasing list $(5, 3, 3, 1)$ represents the partition $5+3+3+1$ of 12; its Ferrers diagram is in Figure 4.1.

Conjugate Partitions

By flipping a Ferrers diagram over the (downward sloping) 45° line, i.e. over its "main diagonal," we get a new diagram whose rows are the columns of the

Figure 4.1

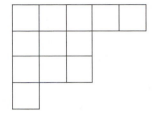

old one. The partition to which this new diagram corresponds is called the *conjugate* of the old one. For example, the partition shown in Figure 4.2 is conjugate to the partition $(5, 3, 3, 1)$ shown in Figure 4.1. We use $(5, 3, 3, 1)^*$ to denote the partition, shown in Figure 4.2, conjugate to $(5, 3, 3, 1)$. Thus $(5, 3, 3, 1)^* = (4, 3, 3, 1, 1)$.

Figure 4.2

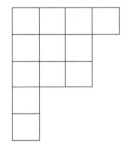

By studying Figure 4.2, we see that a natural formal definition of a conjugate would read:

> the decreasing list which is ***conjugate*** to $(m_1, m_2, \ldots, m_n)$ is the list $(r_1, r_2, \ldots, r_s)$ in which r_i is the number of m_j's larger than or equal to i.

In geometric terms, the number of squares in row i of P^* is equal to the number of rows of P with i or more squares. From the geometric description we can see that the s in the formal definition is actually equal to m_1. The geometric definition of the conjugate is easier for most people to remember than the formal definition and is equivalent to the formal definition. However, the formal definition is sometimes handy to have around for the purpose of keeping proofs from getting too cumbersome. The notion of a conjugate gives us additional information about the number of partitions of an integer into n parts.

Theorem 4.3. The number of partitions of m into n parts is equal to the number of partitions of m into parts the largest of which is n.

Proof. A partition has n parts if and only if its conjugate has its largest part equal to n. Since each partition is the conjugate of some partition, the theorem follows. ∎

Notice the way in which Theorem 4.3 uses our observation that $s = m_1$ in the formal defintion.

The Total Number of Partitions of m

There are no known formulas that compute $P(m, n)$ directly from m and n as there are for $C(n; k)$ and $S(m, n)$ (see the next chapter for a formula for $S(m, n)$). As a result, the numbers $P(m, n)$ have been the subject of intensive study. So has the total number $P(m)$ of partitions of m. In fact, the total number of distributions of m identical objects to m or more identical recipients is the total number $P(m)$ of partitions of m. Explicit formulas for $P(m)$ are also not known, but there are techniques we will study later that provide information about $P(m)$.

There are elementary results connecting the total number of partitions with the number of partitions whose smallest part is some integer. A typical example is the following theorem.

Theorem 4.4. The number of partitions of m is equal to the number of partitions of $m + 1$ whose smallest part is 1.

Proof. Suppose we are given a partition of m of type $(j_1, j_2, \ldots, j_m)$. The partition of type $(j_1 + 1, j_2, \ldots, j_m)$ is a partition of $m + 1$ whose smallest part has size 1. Conversely, if $(h_1, h_2, \ldots, h_{m+1})$ is a partition of $m + 1$ with $h_1 > 0$, then $h_{m+1} = 0$ and $(h_1 - 1, h_2, \ldots, h_m)$ is the type representation of a partition of m. ∎

We will find that the numbers $P(m, n)$ and $P(m)$ can be studied fruitfully by means of the technique of generating functions introduced in the next chapter. We have now carried our study of distributions of objects among recipients as far as is practical without the use of the algebraic methods of Chapter 3.

Our third table of distribution problems, Table 4.2, shows how the table can be filled in by using the quantities we have studied. Even though we do not have formulas expressing all these quantities in terms of more familiar quantities, we do have reasonably efficient computational methods using recursive formulas, methods that require no more work than typical formulas do.

Table 4.2   Third Table of Distribution Problems

Domain (size k) Range (size n)	May receive 0	Each receives ≥ 1
Distinct Distinct	Functions n^k	Surjections $n!S(k,n)$
Distinct Distinct (Each receives ≤ 1)	Injections $(n)_k$	Bijections $\begin{cases} n! \text{ if } n = k \\ 0 \text{ otherwise} \end{cases}$
Distinct Distinct (Order received matters)	Ordered distributions $(n+k-1)_k$	$k!\binom{k-1}{n-1} = k!\binom{k-1}{k-n}$
Distinct Identical	$\sum_{i=0}^{n} S(k,i)$ Bell # B_k if $k \leq n$	Stirling number $S(k,n)$
Distinct Identical (Each receives ≤ 1)	1 if $k \leq n$ 0 otherwise	1 if $k = n$ 0 otherwise
Distinct Identical (Order received matters)	$\sum_{i=0}^{n} L(k,i)$	Broken permutations $L(k,n) = \binom{k}{n}(k-1)_{k-n}$
Identical Distinct	Multisets $\binom{n+k-1}{k}$	$\binom{k-1}{n-1} = \binom{k-1}{k-n}$
Identical Distinct (Each receives ≤ 1)	Subsets $\binom{n}{k}$	1 if $n = k$ 0 otherwise
Identical Identical	$\sum_{i=0}^{n} P(k,i)$ if $n < k$ $P(k)$ if $k \leq n$	$P(k,n)$
Identical Identical (Each receives ≤ 1)	1 if $k \leq n$ 0 otherwise	1 if $k = n$ 0 otherwise

EXERCISES

1. In how many ways can seven identical pieces of candy be placed into three identical bags so that no bag is empty? What if empty bags are allowed?

2. Complete the table of values of $P(m,n)$ through $m = 10$ and $n = 10$.

3. Draw the Ferrers diagrams for partitions with the following type vector representations. What integer is being partitioned?
   (a) $(2, 1, 3, 0, 1, 0, 0, 2)$
   (b) $(0, 0, 2, 2, 2)$
   (c) $(2, 0, 1, 0, 2)$

4. Draw the Ferrers diagrams of partitions with the following decreasing list representations. What integer is being partitioned?
   (a) $(6, 5, 5, 3)$
   (b) $(5, 4, 3, 2, 1)$
   (c) $(8, 6, 6, 4, 2, 2)$

5. Find the conjugates of the partitions in Exercise 3 and draw their Ferrers diagrams.

6. Find the conjugates of the partitions in Exercise 4 and draw their Ferrers diagrams.

7. Show that the number of partitions of m into at most n parts is equal to the number of partitions of m into parts of size at most n.

8. Explain why the number of partitions of m into even parts is the number of partitions of m into parts of even multiplicity (that is, the number of partitions in which each part occurs an even number of times).

9. Find $P(20, 10)$. (Hint: Exercise 2 simplifies this problem.)

10. Find $P(15, 8)$.

11. Show that the number of partitions of m into two parts is $\frac{m}{2}$ if m is even and $\frac{m-1}{2}$ if m is odd.

12. In how many ways can 15 identical apples be placed into five bags so that each bag has at least 2 apples?

13. Show that $P(m) - P(m-1)$ is the number of partitions of m into parts greater than 1.

14. Using the result of Exercise 13, show that

$$P(m+2) + P(m) \geq 2P(m+1).$$

15. Show that $P(m, n) \geq \frac{1}{n!} \binom{m-1}{n-1}$.

16. Prove that the number of partitions of m into unequal odd parts is equal to the number of partitions of m conjugate to themselves. (Such partitions are called *self-conjugate partitions*.) (Hint: Draw the Ferrers diagram for the following pairs of partitions of 9: (9) and $(5, 1, 1, 1, 1)$, $(5, 3, 1)$ and $(3, 3, 3)$. For each pair, show that the Ferrers diagram of the first can be obtained by lining up all squares from the first row *and* column of the second, then all *remaining* squares from the second row and column of the second, then all remaining squares of the third row and column, etc.)

17. Show that the number of partitions of m is equal to the number of partitions of $2m$ into m parts.

18. By subtracting $P(m-1, n-1)$ from $P(m, n)$ and applying Corollary 4.2, derive a recursion formula for $P(m, n)$ with two terms, one involving $P(m-1, n-1)$ and the other involving $P(m-n, n)$.

*19. Show that the number of partitions of m into unequal parts is the number of partitions of m into odd parts.

20. The number of partitions of m into n *distinct* or unequal parts as in the previous exercise is usually denoted by $Q(m, n)$. Prove that

$$Q(m, n) = Q(m-n, n) + Q(m-n, n-1).$$

21. Using $Q(m, n)$ as in the previous exercise, show that

$$Q(m, n) \leq \frac{1}{n!} \binom{m-1}{n-1}.$$

22. Prove that the number of partitions of m into at most n parts is

$$Q\left(m + \binom{n+1}{2}\right), n).$$

What does this tell you about $Q(k, n)$ in terms of k and some function of n and k?

23. Discuss the relationship between partitions into distinct parts and partitions into (a multiset of) consecutive parts, that is, parts that form a list of consecutive, perhaps repeated, numbers, as $1 + 2 + 3 = 6$ and $2 + 2 + 2 = 6$ are for 6.

24. Discuss self-conjugate partitions of m into distinct parts.

25. Show that the total number of ones among all the partitions of n is equal to the total number of distinct parts among all the partitions of n. In other words, if $a(\pi)$ is the number of ones in π, and $b(\pi)$ is the number of distinct parts of π (so that $b(4, 3, 3, 3, 3)$ is 2), show that the sum $A(n)$ of $a(\pi)$ over all partitions π of n equals the sum of $b(\pi)$ over all partitions π of n. (Hint: Establish a bijection between the set of all partitions of numbers less than or equal to n and each of the sets $\{\{\pi\} \times \{1, 2, \ldots, a(\pi)\}\}$ and $\{\{\pi\} \times \{1, 2, \ldots, b(\pi)\}\}$, where, in both sets, π ranges over all partitions of n.)

*26. (Euler) Show that if $m \neq \frac{3k^2 \pm k}{2}$, then the number of partitions of m into an even number of distinct parts is the same as the number of partitions of m into an odd number of distinct parts. Show that in the exceptional case the first number is the second plus $(-1)^k$.

Suggested Reading

Berge, C. 1971. *Principles of Combinatorics,* chs. 1 and 2. New York: Academic Press.

Comtet, L. 1974. *Advanced Combinatorics.* Dordrecht, Holland and Boston: D. Reidel.

Feller, W. 1968. *An Introduction to Probability Theory and Its Applications.* 3d ed. New York: Wiley.

Hardy, G. H., and E. M. Wright. 1960. *An Introduction to the Theory of Numbers.* 4th ed. London: Oxford University Press.

Knuth, D. E. 1973. *The Art of Computer Programming, Fundamental Algorithms,* vol. 1, sec. 1.2.6, 2d ed. Reading, MA: Addison Wesley.

Larsen, L., and B. Hansen. 1986. Solutions to 15th USA Math Olympiad. *Mathematics Magazine* 59 (December): 310–311.

Lovász, L. 1979. *Combinatorial Problems and Exercises,* ch.1. Amsterdam: North-Holland.

Niven, I., and Zuckerman, H. S. 1960. *An Introduction to the Theory of Numbers,* ch. 10. New York: Wiley.

Riordan, J. 1958. *An Introduction to Combinatorial Analysis,* chs. 1, 5, and 6. New York: Wiley.

Tomescu, Ioan. 1975. *Introduction to Combinatorics.* London: Collet's.

3
Algebraic Counting Techniques

Section 1 The Principle of Inclusion and Exclusion

The Size of a Union of Three Overlapping Sets

The sum principle tells us that the size of the union of a family of disjoint sets is the sum of the sizes of the sets. To determine the size of a union of overlapping sets, we clearly must use information about the way the sets overlap—but how? Figure 1.1 shows the Venn diagram of two overlapping sets.

Figure 1.1

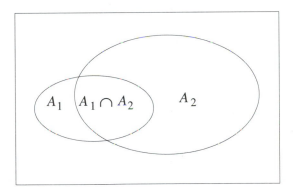

Note that if we add $|A_1|$ and $|A_2|$, we include the size of $A_1 \cap A_2$ twice in this sum, so

$$|A_1| + |A_2| = |A_1 \cup A_2| + |A_1 \cap A_2|.$$

Thus, we get the formula

$$|A_1 \cup A_2| = |A_1| + |A_2| - |A_1 \cap A_2|.$$

Figure 1.2

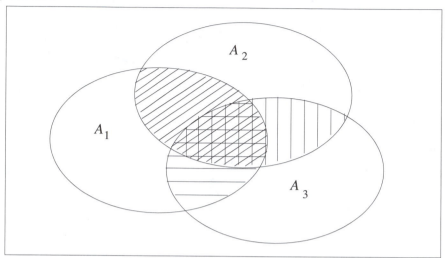

Example 1.1. Find a formula similar to the preceding one for the size of a union of three sets.

Figure 1.2 shows a picture of three overlapping sets. In the sum $|A_1| + |A_2| + |A_3|$, an element of $A_1 \cap A_2$ is included in at least two terms; an element of $A_1 \cap A_2 \cap A_3$ is included in all three terms. In the sum

$$|A_1| + |A_2| + |A_3| - |A_1 \cap A_2| - |A_1 \cap A_3| - |A_2 \cap A_3|, \qquad (1.1)$$

an element in A_1, but not A_2 or A_3, is included in only one term. An element in A_1 and A_2, but not in A_3, is included in two positive terms and one negative term, so it is counted exactly once in this formula. An element in A_1, A_2, and A_3 is included in three positive terms and three negative terms, so it is not counted at all in expression (1.1). Therefore, the expression in (1.1) counts once each element of $A_1 \cup A_2 \cup A_3$ except for the elements in $A_1 \cap A_2 \cap A_3$. Thus,

$$\begin{aligned} |A_1 \cup A_2 \cup A_3| = \ &|A_1| + |A_2| + |A_3| - |A_1 \cap A_2| - |A_1 \cap A_3| \qquad (1.2) \\ &- |A_2 \cap A_3| + |A_1 \cap A_2 \cap A_3|. \quad \blacksquare \end{aligned}$$

A formula such as (1.2) is called an inclusion–exclusion formula, and the argument that led to it is called an inclusion–exclusion argument. We shall formalize the principle of inclusion and exclusion as a general counting principle on the basis of our next example.

The Number of Onto Functions

Recall that a function $f : M \rightarrow N$ is *onto* if each element of N occurs as the image $f(x)$ of some element x of M. We have not yet been able to compute the number of functions from an m-element set M onto a n-element set N. When N has only a few elements, the computation can be made directly.

Example 1.2. Find a formula for the number of functions from an m-element set onto the n-element set $\{1, 2, \ldots, n\}$.

If, for example, $N = \{1\}$, then there is one function from M to N and it is onto. If $N = \{1, 2\}$, there are 2^m functions from M to N. Of these, one function skips 1 because it maps everything to 2, and another function skips 2 because it maps everything to 1. Thus, there are $2^m - 2$ functions from M onto N.

Now if $N = \{1, 2, 3\}$, there are 3^m functions from M to N. Of these, 2^m functions skip 1 because they map into $\{2, 3\}$. In fact, for each of the $\binom{3}{1}$ sets $\{i\}$, there are 2^m functions that skip the element of $\{i\}$. However, the difference

$$3^m - \binom{3}{1} 2^m \tag{1.3}$$

is smaller than the number of onto functions because some functions that skip $\{1\}$ skip $\{2\}$ as well and thus are subtracted twice. In fact, each of the $\binom{3}{2}$ sets $\{i, j\}$, is skipped by exactly one function. This shows that we have subtracted $\binom{3}{2} \cdot 1$ too many functions in the difference (1.3). The number of functions from M onto N is then

$$3^m - \binom{3}{1} 2^m + \binom{3}{2} \cdot 1. \tag{1.4}$$

The same kind of argument with $N = \{1, 2, 3, 4\}$ gives the following. There are 4^m functions from M to N. These functions could skip 1, 2, or 3 (but not all 4) elements of N. The number skipping each of the $\binom{4}{1}$ sets $\{i\}$ is 3^m. In the difference $4^m - \binom{4}{1} 3^m$, we subtract the functions that skip each of the $\binom{4}{2}$ two-element sets $\{i, j\}$ at least twice. Let us examine the quantity

$$4^m - \binom{4}{1} 3^m + \binom{4}{2} 2^m. \tag{1.5}$$

The functions that skip exactly one element are excluded from the quantity exactly once. The functions that skip exactly two elements are excluded from the quantity twice, but they are included back in the $\binom{4}{2} \cdot 2^m$ term exactly once and so are excluded exactly once by the quantity in (1.5). However, the functions that skip the three-element set $\{i, j, h\}$ are excluded three times in the negative term and are included three times (once for each two-element subset of $\{i, j, h\}$) in the term $\binom{4}{2}2^m$. To deal with these functions, note that there are $\binom{4}{3} \cdot 1$ such functions still to be excluded. Thus, we have

$$4^m - \binom{4}{1}3^m + \binom{4}{2}2^m - \binom{4}{3}1^m \tag{1.6}$$

functions from an m-element set onto a four-element set. Formulas (1.4) and (1.6) suggest immediately that the number of functions from an m-element set onto an n-element set is

$$n^m - \binom{n}{1}(n-1)^m + \cdots \pm \binom{n}{n-1}1^m = \sum_{i=0}^{n-1}(-1)^i\binom{n}{i}(n-i)^m$$

$$= \sum_{i=0}^{n}(-1)^i\binom{n}{i}(n-i)^m.$$

(The change from $n-1$ to n in the last equality is simply because $(n-n)^m = 0^m = 0$ when $m > 0$, corresponding to the fact that no functions skip all of N.) We shall prove the formula works later. ∎

Counting Arrangements with or without Certain Properties

Rather than trying to prove each formula we get by an inclusion–exclusion argument, let us try to find a general description of the kind of problem we are solving and state a general principle that can be applied to such problems. Each problem is a problem of counting arrangements. In Example 1.1, our arrangements were just the elements of the sets A_i. In Example 1.2, our arrangements were the functions from M to N. Our arrangements had certain properties that were interesting to us. In Example 1.1, we had a property $\boldsymbol{p}_i$ for each set A_i, the property

"The element x is in set A_i."

In Example 1.2, we had a property $\boldsymbol{p}_j$ for each element j of N, the property

"The function f skips the element j of N."

In Example 1.1, we wanted to know how many arrangements had at least one of the properties. (Note that the union of A_1, A_2, and A_3 is the set of elements in at least one of the sets.) In Example 1.2, we wanted to know how many arrangements had exactly none of the properties.

What kind of data were we given in the two examples? In Example 1.1, the sizes of A_i, $A_i \cap A_j$, and $A_1 \cap A_2 \cap A_3$ are known. The size of A_i is the number of elements with property p_i (and maybe some more properties as well), the size of $A_i \cap A_j$ is the number of elements with properties p_i and p_j, etc. In Example 1.2, we computed the number of functions skipping i; this is the number of functions with at least property p_i. We also computed the number of functions skipping i and j; this is the number of functions with at least properties p_i and p_j.

The Basic Counting Functions $\mathbf{N}_\geq$ and $\mathbf{N}_=$

Thus the ingredients we should expect to deal with in our principle will be a set of arrangements, a set P of properties the arrangements might or might not have, and for each set S of properties the number $\mathbf{N}_\geq(S)$ of arrangements having the properties in S (and perhaps other properties as well). (The notation $\mathbf{N}_\geq$ was suggested by Bender and Goldman to remind us that the numbers we are dealing with are numbers of arrangements with a certain set of properties and *perhaps other* properties as well.) In Example 1.1, $\mathbf{N}_\geq(\{p_i\}) = |A_i|$, $\mathbf{N}_\geq(\{p_i, p_j\}) = |A_i \cap A_j|$, etc. In Example 1.2, with $N = \{1, 2, 3, 4\}$

$$\mathbf{N}_\geq(\emptyset) = 4^m, \quad \mathbf{N}_\geq(\{p_i\}) = 3^m,$$
$$\mathbf{N}_\geq(\{p_i, p_j\}) = 2^m, \quad \mathbf{N}_\geq(\{p_i, p_j, p_k\}) = 1^m,$$
$$\mathbf{N}_\geq(\{p_1, p_2, p_3, p_4\}) = 0.$$

Example 1.3. Charles, Mary, Alice, and Pat are sitting in the four chairs at a library table. They get up to go to class and return to sit at the same table. In how many ways may they sit so that Pat is in the same chair she had before? Using a person's initial for the property that that person sits in his or her previous chair, what is $\mathbf{N}_\geq(S)$ for a set S of properties?

Notice that if Pat sits in her previous chair, then there are three chairs left for the other three people and 3! ways for them to sit in these chairs. We just computed that $\mathbf{N}_\geq(\{P\}) = 3!$. If S is a set with just one initial, then $\mathbf{N}_\geq(S) = 3!$ just as it did with $S = \{P\}$. If S has two elements, then the two people whose initials are in S will sit in their previous places leaving the other two people to sit in $2! = 2$ ways, so that $\mathbf{N}_\geq(S) = 2$. If S has three elements then once three people sit in their previous sets, so must the fourth person, so $\mathbf{N}_\geq(S) = 1$, as it will if S has four elements. Finally, if S is empty, we get $\mathbf{N}_\geq(S) = 4!$. We can summarize by saying $\mathbf{N}_\geq(S) = (4 - |S|)!$. ∎

Example 1.4. In the previous example, what does $\mathbf{N}_=(\{P\})$ stand for? (We will compute it in the next example.) What does $\mathbf{N}_=(S)$ stand for when S is an arbitrary set? How is $\mathbf{N}_{\geq}(\{P\})$ related to $\mathbf{N}_{=}(S)$ for subsets S of our properties?

$\mathbf{N}_{=}(\{P\})$ stands for the number of seating arrangements in which Pat sits in her previous seat *but* nobody else does. $\mathbf{N}_{=}(S)$ stands for the number of seating arrangements in which all the people in S sit in their previous seats but nobody else does. Since $\mathbf{N}_{\geq}(\{P\})$ stands for the ways that Pat can sit in her previous seat and perhaps some other people may sit in their previous seats as well, we have

$$\mathbf{N}_{\geq}(\{P\}) = \mathbf{N}_{=}(\{P\}) + \mathbf{N}_{=}(\{P,C\}) + \mathbf{N}_{=}(\{P,M\}) + \mathbf{N}_{=}(\{P,A\})$$
$$+ \mathbf{N}_{=}(\{P,C,M\}) + \mathbf{N}_{=}(\{P,C,A\}) + \mathbf{N}_{=}(\{P,M,A\}) + \mathbf{N}_{=}(\{P,C,M,A\}).$$

This last equation is a good illustration of the relationship betweeen $\mathbf{N}_{\geq}$ and $\mathbf{N}_{=}$ and is a good place to come back to if you find a particular use of the notation confusing. ∎

The principle of *inclusion and exclusion* tells us how to find the number of arrangements that have exactly a certain set of properties. For example, in Example 1.2, we want to know how many functions have exactly the empty set of skips. We use $\mathbf{N}_{=}(S)$ to stand for the number of arrangements with exactly the properties in S. As in Example 1.2, we shall most often be interested in $\mathbf{N}_{=}(\emptyset)$. From our examples, we expect the principle to tell us that to find $\mathbf{N}_{=}(S)$, we start with $\mathbf{N}_{\geq}(S)$, subtract the number of arrangements with at least one more property besides the properties in S, add the number with at least two more properties, and so on. That is exactly what the summation notation in Theorem 1.1 tells us to do. The symbol J is a dummy variable that takes on as its values all possible sets that are subsets of P and contain S as a subset. Except for the fact that J, S, and P are sets, we use and read the summation like ordinary summation notation. Example 1.5 shows exactly how the notation is used.

The Principle of Inclusion and Exclusion

Theorem 1.1. (Principle of Inclusion and Exclusion). Suppose A is a set of arrangements and suppose P is a set of properties that the arrangements may have. For each $S \subseteq P$, let $\mathbf{N}_{\geq}(S)$ stand for the number of arrangements with at least the properties in S and let $\mathbf{N}_{=}(S)$ stand for the number of arrangements with the properties in S and no others. Then for each $S \subseteq P$,

$$\mathbf{N}_{=}(S) = \sum_{J=S}^{P} (-1)^{|J|-|S|} \mathbf{N}_{\geq}(J).$$

Proof. Suppose an arrangement has only the properties in S. Then it appears in only one term of the sum. If it has one more property p_j in addition to the properties in S, it is counted once in $\mathbf{N}_{\geq}(S)$ and once in $\mathbf{N}_{\geq}(S \cup \{p_j\})$ with opposite sign and so is not counted. If an arrangement has i properties not in S, it appears $\binom{i}{k}$ times in terms of the form

$$\mathbf{N}_{\geq}(S \cup \{p_{i_1}, p_{i_2}, \ldots, p_{i_k}\}),$$

and the sign alternates according to whether k is even or odd. Thus the total number of times this arrangement is counted is

$$1 - \binom{i}{1} + \binom{i}{2} - \binom{i}{3} + \cdots + (-1)^i \binom{i}{i} = (-1+1)^i$$

by the binomial theorem. Therefore, an arrangement with more properties than those in S is counted 0 times. We noted that an arrangement with exactly the properties in S is counted exactly once. Thus, the sum counts exactly the desired arrangements. ∎

Example 1.5. In Examples 1.3 and 1.4, in how many ways may Charles, Mary, Alice, and Pat return to their seats so that Pat sits in her previous seat and nobody else does? In how many ways may they return to their seats so that nobody sits in his or her previous seat? In how many ways may they return to their seats so that somebody sits in his or her previous seat?

By our theorem

$$\mathbf{N}_{=}(\{P\}) = \sum_{S=P}^{\{C,M,A,P\}} (-1)^{|S|-|\{P\}|} \mathbf{N}_{\geq}(S)$$

$$= \mathbf{N}_{\geq}(\{P\}) - \mathbf{N}_{\geq}(\{P,C\}) - \mathbf{N}_{\geq}(\{P,M\}) - \mathbf{N}_{\geq}(\{P,A\})$$
$$+ \mathbf{N}_{\geq}(\{P,C,M\}) + \mathbf{N}_{\geq}(\{P,C,A\}) + \mathbf{N}_{\geq}(\{P,M,A\})$$
$$- \mathbf{N}_{\geq}(\{P,C,M,A\}).$$

Substituting the values from the formula we computed in Example 1.3, we get

$$\mathbf{N}_{=}(\{P\}) = 3! - 2! - 2! - 2! + 1! + 1! + 1! - 0!$$
$$= 6 - 2 - 2 - 2 + 1 + 1 + 1 - 1$$
$$= 2.$$

The number of ways for them to return to the table so that nobody sits in his or her previous seat is $\mathbf{N}_=(\emptyset)$, the number of ways in which exactly the empty set of people take their own seats. By our theorem

$$\mathbf{N}_=(\emptyset) = \sum_{S=\emptyset}^{\{C,M,A,P\}} (-1)^{|S|} \mathbf{N}_\geq(S)$$

$$= \mathbf{N}_\geq(\emptyset) - \mathbf{N}_\geq(\{C\}) - \mathbf{N}_\geq(\{M\}) - \mathbf{N}_\geq(\{A\}) - \mathbf{N}_\geq(\{P\})$$
$$+ \mathbf{N}_\geq(\{C,M\}) + \mathbf{N}_\geq(\{C,A\}) + \cdots - \mathbf{N}_\geq(\{C,M,A\}) - \cdots$$
$$+ \mathbf{N}_\geq(\{C,M,A,P\})$$
$$= 4! - 3! - 3! - 3! - 3! + 2! + 2! + 2! + 2! + 2! + 2!$$
$$- 1! - 1! - 1! - 1! + 0!$$
$$= 24 - 4 \cdot 6 + 6 \cdot 2 - 4 \cdot 1 + 1$$
$$= 9.$$

Notice the way in which the binomial coefficients $\binom{4}{3} = 4$, $\binom{4}{2} = 6$, and $\binom{4}{1} = 4$ appeared in our sum. This is because we had $\binom{4}{3}$ three-element sets for which we subtracted 3!, then $\binom{4}{2}$ two-element subsets for which we added 2!, and $\binom{4}{1}$ one-element sets for which we subtracted 1!. (By now you can probably see $\binom{4}{4}$ and $\binom{4}{0}$ could have been made explicit as well.) We can collect terms in this way whenever $\mathbf{N}_=(S)$ depends only on the size of S; knowing in advance that we can do this will make it easier to write our computations down.

Finally, the number of arrangements in which someone sits where he or she sat before is the total number of seating arrangements *minus* the number of arrangements in which *nobody* sits where they sat before, or $4! - 9 = 15$. ∎

As another example of how to use Theorem 1.1, we prove the formula from Example 1.2.

Onto Functions and Stirling Numbers

Theorem 1.2. The number of functions from an m-element set onto a n-element set is equal to

$$\sum_{i=0}^{n} (-1)^i \binom{n}{i} (n-i)^m = n^m - \binom{n}{1}(n-1)^m + \cdots + (-1)^{n-1} \binom{n}{n-1} 1^m.$$

Proof. As described before, property p_j is "the function f skips the element j of N." $\mathbf{N}_\geq(p_{j_1}, p_{j_2}, \ldots, p_{j_i})$ is the number of functions skipping at least the i elements in $\{j_1, j_2, \ldots, j_i\}$. This is the number of functions mapping into an $(n - i)$-element set, which is the number $(n-i)^m$. As noted before, we want to compute $\mathbf{N}_=(\emptyset)$. By Theorem 1.1,

$$\mathbf{N}_=(\emptyset) = \sum_{I=\emptyset}^{N} (-1)^{|I|}(n - i)^m.$$

Using the fact that there are $\binom{n}{i}$ subsets I of N of size i, we get

$$\mathbf{N}_=(\emptyset) = \sum_{i=0}^{n} (-1)^i \binom{n}{i}(n - i)^m. \qquad \blacksquare$$

Corollary 1.3. The number of partitions of an m-element set into n classes is given by

$$S(m, n) = \frac{1}{n!} \sum_{i=0}^{n} (-1)^i \binom{n}{i}(n - i)^m.$$

Proof. See Theorem 3.3 of Chapter 2. $\blacksquare$

Examples of Using the Principle of Inclusion and Exclusion

Example 1.6. A used car dealer has 18 cars on the lot. Nine of them have automatic transmissions, 12 have power steering, and 8 have power brakes. Seven have both automatic transmissions and power steering, four have automatic transmissions and power brakes, and five have power steering and power brakes. Three cars have power steering and power brakes and automatic transmissions. How many cars have automatic transmission only? How many cars are "stripped"?

To illustrate the principle of inclusion and exclusion, we let our set of arrangements be the cars and our set of properties be $\{AT, PS, PB\}$. For any set X of properties, let $\mathbf{N}_\geq(X)$ be the number of vehicles with the properties in the set X and perhaps some other properties as well. Let $\mathbf{N}_=(X)$ be the number of vehicles with exactly the properties in X. The problem gives us the values of $\mathbf{N}_\geq$ and asks for $\mathbf{N}_=(\{AT\})$ and $\mathbf{N}_=(\emptyset)$. By the principle of inclusion and exclusion

$$\begin{aligned}
\mathbf{N}_=(\{AT\}) &= \mathbf{N}_\geq(\{AT\}) - \mathbf{N}_\geq(\{AT, PS\}) \\
&\quad - \mathbf{N}_\geq(\{AT, PB\}) + \mathbf{N}_\geq(\{AT, PS, PB\}) \\
&= 9 - 7 - 4 + 3 = 1,
\end{aligned}$$

and

$$\begin{aligned}
\mathbf{N}_=(\emptyset) &= \mathbf{N}_\geq(\emptyset) - \mathbf{N}_\geq(\{AT\}) - \mathbf{N}_\geq(\{PS\}) - \mathbf{N}_\geq(\{PB\}) \\
&\quad + \mathbf{N}_\geq(\{AT, PS\}) + \mathbf{N}_\geq(\{AT, PB\}) \\
&\quad + \mathbf{N}_\geq(\{PS, PB\}) - \mathbf{N}_\geq(\{AT, PB, PS\}) \\
&= 18 - 9 - 12 - 8 + 7 + 4 + 5 - 3 = 2.
\end{aligned}$$

Figure 1.3

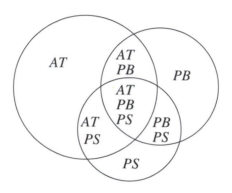

This computation is especially easy to visualize by examining the Venn diagram in Figure 1.3. Think of the area of the entire rectangle as representing the total number of cars. When we subtract the number of cars with each property from the total number of cars, we subtract out the areas labeled AT, PB; PB, PS; and AT, PS twice. Further, we subtract the area labeled AT, PB, PS three times. When we correct for the oversubtraction by adding in the numbers corresponding to the areas AT, PB; PB, PS; and AT, PS, we unfortunately add in the area labeled AT, PB, PS just as often as we subtracted it before. By subtracting out the number corresponding to this last area one more time, we find the number of items corresponding to the area which is inside the rectangle but which is outside all of the circles. ∎

Example 1.7. A group of couples sits around a circular table for a group discussion of marital problems. In how many ways can the group be seated so that no husband and wife sit together?

In this case, our arrangements are seating arrangements and our properties are "husband and wife i sit together." Note that there is no requirement that the seating arrangements alternate sex. We assume the couples are numbered as couple 1 through couple n. We let $N = \{1, 2, \ldots, n\}$. Note that this N is a set and not the $\mathbf{N}_=$ or $\mathbf{N}_\geq$ of Theorem 1.1. We let $\mathbf{N}_\geq(I)$ be the number of arrangements in which the couples numbered by the integers in I sit together as couples, and let $\mathbf{N}_=(I)$ be the number of arrangements in which couples numbered by integers in I—*and no other couples*—sit together. We wish to know $\mathbf{N}_=(\emptyset)$. Theorem 1.1 tells us to compute $\mathbf{N}_\geq(I)$ for every subset I of N.

To compute $\mathbf{N}_\geq(I)$, we note that by seating the i couples first and allowing the other $2n - 2i$ people to sit down at the remaining places, we are in effect arranging $2n - i$ units (think of $2n - 2i$ people and i couples tied together at the ankles and wrists, each couple requiring two chairs) around a circular table. We know that there are $(2n - i - 1)!$ such arrangements. Now each couple can occupy its two chairs in two ways (after untying the ropes in the imaginary description). Therefore, after the $2n - i$ units are assigned to places, there are 2^i ways for the units to be placed in their assigned places. Thus, there are $2^i(2n - i - 1)!$ ways to seat the people so that couples labeled by elements of the i-element set I sit together. Thus, $\mathbf{N}_\geq(I) = 2^i(2n - i - 1)!$, and by applying the principle of inclusion and exclusion, we get

$$\mathbf{N}_=(\emptyset) = \sum_{I=\emptyset}^{N}(-1)^{|I|}\mathbf{N}_\geq(I) = \sum_{i=0}^{n}(-1)^i\binom{n}{i}2^i(2n - i - 1)! \ . \qquad \blacksquare$$

Example 1.8. A bookcase has five shelves each with 10 books on it. Each shelf contains only books on a given one of five different subjects, and each shelf has a different subject. In how many ways can the books be removed for dusting and returned to the shelves so that each subject still has a shelf of its own, even though no shelf has a book that was previously on it? In how many ways can this be done so that at least one shelf has the books that were previously on it?

The arrangements of interest here are arrangements of books on shelves. Once we decide which shelf a given set of 10 books goes on, there are 10! ways to arrange it on this shelf. If we take as property j, "shelf j gets the same subject it had last time," then for a given set I of i properties (shelves), there are $(10!)^i$ ways to fill the shelves specified by I. Next there are $(5 - i)$ other shelves and there are $(5 - i)!$ ways to assign subjects to shelves so that the shelves specified by I—and perhaps some others—get their original subjects back. Then there are $(10!)^{5-i}$ ways to assign the books to places on these shelves. Thus, we have $(10!)^5(5 - i)!$ arrangements having at least the properties in I, so $\mathbf{N}_{\geq}(I) = (10!)^5(5-i)!$. Since we want $\mathbf{N}_{=}(\emptyset)$, we apply the principle of inclusion and exclusion to get (with $N = \{1,2,3,4,5\}$)

$$\mathbf{N}_{=}(\emptyset) = \sum_{I=\emptyset}^{N} (-1)^i (10!)^5 (5-i)!$$

$$= \sum_{i=0}^{5} (-1)^i (10!)^5 \binom{5}{i} (5-i)!$$

$$= 5!(10!)^5 \sum_{i=0}^{5} \frac{(-1)^i}{i!}.$$

Had we had k shelves rather than five, each occurrence of five would be replaced by k.

To answer the second question, notice that either no shelf has a book previously on it or at least one shelf has the books previously on it. Thus, we are asking for the size of the complement of the set whose size we just computed. This is the size $5!(10!)^5$ of the set of all arrangements minus the size of the set of objects having none of the properties, giving us

$$5!(10!)^5 - 5!(10!)^5 \sum_{i=0}^{5} \frac{(-1)^i}{i!} = 5!(10!)^5 \sum_{i=1}^{5} \frac{(-1)^{i+1}}{i!}. \qquad \blacksquare$$

Although the second question we asked in the example may seem rather specialized, it turns out to be a typical kind of useful question. For example, in Example 1.1, property i was the property that an element is in set i. Asking for the size of the union is, as we pointed out earlier, the same as asking, "How many of the objects have at least one of the properties?" A general formula can be derived from Theorem 1.1 to solve such problems; we state it as a corollary for ease of reference. In the following corollary, we use the suggestive notation $\sum_{J\neq S}^{P} f(J)$ to stand for the sum, over all sets J strictly containing S but subsets of P, of $f(J)$.

Corollary 1.4. Suppose A is a set of arrangements and suppose P is a set of properties that the arrangements may have. For each $S \subseteq P$, let $\mathbf{N}_{\geq}(S)$ stand for the number of arrangements with at least the properties in S. Then the number of arrangements with at least one of the properties is

$$N_{\geq}(\emptyset) - N_{=}(\emptyset) = N_{\geq}(\emptyset) - \sum_{J=\emptyset}^{P} (-1)^{|J|} N_{\geq}(J) = \sum_{J\neq\emptyset}^{P} (-1)^{|J|+1} N_{\geq}(J).$$

Proof. We subtract the number of arrangements with none of the properties from the total number of arrangements. ∎

Example 1.9. Use the corollary to do Example 1.1.

Recall that property i is being a member of A_i. The size of the union is the number of elements with at least one of the properties, so

$$|A_1 \cup A_2 \cup A_3| = \sum_{\substack{I \neq \emptyset}}^{\{p_1, p_2, p_3\}} (-1)^{|I|+1} \mathbf{N}_{\geq}(I)$$

$$= \mathbf{N}_{\geq}(\{p_1\}) + \mathbf{N}_{\geq}(\{p_2\}) + \mathbf{N}_{\geq}(\{p_3\}) - \mathbf{N}_{\geq}(\{p_1, p_2\})$$
$$- \mathbf{N}_{\geq}(\{p_1, p_3\}) - \mathbf{N}_{\geq}(\{p_2, p_3\}) + \mathbf{N}_{\geq}(\{p_1, p_2, p_3\})$$
$$= |A_1| + |A_2| + |A_3| - |A_1 \cap A_2| - |A_1 \cap A_3|$$
$$- |A_2 \cap A_3| + |A_1 \cap A_2 \cap A_3|.$$
∎

Derangements

A classic problem in the application of the principle of inclusion and exclusion is the "problem of derangements." A **derangement** of the set $N = \{1, 2, \ldots, n\}$ is a one-to-one function $f : N \to N$ such that $f(i) \neq i$ for any i. In more colorful language, the derangement problem is called the hat check problem (see Exercise 17). To solve the derangement problem, we use as our objects the one-to-one functions from N to N and as our property p_i the property that $f(i) = i$. Then the number of derangements is $\mathbf{N}_{=}(\emptyset)$. We leave the details of developing the formula for Exercise 17.

*Level Sums and Inclusion–Exclusion Counting

In Example 1.7, we might instead be asking ourselves, "What is the probability that no husband and wife sit together?" Since there are $(2n-1)!$ possible seating arrangements, we divide $\mathbf{N}_{=}(\emptyset)$ by $(2n-1)!$ to find this probability. In Example 1.8, we might have wanted the probability that "no shelf has a book previously on it." From this point of view, it is equally interesting to ask, "What is the probability that exactly i couples sit together?" or "What is the probability that exactly i shelves have the books that were previously on them?" To compute the probability that i couples sit together, we need the number of arrangements in which exactly i husbands and wives are side by side; note that this is *not* $\mathbf{N}_{=}(I)$ for some set I, but rather the sum over *all sets* I of size i of $\mathbf{N}_{=}(I)$. Similarly for the bookshelves, we would want the sum over all sets I of size i of $\mathbf{N}_{=}(I)$. This kind of sum makes sense for any inclusion–exclusion problem and represents the number of arrangements that have exactly i properties. In Example 1.7 or 1.8, we could have computed this sum by multiplying one value of $\mathbf{N}_=(I)$ by an appropriate binomial coefficient (which one?), since all values $\mathbf{N}_=(I)$ for all sets of size i are equal. On the other hand, in a problem such as Example 1.6, $\mathbf{N}_=(I)$ depends on more than just the size of I.

It turns out that the sum of $\mathbf{N}_=$ values just described is related to the sum of the $\mathbf{N}_\geq$ values in a way that reduces the sum over subsets in the inclusion–exclusion theorem to a sum over integers. In order to see what this relationship is in any inclusion–exclusion situation, we introduce two new functions related to the two functions $\mathbf{N}_=$ and $\mathbf{N}_\geq$. These two new functions are called **level sums** of the old functions. They are defined by

$$\mathbf{N}^+_=(i) = \sum_{I:|I|=i} \mathbf{N}_=(I)$$

and

$$\mathbf{N}^+_\geq(i) = \sum_{I:|I|=i} \mathbf{N}_\geq(I).$$

In other words, $\mathbf{N}^+_=(i)$ is the sum of all values of $\mathbf{N}_=(I)$ for sets I of size i and $\mathbf{N}^+_\geq(i)$ is the sum of all values of $\mathbf{N}_\geq(I)$ for sets I of size i. Then by substitution of the formula of Theorem 1.1 for $\mathbf{N}_=(I)$,

$$\mathbf{N}^+_=(i) = \sum_{I:|I|=i} \mathbf{N}_=(I) = \sum_{I:|I|=i} \sum_{S=I}^{P} (-1)^{|S|-|I|} \mathbf{N}_\geq(S)$$

$$= \sum_{S:|S|\geq i} \sum_{\substack{I:I\subseteq S \\ \text{and}\,|I|=i}} (-1)^{|S|-|I|} \mathbf{N}_\geq(S)$$

$$= \sum_{S:|S|\geq i} \binom{|S|}{i} (-1)^{|S|-i} \mathbf{N}_\geq(S)$$

$$= \sum_{j=i}^{p} \sum_{S:|S|=j} \binom{j}{i} (-1)^{j-i} \mathbf{N}_\geq(S)$$

$$= \sum_{j=i}^{p} \binom{j}{i} (-1)^{j-i} \sum_{S:|S|=j} \mathbf{N}_\geq(S)$$

$$= \sum_{j=i}^{p} (-1)^{j-i} \binom{j}{i} \mathbf{N}^+_\geq(j).$$

This proves the following theorem.

Theorem 1.5. If $\mathbf{N}_=$ and $\mathbf{N}_\geq$ are related as in Theorem 1.1, and $\mathbf{N}_=^\pm$ and $\mathbf{N}_\geq^\pm$ are the "level sums" given by

$$\mathbf{N}_=^\pm(i) = \sum_{I:|I|=i} \mathbf{N}_=(I)$$

$$\mathbf{N}_\geq^\pm(i) = \sum_{I:|I|=i} \mathbf{N}_\geq(I),$$

then

$$\mathbf{N}_=^\pm(i) = \sum_{j=i}^{p} (-1)^{j-i} \binom{j}{i} \mathbf{N}_\geq^\pm(j).$$

Proof. Already given. ∎

*Examples of Level Sum Inclusion and Exclusion

Example 1.10. In Example 1.6, how many cars have exactly one feature? We will need to know $\mathbf{N}_\geq^\pm(1)$, $\mathbf{N}_\geq^\pm(2)$, and $\mathbf{N}_\geq^\pm(3)$. By inspection and addition, $\mathbf{N}_\geq^\pm(1) = 9 + 12 + 8 = 29$, $\mathbf{N}_\geq^\pm(2) = 7 + 4 + 5 = 16$, and $\mathbf{N}_\geq^\pm(3) = 3$. Then

$$\mathbf{N}_=^\pm(1) = \sum_{j=1}^{3} (-1)^{j-1} \binom{j}{1} \mathbf{N}_\geq^\pm(j)$$

$$= (-1)^0 \binom{1}{1} \cdot 29 + (-1)^1 \binom{2}{1} \cdot 16 + (-1)^2 \binom{3}{1} \cdot 3$$

$$= 29 - 32 + 9 = 6. \qquad \blacksquare$$

Example 1.11. Use Theorem 1.5 to do Example 1.8. For any integer i, $\mathbf{N}_\geq^\pm(i)$ is the sum over all sets I of size i of $\mathbf{N}_\geq(I)$. Thus, $\mathbf{N}_\geq^\pm(i) = \binom{5}{i}(10!)^5(5-i)! = \frac{5!}{i!}(10!)^5$.
Therefore

$$\mathbf{N}_=^\pm(0) = \sum_{i=0}^{5} (-1)^i \frac{5!}{i!}(10!)^5 = 5!(10!)^5 \sum_{i=0}^{5} \frac{(-1)^i}{i!}. \qquad \blacksquare$$

EXERCISES

1. How many functions are there from a five-element set onto a three-element set?

2. What is the Stirling number $S(5,3)$?

3. In how many ways can 12 distinct pieces of candy be passed out to six children so that each child gets a piece?

4. In how many ways can 12 distinct pieces of candy be placed into six identical bags with no bag left empty?

5. In Example 1.6, how many cars have only power brakes? How many cars with power brakes do not have automatic transmission?

6. In an experiment on the effects of fertilizer on 27 plots of a new breed of lawn grass, 8 plots are given nitrogen, phosphorus, and potash fertilizers, 12 plots are given at least nitrogen and phosphorus, 12 plots are given at least phosphorus and potash, and 12 plots are given at least nitrogen and potash. Eighteen plots receive nitrogen, 18 receive phosphorus, and 18 receive potash. How many plots are left unfertilized?

7. In Exercise 6, how many plots are given exactly one of the three nutrients?

8. In a college class of 28 people, nine are women from the East, five are Easterners over 20, and seven are women over 20. Sixteen people in the class are Easterners, 16 are women, and 12 are over 20. There are two Eastern women over 20 in the class. How many men of age 20 or under are in the class? How many of these men are from the East? How many of these men are not Easterners? (This problem is probably best done by Venn diagrams like Figures 1.1 and 1.2, because the questions are not all direct applications of formulas.)

9. Experimental garden plots were treated with four substances: lime, potash, urea, and phosphate. Some were treated with each possible combination of the chemicals. The following table shows how many received each combination of substances. Among the 36 plots used for the experiment, some were control plots that received none of the chemicals. How many control plots were there?

Substances	L	P	U	F	LP	LU	LF	PU	PF	UF	LPU	LUF	LPF	PUF	LPUF
Number	20	19	17	16	14	14	12	12	11	10	10	8	9	7	5

10. The design for an experiment calls for treating experimental garden plots with lime, potash, urea, and phosphate. According to the design, 32 plots are to be treated with each of the individual chemicals and perhaps some others, 16 plots are to be treated with each pair of

chemicals and perhaps some others, 8 are to be treated with three of the chemicals and perhaps another, and 4 are to be treated with all four chemicals. How many plots are needed if none are to receive no treatment at all?

11. In the previous exercise, suppose we wanted to change the 16 plots treated with each two chemicals to 29, increase the 8 plots treated with three chemicals to 12, and increase the number of plots treated with all four chemicals to six. Why is this not possible?

12. In how many ways can k distinct pieces of fruit be passed out to n children if each child gets at least one?

13. Show that

$$
\begin{aligned}
|A_1 \cup A_2 \cup \cdots \cup A_n| = {}& |A_1| + |A_2| + \cdots + |A_n| \\
& - |A_1 \cap A_2| - \cdots - |A_i \cap A_j| - \cdots \\
& + |A_1 \cap A_2 \cap A_3| + \cdots + |A_i \cap A_j \cap A_k| + \cdots \\
& \qquad\qquad\qquad \vdots \\
& - (-1)^n |A_1 \cap A_2 \cap \cdots \cap A_n|.
\end{aligned}
$$

14. In Example 1.7, in how many seating arrangements are exactly i couples seated side by side?

15. You are to make a necklace with n different pairs of beads. The beads in a pair have the same shape but different colors. In how many ways can you make the necklace so that no identically shaped beads are side by side?

16. In Exercise 15, what is the number of necklaces with exactly one pair of identically shaped beads side by side?

17. A hat check person discovers that n people's hats have been mixed up and returns these hats to the owners at random.
    (a) In how many ways can the hats be returned so that all of the owners get someone else's hat?
    (b) What proportion of the total number of distributions do these represent?
    (c) If you have had calculus, what limiting value does this proportion have? (Hint: the limiting value involves the base e of the natural logarithms and is best found by using power series.)

18. In how many ways can the hats of the previous exercise be passed out so that exactly i people receive their own hats?

19. In how many ways can eight pieces of identical candy be passed out to three children so that each child gets at least one piece but no child gets more than four pieces?

20. In how many ways can k identical pieces of candy be passed out to n children so that no child gets m or more pieces?

21. In how many ways can n distinct books be arranged on k shelves so that no shelf gets m or more books?

22. Each person attending a card party is to bring a prize. The person in charge has arranged to give out exactly as many prizes as there are people, but anyone can win any number of prizes. If there are n attendees, in how many ways can the prizes be passed out so that no one leaves with the prize he or she brought?

23. A group of n people is attending a seminar in a room with m chairs. In the middle of the seminar there is a break and everyone leaves the room. In how many ways can the group sit down after the break if no one is in the same chair as before?

24. In a nursery school, m children take off their wet socks and set them on the radiator to dry. By the end of the day, after the socks have become hopelessly mixed up, each child takes two socks.
    (a) In how many ways can the children prepare to go home in such a way that each child has at least one sock belonging to someone else?
    *(b) In how many ways can they go home with each child wearing two socks belonging to someone else?

25. In Example 1.7, if we require that the couples be seated in the order man, woman, man, woman,... (i.e., alternating gender), then we have a classic problem in combinatorics called the "menage problem." The main difference in the two problems lies in the computation of $\mathbf{N}_{\geq}(I)$. Of course, if I is empty, $\mathbf{N}_{\geq}(I) = \mathbf{N}_{\geq}(\emptyset)$ is just $(n-1)!n!$. If I has some couples in it, pick the first couple (say, in alphabetical order) and seat them. There are $2n$ ways to choose places for them and two ways to seat them in these two places. Since the seating pattern must alternate in gender, this determines for the entire table which places are for men and which are for women. Once you choose two side-by-side places for any other couple, you can seat them in these places in just one way. Now to fill in the rest of the table, make the same kind of argument we made before using $i-1$ "tied together couples" and $2n-2i$ people (consistently with their gender). Observe that each arrangement is equivalent to $2n$ other arrangements this time. From here on, the argument is parallel. Solve the menage problem.

*26. Do Exercises 15 and 16 for the case where the identically shaped beads have the same color. (There is a troublesome symmetry problem you must avoid here.)

27. A group of n people is seated around a round table. The group leaves the table for a break and then returns. In how many ways can the people sit down so that no one is to the right of the same person as in the previous seating?

*28. A group of n people is seated around a round table. The group leaves the table for a break and then returns. In how many ways can the people sit down so that no one is next to the same person in both seatings? (Note that a person cannot be to the right of someone else in one seating and to the left in the other.)

Section 2 The Concept of a Generating Function

Symbolic Series

Many combinatorial problems can be interpreted as problems of combining arrangements according to certain rules to make new arrangements. Problems in algebra often involve following algebraic rules in combining expressions into new ones. While the operations we perform to do the combining *are* different from those we perform in algebra, there is enough similarity to allow us to transform certain combinatorial problems into algebraic ones. It is this kind of translation that leads to the theory of generating functions. We begin our study of the theory with examples designed to illustrate the translation process; then we will be able to develop theorems to apply in practice as shortcuts to the translation process.

Example 2.1. Suppose we have two apples, three nectarines, and four plums. We shall try to give a compact description of all logically possible fruit selections with at least one of each fruit. In the next example, we will see how this symbolic representation of fruit selections helps us compute the number of ways to select k pieces of fruit.

If we think of A as standing for "we take an apple," N as standing for "we take a nectarine," and P as standing for "we take a plum," then it is natural to use ANP to stand for "we take an apple, a nectarine, and a plum" and to use

$$ANNPPP$$

to stand for "we take an apple, two nectarines, and three plums." Algebra suggests the more compact shorthand

$$AN^2P^3$$

for the second selection. Although we have listed the symbols for the fruits in alphabetical order, note that PNA stands for the same fruit selection as ANP. Our snack could be ANP or ANP^2, but it couldn't be both. This is what we mean by the *excluxive or* of ANP and ANP^2. If we use $\oplus$ to stand for "exclusive or" (for which $\oplus$ is a common symbol in electrical engineering and computer science), then the statement that we take an apple, a nectarine, and a plum or else we take an apple, a nectarine, and two plums is represented by

$$ANP \oplus ANP^2,$$

and the statement that we take an apple, a nectarine, and a plum or one each of two and two of the third, or two each of two and one of the third, and so on is represented by

$$ANP \oplus ANP^2 \oplus AN^2P \oplus A^2NP \oplus AP^2N^2 \oplus \cdots \oplus A^2N^3P^4. \qquad (2.1)$$

Notice that we are making use of all possible ways of writing down a monomial using A or A^2, using N, N^2, or N^3 and using P, P^2, P^3, or P^4 as the factors of the monomial. Adding all these monomials gives the same result as using the commutative, distributive, and associative laws to expand the "product"

$$(A \oplus A^2)(N \oplus N^2 \oplus N^3)(P \oplus P^2 \oplus P^3 \oplus P^4). \tag{2.2}$$

We refer to this product as a **symbolic series** describing all of the possible fruit selections. ∎

How does this description of the possible fruit selections help us count the number of selections with, say, six pieces of fruit? The selections with six pieces of fruit correspond to monomials such as AN^2P^3 or $A^2N^2P^2$ whose exponents add to six. Thus we want to know the *number* of terms with total exponent 6. If we were to replace each of A, P, N by x in

$$ANP \oplus ANP^2 \oplus AN^2P \oplus A^2NP \oplus AP^2N^2 \oplus \cdots \oplus A^2N^3P^4$$

(which is (2.1)), then every term with total exponent six becomes x^6 and no other terms become x^6. Thus, the number of times x^6 appears is the number of fruit selections with six pieces of fruit. This number will be the coefficient we get for x^6 by using algebra to collect terms. We remarked that we use only the commutative, distributive, and associative laws in expanding (2.2) to get the expression (2.1). Since these laws apply to ordinary algebra as well, we may substitute x for A, N, and P in (2.2) and expand the resulting polynomial by ordinary algebra in order to get the result of making the substitution in (2.1).

Example 2.2. Show the result of substituting x for A, N, and P and $+$ for $\oplus$ in (2.2) and simplifying so that the coefficient of x^k (or in particular, x^6) is the number of fruit selections with k (or 6) pieces of fruit, taking at least one of each kind.

The substitution gives us

$$(x + x^2)(x + x^2 + x^3)(x + x^2 + x^3 + x^4)$$
$$= x^3(1 + x)(1 + x + x^2)(1 + x + x^2 + x^3)$$
$$= x^3(1 + 2x + 2x^2 + x^3)(1 + x + x^2 + x^3)$$
$$= x^3(1 + 3x + 5x^2 + 6x^3 + 5x^4 + 3x^5 + x^6)$$
$$= x^3 + 3x^4 + 5x^5 + 6x^6 + 5x^7 + 3x^8 + x^9.$$

Thus, there are six ways to select six pieces of fruit, five ways to select five or seven pieces of fruit, and so on. ∎

Suppose that instead of wanting to know how many selections have a certain number of fruits, we want to know how many selections cost a certain amount or have a certain number of calories. It turns out that by making different substitutions into the symbolic series, we can answer these questions as well.

Example 2.3. Suppose that a plum has 20 calories, a nectarine has 40 calories, and an apple has 60 calories. Show that if we substitute x^{20} for P, x^{40} for N, and x^{60} for A into the symbolic series of Example 2.1, then the coefficient of x^n in the polynomial that results is the number of fruit selections with n calories. Make the substitution and find out how many selections with at least one of each kind of fruit have 200 calories.

The fruit selection $A^i N^j P^k$ has $60i + 40j + 20k$ calories. Substituting x^{60} for A, x^{40} for N and x^{20} for P gives us a term of the form $x^{60i} x^{40j} x^{20k} = x^{60i+40j+20k}$, so the exponent of the power of x that results from the substitution into a monomial representing a fruit selection is the number of calories that the selection has. The number of monomials with $60i + 40j + 20k = n$ is the number of fruit selections with n calories and is also the coefficient of x^n. Thus, the coefficient is the number of selections with that number of calories.

We make the substitution into (2.2), multiplying out and collecting terms to get

$$\begin{aligned}
(x^{60} &+ x^{120})(x^{40} + x^{80} + x^{120})(x^{20} + x^{40} + x^{60} + x^{80}) \\
&= x^{120}(1 + x^{20} + 2x^{40} + 3x^{60} + 3x^{80} + 4x^{100} \\
&\qquad\qquad + 3x^{120} + 3x^{140} + 2x^{160} + x^{180} + x^{200}) \\
&= x^{120} + x^{140} + 2x^{160} + 3x^{180} + 3x^{200} + 4x^{220} \\
&\qquad\qquad + 3x^{240} + 3x^{260} + 2x^{280} + x^{300} + x^{320} \, .
\end{aligned}$$

From this we see that three selections of fruit have 200 calories. We see also that more selections have 220 calories than any other amount and that the number of selections with a given calorie count decreases symmetrically as we increase or decrease the calorie count from 220. The term-by-term multiplication and collection of terms can be a painstaking task; although it is necessary and useful in some cases, many problems that arise in practice have enough regularity to allow us to use some general algebraic techniques that we shall introduce later. ∎

From our two examples we see that by making an appropriate substitution into a symbolic series we can convert the process of *exhibiting* in shorthand all possible solutions to a problem into a process of *counting* all solutions of a problem according to a specific numerical criterion that we choose. Of course, the numerical criterion cannot be arbitrary; it must be computable by taking weighted sums of the exponents in the series. In our case, this means our criteria have to involve something we can compute in terms of the number of fruits of each kind in our selection. How do we choose what powers of x to substitute for the symbols? The exponent we choose for x in substituting for a certain symbol will be the numerical weight associated with that symbol in the weighted sum; thus we chose the number of calories for the exponent on x when we were interested in the number of calories in a fruit selection.

Example 2.4. If an apple costs 40 cents, a nectarine costs 40 cents, and a plum costs 20 cents, what powers of x should we substitute into the symbolic series of Example 2.1 so that the coefficient of x^n in the resulting expression is the number of fruit selections that cost n cents? How many fruit selections cost \$2.00?

If we substitute x^{40} for A, x^{40} for N, and x^{20} for P, then the sum $40i + 40j + 20k$ resulting from the monomial $A^i N^j P^k$ will be the cost of this fruit selection. Since the sum is the exponent of x in the expression that results, the coefficient of this power of x will be the number of fruit selections with this cost.

Making the substitution, multiplying out and collecting terms gives us

$$
\begin{aligned}
(x^{40} + x^{80})&(x^{40} + x^{80} + x^{120})(x^{20} + x^{40} + x^{60} + x^{80}) \\
&= x^{100}(1 + x^{20} + 3x^{40} + 3x^{60} + 4x^{80} + 4x^{100} \\
&\qquad\qquad + 3x^{120} + 3x^{140} + x^{160} + x^{180}) \\
&= x^{100} + x^{120} + 3x^{140} + 3x^{160} + 4x^{180} + 4x^{200} \\
&\qquad\qquad + 3x^{220} + 3x^{240} + x^{260} + x^{280}.
\end{aligned}
$$

Thus, there are four ways to select \$2.00 worth of fruit, taking at least one of each kind. ∎

The symbolic series we developed in Example 2.1 for fruit selections assumes both a lower and an upper limit of the number of pieces of each kind of fruit we take. There is also a natural symbolic series that we can write down to illustrate the process of making fruit selections from an unlimited supply.

Example 2.5. Let us use the preceding techniques to illustrate the possible fruit selections from apples, nectarines, and plums with no restrictions on how many of each fruit we take. When we attempt to write down the result of forming all monomials and using the circled plus sign to stand for exclusive *or*, we have no last term. We also have terms representing the possibility of taking none of a certain kind of fruit. A^0 is a convenient notation for "take no apples." By beginning with the possibility of taking none of any kind of fruit, we obtain, in place of (2.1),

$$A^0 N^0 P^0 \oplus A N^0 P^0 \oplus \cdots \oplus ANP \oplus ANP^2 \oplus \cdots \oplus A^i N^j P^k \oplus \cdots . \quad (2.3)$$

The statement that we may take zero, one, two, and so on through any number of apples may be represented symbolically as

$$A^0 \oplus A^1 \oplus A^2 \oplus \cdots \oplus A^i \oplus \cdots . \quad (2.4)$$

Similarly the statements that we may take any number of nectarines or plums may be represented by

$$N^0 \oplus N^1 \oplus N^2 \oplus \cdots \oplus N^j \oplus \cdots . \quad (2.5)$$

and

$$P^0 \oplus P^1 \oplus P^2 \oplus \cdots \oplus P^k \oplus \cdots . \quad (2.6)$$

As with the expression in (2.1), we may think of expression (2.3) as the result of repeatedly choosing one term each from (2.4), (2.5), and (2.6), multiplying these terms together, and adding the product into a running sum. If we think of applying the commutative, distributive, and associative laws from left to right in the "product"

$$(A^0 \oplus A^1 \oplus \cdots \oplus A^i \oplus \cdots)(N^0 \oplus N^1 \oplus \cdots \oplus N^j \oplus \cdots)(P^0 \oplus P^1 \oplus \cdots \oplus P^k \oplus \cdots) \quad (2.7)$$

then the terms that we get from (2.7) are exactly the terms of (2.3), so we may think of (2.3) and (2.7) as two different symbolic representations of the fruit selection process. These representations are connected by our use of the commutative, distributive, and associative laws; in fact, the notation is chosen so as to encourage us to use these laws without even needing to think of them. ∎

Example 2.6. What should we substitute for A, N, and P in (2.7) so that the coefficient of x^n in the expression that results will be the number of selections of fruit with n pieces of fruit? What is the expression that results?

If we substitute x for each of A, N, and P, then in the resulting expression each $A^i N^j P^k$ term will become x^{i+j+k}, and only terms with $i + j + k = n$ will become x^n. Therefore, the coefficient of x^n will be the number of fruit selections with exactly n pieces of fruit. Since each of (2.4), (2.5), and (2.6) becomes $1 + x + x^2 + \cdots + x^i + \cdots$, (2.7) becomes

$$(1 + x + x^2 + \cdots + x^i + \cdots)^3 .$$

■

Power Series

The reader acquainted with the notion of power series will notice that the symbolic multiplications we performed in the last example are ordinary multiplication of power series. A **power series** is an expression of the form

$$\sum_{i=0}^{\infty} a_i x^i = a_0 x^0 + a_1 x^1 + a_2 x^2 + \cdots$$

$$= a_0 + a_1 x + a_2 x^2 + \cdots$$

where the a_i's are numbers. Power series are multiplied in the same way as polynomials. For example, the power series

$$1 + y + y^2 + y^3 + \cdots + y^i + \cdots$$

times the polynomial $1 - y$ gives the product

$$
\begin{array}{l}
1 + y + y^2 + y^3 + \cdots + y^i + \cdots \\
\underline{\hspace{2em} 1 - y \hspace{2em}} \\
-y - y^2 - y^3 - y^4 - \cdots \qquad - y^{i+1} - \cdots \\
\underline{1 + y + y^2 + y^3 + y^4 + \cdots + y^i + y^{i+1} + \cdots} \\
1
\end{array}
.
$$

Thus, $(1 - y)(1 + y + y^2 + \cdots) = 1$, or after division by $1 - y$,

$$\sum_{i=0}^{\infty} y^i = 1 + y + y^2 + y^3 + \cdots = \frac{1}{1 - y}. \tag{2.8}$$

This is the formula for the sum of a *geometric sequence* often presented in algebra courses. The series is called a **geometric series**.

We proceeded intuitively in the preceding multiplication. If you examine how you would apply the distributive, associative, and commutative laws to the product

$$\sum_{i=0}^{\infty} a_i x^i \sum_{j=0}^{\infty} b_j x^j,$$

you will see that the appropriate definition for the product of two power series is

$$\sum_{i=0}^{\infty} a_i x^i \sum_{j=0}^{\infty} b_j x^j = \sum_{k=0}^{\infty} \Big(\sum_{i=0}^{k} a_i b_{k-i} \Big) x^k. \tag{2.9}$$

What Is a Generating Function?

The power series we get by substitution into symbolic series provide us with a way of "generating" information about the number of solutions of certain kinds of combinatorial problems. For this reason, we define the *generating series* or **generating function** for the sequence c_i of numbers to be the power series $\sum_{i=0}^{\infty} c_i x^i$. For example, the generating function for the sequence $3i^2 + 4$ is

$$\sum_{i=0}^{\infty} (3i^2 + 4) x^i.$$

Similarly, the power series

$$\sum_{i=0}^{\infty} \binom{n}{i} x^i = \sum_{i=0}^{n} \binom{n}{i} x^i = (1 + x)^n$$

(where we know n is a constant since it is not the dummy variable that is shown in the lower limit of the sum, we know the first equality because $\binom{n}{i}$ is 0 when $i > n$, and we know the second equality because of the binomial theorem) is the generating function for $\binom{n}{i}$. We might give any of those three forms of the generating function when we talk about the generating function for $\binom{n}{i}$.

In algebra and calculus we learn how a power series in x can often be used to represent a function of x for appropriate x values. This is why generating series have come to be called generating functions. In combinatorial mathematics, we think of generating functions as a convenient notation to use with sequences of numbers related to structures we build by combining simpler ones. The power of generating functions in combinatorial mathematics lies in part in their ability to mirror the associative, distributive, and perhaps commutative laws satisfied by common methods of combining arrangements. Thus power series that do not represent functions can still be useful in combinatorics.

At times various calculus theorems about power series can be of value. For example, if we know that the generating function for a sequence of numbers is an infinitely differentiable function f, we can apply Taylor's theorem (from calculus) to compute the numbers c_i. Note, however, that we have defined power series as certain kinds of expressions, not in the way that (convergent) power series are defined in a calculus course. In this context, power series are often called *formal power series*. We leave to the reader familiar with the calculus of power series the task of verifying that the conclusions we make when applying calculus to power series that represent differentiable functions are appropriate. In fact, one can arrange things so as to use a variation of calculus with power series that do not represent differentiable functions! Since we do not use this subject, we do not develop it.

In a typical application of generating functions, we have a sequence a_i such that a_i is the number of objects in a certain set that have a "value" of some sort equal to i. For example, our objects might be piles (multisets) of apples. Then the value of interest might be the size of the pile, the monetary price of the pile, or the number of calories in the pile. If we assume that we have an empty pile, a pile with one apple, a pile with two, and so on, then a_i is 1 if i is a possible value of the size, price, or calorie content, respectively, and otherwise a_i is 0. Multiplying generating functions corresponds to adding values in the sense described in our next theorem, the **product principle for generating functions**.

The Product Principle for Generating Functions

Theorem 2.1. Let v and w be nonnegative integer-valued functions defined on sets S and T. Let a_i be the number of objects s in S with $v(s) = i$ and b_i be number of objects t in T with $w(y) = i$. Then

$$\left(\sum_{i=0}^{\infty} a_i x^i \right) \left(\sum_{i=0}^{\infty} b_i x^i \right)$$

is the generating function for the sequence c_j, where c_j is the number of ordered pairs $(s, t) \in S \times T$ with $v(s) + w(t) = j$.

Proof. By Equation (2.9) for multiplying power series,

$$c_j = \sum_{i=0}^{j} a_i b_{j-i}.$$

By the product principle for sets, $a_i b_{j-i}$ is the number of ordered pairs (s, t) with $v(s) = i$ and $w(t) = j - i$. Since the ordered pairs (s, t) with $v(s) = i$ and $w(t) = j - i$ are disjoint from those with $v(s) = i'$ and $v(t) = j - i'$, the sum principle says that adding $a_i b_{j-i}$ over all conceivable values of i will give us the total number of ordered pairs with total value j. However, the only values of i that could yield $v(s) + w(t) = j$ are those for which $v(s)$ is $0, 1, 2, \ldots, j$. The equation for c_j, and thus the assertion of the theorem, follows. ∎

Example 2.7. In how many ways can we select a multiset of eight candy bars that are either pure milk chocolate or milk chocolate almond bars?

We can choose $0, 1, 2, \ldots$, etc. pure chocolate bars. Thus, if we let $a_i = 1$ for each i, then a_i represents the number of ways to choose i pure chocolate candy bars, and $\sum_{i=0}^{\infty} x^i$ is the generating function for the number of ways to choose a multiset of i pure chocolate candy bars. (Here S is the set containing the empty set of pure chocolate bars, one one-element set of them, one two-element multiset of them, one three-element multiset of them, and so on.) The generating function $\sum_{i=0}^{\infty} x^i$ is also the generating function for the number of ways to select i almond candy bars. (What is the set T?) Thus, by Theorem 2.1,

$$\sum_{i=0}^{\infty} x^i \cdot \sum_{i=0}^{\infty} x^i = \sum_{j=0}^{\infty} c_j x^j$$

is the generating function for the number of ways to select a total of j candy bars, and the coefficient of x^8 is $\sum_{i=0}^{8} 1 \cdot 1 = 9$. ∎

Thus, there are nine ways to choose our eight-element multiset of two kinds of candy bars; notice that $9 = \binom{8+2-1}{8} = \binom{9}{8}$, as we learned in Chapter 2.

Corollary 2.2. Let $v_1, v_2, \ldots v_n$ be nonnegative integer valued functions defined on sets $S, T, \ldots U$. Let a_i be the number of objects s in S with $v_1(s) = i$, b_i be the number of objects t in T with $v(t) = i, \ldots$, c_i be the number of objects u in U with $v(u) = i$. Then

$$\sum_{i=0}^{\infty} a_i x^i \sum_{i=0}^{\infty} b_i x^i \cdots \sum_{i=0}^{\infty} c_i x^i$$

is the generating function for the sequence d_j where d_j is the number of n-tuples $(s, t, \ldots, u)$ in $S \times T \times \cdots \times U$ with $v_1(s) + v_2(t) + \cdots + v_n(u) = j$.

Proof. Apply mathematical induction using Theorem 2.1. ∎

The Generating Function for Multisets

Theorem 2.3. The generating function for the number $C(n+k-1; k)$ of k-element multisets of an n-element set is $(1-x)^{-n}$.

Proof. A k-element multiset of an n-element set is described by a multiplicity function that is essentially an n-tuple of nonnegative integers that add up to k. Thus, the generating function for the number of n-tuples of nonnegative integers whose values add up to k is also the generating function for k-element multisets of an n-element set, namely $\sum_{k=0}^{\infty} C(n+k-1;k)x^k$. Now

$$1 + x + x^2 + \cdots$$

is the generating function for the number of ways to select a single nonnegative integer whose value is j, because there is exactly one way to select an integer whose value is j. From Corollary 2.2 applied to the generating function $\sum_{k=0}^{\infty} C(n+k-1;k)x^k$ we found for n-tuples, we see that $C(n+k-1;k)$ is the coefficient of x^k in the product of the n equal generating functions

$$1 + x + x^2 + \cdots = (1-x)^{-1}$$

for the number of ways to select a single nonnegative integer whose value is j. This product of n equal series is $(1-x)^{-n}$. ∎

Corollary 2.4. $(1-x)^{-n} = \sum_{k=0}^{\infty} C(n+k-1;k)x^k$ for each nonnegative integer n.

Note the similarity between Corollary 2.4 and the binomial theorem, which states that $(1+x)^n = \sum_{k=0}^{n} C(n;k)x^k$.

Polynomial Generating Functions

In our first three examples, the generating function that occurred turned out to be simply a polynomial—a particularly elementary kind of power series. Of course, Theorem 2.1 applies to polynomial generating functions.

Example 2.8. If in Example 2.7 there are six almond bars and five pure chocolate bars, then in how many ways can we choose a total of eight candy bars?

In this case we can choose 0, 1, 2, 3, 4, or 5 pure chocolate bars. Thus, we see that the generating function for pure chocolate bars is $1 + x + x^2 + x^3 + x^4 + x^5$. Similarly, the generating function for almond bars is $1 + x + x^2 + x^3 + x^4 + x^5 + x^6$. Finally, by the product principle, the generating function for ordered pairs of selections of candy bars is

$$(1 + x + x^2 + x^3 + x^4 + x^5)(1 + x + x^2 + x^3 + x^4 + x^5 + x^6)$$
$$= 1 + 2x + 3x^2 + 4x^3 + 5x^4 + 6x^5 + 6x^6 + 5x^7 + 4x^8 + 3x^9 + 2x^{10} + x^{11}$$

Thus, there are four ways to select eight candy bars from among the two types. ∎

In the same way, if we had n different candy bars and wanted to select k of them, choosing zero or one of each kind, our generating function for the number of choices would, by Theorem 2.1, be a product of n factors:

$$(1+x)(1+x)\cdots(1+x) = (1+x)^n.$$

However, by making such a selection, we select k different elements from our n-element set. We have in essence used Theorem 2.1 to rederive the binomial theorem, that, using $C(n;k)$ to stand for the number of ways to choose a k-element subset of an n-element set, $(1+x)^n = \sum_{k=0}^n C(n;k)x^k$. Thus we say that $(1+x)^n$ is the generating function for subsets of an n-element set.

Extending the Definition of Binomial Coefficients

By extending the definition of $C(n;k)$ to negative values of n, we can relate this generating function for subsets to the generating functions for multisets. One way to write $\binom{n}{k}$ is $\frac{(n)_k}{k!}$, where $(n)_k = n(n-1)\cdots(n-k+1)$.

It is interesting to note that when n is a negative number, $n = -m$, if we *define*

$$\binom{-m}{k} = \frac{(-m)_k}{k!} = \frac{(-m)(-m-1)\cdots(-m-k+1)}{k!},$$

then

$$\binom{-m}{k} = (-1)^k \frac{(m+k-1)_k}{k!}$$

$$= (-1)^k C(m+k-1;k).$$

Thus, applying Corollary 2.4 to the power series

$$(1+x)^{-m} = (1-(-x))^{-m},$$

we get

$$(1+x)^{-m} = \sum_{k=0}^\infty C(m+k-1;k)(-1)^k x^k$$

$$= \sum_{k=0}^\infty \binom{-m}{k} x^k.$$

The Extended Binomial Theorem

In other words, with the natural definition of $\binom{n}{k}$ for negative values of n, we now have for all integers n

$$(1+x)^n = \sum_{k=0}^\infty \binom{n}{k} x^k,$$

and for any positive or negative integer n, $(1+x)^n$ is the generating function for the binomial coefficients $\binom{n}{k}$.

Theorem 2.5. For any integer n,

$$(x+y)^n = \sum_{k=0}^{\infty} \binom{n}{k} x^{n-k} y^k.$$

Proof. We observe that $(x+y)^n = x^n(1+\frac{y}{x})^n$ and apply the preceding formula to the second term of this product. ∎

EXERCISES

1. Using the letter B for bananas, write the symbolic series that says that someone chooses one, two, or three bananas.

2. What is the symbolic series that says that someone takes between one and three each of bananas, pears, apples, tangerines, and nectarines?

3. Write down the polynomial that results from substituting x for all the letters and changing from logical to algebraic notation in Exercise 2.

4. If someone is to choose between one and three each of bananas, pears, apples, tangerines, and nectarines, in how many ways can this person choose seven pieces of fruit?

5. Assume that bananas and apples each cost 20 cents, pears and tangerines each cost 30 cents, and nectarines cost 25 cents each. What power of x should be substituted for each of the fruit symbols in Exercise 2 so that the coefficient of x is the number of fruit selections whose cost is the exponent of x? What polynomial results? In how many ways can someone select $1.00 worth of fruit in the preceding exercises? In how many ways may someone select $2.50 worth of fruit?

6. A "snack pack" has three packages each of potato chips, corn chips, nacho chips and cheese twists. Show symbolically how we may choose snacks if we try at least one of each snack. (Use initials P, C, N, and T.) Make the substitution of x for each initial and show the polynomial that arises. In how many ways can we choose a total of seven packages of snacks?

7. Redo the previous exercise allowing us to take none of any snack as well. Use P^0, for example, to say we take no potato chips.

8. Using P for pennies, N for nickels, D for dimes, and Q for quarters, write the symbolic series for making change using from zero to five of each kind of coin. What should you substitute for each letter so that

the coefficient of x^n in the result is the number of ways to make n cents using from zero to five of each coin? What is the polynomial that results?

9. A bake shop makes four kinds of candies that weigh one ounce, three kinds of fruit bars that weigh two ounces, and one kind of brownie that weighs four ounces. Using A, B, C, and D to stand for the candies, E, F, and G for the fruit bars, and H for the brownies, write a symbolic series for the ways to use between 0 and 4 of each variety to make a box of goodies. What should be substituted for each letter to answer the question "in how many ways can we choose a 1-pound box?" What is the polynomial that results?

10. The symbolic series corresponding to taking no apples, two apples, four apples, etc. from an unlimited supply is $A^0 \oplus A^2 \oplus A^4 \oplus \cdots$. What is the generating function for the number of ways to choose an even number of apples from an unlimited supply?

11. Write down the symbolic series and then the corresponding generating function for the number of ways to choose an even number of apples and any number of tangerines from unlimited supplies.

12. What is the sequence whose generating function is the power series $x + 2x^2 + 3x^3 + 4x^4 + \cdots$?

13. Find the product of the polynomial $1 + x$ and the power series $1 - x + x^2 - x^3 + x^4 - x^5 + \cdots$.

14. Find the product of the polynomial $1 - x$ and the power series $x + x^2 + x^3 + x^4 + \cdots$.

15. Using $F_1, F_2, \ldots, F_n$ to stand for n different kinds of fruit, explain why

$$\prod_{i=1}^{n} (F_i^0 + F_i^1 + F_i^2 + F_i^3)$$

is a symbolic series that represents the number of ways of choosing between zero and three of n different kinds of fruit. What is the generating function for the number of ways to select fruit, taking between zero and three of each kind?

16. Modify the previous exercise by allowing someone to take any even number between zero and four of any of the n fruits. Write down the appropriate symbolic series and generating function.

17. What is the generating function in which the coefficient of x^i is the number of ways the sum of the tops can be i when you roll a die n times?

18. The equation $c_j = \sum_{i=0}^{j} a_i b_{j-i}$ given in the proof of Theorem 2.1 actually defines the meaning of $f(x) \cdot g(x)$. Write down the expression this equation gives for $(f(x) \cdot g(x)) \cdot h(x)$ and the expression this equation gives for $f(x) \cdot (g(x) \cdot h(x))$. Use this to explain why the associative law holds for the multiplication of power series.

19. Read the previous exercise. Write down the expressions given by the equation defining c_j in the previous exercise for $f(x)g(x) + h(x)g(x)$ and $(f(x) + h(x))g(x)$. Use this to explain why the distributive law holds for multiplication and addition of power series.

20. (a) Use the fact that $(1 - x)(1 + x + x^2 + \cdots + x^i + \cdots) = 1$ to find the value of
$$1 + \frac{1}{2} + \frac{1}{4} + \cdots + \frac{1}{2^i} + \cdots .$$

    (b) What is the product $(1 + x)(1 - x + x^2 - x^3 + \cdots + (-1)^i x^i + \cdots)$? What does this tell you about the value of $1 - \frac{1}{2} + \frac{1}{4} - \frac{1}{8} + \cdots$?

    (c) Try to explain what goes wrong in parts (a) and (b) if you replace $\frac{1}{2}$ by 2 and in this way attempt to compute $1 + 2 + 4 + \cdots$ or $1 - 2 + 4 - 8 + \cdots$.

21. Find the coefficient of x^7 in $(1 + x + x^2 + \cdots)^9$. (Try Corollary 2.3.)

22. Find the coefficient of x^k in $(1 + x + x^2 + \cdots)^9$.

23. Find the coefficient of x^7 in $(1 + x + x^2 + \cdots)^n$.

24. Find the coefficient of x^7 in $(x + x^2 + \cdots)^n$.

25. Find the coefficient of x^{10} in $(1 + x^2 + x^4 + x^6 + \cdots)^6$.

26. Find the coefficient of x^{10} in $\frac{1}{1-x} \cdot \frac{1}{1-x^2}$.

27. Extend Example 2.7 by allowing bittersweet chocolate candy bars as well. In how many ways can we select eight candy bars? (Hint: The sum in the example was really the sum of $1^i 1^j$ over all i and j such that $i + j = 8$. The condition $i + j = 8$ converted it into a single sum; here you can convert to a double sum. Using Corollary 2.3 might be easier!)

28. Two dice (a red one and a white one) are thrown. Write a generating function for the number of outcomes in which the two faces showing on top add up to i. Using Theorem 2.1, explain why this generating function can be factored and what its factors mean.

29. A penny, nickel, dime, and quarter are tossed. Write the generating function for the number of ways that i heads can occur. Explain why this polynomial can be factored and explain what the factors mean.

30. Repeat Exercise 28 with three dice.

31. What is the generating function for the sum of the faces if n distinct dice are thrown?

32. Extend Example 2.8 by allowing four bittersweet candy bars. In how many ways can we select eight candy bars in this case?

33. What is the generating function for the number of ways of distributing identical pieces of candy to Joe and Mary so that Joe gets a nonzero even number of pieces and Mary gets an odd number of pieces? What if no more than 11 pieces are to be handed out?

34. Write the generating function for the number of ways to choose a snack of n items chosen to include zero or one each of s distinct candy bars, zero or two each of r different kinds of fruit, and one or two soft drinks chosen from among t varieties.

35. What is the generating function for the number of ways to pass out apples (from a potentially infinite supply) to five children? What if we require that each child gets at least one apple? In both cases, how many distributions use 15 apples?

36. What is the generating function for the number of ways to pass out an even number of oranges to each of k children so that each child gets something? How many distributions use $2n$ oranges?

37. What is the generating function for the number of multisets that can be formed using the letter a an even number of times and the letter b any number of times?

38. What is the generating function for the number of multisets that can be chosen from an m-element set A and a (disjoint) k-element set B so that elements chosen from B are used an even number of times?

39. Show by means of generating functions that the number of multisets of k objects chosen from an n-object set such that each object appears at least once is $C(k-1; k-n)$.

40. If $f(x)$ is the generating function for the sequence a_i, and f is infinitely differentiable in a neighborhood of $x = 0$, what does Taylor's theorem tell us about the value of the a_i's in terms of f?

41. Using calculus, take r derivatives of $\frac{1}{1-x}$ and from the result compute the generating function for the falling factorial numbers $(n)_r$.

42. Expand $(1+x)^n(1+x)^n$ and use the resulting expression to prove that

$$\binom{2n}{n} = \sum_{k=0}^{n} \binom{n}{k}^2.$$

43. From their generating function for a fixed n, what can you conclude about the sum of the Stirling numbers $s(n, j)$ of the first kind?

44. Using calculus and generating functions, determine the value of the sum

$$\binom{n}{1} + 2\binom{n}{2} + 3\binom{n}{3} + \cdots + n\binom{n}{n}.$$

*45. Using generating functions, show that

$$\sum_{i=0}^{k} \binom{m}{i}\binom{n}{k-i} = \binom{m+n}{k}.$$

*46. Fractional powers also satisfy the binomial theorem. In particular, for any real number r, we define

$$\binom{r}{k} = \frac{(r)_k}{k!} = \frac{r(r-1)\cdots(r-k+1)}{k!}$$

so that the statement of the binomial theorem, namely

$$(1+x)^r = \sum_{k=0}^{\infty} \binom{r}{k} x^k$$

makes sense for any real number r (whether or not it is true).

(a) Write explicit expressions for $\binom{\frac{1}{2}}{}$ and $\binom{-\frac{1}{2}}{}$.

(b) Show that fractional powers of the binomial $(1 + x)$ obey the binomial theorem.

(c) Show that $(1 - 4x)^{-\frac{1}{2}}$ is the generating function for $\binom{2n}{n}$.

47. Restate Theorem 2.1 as a theorem about n-tuples and prove this new version of the theorem.

Section 3 Applications to Partitions and Inclusion–Exclusion

Pólya's Change-Making Example

Much of what is known about the number of partitions of an integer has been found or proved through the use of generating functions. To build our intuition we begin our study of these applications with another example involving symbolic series.

Example 3.1.† Suppose we have a pile of nickels, dimes, and quarters, and we wish to make change for some amount of money, for example, a dollar. Find a symbolic series that describes all ways to do this.

Using Q for quarters, D for dimes, and N for nickels, we can visualize the process of selecting coins by constructing three symbolic strings

$$N^0 \oplus N^1 \oplus N^2 \oplus \cdots \oplus N^i \oplus \cdots \tag{3.1}$$

$$D^0 \oplus D^1 \oplus D^2 \oplus \cdots \oplus D^j \oplus \cdots \tag{3.2}$$

$$Q^0 \oplus Q^1 \oplus Q^2 \oplus \cdots \oplus Q^k \oplus \cdots \tag{3.3}$$

to stand for the fact that we can take 0, 1, 2, 3, and so on nickels; 0, 1, 2, 3, and so on dimes; and 0, 1, 2, 3, and so on quarters. The selection of i nickels, j dimes, and k quarters yields the expression $N^i D^j Q^k$. This expression can be derived by multiplying N^i from the first symbolic series, D^j from the second, and Q^k from the third. Since this is a typical term in the product of the three series (3.1), (3.2), and (3.3), we obtain

$$(N^0 \oplus N^1 \oplus \cdots \oplus N^i \oplus \cdots)(D^0 \oplus D^1 \oplus \cdots \oplus D^j \oplus \cdots)(Q^0 \oplus Q^1 \oplus \cdots + Q^k + \cdots) \tag{3.4}$$

as the desired symbolic series. ∎

Of course, not all terms $N^i D^j Q^k$ in the product correspond to change for a dollar; we get a dollar if and only if $5i + 10j + 25k = 100$. This suggests replacing N by x^5, D by x^{10}, and Q by x^{25} to get the generating function for the number of ways to make change for some number of cents. Making these substitutions gives the same series as applying Theorem 2.1 with the value function that assigns $5i$ as the value of a multiset of i nickels, $10j$ as the value of a multiset of j dimes, and $25k$ as the value of a multiset of

---

† This example, the material preceding Theorem 2.1, and to some extent all examples involving symbolic series are based on George Pólya's highly readable and illuminating paper "Picture writing." (See Suggested Reading.)

k quarters. Thus, the number of ways to make change for a dollar is the coefficient of x^{100} in the product of the power series

$$(x^0 + x^5 + x^{10} + \cdots)(x^0 + x^{10} + x^{20} + \cdots)(x^0 + x^{25} + x^{50} + \cdots) \qquad (3.5)$$

$$= \sum_{i,j,k} x^{5i} x^{10j} x^{25k} = \sum_{i,j,k} x^{5i+10j+25k}.$$

Example 3.2. Use the formula for the sum of a geometric series to simplify the generating function (3.5) for change making.

By making the substitutions x^5, x^{10}, and x^{25} for y in Equation 2.8 of the previous section, we see that

$$1 + x^5 + x^{10} + \cdots = \sum_{i=0}^{\infty} x^{5i} = (1 - x^5)^{-1}$$

$$1 + x^{10} + x^{20} + \cdots = \sum_{j=0}^{\infty} x^{10j} = (1 - x^{10})^{-1}$$

and

$$1 + x^{25} + x^{50} + \cdots = \sum_{k=0}^{\infty} x^{25k} = (1 - x^{25})^{-1}.$$

Substituting these into (3.5), we see that the number of ways to make change for a dollar is the coefficient c_{100} of x^{100} in

$$\sum_{i=0}^{\infty} c_i x^i = (1 - x^5)^{-1} \cdot (1 - x^{10})^{-1} \cdot (1 - x^{25})^{-1} = \frac{1}{(1 - x^5)(1 - x^{10})(1 - x^{25})}. \ \blacksquare$$

Systems of Linear Recurrences from Products of Geometric Series

Although we have translated the change-making problem into a problem that looks quite different, we still have not solved it. How, after all, are we to go about finding the coefficient c_{100}? It turns out that we can work out a table of values of c_i from what we already know. Let us try to develop an approach to finding the values of c_i by starting with a simpler problem. If we had a pile of nickels and dimes only, then the number of ways to make change for a dollar would be the coefficient b_{100} in the product of two series

$$(1 - x^5)^{-1}(1 - x^{10})^{-1} = \sum_{i=0}^{\infty} b_i x^i, \qquad (3.6)$$

and if we had nickels only, the number of ways to make change for a dollar would be the coefficient a_{100} in

$$(1 - x^5)^{-1} = \sum_{i=0}^{\infty} a_i x^i.$$

We already know that $a_i = 1$ if $i = 0, 5, 10, \ldots$, and $a_i = 0$ if i is not a multiple of 5. If we multiply Equation (3.6) for $\sum b_i x^i$ by $1 - x^{10}$, we get

$$(1 - x^5)^{-1} = (1 - x^{10}) \sum_{i=0}^{\infty} b_i x^i$$

or

$$\sum_{i=0}^{\infty} a_i x^i = \sum_{i=0}^{\infty} b_i (x^i - x^{i+10})$$

$$= \sum_{i=0}^{\infty} b_i x^i - \sum_{i=0}^{\infty} b_i x^{i+10}.$$

By letting $b_{-10} = b_{-9} = \cdots = b_{-1} = 0$, we may rewrite this as

$$\sum_{i=0}^{\infty} a_i x^i = \sum_{i=0}^{\infty} b_i x^i - \sum_{i=0}^{\infty} b_{i-10} x^i. \tag{3.7}$$

Now we already know that b_0 must be one, because $a_0 b_0$ is the coefficient of x^0 in the product $(1 - x^5)^{-1}(1 - x^{10})^{-1}$ and 1 is the coefficient of x^0 in the product $(1 + x^5 + x^{10} + \cdots)(1 + x^{10} + x^{20} + \cdots)$. By combining the two summations on the right hand side of Equation 3.7, we see that $a_i = b_i - b_{i-10}$ for $i \geq 0$. Thus,

$$b_i = a_i + b_{i-10}.$$

This equation is called a linear recurrence in two variables. How does the equation help us? First, b_{i-10} is 0 if i is less than 10, so $b_i = a_i$ if $i < 10$. To get b_{10}, note that

$$b_{10} = a_{10} + b_0 = 1 + 1 = 2.$$

To get b_{11}, note that

$$b_{11} = a_{11} + b_1 = 0 + 0 = 0.$$

To get b_{20}, note that

$$b_{20} = a_{20} + b_{10} = 1 + 2 = 3.$$

Now whenever i is not a multiple of 5, a_i is 0. Thus,

$$b_i = a_i + b_{i-10} = b_{i-10}$$

and by induction, it is clear that $b_i = 0$ when i is not a multiple of 5. In other words, we need only compute b_i by the technique shown above when i is a multiple of 5. This is how row 2 was created in Table 3.1.

Table 3.1

$i =$	0	5	10	15	20	25	30	35	40	45	50	55	60	65	70	75	80	85	90	95	100
$a_i =$	1	1	1	1	1	1	1	1	1	1	1	1	1	1	1	1	1	1	1	1	1
$b_i =$	1	1	2	2	3	3	4	4	5	5	6	6	7	7	8	8	9	9	10	10	11
$c_i =$	1	1	2	2	3	4	5	6	7	8	10	11	13	14	16	18	20	22	24	26	29

What does this suggest for computing the numbers c_i? Recall that

$$(1 - x^5)^{-1}(1 - x^{10})^{-1}(1 - x^{25})^{-1} = \sum_{i=0}^{\infty} c_i x^i,$$

so that, by substitution from (3.6),

$$\left(\sum_{i=0}^{\infty} b_i x^i \right)(1 - x^{25})^{-1} = \sum_{i=0}^{\infty} c_i x^i.$$

By multiplying both sides by $1 - x^{25}$, we get

$$\sum_{i=0}^{\infty} b_i x^i = (1 - x^{25}) \sum_{i=0}^{\infty} c_i x^i.$$

If we let $c_{-25} = c_{-24} = \cdots = c_{-1} = 0$, then we get

$$\sum_{i=0}^{\infty} b_i x^i = \sum_{i=0}^{\infty} (c_i - c_{i-25}) x^i,$$

so that $c_i - c_{i-25} = b_i$ or $c_i = b_i + c_{i-25}$. Thus, $c_0 = b_0 + 0 = 1$ and $c_{25} = b_{25} + c_0 = 3 + 1 = 4$. Of course, $c_i = 0$ if i is not a multiple of 5 (why?) and we can compute line 3 of Table 3.1 from line 2 by applying the formula $c_i = b_i + c_{i-25}$.

From the table, we see that there are 29 ways to make change for a dollar using nickels, dimes, and quarters.

Now suppose we had asked for the number of ways to make change using nickels, dimes, quarters, and half-dollars. The same considerations would lead us to a product of four geometric series rather than three. If this product were rewritten as $\sum_{i=0}^{\infty} d_i x^i$, we would get the d_i's from the c_j's in the way we get the c_i's from the b_j's. This would add a new row to Table 3.1. If we wanted to include pennies as well, then the series $\sum a_i x^i$ would be the product of a geometric series for pennies and our previous product of three geometric series. It appears that we would have to compute the a_i's much as we computed the b_i's and then we would end up entirely rebuilding Table 3.1 in order to find the number of ways to make change for a dollar. In fact we could simply add a new row for pennies by filling in zeros in our current table for the number of ways to make change for 1, 2, 3, 4, 6, 7, 8, 9, 11, etc. cents and applying our next theorem.

Theorem 3.1. Suppose the generating function for the sequence a_n satisfies the equation

$$\sum_{i=0}^{\infty} a_i x^i = \frac{1}{1 - bx^n} \sum_{i=0}^{\infty} c_i x^i,$$

then the sequence a_i may be computed from the sequence c_i (and appropriate values of a_i for $i < 0$) by the recurrence

$$a_i = ba_{i-n} + c_i.$$

Proof. Writing out the proof, which is similar to the computations in the preceding examples, is Exercise 20. ∎

Generating Functions for Integer Partitions

The problem of making change calls for finding a multiset of fives, tens, and twenty-fives that add to a certain number. Thus, in the terminology of Chapter 2, we are finding a partition of that number into parts of size five, ten, and twenty-five. It should therefore not be a surprise that generating functions are a useful tool for the study of partitions of integers.

Example 3.3. As defined in Chapter 2, a *partition* of an integer n is a multiset of positive integers that add up to n. (The empty multiset of positive numbers is a partition of the integer zero by definition.) What is the generating function for partitions of integers into parts of size 1, 2, 3, 4, or 5?

The number of ways of choosing a multiset of 1's that add up to n is 1 (our only choice is a multiset of n ones), so the generating function for partitions of n all of whose parts have size 1 is

$$1 + x + x^2 + x^3 + \cdots = \frac{1}{1-x}.$$

The number of ways of choosing a multiset of 2's that add up to n is 1 if n is even and 0 if n is odd. Thus, the generating function for partitions of n all of whose parts have size 2 is

$$1 + x^2 + x^4 + x^6 + \cdots = \frac{1}{1-x^2}.$$

Similarly, the generating functions for the number of partitions of n with all parts of size 3, 4, and 5 are, respectively,

$$\frac{1}{1-x^3}, \quad \frac{1}{1-x^4}, \quad \text{and} \quad \frac{1}{1-x^5}.$$

Thus by Theorem 2.1, the generating function for the number of partitions of n into parts of size 1, 2, 3, 4, or 5 is

$$\frac{1}{1-x} \cdot \frac{1}{1-x^2} \cdot \frac{1}{1-x^3} \cdot \frac{1}{1-x^4} \cdot \frac{1}{1-x^5}. \qquad \blacksquare$$

If we apply Theorem 2.1 in the same way to partitions into parts of size $1, 2, \ldots$ and so on up to m, we obtain the following theorem.

Theorem 3.2. The coefficient of x^n in the product

$$\frac{1}{1-x} \cdot \frac{1}{1-x^2} \cdot \frac{1}{1-x^3} \cdots \frac{1}{1-x^m} = \prod_{i=1}^{m} \frac{1}{1-x^i}$$

is the number of partitions of the integer n, so long as $m \geq n$.

Proof. As in our Example 3.3, the generating function for partitions of integers into parts of size $1, 2, \ldots, m$ is the product shown. However, if $m \geq n$, then all parts of every partition of n are of size less than or equal to m. $\blacksquare$

Now recall that

$$\frac{1}{1-x^i} = 1 + x^i + x^{2i} + \cdots,$$

giving us two ways to write the symbolic product

$$\prod_{i=1}^{\infty} \frac{1}{1-x^i} = (1+x+x^2+\cdots)(1+x^2+x^4+\cdots)\cdots(1+x^i+x^{2i}+\cdots)\cdots,$$

which is a product of infinitely many power series. We may interpret this product as a power series. This power series is the sum of the products $t_1 t_2 \cdots t_i \cdots$, with one term from each series, in which only finitely many terms t_i are different from 1. (For example, we might choose x^3 from the first factor, 1 from the second factor, x^6 from the third factor, and 1 from each of the remaining factors to give us $x^3 \cdot 1 \cdot x^6 \cdot 1 \cdots 1 \cdots = x^9$. Here, only two of the chosen terms are different from 1.) How do we get the coefficient of x^m in such an infinite product? In this product, the only terms that can be multiplied together to yield x^m will be terms of the form x^j with $j \leq m$. Thus, the coefficient of x^m in this infinite product is the same as the coefficient of x^m in the product of the first m series. This gives us the generating function for all partitions of integers.

> **Theorem 3.3.** The generating function in which the coefficient of x^n is the number of partitions of the integer n is
>
> $$\prod_{i=1}^{\infty} \frac{1}{1-x^i}.$$

Many of the interesting results about partitions show that the numbers of partitions in two different families are the same. By interpreting symbolic series in different ways, by algebraic manipulations, or by using different value functions, we can sometimes show that the generating function for the number of partitions of one kind is equal to the generating function for the number of partitions of a second kind.

Example 3.4. Write down the generating function for the number of ways to partition an integer into parts of size no more than m, each used an odd number of times. Write down the generating function for the number of partitions of an integer into parts of size no more than m, each used an even number of times. Are the two numbers for which we gave generating functions related?

If we are to partition an integer using only the part i and using it an odd number of times, then we can partition only the numbers i, $3i$, $5i$, and so on (and do so only by choosing the appropriate number of i's). Therefore, the generating function for partitions whose only part is i, used an odd number of times, is

$$x^i + x^{3i} + x^{5i} + \cdots + x^{(2j+1)i} + \cdots .$$

Thus, by the product principle for generating functions, the generating function for partitions with parts of size no more than m, each used an odd number of times, is

$$\prod_{i=1}^{m}(x^i + x^{3i} + x^{5i} + \cdots x^{(2j+1)i} + \cdots)$$

$$= \prod_{i=1}^{m}\sum_{j=0}^{\infty} x^{(2j+1)i}$$

$$= x^{1+2+\cdots+m}\prod_{i=1}^{m}\sum_{j=0}^{\infty} x^{2ji} = x^{1+2+\cdots+m}\prod_{i=1}^{m}\frac{1}{1-x^{2i}} .$$

To get the generating function for partitions into parts of size no more than m each used an even number of times, observe that the generating function for partitions using only the part i and using it an even number of times is $x^0 + x^{2i} + x^{4i} + \cdots$. Thus, by the product principle for generating functions, the generating function for partitions into parts of size no more than m, each used an even number of times, is

$$\prod_{i=1}^{m}\sum_{j=0}^{\infty} x^{2ij}.$$

Since

$$1 + 2 + \cdots + m = \binom{m+1}{2},$$

the coefficient of x^k in the second generating function is the coefficient of $x^{k+\binom{m+1}{2}}$ in the first. Therefore, the number of partitions of n into parts that are of size no more than m, each used an even number of times, equals the number of partitions of $n + \binom{m+1}{2}$ into parts of size no more than m, each used an odd number of times. ∎

There is a second interpretation of the second generating function which, although natural to an experienced user of generating functions, is probably not obvious to the novice. Can you find it?

Generating Functions Sometimes Replace Inclusion–Exclusion

As in Exercise 20 of Section 1 of this chapter, it is possible to apply the principle of inclusion and exclusion to determine the number of ways to hand out k identical pieces of candy to n children so that no child gets m or more pieces. (We use, as property i, "child i gets more than $m-1$ pieces." To find the number of distributions with at least a set of size j of the properties, we observe, as in Exercise 3 of Section 2 of Chapter 2, that there is one way to first pass out m pieces of identical candy to each of the j children and then the number of ways to pass out the remaining candy as we choose is $\binom{n-jm+k-1}{n-1}$.) Generating functions make such a problem easier.

Example 3.5. In how many ways can we distribute 10 pieces of candy to three children so that no child gets more than 4 pieces?

Each child could get 0, 1, 2, 3, or 4 pieces, so the generating function for the number of ways to give n pieces of candy to child i is $1+x+x^2+x^3+x^4$. (Since all the pieces of candy are identical, there is one way to pass out each possible number of pieces.) Thus, the generating function for the number of ways to pass out n pieces of candy to three children is $(1+x+x^2+x^3+x^4)^3$. (In other words, the coefficient of x^n in this product is the number of ways to pass out a total of n pieces to the three children.) The number of ways to pass out 10 pieces is the coefficient of x^{10} in this power of a multinomial. This coefficient can be computed using the multinomial theorem. It can also be computed more easily by using some clever algebra. One of the standard factorizations we learn in algebra is

$$(1-x)(1+x+x^2+x^3+x^4) = 1-x^5,$$

so

$$(1+x+x^2+x^3+x^4) = \frac{1-x^5}{1-x}.$$

Thus,

$$\begin{aligned}
(1+x+x^2+x^3+x^4)^3 &= \frac{(1-x^5)^3}{(1-x)^3} \\
&= (1-x^5)^3(1-x)^{-3} \\
&= (1-3x^5+3x^{10}-x^{15})\sum_{i=0}^{\infty}\binom{3+i-1}{i}x^i \\
&= (1-3x^5+3x^{10}-x^{15})\sum_{i=0}^{\infty}\binom{2+i}{i}x^i.
\end{aligned}$$

In this product, the term involving x^{10} is

$$\binom{2+10}{10}x^{10} - 3x^5\binom{2+5}{5}x^5 + 3x^{10}\binom{2+0}{0}x^0$$

$$= \left[\binom{12}{10} - 3\binom{7}{5} + 3\right]x^{10}$$

$$= (66 - 63 + 3)x^{10} = 6x^{10}.$$

Thus, there are six ways to pass out the candy. ∎

The method used in the preceding example could be generalized to deal with an arbitrary number n of children—in which case we would have $(1 - x^5)^n(1 - x)^{-n}$; or to deal with an arbitrary upper bound m on the number of pieces of candy—5 would be replaced by $m + 1$. The example already shows us essentially how to deal with an arbitrary number of pieces of candy—this number just tells us what power of x to examine. Note how the example mirrors the use of inclusion–exclusion: the coefficients of $1 - 3x^5 + 3x^{10} - x^{15}$ are binomial coefficients and they alternate in sign.

*Generating Functions and Inclusion–Exclusion on Level Sums

These are exactly the sorts of coefficients that we found in the relationship between $\mathbf{N}^{\pm}_{=}$ and $\mathbf{N}^{+}_{\geq}$ in Theorem 1.5 of this chapter. In fact, Theorem 1.5 can be used to give a relationship between the generating functions for the numerical functions $\mathbf{N}^{\pm}_{=}$ and $\mathbf{N}^{+}_{\geq}$.

In particular, assume as in Section 1 that $\mathbf{N}_{=}$ and $\mathbf{N}_{\geq}$ count arrangements that have exactly the properties and at least the properties, respectively, in one of the subsets of a set P of properties. We define

$$\mathbf{N}^{+}_{=}(i) = \sum_{T:|T|=i} \mathbf{N}_{=}(T)$$

and

$$\mathbf{N}^{+}_{\geq}(i) = \sum_{T:|T|=i} \mathbf{N}_{\geq}(T).$$

According to Theorem 1.5,

$$\mathbf{N}^{+}_{=}(i) = \sum_{j=i}^{p}(-1)^{j-i}\binom{j}{i}\mathbf{N}^{+}_{\geq}(j).$$

Note that $\mathbf{N}_{\geq}^{+}(j) = 0$ if $j > p$, and $\binom{j}{i} = 0$ if $j < i$. Thus, we can rewrite the sum as

$$\mathbf{N}_{=}^{+}(i) = \sum_{j=0}^{\infty}(-1)^{j-i}\binom{j}{i}\mathbf{N}_{\geq}^{+}(j).$$

This form is well suited for relating the power series for $\mathbf{N}_{=}^{\pm}$ and $\mathbf{N}_{\geq}^{+}$. In particular,

$$\sum_{i=0}^{\infty}\mathbf{N}_{=}^{+}(i)x^i = \sum_{i=0}^{\infty}\left(\sum_{j=0}^{\infty}(-1)^{j-i}\binom{j}{i}\mathbf{N}_{\geq}^{+}(j)\right)x^i$$

$$= \sum_{j=0}^{\infty}\left(\sum_{i=0}^{\infty}(-1)^{j-i}\binom{j}{i}x^i\right)\mathbf{N}_{\geq}^{+}(j)$$

$$= \sum_{j=0}^{\infty}\mathbf{N}_{\geq}^{+}(j)\sum_{i=0}^{\infty}(-1)^{j-i}\binom{j}{i}x^i$$

$$= \sum_{j=0}^{\infty}\mathbf{N}_{\geq}^{+}(j)(x-1)^j.$$

In summary,

> ***Theorem 3.4.*** Suppose $\mathbf{N}_{=}$ and $\mathbf{N}_{\geq}$ are defined on subsets of a set P and $\mathbf{N}_{\geq}(T) = \sum_{S=T}^{P}\mathbf{N}_{=}(S)$. Then the generating function for the associated numerical function $\mathbf{N}_{=}^{\pm}$ may be obtained from the generating function for $\mathbf{N}_{\geq}^{+}$ by replacing x^j with $(x-1)^j$.
>
> *Proof.* Given before the statement of the theorem. ∎

Example 3.6. Rework Example 3.5 using Theorem 3.4. The property of interest here is "child i gets at least five pieces of candy." Thus, for a given set I of children (properties), we give each child in I five pieces of candy (in one way) and then pass out the remaining pieces in

$$\binom{10 - 5i + 3 - 1}{3 - 1}$$

ways. Thus, $\mathbf{N}_{\geq}(I)$ depends only on the size of I, so

$$\mathbf{N}_{\geq}^{+}(i) = \binom{3}{i}\cdot\mathbf{N}_{\geq}(I) = \binom{3}{i}\binom{12 - 5i}{2}.$$

Since we don't want any child to get more than four pieces, we must find the number of arrangements in which no (i.e., zero) children get more than four pieces of candy. However, this number of arrangements is $N^+(0)$, which is the coefficient of x^0 in

$$\sum_{i=0}^{\infty} N^+_=(i)x^i = \sum_{j=0}^{\infty} \binom{3}{j} \binom{12 - 5j}{2}(x-1)^j.$$

However, in a power series, the coefficient of x^0 is $f(0)$. Thus,

$$N^\pm_=(0) = \sum_{j=0}^{\infty} \binom{3}{j} \binom{12 - 5j}{2}(-1)^j = \sum_{j=0}^{3} \binom{3}{j} \binom{12 - 5j}{2}(-1)^j$$

$$= \binom{3}{0}\binom{12}{2} - \binom{3}{1}\binom{7}{2} + \binom{3}{2}\binom{2}{2} - 0$$

$$= 6. \qquad\blacksquare$$

Note that we have the same alternating series as before, although it arose from a much different sequence of operations with power series.

EXERCISES

1. What is the generating function in which the coefficient of x^n is the number of ways of choosing an even number of nickels, an even number of dimes, and an even number of quarters with total value n cents?

2. What is the generating function for the number of ways to make n cents using at least one nickel, at least one dime, and at least one quarter?

3. Extend the change-making example (with which we began the section) by adding 50-cent pieces. In how many ways can we make change for a dollar in this case?

4. Extend the change-making example (with which we began the section) by allowing pennies. (You need only include the columns of the table corresponding to multiples of five cents.) In how many ways can we make change for a dollar in this case?

5. Extend the change-making example (with which we began the section) by allowing both pennies and 50-cent pieces. In how many ways can we make change for a dollar in this case?

6. Explain why one of the generating functions in Example 3.4 is the generating function for the number of partitions of an integer into even parts of size no more than $2m$. How is the number of such partitions related to the number of partitions of an integer into parts, each of which is used an even number of times?

7. What is the generating function for the number of partitions of an integer such that each part is used an even number of times?

8. What is the generating function for the number of partitions of an integer into parts all of which are even numbers?

9. On the basis of the preceding two exercises, what can we say about the number of partitions of an integer n into even parts?

10. What is the generating function for the number of partitions of the integer n into distinct parts, that is, into parts such that no two are equal? (Hint: Each part is used at most once, so the relationship between this generating function and the ones we have studied should be similar to the relation between the generating function for subsets and the generating function for multisets.)

11. Make the substitution $\dfrac{1 - x^{2i}}{1 - x^i}$ for $1 + x^i$ in the product

$$\prod_{i=1}^{\infty}(1 + x^i).$$

Note that all terms in the numerator cancel out, leaving only every other term in the denominator. What can you conclude about the relationship between the number of partitions of n into distinct parts and the number of partitions of n into odd parts?

12. What is the generating function for the number of partitions of the integer n into distinct even parts?

13. What is the generating function for the number of partitions of the integer n into parts, at most i of which are of size i?

14. (a) What is the generating function for the number of ways of making an n-ounce candy assortment (with n an integer) using three different kinds of candy weighing one-half ounce, four different kinds of candy weighing one ounce, two different kinds weighing two ounces, and one kind of candy weighing four ounces?

    *(b) What can you say about the corresponding generating function for an n-pound candy assortment? Why is there a difference between the processes of finding the generating function for an integral number of ounces and that of finding the generating function for an integral number of pounds?

15. For problems involving partitions of an integer into a specified number of parts, it is often convenient to work with a generating function involving two variables, one relating to the number of parts and one relating to the number being partitioned. Explain why the coefficient of $t^m x^n$ in

$$\prod_{i=1}^{\infty}\frac{1}{1 - tx^i} = g(t, x)$$

is the number of partitions of the integer n into m parts.

*16. Explain on the basis of Exercise 15 why the generating function (of one variable) for the number $P(n, m)$ of partitions of the integer n into m parts is equal to

$$\frac{x^m}{(1-x)(1-x^2)\cdots(1-x^m)}.$$

17. Use generating functions to show that the number of partitions of $2n$ into n parts (where we think of m as fixed and n as the dummy variable in the power series) is the number of partitions of n into any number of parts.

18. On the basis of Exercise 15, show that the number of partitions of n with parts no more than m is equal to the number of partitions of $m+n$ with exactly m parts.

19. (a) Discuss how the exponents in the generating function in Exercise 15 relate to such parameters of the Ferrers diagram of a partition as the number of rows, the number of columns, and the number of squares.
    (b) If we replace the infinity by the integer k, what parameter of the Ferrers diagram relates to k? How does it relate? By what must we multiply to ensure that k equals this parameter?

20. Prove Theorem 3.1.

21. Show that if $\sum_{i=0}^{\infty} a_i x^i = \frac{1}{p(x)} \sum_{i=0}^{\infty} b_i x^i$ and $p(x) = \sum_{i=0}^{k} c_i x^i$ with $c_0 \neq 1$, then the sequence a_i can be computed from the sequence b_i by the recurrence

$$a_n = \frac{1}{c_0}\left(b_n - \sum_{i=1}^{k} c_i a_{n-i}\right).$$

22. Use generating functions to determine how many six-element multisets can be formed by using up to three a's, up to five s's, up to two e's, and up to three c's? What if each letter must be used once?

23. Use generating functions to determine in how many ways you can make change for a dollar using no more than 10 nickels, no more than 5 dimes, and no more than 4 quarters.

24. Use generating functions to determine in how many ways you can pass out ten identical pieces of candy to three children so that each child gets between two and four pieces? Use generating functions to answer the question for between three and five pieces.

25. Show that the generating function for the number of k-element multisets of an n-element set such that no element appears more than m times is $(1 - x^{m+1})^n / (1 - x)^n$.

26. Modify the generating function in the previous exercise to count the number of k-element multisets of an n-element set such that each element appears at least j times and no more than m times.

27. What is the generating function for partitions of n such that each number between 1 and k is used as a part at least j times and at most m times?

*28. Redo the "hat check" problem (Exercise 17, Section 1) using Theorem 3.4.

*29. Redo Exercise 20, Section 1, using generating functions.

*30. Redo Exercise 21, Section 1, using generating functions.

*31. Redo Exercise 22, Section 1, using generating functions.

*32. Redo Exercise 23, Section 1, using generating functions.

*33. Find the number of functions from a k-element set onto an n-element set by using generating functions.

Section 4   Recurrence Relations
and Generating Functions

The Idea of a Recurrence Relation

One of the reasons why generating functions are an important tool is that they allow us to manipulate and sometimes explicitly find sequences of numbers that satisfy rules such as

$$a_n = 2a_{n-1},$$

or

$$a_n = 2a_{n-1} - a_{n-2},$$

or

$$a_n = a_{n-1} + 2.$$

Such rules are called *linear recurrence relations* or *linear difference equations*.

(When referring to these equations as difference equations, we would normally rewrite them in a different notation as

$$2a_n - 2a_{n-1} = a_n, \quad \text{or, in new notation,} \quad 2\Delta a_n = a_n,$$

or

$$a_n - 2a_{n-1} + a_{n-2} = 0, \quad \text{or, in new notation,} \quad \Delta^2 a_n = 0,$$

or

$$a_n - a_{n-1} = 2, \quad \text{or, in new notation,} \quad \Delta a_n = 2 \text{ .)}$$

To be specific, a **recurrence relation** or **recurrence** is an equation that expresses the nth term a_n of a sequence in terms of values a_k with $k < n$ and perhaps the value b_n of some other sequence as well. For example, in $a_n = 3a_{n-1} - a_{n-2} + 2^n$ we have expressed a_n in terms of the earlier values a_{n-1} and a_{n-2} as well as the sequence $b_n = 2^n$. When each term is a multiple of some a_i, the recurrence is called *homogeneous*. Thus the first two of our original three examples are homogeneous. We say the recurrence has *order* k if it involves a_n through a_{n-k} or, equivalently, a_{n+k} thro ugh a_n. Our first and third original examples are of first order while the second one is a second-order recurrence. We have already used such recurrences to compute tables of values. Equations like those above are rather analogous to differential equations, and the use of generating functions to find numbers a_i satisfying these equations is similar to the use of power series in solving differential equations. The equation $a_n = 2a_{n-1}$ arises in the study of subsets of a set—if a_n is the number of subsets of $\{1, 2, \ldots, n\}$, then a_n is the number of subsets not containing n plus the number of subsets containing n, and both of these numbers are a_{n-1}. Of course, we have already shown that since $a_0 = 1$, it follows that $a_n = 2^n$. Let us also derive this result using generating functions, showing each detail explicitly.

How Generating Functions Are Relevant

Example 4.1. Find the generating function $\sum_{i=0}^{\infty} a_i x^i$ in which a_i is the number of subsets of an i-element set.

Since we know that $a_{i+1} = 2a_i$, we may multiply both sides of the equation by x^{i+1} and sum over all i to get

$$\sum_{i=0}^{\infty} a_{i+1} x^{i+1} = \sum_{i=0}^{\infty} 2a_i x^{i+1} \tag{4.1}$$

or

$$\sum_{i=0}^{\infty} a_{i+1} x^{i+1} = 2x \sum_{i=0}^{\infty} a_i x^i. \tag{4.2}$$

If we let $j = i + 1$, then we have $a_j x^j$ on the left-hand side and $a_i x^i$ on the right-hand side. This lets us "solve" the equation for the power series $\sum_{i=0}^{\infty} a_i x^i$. We now proceed to do so:

$$\sum_{j=0}^{\infty} a_j x^j - a_0 = 2x \sum_{i=0}^{\infty} a_i x^i,$$

or, changing the "dummy" variable j to i and rearranging terms,

$$\sum_{i=0}^{\infty} a_i x^i - 2x \sum_{i=0}^{\infty} a_i x^i = a_0,$$

so

$$(1 - 2x) \sum_{i=0}^{\infty} a_i x^i = a_0.$$

By the formula for the sum of an infinite geometric series,

$$\sum_{i=0}^{\infty} a_i x^i = \frac{a_0}{1 - 2x} = a_0(1 + 2x + (2x)^2 + (2x)^3 + \cdots)$$

$$= a_0 \sum_{i=0}^{\infty} 2^i x^i.$$

Thus, $a_i = a_0 2^i$, and since $a_0 = 1$, $a_i = 2^i$. ∎

In the change from (4.1) to (4.2), we factored $2x$ out of the sum. Had our recurrence been $a_{i+1} = ba_i$, we could have factored the bx similarly. If, however, we had $a_{i+1} = ia_i$ or $a_{i+1} = i^2 a_i$, the factorization would not work. (There is, however, a method for solving such recurrences involving a slightly different kind of generating function, introduced in Section 5 of this chapter.) On the other hand, this technique does let us solve a wide variety of recurrences that arise in practice. In the theorem below d_i corresponds to a swquence we already know, while a_i is a sequence whose values we wish to find in terms of d_i.

Theorem 4.1. If the sequence a_n satisfies the recurrence

$$a_{i+1} = ba_i + d_{i+1},$$

then if $b \neq 0$, the generating function for a_i is

$$\sum_{i=0}^{\infty} a_i x^i = \frac{a_0 - d_0 + \sum_{j=0}^{\infty} d_j x^j}{1 - bx}$$

and

$$a_n = b^n \left(a_0 + \sum_{i=1}^{n} b^{-i} d_i \right).$$

Proof. By multiplying both sides of the recurrence by x^{i+1} and solving for $\sum_{i=1}^{\infty} a_i x^i$, we get the generating function given. Finding the coefficient of x^i on both sides of the equation gives us the preceding formula for a_i. ∎

Example 4.2. A hypothetical radioactive compound decays in 1 year to a mixture half of which is the original compound and half of which is nonradioactive. Let us suppose a company produces this substance as a by-product of its operations and at the end of each year adds 100 kilograms of it to material already in storage. If they start with no radioactive material, how much radioactive material will be in storage after n years?

We may use the recurrence $a_n = .5a_{n-1} + 100$ to describe the process. Then by Theorem 4.1,

$$a_n = \left(\frac{1}{2} \right)^n \left(0 + \sum_{i=1}^{n} \left(\frac{1}{2} \right)^{-i} \cdot 100 \right)$$

$$= 100 \cdot \left(\frac{1}{2} \right)^n \sum_{i=1}^{n} 2^i = 100 \cdot \left(\frac{1}{2} \right)^n \left(\frac{2^{n+1} - 1}{2 - 1} - 1 \right)$$

$$= 100 \left(2 - 2 \left(\frac{1}{2} \right)^n \right) = 200 \left(1 - \left(\frac{1}{2} \right)^n \right).$$

Notice that the total amount of radioactive material stored gets close to, but never reaches, 200 kilograms. ∎

Second-Order Linear Recurrence Relations

A classic example of how recurrence relations arise is Fibonacci's problem; we present three variations on Fibonacci's problem culminating in Fibonacci's original example. The problems all deal with an imaginary population of rabbits. Similar problems arise in the study of leaf and tip growth in plants, in the analysis of algorithms, and in geometry.

Example 4.3. In variation 1, we have a population of rabbits who reproduce in pairs. Each pair of rabbits born in a particular month produces a pair of offspring in the following month and then becomes infertile by the end of that second month. Over the period of our observations, no rabbits die. Thus, if a_n denotes the number of pairs of rabbits present at the end of n months, the number of pairs of baby rabbits present at the end of month $n + 2$ is the number of pairs of rabbits that have not yet reproduced, namely $a_{n+1} - a_n$. Adding this number to a_{n+1} gives the total number of pairs of rabbits present at the end of month $n + 2$, so that

$$a_{n+2} = 2a_{n+1} - a_n.$$

We shall find a generating function for a_n. From this we will find a_n.

Substituting the equation for a_{n+2} into the term a_{n+2} of the generating function for a_n and simplifying yields

$$\sum_{n=0}^{\infty} a_{n+2}x^{n+2} = 2\sum_{n=0}^{\infty} a_{n+1}x^{n+2} - \sum_{n=0}^{\infty} a_n x^{n+2},$$

$$\sum_{n=0}^{\infty} a_{n+2}x^{n+2} = 2x\sum_{n=0}^{\infty} a_{n+1}x^{n+1} - x^2\sum_{n=0}^{\infty} a_n x^{n},$$

$$\sum_{n=2}^{\infty} a_n x^{n} = 2x\sum_{n=1}^{\infty} a_n x^{n} - x^2\sum_{n=0}^{\infty} a_n x^{n},$$

$$\sum_{n=0}^{\infty} a_n x^{n} - a_1 x - a_0 = 2x\left(\sum_{n=0}^{\infty} a_n x^{n} - a_0\right) - x^2\sum_{n=0}^{\infty} a_n x^{n}.$$

Rearranging terms and placing all occurrences of the generating function on the left gives

$$\sum_{n=0}^{\infty} a_n x^{n}(1 - 2x + x^2) = (a_1 - 2a_0)x + a_0$$

or

$$\sum_{n=0}^{\infty} a_n x^n = \frac{a_0 + (a_1 - 2a_0)x}{(1-x)^2}.$$

However, by the extended binomial theorem (Theorem 2.2),

$$(1-x)^{-2} = \sum_{k=0}^{\infty} C(2+k-1;k)x^k$$

$$= \sum_{k=0}^{\infty} C(k+1;k)x^k = \sum_{k=0}^{\infty} (k+1)x^k,$$

so by substitution,

$$\sum_{n=0}^{\infty} a_n x^n = (a_0 + (a_1 - 2a_0)x) \sum_{k=0}^{\infty} (k+1)x^k$$

$$= a_0 \sum_{k=0}^{\infty} (k+1)x^k + (a_1 - 2a_0) \sum_{k=0}^{\infty} (k+1)x^{k+1}$$

$$= a_0 + a_0 \sum_{k=1}^{\infty} (k+1)x^k + (a_1 - 2a_0) \sum_{k=1}^{\infty} kx^k$$

$$= a_0 + \sum_{k=1}^{\infty} \big[k(a_1 - a_0) + a_0 \big] x^k.$$

Thus, $a_n = n(a_1 - a_0) + a_0$. Note that if we started out so that $a_1 = a_0$, then the end-of-the-month population would always be a_0; if a_1 were greater than a_0, then the population would grow linearly; and if a_1 were less than a_0, our population would die out (the assumptions we use in describing the recurrence relation would be invalid, though, because the reasoning we used to find the recurrence implies that any rabbits present at the end of month n but not at the beginning of the month were born during month n and no rabbits are removed during month n). The values of a_0 and a_1 represent the initial conditions of the experiment. If in month 0 we have a pair of rabbits that have had babies once, then a_1 would be a_0. However, if in month 0 we have rabbits born during month 0, then $a_1 = 2a_0$. (Note: By month 0 we mean the month at the end of which we begin recording the population.) This problem is one in which a solution could have been found without generating functions; however, the methods used are quite general. ∎

In outline, to solve a linear recurrence relation of order k—that is, a linear recurrence of the form

$$a_{n+k} = \sum_{i=0}^{k-1} b_i a_{n+i},$$

we substitute the relation for a_{n+k} into the generating function or equivalently, we multiply both sides of the recurrence by x^{n+k} and add over all n. We then reexpress each infinite series in terms of $\sum_{n=0}^{\infty} a_n x^n$ and solve for this power series. The result always turns out to be a quotient of two polynomials and depends on the numbers $a_0, a_1, \ldots, a_{k-1}$. If we know these numbers in advance, they can be used instead of the symbols a_i; this simplifies the arithmetic. As we shall see, a quotient of two polynomials can be reexpressed as a product of a polynomial and a power series, in particular, as a product of a polynomial and some number of geometric series. Let us illustrate with a second rabbit problem.

Example 4.4. Now assume that a pair of rabbits requires a maturation period of 1 month before they can produce offspring (in other words they are not mature at the end of the month in which they are born and so do not produce offspring in the next month); each pair of mature rabbits present at the end of 1 month produces two new pairs of rabbits by the end of the next month and again by the end of each of the following months. (We still assume that no rabbits die during our observations.) The following recurrence relation states that the rabbit population (measured in pairs) at the end of 1 month consists of all rabbits present at the end of the previous month plus new offspring produced by rabbits that are mature:

$$a_{n+2} = a_{n+1} + 2a_n.$$

Substituting this into the power series gives us

$$\sum_{n=0}^{\infty} a_{n+2} x^{n+2} = \sum_{n=0}^{\infty} a_{n+1} x^{n+2} + 2 \sum_{n=0}^{\infty} a_n x^{n+2},$$

which, after appropriate manipulation, yields

$$\left(\sum_{n=0}^{\infty} a_n x^n \right) (1 - x - 2x^2) = a_1 x + a_0 - a_0 x,$$

so that

$$\sum_{n=0}^{\infty} a_n x^n = \frac{(a_1 - a_0)x + a_0}{1 - x - 2x^2} = \frac{a_0 + (a_1 - a_0)x}{(1 - 2x)(1 + x)} \tag{4.3}$$

$$= (a_0 + (a_1 - a_0)x) \left(\sum_{i=0}^{\infty} (2x)^i \right) \sum_{j=0}^{\infty} (-x)^j$$

$$= (a_0 + (a_1 - a_0)x) \left(\sum_{i=0}^{\infty} 2^i x^i \right) (\sum_{j=0}^{\infty} (-1)^j x^j)$$

$$= (a_0 + (a_1 - a_0)x) \sum_{k=0}^{\infty} \left(\sum_{i=0}^{k} 2^i (-1)^{k-i} \right) x^k$$

$$= (a_0 + (a_1 - a_0)x) \sum_{k=0}^{\infty} (-1)^k \left(\sum_{i=0}^{k} (-2)^i \right) x^k.$$

By applying the formula for the sum of a geometric series, we may show that the coefficient of x^k is $\frac{2^{k+1}}{3} + \frac{(-1)^k}{3}$. Thus,

$$\sum_{n=0}^{\infty} a_n x^n = \frac{a_0 + (a_1 - a_0)x}{(1 - 2x)(1 + x)}$$

$$= \frac{a_0 + (a_1 - a_0)x}{3} \sum_{i=0}^{\infty} (2^{i+1} + (-1)^i)x^i,$$

so that for $i > 0$

$$a_i = \frac{a_1 - a_0}{3}(2^i + (-1)^{i-1}) + \frac{a_0}{3}(2^{i+1} + (-1)^i)$$

$$= \frac{a_1 + a_0}{3} \cdot 2^i + \frac{2a_0 - a_1}{3} \cdot (-1)^i.$$

Thus, since a_0 and a_1 are nonnegative, if either is nonzero the population will grow essentially exponentially. If we start with mature rabbits at time 0, then $a_1 = 3a_0$; if we start with immature rabbits at time 0, then $a_1 = a_0$.

Although the algebra in this example is not difficult, neither is it pleasant. As we shall see later, the somewhat more sophisticated algebraic concept of partial fractions will trim the work needed in situations similar to this example. By generalizing the result of this computation, the next theorem allows us to bypass most of the algebraic computation. ∎

We summarize the technique of the example in a theorem.

Theorem 4.2. If the polynomial x^2+bx+c has distinct roots r_1 and r_2, then the generating function for the solution to the recurrence relation

$$a_{n+2} + ba_{n+1} + ca_n = 0 \qquad (4.4)$$

is given by

$$\sum_{i=0}^{\infty} a_i x^i = \frac{a_0 + (a_1 + ba_0)x}{(1 - r_1 x)(1 - r_2 x)} = (a_0 + (a_1 + ba_0)x) \sum_{i=0}^{\infty} \frac{r_2^{i+1} - r_1^{i+1}}{r_2 - r_1} x^i$$

and

$$a_n = \frac{a_0(r_2 + b) + a_1}{r_2 - r_1} r_2^n + \frac{a_0(r_1 + b) + a_1}{r_1 - r_2} r_1^n,$$

so that there are constants c_1 and c_2 with

$$a_n = c_1 r_1^n + c_2 r_2^n.$$

Proof. When we multiply both sides of (4.4) by x^{n+2}, sum, and solve for the generating function for a_n, we get (as Exercise 13 asks you to show)

$$\sum_{n=0}^{\infty} a_n x^n = \frac{(a_1 + ba_0)x + a_0}{1 + bx + cx^2}.$$

Our assumption that r_1 and r_2 are roots of $x^2 + bx + c$ implies that

$$1 + bx + cx^2 = (1 - r_1 x)(1 - r_2 x).$$

(Since $(x - r_1)(x - r_2) = x^2 + bx + c$, we have that $c = r_1 r_2$, and $b = -r_1 - r_2$. When we use these facts to expand $(1 - r_1 x)(1 - r_2 x)$, we see that it is $1 + bx + cx^2$.) As Exercise 14 asks you to show, the coefficient of x^k in $(1 - r_1 x)^{-1}(1 - r_2 x)^{-1}$ is

$$\sum_{i=0}^{k} r_1^i r_2^{k-i} = r_2^k \sum_{i=0}^{k} \left(\frac{r_1}{r_2}\right)^i.$$

But

$$r_2^k \sum_{i=0}^{k} \left(\frac{r_1}{r_2}\right)^i = r_2^k \frac{1 - \left(\frac{r_1}{r_2}\right)^{k+1}}{1 - \frac{r_1}{r_2}} = \frac{r_2^{k+1} - r_1^{k+1}}{r_2 - r_1},$$

so that

$$\sum_{n=0}^{\infty} a_n k x^n = \frac{(a_1 + ba_0)x + a_0}{(1 - r_1 x)(1 - r_2 x)} = (a_0 + (a_1 + ba_0)x) \sum_{i=0}^{\infty} \frac{r_2^{i+1} - r_1^{i+1}}{r_2 - r_1} x^i.$$

Thus

$$\begin{aligned}
a_n &= \frac{r_2^{n+1} - r_1^{n+1}}{r_2 - r_1} a_0 + (a_1 + ba_0) \frac{r_2^n - r_1^n}{r_2 - r_1} \\
&= \frac{r_2 a_0 + a_1 + ba_0}{r_2 - r_1} r_2^n - \frac{r_1 a_0 + a_1 + ba_0}{r_2 - r_1} r_1^n,
\end{aligned}$$

which is the formula in the statement of the theorem since $r_1 - r_2 = -(r_2 - r_1)$. ∎

Because of its central role in describing the solution to the recurrence relation $a_{n+2} + ba_{n+1} + ca_n = 0$, the polynomial $p(x) = x^2 + bx + c$ is called the **characteristic polynomial** of the recurrence relation.

The Original Fibonacci Problem

Fibonacci's original problem is no more difficult theoretically; however, a slightly unexpected square root appears.

Example 4.5. Now, as in Example 4.3, our rabbits can reproduce after one month maturation; however, each pair of mature rabbits present at the end of one month produces exactly one pair of baby rabbits during (and before the end of) the next month and each succeeding month. Again, we assume no rabbits die during our observations. (This also describes a somewhat oversimplified model of plant growth and branching; the rabbits correspond to branches and new branches grow from those that have matured over at least one growing season.) Then if a_n is the number of pairs of rabbits at the end of month n, our recurrence relation will be

$$a_{n+2} = a_{n+1} + a_n.$$

Substituting into the generating function as before and using the facts that b and c are 1, we get

$$\sum_{i=0}^{\infty} a_i x^i = \frac{a_0 + (a_1 - a_0)x}{1 - x - x^2}.$$

Note that $-x^2 - x + 1$ has the roots $\frac{1+\sqrt{5}}{2}$ and $\frac{1-\sqrt{5}}{2}$.

It is traditional to start with one pair of baby rabbits; thus $a_0 = a_1 = 1$ (pair). Theorem 4.2 gives us

$$a_n = \frac{1}{\sqrt{5}} \left(\frac{1+\sqrt{5}}{2} \right)^{n+1} - \frac{1}{\sqrt{5}} \left(\frac{1-\sqrt{5}}{2} \right)^{n+1}.$$

The numbers a_n are called the **Fibonacci numbers** and are usually denoted by F_n. Note that for large values of n, the $\left(\frac{1+\sqrt{5}}{2} \right)^{n+1}$ term will be large in absolute value while the second power will be less than one; thus the Fibonacci numbers a_n are essentially $\left(\frac{1+\sqrt{5}}{2} \right)^{n+1}$.

Notice that from the recurrence relation $F_{n+2} = F_{n+1} + F_n$ we see that for $n = 0, 1, 2, 3, 4, 5, 6$, we have $F_n = 1, 1, 2, 3, 5, 8, 13$. Thus the square roots in the preceding formula for a_n must disappear! Can you explain why? ∎

General Techniques

Theorem 4.2 lets us write down solutions to any homogeneous linear recurrence relation of degree 2, as long as the characteristic polynomial has distinct roots. If the characteristic polynomial has only one root and so may be factored as $(x - r)^2$, then the extended binomial theorem gives us the generating function for a_n in a form suitable for writing down a formula for a_n. The result will always be a sum of a multiple of r^n and a multiple of nr^n.

Similar techniques let us solve any recurrence of the form

$$a_n + ba_{n-1} + ca_{n-2} = d_n.$$

The generating function is a sum of two functions, the generating function we have found for the case $d_n = 0$ plus

$$\frac{1}{1 + bx + cx^2} \sum_{n=2}^{\infty} d_n x^n.$$

For recurrences of higher order, there is an analogous characteristic polynomial and the solutions are combinations of powers of roots of this polynomial with some added terms involving n, n^2, etc. if the characteristic polynomial has repeated roots. To prove this (or to write down a formula for a_n) as we proved Theorem 4.2 would involve tedious computation involving sums of geometric and binomial series.

The generating function approach often works for other variations on our examples as well, and so it is better to practice working directly with the generating functions rather than memorize the formulas of Theorem 4.2.

One important technique for dealing with generating functions is the method of partial fractions, which we now illustrate by giving a second solution to the recurrence of Example 4.4. This method, which lets us replace a product by a sum, is sometimes used in calculus for integration of quotients of polynomials. The basic idea is that we can find numbers r and s with

$$\frac{1}{(ax+b)(cx+d)} = \frac{r}{ax+b} + \frac{s}{cx+d}$$

if $(ax+b)$ and $(cx+d)$ aren't multiples of each other. Also, we can find numbers r, s, and t with

$$\frac{1}{(ax+b)^2(cx+d)} = \frac{r}{(ax+b)^2} + \frac{s}{ax+b} + \frac{t}{cx+d}.$$

The idea can be extended to larger numbers of terms or higher powers.

In our example we write the equation

$$\frac{1}{(1-2x)(1+x)} = \frac{r}{(1-2x)} + \frac{s}{(1+x)},$$

which gives us

$$\frac{1}{(1-2x)(1+x)} = \frac{r+rx+s-2sx}{(1-2x)(1+x)}$$

so that $r + s = 1$ and $rx - 2sx = 0$. Dividing the second equation by x (or replacing x by the possible value 1) yields $r - 2s = 0$, so $3s = 1$ and $3r = 2$. Thus,

$$\frac{1}{(1-2x)(1+x)} = \frac{2}{3} \cdot \frac{1}{1-2x} + \frac{1}{3} \cdot \frac{1}{1+x}$$

$$= \frac{2}{3} \sum_{i=0}^{\infty} (2x)^i + \frac{1}{3} \sum_{i=0}^{\infty} (-x)^i$$

$$= \frac{2}{3} \sum_{i=0}^{\infty} 2^i x^i + \frac{1}{3} \sum_{i=0}^{\infty} (-1)^i x^i.$$

Collecting terms and multiplying by $a_0+(a_1-a_0)x$ yields from Equation (4.3)

$$\sum_{i=0}^{\infty} a_i x^i = \frac{a_0 + (a_1 - a_0)x}{(1-2x)(1+x)}$$

$$= \frac{a_0 + (a_1 - a_0)x}{3} \sum_{i=0}^{\infty} (2^{i+1} + (-1)^i) x^i,$$

so that for $i > 0$

$$a_i = \frac{a_1 - a_0}{3}(2^i + (-1)^{i-1}) + \frac{a_0}{3}(2^{i+1} + (-1)^i)$$

$$= \frac{a_1 + a_0}{3} \cdot 2^i + \frac{2a_0 - a_1}{3} \cdot (-1)^i.$$

By using the method of partial fractions, we can always rewrite our generating function in a form suitable for giving a formula for a_n, because every polynomial can be written as a product of linear factors (perhaps involving complex numbers in the coefficients as in $(x - i)(x + i) = x^2 + 1$).

EXERCISES

1. (a) What is a simpler form for $\sum_{i=0}^{\infty}(\frac{1}{3})^i x^i$, $\sum_{i=0}^{\infty}(-\frac{1}{3})^i x^i$?

   (b) To which recurrence relations and what condition of a_0 do these generate the solutions? (Hint: This question requires that you reason in the "opposite direction" from the way we reasoned in the text.)

2. Use generating functions to solve the recurrence relation $a_n = 3a_{n-1}$.

3. Redo Example 4.3 assuming *in advance* that $a_0 = 1$ and $a_1 = 2$.

4. Redo Example 4.4 assuming *in advance* that $a_1 = a_0 = 1$.

5. Use generating functions to solve the recurrence relation $a_{n+2} = 4a_{n+1} - 4a_n$. First consider $a_0 = a_1 = 1$; next consider $a_0 = 1$, $a_1 = 2$ and $a_0 = 0$, $a_1 = 1$.

6. Use generating functions to solve the recurrence relation $a_{n+3} = 3a_{n+2} - 3a_{n+1} + a_n$ in the case with $a_0 = a_1 = a_2 = 1$. Next consider the case $a_0 = 1$, $a_1 = 3$, and $ka_2 = 6$.

7. Solve the following variant of Fibonacci's problem. Each mature pair of rabbits present at the end of any given month produces three more pairs of rabbits during the month following; further, the baby rabbits take a month to mature. Assuming no rabbits die, how many rabbits are present after n months, assuming we start with 10 baby rabbits?

8. Solve the following variant of Fibonacci's problem. Rabbits require 1 month of maturation before they can produce babies; each pair of mature rabbits present at the beginning of a month produces one pair of rabbits during that month; rabbits die after their second offspring are produced. (This gives a linear recurrence of order 3.) How many rabbits are present at the end of n months, assuming we start with 10 baby rabbits?

9. One natural model for the growth of a tree is that all new wood grows from the previous year's growth and is proportional to it in amount. To

be more precise, let us assume that the new growth in a given year is the constant c times the amount of new growth in the previous year. Assuming we are measuring new growth as the sum of the lengths of the new branches, write a recurrence relation for the total length a_n of all the branches of the tree at the end of growing season n. Find the general solution to your recurrence relation. Assuming we begin with a 1-meter-long cutting which is one half new wood and which grows to a total length of 2 meters by the end of year 1, what will the total length of the tree be in year n?

10. The linear recurrence $a_{n+1} = a_n + 1$ has the (almost obvious) solution $a_n = a_0 + n$. Using generating functions, show how this arises.

11. Find the generating function for the solution of the recurrence relation $a_{n+1} = ba_n + d$. Give a formula for a_n in terms of a_0, b, and d.

12. The proof given for Theorem 4.1 was actually a brief outline of a proof. Write out a complete proof.

13. In the first sentence of the proof of Theorem 4.2, we asserted that the generating function for a_n could be written in the form

$$\sum_{n=0}^{\infty} a_n x^n = \frac{(a_1 + ba_0)x + a_0}{1 + bx + cx^2}.$$

Show that this is the case.

14. In the proof of Theorem 4.2, we stated that the coefficient of x^k in $(1 - r_1 x)^{-1}(1 - r_2 x)^{-1}$ is $\sum_{i=0}^{k} r_1^i r_2^{k-i} = r_2^k \sum (\frac{r_1}{r_2})^i$. Show that this is the case.

15. The recurrence relation $a_{n+2} - 2ra_{n+1} + r^2 a_n = 0$ has the most general form that a homogeneous degree 2 recurrence can have if its characteristic polynomial has one repeated root. Find a general formula for its solution in terms of a_0 and a_1, thus proving an analog of Theorem 4.2 for the case of repeated roots.

16. The "tower of Hanoi" is a puzzle consisting of three vertical posts (mounted on a board) and some number n of rings of different diameters. In standard form, the rings are all stacked on one post in decreasing order of size from bottom to top (Figure 4.1). A solution to the puzzle consists of first choosing a second post on which the rings are to be stacked; then moving rings from post to post in such a way that a larger ring is never placed on a smaller ring with the goal of getting all the rings on the chosen post. If a_n is the minimum number of moves to solve a puzzle with n rings, then $a_{n+1} = 2a_n + 1$. (First solve the

problem of getting all but the biggest ring onto the third post, next move the biggest ring to the chosen post, then solve the problem of moving all but the biggest ring to the chosen post.) Find the number of moves needed with n rings. In particular, what if $n = 5$?

Figure 4.1

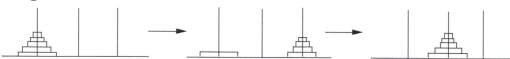

17. Write a recursive program that uses the idea underlying the recurrence relation of Exercise 16 to give a computer solution of the tower of Hanoi problem for any number of rings.

18. Solve the recurrence relation $a_{k+1} = a_k + 2^k$ with $a_0 = 2$.

19. Solve the recurrence relation $a_{n+2} = 4a_{n+1} - 4a_n + 2^n$ with $a_0 = 2$, $a_1 = 4$.

20. Find the generating function for the solution to $a_{n+2} = 3a_{n+1} + 2^n$ for $a_0 = 1$.

21. Find the generating function for the solution to $a_{n+2} = a_{n+1} + 2a_n + 2^n$ for $a_0 = 2$, $a_1 = 4$.

*22. A merge sort of a list of numbers can be described as follows. If the list has only one element, do nothing to that one-element list. Otherwise, split the list in half, apply merge sort to each half, then merge the two sorted lists in increasing order. Let a_n be the number of comparisons made by a merge sort on an n-element list. For $n = 1, 2, 4$, figure out by experiment how many comparisons you use (note that you must make some comparisons as you merge). Assuming n is a power of 2, write a recurrence relation for the numbers a_n. Since this recurrence involves $a_{n/2}$, it is not linear, and the merging keeps it from being homogeneous. There is a solution to this recurrence involving $n \log_2(n)$. One way to find it is to make the substitution $n = 2^k$. Make this substitution, solve the resulting recurrence, and convert back from k to n to get a formula for a_n. This kind of recurrence arises in analyzing many of the "divide and conquer" algorithms used to solve computing problems.

*23. Find a recurrence relation for the number of ways to divide a regular n-gon into triangles by means of nonintersecting diagonals. Solve this recurrence relation. Explain the relationship of this exercise to Catalan numbers.

24. Let a_n be the Fibonacci numbers of Example 4.5. Find all n with $a_n = n$. Find all n with $a_n = n^2$.

25. Show that $\sum_{k=0}^{n} \binom{n-k}{k}$ is a solution to the Fibonacci recurrence relation. Is it equal to one of the Fibonacci numbers a_m of Example 4.5?

*26. Show that if $F_0, F_1, F_2, \ldots, F_n \ldots$ is the sequence of Fibonacci numbers of Example 4.5, then for each m and n

$$b_{m,n} = \sum_{k=0}^{n} \binom{n}{k} F_{m+k}$$

is also a Fibonacci number F_j for some value of j.

Section 5 Exponential Generating Functions

Indicator Functions

While powers of binomials arise frequently in the use of generating functions, many other kinds of polynomials and series arise as well. Recall that the Stirling numbers relate falling factorial polynomials to ordinary polynomials. In particular, since

$$(x)_n = x(x-1)\cdots(x-n+1) = \sum_{k=0}^{n} s(n,k)x^k,$$

the falling factorial polynomial of degree n is the generating function for the Stirling numbers of the first kind. It is natural in this situation to ask if the similar polynomial equation involving the Stirling numbers of the second kind gives a generating function for them. This polynomial equation is

$$x^n = \sum_{k=0}^{n} S(n,k)(x)_k.$$

Note that we no longer have powers of x in the sum, and so this is not a generating function in the usual sense. For this reason, we define generating functions relative to a set of "indicator functions." For our purposes, a set I of polynomials will be called a set of *indicator functions* if for each nonnegative integer n there is exactly one polynomial $p_n(x)$ of degree n in I. A straightforward proof by induction shows that any polynomial can then be represented in one and only one way as a sum of numerical multiples of indicator polynomials in I. (In the language of linear algebra, I is a basis for the vector space of polynomials.) At times even more general indicator functions can be described, but our description includes all the usual types of such polynomials.

We define the *generating function for the sequence a_n relative to I* to be the series

$$\sum_{n=0}^{\infty} a_n p_n(x).$$

Thus, x^n is the generating function for the Stirling numbers of the second kind relative to the indicator family of falling factorial polynomials.

Exponential Generating Functions

One other indicator family is suggested by an analysis of the binomial theorem. Recall that $(n)_j = \frac{n!}{(n-j)!}$ is the number of one-to-one functions from a j-element set to an n-element set. The binomial theorem can be rewritten to give $(n)_j$ explicitly:

$$(x+1)^n = \sum_{j=0}^{n} \binom{n}{j} x^j = \sum_{j=0}^{n} \frac{(n)_j}{j!} x^j = \sum_{j=0}^{n} (n)_j \frac{x^j}{j!}.$$

Thus, the generating function for the number of one-to-one functions from a j-element set into an n-element set *relative to* the indicator functions $\frac{x^j}{j!}$ is $(x+1)^n$. Note that $(n)_j$ is 0 if $j > n$, so

$$\sum_{j=0}^{n} \binom{n}{j} x^j = \sum_{j=0}^{\infty} (n)_j \frac{x^j}{j!}.$$

The importance of one-to-one functions in our work so far suggests that generating functions relative to the indicator family $\frac{x^j}{j!}$ might arise in other situations as well, and in fact they do. Since the exponential function e^x is the generating function for the all ones $(1, 1, 1, 1, \ldots)$ sequence relative to $\frac{x^j}{j!}$, it is customary to call

$$\sum_{j=0}^{\infty} a_j \frac{x^j}{j!}$$

the **exponential generating function** for the sequence a_i.

Products of Exponential Generating Functions

Many of our important results on generating functions were based on multiplying generating functions. We began our study of generating functions with the study of symbolic series. From these examples, we were led to understand what happens when we multiply generating functions. First, if we have two generating functions $\sum_{i=0}^{\infty} a_i x^i$ and $\sum_{i=0}^{\infty} b_i x^i$, we interpret a_i as being the number of arrangements s in some set S whose "value" $v_1(s)$ is the integer i and interpret b_j as being the number of arrangements t in some set T whose "value" $v_2(t)$ is the integer j. Then the coefficient c_k of x^k in the product

$$\sum_{k=0}^{\infty} c_k x^k = \sum_{i=0}^{\infty} a_i x^i \cdot \sum_{j=0}^{\infty} b_j x^j$$

is equal to the number of ordered pairs (s, t) with s in S and t in T and $v_1(s) + v_2(t) = k$.

This result does not apply to the a_i, b_i, and c_i in exponential generating functions. However, there is a similar interpretation of the coefficients c_k of $\frac{x^k}{k!}$ in the product

$$\sum_{i=0}^{\infty} c_k \frac{x^k}{k!} = \sum_{i=0}^{\infty} a_i \frac{x^i}{i!} \cdot \sum_{j=0}^{\infty} b_j \frac{x^j}{j!}$$

of exponential generating functions.

To lead into this interpretation, we begin with another example. Suppose we have two bookshelves, shelf S and shelf T. We will arrange books on these shelves according to some prescribed rules such as, perhaps, no shelf is empty or the books are in alphabetical order by title or, perhaps, arranged totally at random.

A given arrangement s of books on shelf S uses a certain set $v_S(s) = B_s$ of books, while a given arrangement t of books on shelf T uses a certain set $v_T(t) = B_t$ of books. Of course, B_s and B_t have no books in common, so $B_s \cap B_t$ is empty and $B_s \cup B_t$ is the entire set B of books used on both shelves. We will assume that the number of prescribed arrangements of a set I of books on shelf S depends only on the size i of the set I and thus is some number a_i. We assume the same thing about the number of arrangements b_j of a set J of j books on shelf T. (Note, we aren't assuming that $a_i = b_i$; for example, for some reason we might be allowed to place the books on shelf S at random but on shelf T only in the single arrangement of alphabetical order by title.)

In this context, is there a natural interpretation of the coefficients c_k of the product

$$\sum_{k=0}^{\infty} c_k \frac{x^k}{k!} = \sum_{i=0}^{\infty} a_i \frac{x^i}{i!} \cdot \sum_{j=0}^{\infty} b_j \frac{x^j}{j!} \ ?$$

First, notice that $\frac{c_k}{k!}$ will be the *sum* of all products of the form $\frac{a_i b_j}{i! j!}$ in which $i + j = k$. Thus, $j = k - i$ and c_k is the sum over all values of i of the product

$$\frac{k!}{i!(k-i)!} a_i b_{k-i} = C(k; i) a_i b_{k-i}.$$

If we have a specific set K of k books, then we may interpret $C(k; i)$ as the number of ways to choose i books for shelf S and $k - i$ books for shelf T. Then we may interpret $a_i b_{k-i}$ as the number of ways of arranging this ordered pair of sets of books on the two shelves. To compute c_k, we must add

$$C(k; i) a_i b_{k-i}$$

over all values of i. Thus,

$$c_k = \sum_{i=0}^{k} C(k;i)a_i b_{k-i},$$

which is the total number of ways of dividing the set K of books into an ordered pair of sets, one for shelf S and one for shelf T, and then arranging these books. This is just the number of permissible arrangements of k books on the two shelves.

To summarize, suppose that we are given the exponential generating function $\sum_{i=0}^{\infty} a_i \frac{x^i}{i!}$ for the number of permissible arrangements of a set I of size i on shelf S and the exponential generating function $\sum_{j=0}^{\infty} b_j \frac{x^j}{j!}$ for the number of permissible arrangements of a set J of size j on shelf T. The coefficient c_k of $\frac{x^k}{k!}$ in the product of the two generating functions is the number of permissible arrangements of a set K of size k on the two shelves S and T. We now give an example of what the actual generating functions might be under the condition that both shelf S and shelf T must receive at least one book.

We will then try to formulate these ideas concisely enough to state a theorem that summarizes them.

Example 5.1. What is the exponential generating function for the number of ways to arrange k books on two shelves so that each shelf receives at least one book?

There are no ways to arrange zero books on shelf S (so that it gets at least one book) and $i!$ ways to arrange i books if $i > 0$. Thus, the exponential generating function for the arrangements of books on the first shelf, $\sum_{i=0}^{\infty} a_i \frac{x^i}{i!}$, is $\sum_{i=1}^{\infty} i! \frac{x^i}{i!}$ and the exponential generating function for arrangements of books on the second shelf, $\sum_{j=0}^{\infty} b_j \frac{x^j}{j!}$, is $\sum_{j=1}^{\infty} j! \frac{x^j}{j!}$. However,

$$\sum_{i=1}^{\infty} i! \frac{x^i}{i!} = \sum_{i=1}^{\infty} x^i = \frac{x}{1-x}.$$

Thus, the exponential generating function for the number of ways to arrange the books on the two shelves is

$$\frac{x^2}{(1-x)^2} = x^2 \sum_{i=0}^{\infty} C(2+i-1;i)x^i = \sum_{i=0}^{\infty} \binom{i+1}{i} x^{i+2}$$

$$= \sum_{k=2}^{\infty} \binom{k-1}{k-2} x^k = \sum_{k=2}^{\infty} (k-1)x^k$$

$$= \sum_{k=2}^{\infty} k!(k-1)\frac{x^k}{k!}.$$

Thus, there are $k!(k-1)$ ways to arrange k books on two shelves so that each shelf receives at least one book. ∎

From our example, let us try to find a general principle that explains what is happening. For ordinary generating functions, we had nonnegative *integer*-valued functions v_1 and v_2 defined on sets S_1 and S_2 as our "value functions" and we interpreted the coefficient of x^n in the product of the two relevant power series as the number of ordered pairs (a, b) in $S_1 \times S_2$ of value $v_1(a) + v_2(b) = n$. In the case of exponential generating functions, we assign to each element a of S_1 a *subset* $V_1(a)$ of some relevant set (such as the preceding set of books), and to each element b of S_2 a *subset* $V_2(b)$. Further, we assume the number of elements a with $V_1(a) = I$ is finite and equal to the number of elements a with $V_1(a) = J$ if I and J are finite and of the same size. We make the same assumption about V_2. We let a_i be the number of elements of S_1 whose value is a fixed set I of size i; we let b_j be the number of elements of S_2 whose value is a fixed set J of size j. Then if

$$\sum_{n=0}^{\infty} c_n \frac{x^n}{n!} = \sum_{i=0}^{\infty} a_i \frac{x^i}{i!} \sum_{j=0}^{\infty} b_j \frac{x^j}{j!},$$

we may interpret c_n as the number of ordered pairs whose values are disjoint sets such that their union is a specific set N of size n.

The coefficient of $\frac{x^n}{n!}$ in a product of k generating functions may similarly be interpreted as the number of k-element lists whose k values are disjoint sets such that their union is a specific set N of size n.

The Exponential Generating Function for Onto Functions

Example 5.2. Compute the exponential generating function for the number of functions from an n-element set *onto* the set $\{1, 2\}$.

Let S_1 and S_2 both be the set of all nonempty subsets of the integers. Let the "value" of an element of S_1 or S_2 be that set itself. Then the number of ordered pairs of elements whose values are disjoint sets with union N is the number of *ordered* partitions of N into two parts, or equivalently the number of functions from N onto the set $\{1, 2\}$. In this case, $a_i = b_i = 1$ if $i > 0$, $a_0 = b_0 = 0$. Thus,

$$\sum_{i=0}^{\infty} a_i \frac{x^i}{i!} = \sum_{i=0}^{\infty} b_i \frac{x^i}{i!} = \sum_{i=1}^{\infty} \frac{x^i}{i!} = e^x - 1,$$

so that

$$\sum_{n=0}^{\infty} c_n x^n = (e^x - 1)^2 = e^{2x} - 2e^x + 1$$

$$= \sum_{i=0}^{\infty} 2^i \cdot \frac{x^i}{i!} - \sum_{i=0}^{\infty} 2\frac{x^i}{i!} + 1$$

$$= \sum_{i=1}^{\infty} (2^i - 2)\frac{x^i}{i!}.$$

Thus, the number of functions from an n-element set N onto the set $\{1, 2\}$ is $2^n - 2$ when $n \geq 1$, as we would expect from an elementary analysis of the problem. However, a more impressive fact is that with the same kind of computation, we can see that the exponential generating function for the number of functions that map onto a k-element set is $(e^x - 1)^k$. ∎

We have not yet proved the result just used. A precise statement of this result and its proof follows. Although not difficult, the proof is an intricately condensed version of the computations we made in the bookshelf example. Thus, there is no harm in passing over the proof.

The Product Principle for Exponential Generating Functions

Theorem 5.1. Let S_1 and S_2 be sets and V_1 and V_2 be functions from S_1 and S_2, respectively, to the subsets of a set T. Suppose that, for each i, the number of x in S_i with $V_i(x) = J$ and the number of y in S_i with $V_i(y) = K$ are equal if J and K have the same size. If a_i is the number of x in S_1 with $V_1(x) = I$ for any given set I of size i, and b_j is the number of y in S_2 with $V_2(y) = J$, for any given set J of size j, and if

$$\sum_{k=0}^{\infty} c_k \frac{x^k}{k!} = \sum_{i=0}^{\infty} a_i \frac{x^i}{i!} \sum_{j=0}^{\infty} b_j \frac{x^j}{j!}, \tag{5.1}$$

then c_k is the number of ordered pairs (x, y) in $S_1 \times S_2$ such that $V_1(x)$ and $V_2(y)$ are disjoint sets whose union is a given set K of size k.

Proof. The coefficient of x^n on the right-hand side of (5.1) is

$$\sum_{i=0}^{n} \frac{a_i}{i!} \frac{b_{n-i}}{(n-i)!},$$

so that the coefficient of $\frac{x^n}{n!}$ is

$$\sum_{i=0}^{n} \frac{n!}{i!(n-i)!} a_i b_{n-i} = \sum_{i=0}^{n} \binom{n}{i} a_i b_{n-i},$$

which is the number of ways first to divide an n-element subset N of T into a set I of size i and a (disjoint) set J of size $n - i$, and second to choose an x_1 in S_1 with $V_1(x_1) = I$ and an x_2 in S_2 with $V_2(x_2) = J$. Thus this coefficient, which is c_n, is the number of ways to choose an ordered pair (x_1, x_2) in $S_1 \times S_2$ with $V_1(x_1) \cap V_2(x_2) = \emptyset$ and $V_1(x_1) \cup V_2(x_2) = N$ for any given n-element subset N of T. ∎

There is a version of this theorem for a product of k generating functions; we state it next.

__Theorem 5.2.__ Let $S_1, S_2, \ldots, S_k$ be sets and let V_i be a function from S_i to the subsets of some set T. Suppose further that for all j-element sets J, the number of elements a of S_i such that $V_i(a) = J$ is the same. Let $f_i(x)$ be the exponential generating function in which the coefficient of $\frac{x^j}{j!}$ is the number of elements a of S_i with $V_i(a) = J$ for one particular j-element set J. Then the coefficient of $\frac{x^n}{n!}$ in the product of the k generating functions $f_1, f_2, \ldots, f_k$ is the number of lists $(a_1, a_2, \ldots, a_k)$ of elements $a_i \in S_i$ such that the sets $V(a_1), V(a_2), \ldots, V(a_k)$ are disjoint and their union is a particular n-element subset N of the set T.

Putting Lists Together and Preserving Order

As our example of the bookshelves suggests, exponential generating functions frequently are useful in situations where the arrangements we are counting are ordered in some way. In lists of objects, the order in which the objects appear is important, whereas in sets or multisets the order is unimportant. Thus, results we developed for sets or multisets using ordinary generating functions may have analogs for lists that we could develop using exponential generating functions. In particular, if we multiply the generating functions for multisets chosen from a set A (with certain restrictions) and multisets chosen from a set B (with perhaps other restrictions) and if A and B are disjoint, then we get the generating functions for multisets chosen from $A \cup B$ (subject to all the restrictions). There are many ways to put two lists, one chosen from A and one chosen from B, together to form a new list.

__Example 5.3.__ A state will allow any sequence of four to eight letters and numbers on its license plates, subject to the condition that the letters themselves, in the order in which they appear on the license plate, should not spell any of a list of forbidden words. Thus a license plate can be formed by taking an acceptable list of letters and an arbitrary list of numbers and then mixing the numbers together with the letters without changing the order of either list. Our next definition makes this idea of mixing lists together precise, and our next theorem will show us how to get the generating function for the number of acceptable lists of letters and numbers from the generating functions for the number of acceptable lists of letters and for the number of lists of numbers. (Typically we would get the generating function for the number of n-symbol license plates and then restrict ourself to the coefficients that correspond to values of n between 4 and 8.) ∎

We will say that we *interleave* or *shuffle* together the lists $a_1, a_2, \ldots, a_i$ and $b_1, b_2, \ldots, b_j$ when we form a list of $i + j$ things by choosing i places among $i + j$ where we place the a_k's (in the order of their original list) and by placing the b_k's in the other j places (again in the order of their original list). Thus we have formed a new list $c_1, c_2, \ldots, c_{i+j}$ in which each c_k is either an a_i or b_j, and if we write down just the a's from this new list (preserving order), we get the list of a's, and similarly for the b's.

Theorem 5.3. Suppose A is a set of lists with a_n lists of length n chosen from a set S and B is a set of lists with b_m lists of length m chosen from a set R with $S \cap R = \emptyset$. Then the exponential generating function for the number of lists of length k chosen from $S \cup R$ that may be formed by shuffling together lists from A and B is

$$\sum_{i=0}^{\infty} a_i \frac{x^i}{i!} \cdot \sum_{j=0}^{\infty} b_j \frac{x^j}{j!}.$$

Proof. We may think of a list of length i as occupying a certain set of i positions chosen from $\{1, 2, \ldots, k\}$—for example, the first i positions or the last i positions. Each way of shuffling a list of length i from A together with a list of length $k-i$ from B uses a certain i-element subset I of $\{1, 2, \ldots, k\} = K$ for the lists from A and the complementary set $K - I$ for the lists from B. There are a_i lists that use positions in I and b_{k-i} lists that use positions in $K - I$. The total number c_k of ways to shuffle together lists from A and B to get a list of length k is the number of pairs of lists, one from A and one from B, that together use positions in the set $K = \{1, 2, \ldots, k\}$. By Theorem 5.1, the exponential generating function for c_k is the product of the exponential generating function for a_i and the exponential generating function for b_j. ∎

Example 5.4. The state of Old York forbids 7 three-letter words, 40 four-letter words, 60 five-letter words, 60 six-letter words, 70 seven-letter words, and 70 eight-letter words from its license plates. (In practice, no more than eight letters will appear on a license plate, so we can pretend they forbid no words of length 9 or more.) The state allows any sequence of the digits 3–9 on its license plates. What is the exponential generating function for the number of legal license plates consisting of any mixture of digits and letters that does not contain, in order, the letters of a forbidden word? How would we compute how many seven-symbol license plates are possible?

The exponential generating function for the number of legal words of length i is

$$\sum_{i=0}^{\infty} \frac{26^i}{i!} x^i - 7\frac{x^3}{3!} - 40\frac{x^4}{4!} - 60\frac{x^5}{5!} - 60\frac{x^6}{6!} - 70\frac{x^7}{7!} - 70\frac{x^8}{8!}.$$

The exponential generating function for the number of legal sequences of digits is $\sum_{i=0}^{\infty} \frac{10^i}{i!}$. Thus the exponential generating function for the number of n-symbol license plates (with no restriction on the number of letters and numbers) is

$$\left(\sum_{i=0}^{\infty} \frac{26^i}{i!} x^i - 7\frac{x^3}{3!} - 40\frac{x^4}{4!} - 60\frac{x^5}{5!} - 60\frac{x^6}{6!} - 70\frac{x^7}{7!} - 70\frac{x^8}{8!}\right) \cdot \sum_{i=0}^{\infty} \frac{10^i}{i!}.$$

To compute the number of seven-symbol license plates, we would compute the coefficient of x^7 and multiply it by 7!. ∎

Exponential Generating Functions for Words

Example 5.5. We use the phrase "nonsense word" to mean a string of alphabetical letters that is not necessarily a word of English. How many nonsense words of length n may be made using the letters a, b, c, and d?

For each letter, the number of ways to make a list of j copies of that letter is just one, so that generating function for the number of lists of each individual letter is

$$\sum_{i=0}^{n} 1\frac{x^i}{i!} = e^x.$$

Multiplying the four generating functions together gives

$$\sum_{i=0}^{\infty} a_i x^i = e^{4x} = \sum_{i=0}^{\infty} 4^i \frac{x^i}{i!}$$

for the generating function for words using the four letters, so there are 4^n words of length n. ∎

Solving Recurrence Relations with Other Generating Functions

We have seen how ordinary generating functions allowed us to solve all first-order linear recurrence relations with a constant coefficient. By using a wider variety of indicator functions, we may solve all first-order linear recurrence relations. As an example, let us examine the recurrence for a sequence we know, the recurrence $a_n = na_{n-1}$ for the number of bijections of an n-element set. (Note that the only solution to this recurrence that counts bijections is the one with $a_0 = 1$.)

Example 5.6. Show that the solutions to the recurrence $a_n = na_{n-1}$ are all of the form $a_n = a_0 n!$.

We substitute na_{n-1} in for a_n in the exponential generating function for a_n to get

$$\sum_{n=0}^{\infty} a_n \frac{x^n}{n!} = a_0 + \sum_{n=1}^{\infty} a_n \frac{x^n}{n!}$$

$$= a_0 + \sum_{n=1}^{\infty} na_{n-1} \frac{x^n}{n!}$$

$$= a_0 + x \sum_{i=0}^{\infty} a_i \frac{x^i}{i!}$$

so that

$$(1 - x) \sum_{n=0}^{\infty} a_n \frac{x^n}{n!} = a_0$$

or

$$\sum_{n=0}^{\infty} a_n \frac{x^n}{n!} = \frac{a_0}{1 - x} = a_0 \sum_{i=0}^{\infty} x^i$$

giving us $a_n/n! = a_0$, or

$$a_n = a_0 n! \,. \qquad \blacksquare$$

From the example it appears that the way in which the exponential generating function helped was in its ability to cancel out the coefficient of n of a_{n-1} and yet retain its original form. With a similar computation we may prove a much more general result.

Theorem 5.4. If $a_n = b_n a_{n-1} + d_n$ and for all n, $b_n \neq 0$, then the generating function relative to the indicator family $x^n / \prod_{i=0}^{n} b_i$, written $x^n / \Pi b_n$ for short, is

$$\sum_{n=0}^{\infty} a_n \frac{x^n}{\Pi b_n} = \frac{1}{1 - x} \left(\frac{a_0 - d_0}{b_0} + \sum_{n=0}^{\infty} \frac{d_n x^n}{\Pi b_n} \right),$$

and

$$a_n = \frac{a_0 - d_0}{b_0} \Pi b_n + \sum_{i=0}^{n} d_i \frac{\Pi b_n}{\Pi b_i} .$$

Proof. Similar to Example 5.4 ∎

Example 5.7. An analysis of the computer algorithm known as Quick-sort gives

$$c_{n+1} = \frac{n+2}{n+1} c_n + 2 \frac{n}{n+1}$$

for the expected number of comparisons of elements of the list being sorted.

By letting $b_0 = 1$, $c_0 = d_0 = 0$, $a_n = c_{n+1}$, $b_n = (n+2)/(n+1)$, and $d_n = 2n/(n+1)$, we see that $a_0 = c_1 = 0$ and thus by Theorem 5.4,

$$c_n = 2(n+1) \sum_{i=1}^{n-1} \frac{i}{(i+2)(i+1)} \le 2(n+1)H_n ,$$

where $H_n = \sum_{i=1}^{n} \frac{1}{i}$. The numbers H_n are called the **harmonic numbers** and are between $\log_e(n) + \frac{1}{2}$ and $\log_e(n) + 1$. ∎

Using Calculus with Exponential Generating Functions

Exponential generating functions will help in solving certain kinds of second-order recurrence relations that are not constant coefficient, that is, recurrence relations of the form

$$a_n = b_n a_{n-1} + c_n a_{n-2},$$

in which b_n and c_n are functions of n. The classical example is the problem of derangements (the hat check problem). Let D_n be the number of lists of $\{1, 2, \ldots, n\}$ in which integer i is not in position i for any i. Then some integer k between 1 and $n-1$ is in position n. (Thus, there are $n-1$ choices for k.) Now we examine two cases. In case one, n is in position k; in this case we have D_{n-2} arrangements of the remaining integers. In case two, n is not in position k. Thus, among positions 1 up to $n-1$ we see a list of the numbers $1, 2, \ldots, k-1, k+1, \ldots, n$, such that i is not in position i if $i < n$ and n is not in position k. In this case we have D_{n-1} such arrangements. Since in both cases we had $n-1$ choices for k, we have

$$D_n = (n-1)D_{n-1} + (n-1)D_{n-2}.$$

(From this it is possible to derive by mathematical induction that $D_n = nD_{n-1} + (-1)^n$.)

This may be rewritten as

$$D_{n+2} = (n+1)D_{n+1} + (n+1)D_n.$$

To remove the $(n+1)$ from the recursion, we multiply by $\frac{x^{n+1}}{(n+1)!}$ and sum over all $n \geq 0$. (Note $D_2 = 1$, $D_1 = 0$, so if we set $D_0 = 1$ we may start the sum at $n = 0$ and satisfy the recurrence relation.) We get

$$\sum_{n=0}^{\infty} D_{n+2} \frac{x^{n+1}}{(n+1)!} = \sum_{n=0}^{\infty} D_{n+1} \frac{x^{n+1}}{n!} + \sum_{n=0}^{\infty} D_n \frac{x^{n+1}}{n!}.$$

Setting $D(x) = \sum_{n=0}^{\infty} D_n \frac{x^n}{n!}$ and using primes to denote derivatives, we get

$$\left(\sum_{n=0}^{\infty} D_{n+2} \frac{x^{n+2}}{(n+2)!} \right)' = x \left(\sum_{n=0}^{\infty} D_{n+1} \frac{x^{n+1}}{(n+1)!} \right)' + xD(x)$$

or,

$$(D(x) - D_1 x - D_0)' = x(D(x) - D_0)' + xD(x)$$

so that

$$D'(x) - D_1 = xD'(x) + xD(x),$$

giving

$$D'(x)(1-x) = xD(x),$$

because $D_1 = 0$. This gives us

$$\frac{D'(x)}{D(x)} = \frac{x}{1-x} = \frac{1}{1-x} - 1,$$

and by integration,

$$\ln D(x) = -\ln(1-x) - x + c,$$

and finally by exponentiation,

$$D(x) = \frac{1}{1-x} e^{-x} e^c.$$

Now $D(0) = 1$, so $c = 0$, giving

$$D(x) = \frac{e^{-x}}{1-x} = e^{-x}(1 + x + x^2 + \cdots)$$

$$= \sum_{i=0}^{\infty} (-1)^i \frac{x^i}{i!} \cdot \sum_{j=0}^{\infty} x^j$$

$$= \sum_{i=0}^{\infty} \left(\sum_{j=0}^{i} \frac{(-1)^j}{j!} \right) x^i.$$

Thus, since $D(x)$ is the exponential generating function for D_n, we get

$$D_n = n! \sum_{j=0}^{n} \frac{(-1)^j}{j!}.$$

EXERCISES

1. Show that the ordinary generating function for the number of ways to place a set of n books on two shelves is $\sum_{n=0}^{\infty} (n+1)! x^n$. What if each shelf must receive at least one book?

2. Show that for any real number a, other than 0, $\lim_{n \to \infty} n! a^n$ does not exist because $n! a^n$ tends to infinity. What does this tell you about the convergence (or lack of convergence) of the ordinary generating function in Exercise 1?

3. For what values of x does the exponential generating function for the arrangements in Exercise 1 converge?

4. What is the exponential generating function for the number of ways to place n books on one shelf so that the shelf does not receive more than 10 books?

5. What is the exponential generating function for the number of ways to arrange n books on two shelves so that neither shelf gets more than 10 books?

6. Do Exercise 5 assuming each shelf must receive at least one book.

7. Do the bookshelf example (Example 5.1) at the beginning of this section for n shelves.

8. Do Exercise 5 for k shelves.

9. Do Exercise 5 for k shelves, each of which must receive a book.

10. What is the exponential generating function for the number of functions from a set N onto a set K such that each element of K is the image of at most two elements of N? Assume K is fixed.

11. What does Taylor's theorem tell us about the number a_k in the exponential generating function

$$f(x) = \sum_{k=0}^{\infty} a_k \frac{x^k}{k!}?$$

12. How many nonsense words of six letters may be made using up to three a's, five s's, two e's, and three c's if each letter must be used at least once? What is the generating function for the number of i-letter (nonsense) words that can be made from this selection of letters?

13. What is the exponential generating function for Stirling numbers $S(n, k)$ of the second kind with fixed k? (Hint: How do they relate to onto functions?)

14. What is the exponential generating function for the number of ways to pass out n distinct pieces of candy to three children so that each child gets a piece? In how many ways can 10 pieces be passed out so that each child gets a piece?

15. Redo Exercise 14 assuming each child gets at least two pieces.

16. Redo Exercise 14 assuming no child gets more than four pieces.

17. Find the number of ordered distributions of k distinct objects to n distinct recipients so that everybody gets at least one object.

18. A function from N to K is "doubly onto" if each element of K is the image of at least two elements of N. What is the exponential generating function for the number of doubly onto functions onto a k-element set?

19. For what distribution problem is a power of the hyperbolic cosine the relevant exponential generating function?

20. What is the exponential generating function for words from the 26-letter alphabet in which the vowels (a, e, i, o, or u) must be used an even number of times?

21. Give a proof of Theorem 5.2, either by induction or by modifying the proof of Theorem 5.1

22. Prove Theorem 5.4.

23. Find general solutions to the following recurrences.
    (a) $a_n = n^2 a_{n-1} + 1$.

(b) $a_n = na_{n-1} + n!$.

24. Explain why the total number of k-element permutations of an n-element set (totaled over all k) satisfies the recurrence $a_n = na_{n-1} + 1$. (Don't forget the empty permutation.) What is a_0? Find a formula for a_n by solving the recurrence relation. In retrospect, is the formula surprising?

25. For a fixed rational number $r > 0$, solve the recurrence $a_n = n^r a_{n-1}$.

26. Let $s_n(k)$ stand for the sum of the kth powers of the first n integers. For n fixed, find the exponential generating function for $s_n(k)$.

27. Use the recurrence

$$B_{n+1} = \sum_{i=0}^{n} \binom{n}{i} B_{n-i}$$

to find the exponential generating function for the Bell numbers.

28. Show that the exponential generating function for the solutions to the recurrence $a_n = na_{n-1} + n$ with $a_0 = 0$ is $xe^x/(1-x)$. What does this tell you about a_n?

29. Recall from Section 1 of this chapter that D_n stands for the number of derangements (permutations that fix no elements) of an n-element set. Why is $D_n - nD_{n-1} = -\{D_{n-1} - (n-1)D_{n-2}\}$? Conclude from this that

$$D_n = nD_{n-1} + (-1)^n.$$

30. Use the recurrence relation $D_n = nD_{n-1} + (-1)^n$ to derive the exponential generating function $D(x)$ for D_n without using derivatives.

31. Let R_n be the number of lists of $\{1, 2, \ldots, n\}$ such that $i + 1$ is not immediately to the right of i. Show that R_n satisfies a recurrence relation much like that we found in this section of the text for D_n; find the exponential generating function for R_n, find a relationship between R_n and D_n, and find a formula for R_n.

Suggested Reading

Goulden, I. P., and D. M. Jackson. 1983. *Combinatorial Enumeration.* New York: Wiley.

Hall, Marshall, Jr. 1986. *Combinatorial Mathematics,* 2d ed., chs. 2–4. New York: Wiley.

Hardy, G. H., and E. M. Wright. 1960. *An Introduction to the Theory of Numbers,* 4th ed. London: Oxford University Press.

Knuth, Donald E. 1973. *The Art of Computer Programming: Fundamental Algorithms,* vol. 1, 2d ed. Reading, MA: Addison Wesley

Liu, C. L. 1968. *Introduction to Combinatorial Mathematics*, chs. 2–4. New York: McGraw-Hill.

Lovász, László. 1979. *Combinatorial Problems and Exercises*, chs. 1 and 2. Amsterdam: North-Holland.

Niven, Ivan, and Herbert Zuckerman. 1960. *An Introduction to the Theory of Numbers*, ch. 10. New York: Wiley.

Pólya, George. 1956. Picture writing. *American Mathematical Monthly* (December): 689.

Riordan, John. 1958. *An Introduction to Combinatorial Analysis*, chs. 2, 3, 7, 8. New York: Wiley.

Stanley, Richard P. 1986. *Enumerative Combinatorics.* Monterey, CA: Wadsworth & Brooks-Cole.

Stanley, Richard P. 1978. Generating functions. In *Studies in Combinatorics*, *MAA Studies in Mathematics*, vol. 17, ed. Gian-Carlo Rota. Washington DC: Mathematical Association of America.

Wilf, Herbert S. 1990 *Generatingfunctionology* San Diego, CA: Academic Press.

4
Graph Theory

Section 1 Eulerian Walks and the Idea of Graphs

The Concept of a Graph

In our work so far, we have concentrated on formulas that are useful in computing the number of arrangements of a certain kind. In this chapter, we begin another kind of study, the study of the *properties* of certain kinds of arrangements. In the past, a number of different kinds of arrangements have been called graphs. There is no general agreement among mathematicians on exactly which "graphical arrangements" should be called graphs; however, the terminology introduced here, which is largely the "Michigan" terminology popularized by Frank Harary (while he was a professor at the University of Michigan), seems to be well established. The other natural terminology uses the word "graph" for what we call a "multigraph" in the discussion that follows; then what we call a "graph" is called a "simple graph."

To begin with a concrete example, suppose a company has a distributed computer network with machines in seven cities. Not all of these computers are able to communicate with one another; Table 1.1 gives, for each of the seven cities, the locations of the computers that can communicate with the computer in the given city.

Each day the system administrator runs a program that tests the communications links. The testing consists of sending messages back and forth and comparing transmitted and received messages. If a message could start in one city and be transmitted through the network in such a way as to use each link exactly once and end up in the starting city, this would be the least costly way to check all the links. Is there such a routing? In Figure 1.1 we show a simplified geographic picture of the network, with the circles representing cities and the lines representing communications links.

The "map" in Figure 1.1 could just as well stand for possible airline routes between cities—in this context it is natural to ask if an airplane could

Table 1.1

Computer Location	Potential Communications Links			
Boston	Albany	Atlanta	Cleveland	Denver
Albany	Boston	Cleveland	Atlanta	
Atlanta	Boston	Albany	Cleveland	Dallas
Cleveland	Atlanta	Albany	Boston	Denver	Dallas
Dallas	Atlanta	Denver	Sacramento	Cleveland
Denver	Boston	Dallas	Cleveland	Sacramento
Sacramento	Denver	Dallas		

Figure 1.1

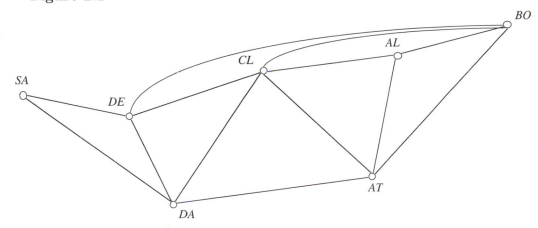

shuttle from city to city, passing through each connection on the route map exactly once. If the figure were a map of a neighborhood and the circles were intersections of streets, we might wish to ask whether a mail delivery truck could enter the neighborhood at one point, deliver mail along each street, and then leave after traversing each street exactly once.

The picture in Figure 1.1 is generally called a *graph*. Informally, a graph is a collection of points, called vertices, together with lines between the points, called edges. A more precise definition is that a **graph** consists of set V, which is called a **vertex set**, and a set E of two-element subsets of V, which is called an **edge set**. The elements of V are called **vertices** and the elements of E are called **edges**. We use the shorthand notation $G = (V, E)$ to say "G is a graph with vertex set V and edge set E."

Example 1.1. Write down the vertex set and edge set of the graph shown in Figure 1.1. By inspection we write

$$V = \{BO, AL, AT, CL, DA, DE, SA\}$$

and

$$E = \{\{BO, AT\}, \{BO, AL\}, \{BO, CL\}, \{BO, DE\}, \{AL, AT\},$$
$$\{AL, CL\}, \{AT, CL\}, \{AT, DA\}, \{CL, DE\}, \{CL, DA\},$$
$$\{DE, DA\}, \{DE, SA\}, \{DA, SA\}\}. \qquad \blacksquare$$

Multigraphs and the Königsberg Bridge Problem

One of the questions that gave rise to modern graph theory was a question that Euler (pronounced "oiler") attributed to the citizens of the old town of Königsberg in Prussia. (Königsberg, now called Kaliningrad, is a port northwest of Moscow on the Baltic sea. In a fascinating twist of history, it is part of Russia but separated from the rest of Russia by Lithuania.) Königsberg consisted of an island where two branches of the river Pregel joined, together with some land along each riverbank; it also had quite a few bridges. A schematic map of Königsberg is shown in Figure 1.2.

Figure 1.2

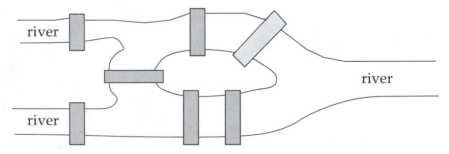

The bars represent bridges crossing rivers; it is possible to walk between any two bridges touching a given land mass. Euler attributed to the townspeople the question, "Is it possible to take a walk through town, starting and ending at the same place, and cross each bridge exactly once?" Euler recognized that the shape of the land masses made no difference to the problem and that the more abstract diagram in Figure 1.3., in which circles represent land masses and lines represent bridges, contained the essence of the problem.

The picture represents a type of arrangement that we will call a multigraph. A **multigraph** consists of a set V, called a *vertex* set, and a

Figure 1.3

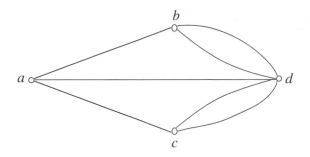

multiset of two-element multisets (chosen from V) called *edges*. We use the notation $G = (V, E)$ and refer to the "multigraph G." Another multigraph is shown in Figure 1.4. The graph shown in Figure 1.1 is also a multigraph; that is, graphs are just special kinds of multigraphs.

Example 1.2. The edge multiset of the multigraph in Figure 1.3 is

$$\{\{a, b\}, \{b, d\}, \{b, d\}, \{c, d\}, \{c, d\}, \{a, c\}, \{a, d\}\}.$$

The edge multiset for Figure 1.4 is

$$\{\{1, 2\}, \{2, 4\}, \{2, 4\}, \{4, 4\}, \{3, 4\}, \{3, 3\}, \{2, 3\}, \{1, 3\}\}. \qquad ∎$$

Figure 1.4

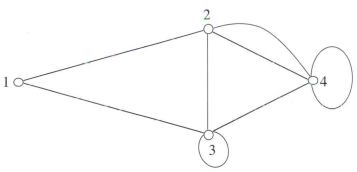

Although our definition of a multigraph is expressed in terms of multisets, we can visualize multigraphs in much the same way we visualize graphs. We start by drawing points on a piece of paper, one for each vertex, connecting two vertices with a line whenever these two vertices are the vertices of an edge, and drawing a line from a vertex to itself whenever the multiset containing that vertex twice is an edge. An edge connecting a vertex to itself is called a *loop*. Notice that a graph has no loops because a loop is not representable as a two-element set. (As mentioned earlier, what we call a multigraph is sometimes called a *graph*; then what we call a graph is called a *simple graph*.)

Walks, Paths, and Connectivity

The questions we have asked in our examples can be asked for any multigraph. Such questions arise so often that special terminology has been developed to help make these ideas easy to discuss. A *walk* in a multigraph is an alternating sequence of vertices and edges

$$v_1 e_1 v_2 e_2 \cdots e_{n-1} v_n$$

such that e_i contains v_i and v_{i+1}.

A *path* is a walk in which no vertex appears twice. For example, the sequence $1\{1,3\}3\{3,2\}2$ is a path from vertex 1 to vertex 2 in Figure 1.4. (Unfortunately, what we call a walk is sometimes called a *path*; then what we call a path is called a *simple path*.) If the first and last vertices of a walk are equal, the walk is called a *closed walk*; a closed walk in which only the first and last vertices are equal is a *cycle*. The sequence $1\{1,3\}3\{3,2\}2\{1,2\}1$ is a cycle in Figure 1.4. See Figure 1.5 for some examples of cycles. (Unless the edge $\{x,y\}$ has multiplicity more than 1, however, we do not call the trivial closed walk $x\{x,y\}y\{x,y\}x$ a cycle.)

Figure 1.5   The multigraphs (a) and (b) are cycles; in (c) the four-edge closed walk from a to b and back twice is not a cycle. However, the four-edge closed walk in (c) is an Eulerian walk.

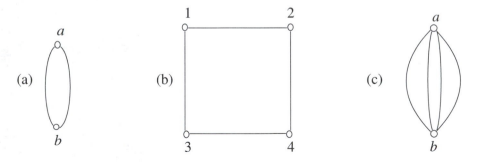

A walk that includes each edge of the multigraph $G = (V, E)$ exactly as many times as it is in E is called an *Eulerian* walk. In our two examples, we asked whether a multigraph had a closed Eulerian walk. It was Euler who noticed that the question of the existence of such a walk could be answered by determining how many edges touch each vertex. The *degree* of a vertex v, denoted by $d(v)$, is the number of two-element edges that touch v plus twice the number of loops that touch v. Thus, the degree of v is the total number of times v appears in edges of the multigraph. Intuitively speaking, it is the number of lines "sticking out of" v.

Note that if a multigraph has an Eulerian walk, then (with the possible exception of the first and last vertices in the walk) each vertex appears an even number of times in edges—once in each edge preceding it in the walk and once more in each edge following it in the walk. Thus, each vertex (with the possible exception of the first and last in the walk) has even degree. Further, if the first and last vertices are the same, it has even degree, and if the first and last vertices are different, then they have odd degree. From this, we conclude there is no Eulerian walk in the multigraph of Figure 1.3. Therefore, the answer to the Königsberg bridge question is "No."

Note also that if a multigraph has an Eulerian walk, then the vertices on the walk are *connected* in the sense that given any two vertices, there is a walk from one to the other. We say that the vertex u **is connected to** the vertex v if there is a walk from u to v. It is simple to verify that the relation "is connected to" is an equivalence relation on the vertices of a multigraph. Thus the vertices are divided into equivalence classes (called **connectivity classes** or **connected components**), so that vertices in the same class are connected but vertices in different classes are not. Every edge is between two vertices in some connectivity class. A vertex that lies in no edges forms a connectivity class by itself; we call such a vertex an **isolated vertex**. A graph with just one connectivity class is called a **connected graph**. If an Eulerian graph has no isolated vertices, then every vertex must be on an Eulerian walk, so that the graph is connected. Figure 1.6 shows two graphs, one connected and one not connected.

Figure 1.6 Graph (a) is connected, but graph (b) has three connected components.

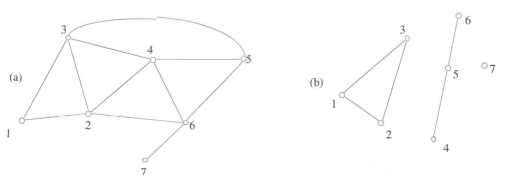

Eulerian Graphs

The two ideas of *degree* and *connectivity* can be used to describe the **Eulerian multigraphs**, the multigraphs with an Eulerian walk.

Theorem 1.1. A multigraph (without isolated vertices) is Eulerian if and only if it is connected and either all or all but two vertices have even degree. If every vertex has even degree, then all Eulerian walks are closed; if two vertices have odd degree, then every Eulerian walk starts at one of these vertices and ends at the other.

Proof. We have already verified that an Eulerian multigraph is connected and that either zero or two of its vertices have odd degree. Further, we have observed that the Eulerian walk is closed if and only if zero vertices have odd degree. Thus we now prove that if a multigraph is connected and has zero or two vertices of odd degree, then it is Eulerian. Our proof is by induction. It will give an inductive description of an algorithm that lets us find Eulerian walks. (Note: This proof has been written out in great detail because it is one of our first examples of induction that does not involve formulas. Its length should not be regarded as a measure of difficulty.) We let S be the set of all integers n so that if a connected multigraph with n edges has zero or two vertices of odd degree, then it has an Eulerian walk.

A connected multigraph with one edge has either two vertices of degree 1 or one vertex of degree 2. Thus, it has either the Eulerian walk $v_1 e v_2$ or the Eulerian walk $v_1 e v_1$. Therefore, 1 is in S.

Now suppose that all nonnegative integers less than n are in S. We assume the multigraph G has n edges, is connected, and has zero or two vertices of odd degree. Let x be a vertex of G, a vertex of odd degree if G has any such vertices. Construct a walk as follows. Let $v_1 = x$, let $e_1 = \{x, y\}$ be an edge containing x, and let $v_2 = y$. We continue on in this way. If v_i touches an edge yet to be used in the walk, denote the edge by e_i and let $v_{i+1} = y$ if $e_i = \{v_i, y\}$. (Note: If an edge is in E more than once, then it may be used as often as it is in E.) Continue this process until we reach a vertex v_r that does not touch any unused edges. If $v_r = x = v_1$, then the vertex x has even degree and the walk is closed. If $v_r \neq v_1$, then v_r and v_1 both have odd degree. All other vertices v_i have even degree *and* are incident with—that is, precede or follow—an even number of edges in the walk. If the walk constructed is an Eulerian walk, then we are done. If the walk is not Eulerian, then we let E' consist of the edges in E but not in the walk. Then we let $G' = (V, E')$.

Figure 1.7 Two connected possibilities for G'. The edges shown as solid lines are in the path beginning at x; the other edges are dashed. To splice "paths" together, go from x to v_2; then follow the dashed lines until you return to v_2; finally, follow the solid lines again.

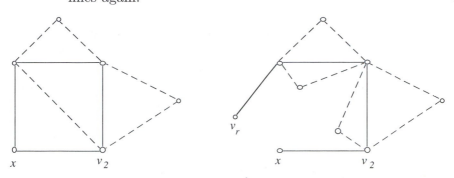

In G', *all* vertices have even degree. (Thus, if G' were connected, we would have by induction a closed Eulerian walk in G' that could be "spliced" with the walk we constructed to get an Eulerian walk in G.) (See Figure 1.7.)

Let $C_1, C_2 \ldots, C_k$ be the connected components of G'. Each of these components contains a vertex of the walk we constructed (why?); say v_{j_i} is in C_i. Now each edge in E' connects two vertices in some C_i. Thus, $E' = E_1 \cup E_2 \cup \cdots \cup E_k$, where E_i connects edges in C_i only. The multigraph (C_i, E_i) is connected and each vertex in it has even degree. Further, (C_i, E_i) has fewer edges than G, and so by the induction hypothesis there is a closed Eulerian walk in (C_i, E_i) starting and ending at v_{j_i}.

Now we can construct an Eulerian walk in G. (See Figure 1.8.) Follow the original walk of v_i's until the first v_{j_i} is reached. Then follow the Eulerian walk in C_i, returning to v_{j_i}. Continue along the original walk to the next v_{j_i} and repeat the process until you have reached v_r. Since each edge is either in an E_i or in the original walk, this walk is Eulerian, so G is Eulerian and $n \in S$. Thus, by the principle of mathematical induction, S contains all positive integers, and this proves the theorem. ∎

From the theorem we may conclude that the only Eulerian walks in the multigraph of Figure 1.4 start and end at either vertex 3 or vertex 4. Thus, the multigraph of Figure 1.4 has no closed Eulerian walks. In

Figure 1.8 Disconnected possibilities for G'. Solid and dashed lines are as in Figure 1.7. Go from x to v_2; follow the dashed lines back to v_2; follow the solid lines to v_{r-1}; follow the dashed lines back to v_{r-1}; and then go on the solid lines again. The curved lines separate the connected components of G'.

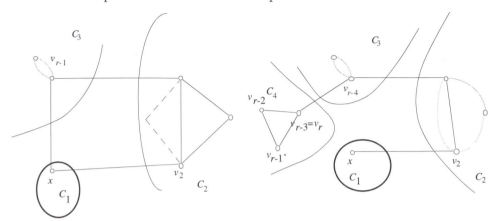

Figure 1.1, only AL and CL have odd degree; therefore, there is an Eulerian walk starting at CL and ending at AL. Although this means we couldn't solve the airplane shuttle problem or the mail delivery problem, some of the additional structure of this particular graph means we can still efficiently test the communications links. Namely, send the test messages from Cleveland to Albany and record what arrives in Albany. Now there is an Eulerian walk on the remainder of the graph from Albany to Albany. When the message last leaves Cleveland, check it against the original message. When the message arrives in Albany for the last time, check it against the recorded message there. If neither check shows an error, then presumably the communications links are functioning properly.

EXERCISES

1. Draw pictures of all the graphs that have the vertex set $\{1, 2, 3\}$. Do not draw more than one picture for any given graph. If two different graphs can be drawn with the same picture, include a picture of only one of them. For each picture, indicate how many different graphs can be drawn with that picture.

2. Draw a picture of all the connected graphs that have the vertex set $\{1, 2, 3, 4\}$. Do not draw more than one picture for any given graph. If two different graphs can be drawn with the same picture, include a picture of only one of them. For each picture, indicate how many different graphs can be drawn with that picture.

3. What is the maximum number of edges possible in a disconnected graph on five vertices?

4. Explain why the relation "is connected to" is an equivalence relation.

5. If two bridges in Königsberg collapse, does Euler's question have an affirmative answer? Does it matter which two collapse? What if you don't have to start and end at the same place?

6. If one bridge in Königsberg collapses, does Euler's question have an affirmative answer? Does it matter which one collapses? What if you don't have to start and end at the same place?

7. Find an Eulerian walk in graph (a) of Figure 1.6.

8. Explain how the mail truck could enter the multigraph of Figure 1.4 at vertex 3 and traverse each street *once in each direction*.

9. Explain why the sum of the degrees of the vertices in a multigraph is an even number. (Hint: How does the sum of the degrees relate to the number of edges?)

10. Prove that in a multigraph, the number of vertices of odd degree is an even number. (Hint: Exercise 9 makes this easier.)

11. Prove that in a graph with n vertices, if two vertices are connected by a walk, then they are connected by a walk with $n-1$ or fewer edges. Is this result true for multigraphs?

12. If a graph consists of a cycle on n vertices (with no extra edges), then how many edges does it have?

13. Prove that a connected multigraph with n vertices has at least $n-1$ edges.

14. Prove that if a graph on n vertices has $\frac{n^2-3n+4}{2}$ or more edges, then it is connected. Is this true for multigraphs?

15. Prove that in a multigraph (with V finite) with exactly two vertices of odd degree, these vertices are connected.

16. Can a graph with 10 vertices have 50 edges? What is the maximum number of edges it can have?

17. A graph is *complete* if there is an edge between every pair of distinct vertices. How many edges does a complete graph with n vertices have?

18. For what values of n is a complete graph (see the previous problem) on n vertices Eulerian?

19. A path in a multigraph is said to be *Hamiltonian* if it includes each vertex exactly once. No easy general condition for checking whether a graph has a Hamiltonian path is known.

(a) Does a complete graph (Exercise 17) always have a Hamiltonian path?

(b) Show that if a graph G has n vertices and if the sum of the degrees of any two vertices is n or more, then G has a Hamiltonian path. (The result holds if the sum of the degrees is $n - 1$ rather than n. You might try to prove this stronger result.)

(c) Give an example that shows (b) is false for multigraphs.

20. A graph is **Hamiltonian** if it has a cycle which contains all the vertices of the graph. Such a cycle is called a *Hamiltonian cycle*. Find a graph with 8 vertices and 13 edges that is not Hamiltonian.

21. Either prove or give a counterexample to the statement "If all vertices of a graph have degree three, then the graph is Hamiltonian." (See the previous problem for the meaning of Hamiltonian.)

22. (a) Give an example of an Eulerian graph that is not Hamiltonian.
    (b) Give an example of a Hamiltonian graph that is not Eulerian.

23. How many graphs are there which have the n-element vertex set $\{v_1, v_2, \ldots, v_n\}$?

24. What is the maximum possible number of edges in a graph which is not connected and why?

25. If a graph has k connected components and n vertices, what are the maximum and minimum number of edges it can have and why?

26. A table such as Table 1.1, called an *adjacency table* for a graph, is a good way to represent a graph for a computer. Write a computer program that makes use of an adjacency table to list the connected component containing a vertex as follows: Start a list of all vertices adjacent to the chosen one; then for each vertex in the list add all those adjacent to it. Modify this program to use as of part of one that uses the method of Theorem 1.1 to find an Eulerian walk.

Section 2 Trees

The Chemical Origins of Trees

Another origin of graph theory was the study of molecules of hydrocarbons. A hydrocarbon is a compound formed from hydrogen atoms and carbon atoms. A molecule of a compound consists of atoms of the constituent parts held together by chemical bonds. A carbon atom can form four such bonds to hydrogen atoms—or other carbon atoms—and a hydrogen atom can form exactly one bond. This leads to a graphical representation of hydrocarbons as multigraphs in which each vertex has degree 4 or 1. Typically, a vertex of degree 4 is labeled with a C and a vertex of degree 1 is labeled with an H. Several graphs of hydrocarbons are shown in Figure 2.1. These hydrocarbons are called saturated because their molecules contain the maximum amount of hydrogen possible for the amount of carbon they contain.

Figure 2.1 Several examples of hydrocarbons.

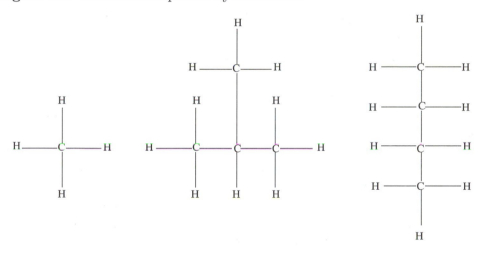

Figure 2.2 shows examples of unsaturated hydrocarbons. Note that the multigraphs in Figure 2.1 are in fact graphs, but the ones in Figure 2.2 have multiple edges. Also, the second hydrocarbon in Figure 2.2 has a closed walk in its multigraph. Further, the graph of each compound is connected, because all the atoms of a molecule are bound into the molecule in some way.

A connected graph with no cycles (closed paths) is called a **tree** (for the geometric reason that any path that can be traced out among the branches of a tree cannot close back upon itself). A graph with no cycles is called a **forest**. (A forest is thus a disjoint union of trees.) Figure 2.3 shows a tree and a forest.

Figure 2.2

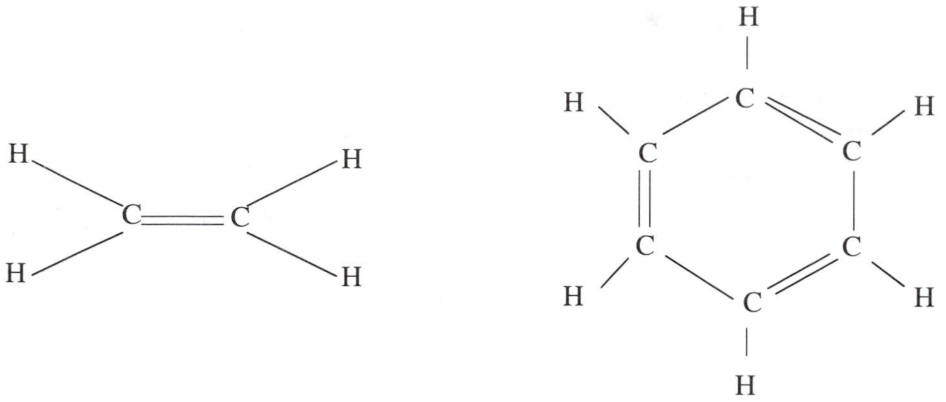

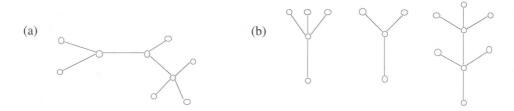

Figure 2.3 The graph labeled (a) is a tree; the graph labeled (b) is a forest but not a tree.

Basic Facts about Trees

There are quite a few alternate descriptions of trees. Here are two relatively simple descriptions.

>*Theorem 2.1.* The following statements about a graph G are equivalent.
>
>(1) G is a tree.
>
>(2) There is a unique path between any two vertices of G.
>
>(3) G is connected, but removing any edge from the edge set E of G leaves a disconnected graph.

Proof. We show that statement (1) implies statement (2), that statement (2) implies statement (3), and finally that statement (3) implies statement (1); in this way we prove they are all equivalent.

To show that statement (1) implies statement (2), suppose G is a tree, and let x and y be two distinct vertices of G. Since a tree is connected, there is a path from x to y. If there are two different paths from x to y, they must have different edge sets, for the edge set of a path from x to y (in any graph) completely determines the path. (There is one edge of the path leaving x; it determines the vertex following x and so on.) Thus, x and y lie on a closed walk in which at least one edge is used only once. Now among all closed walks using at least one edge only once, choose one having no more edges than any other (that is, having as few edges as possible). Let the chosen closed walk be

$$x_0 e_1 x_1 \cdots e_n x_n$$

with $x_0 = x_n$. Now suppose $x_i = x_j$ with $i < j$. If $j - i < n$, there are two closed walks,

$$x_0 e_1 \cdots x_i e_{j+1} \cdots x_n \text{ and } x_i e_{i+1} x_{i+1} \cdots x_{j-1} e_j x_j,$$

one of which uses some edge only once. Thus, the chosen closed walk has $x_i = x_j$ only if $i = 0$ and $j = n$. No edge is used twice, for then a vertex other than x_0 would appear twice. Therefore, the chosen walk is a cycle, which is impossible in a tree. For this reason, there cannot be two different paths from x to y. This shows that statement (1) implies statement (2).

To show that statement (2) implies statement (3), suppose G is a graph in which each pair of points is connected by a unique path. Then G is connected, and if $e = \{x, y\}$ is an edge, xey must be the unique path from x to y, so deleting e from the edge set of G leaves x and y disconnected. This shows that statement (2) implies statement (3).

Now suppose G is a connected graph such that the removal of any edge yields a disconnected graph. Then G can have no cycles, since deleting an edge of a cycle cannot disconnect a graph. Thus, G is a tree. This shows that statement (3) implies statement (1)

Therefore, each statement implies the other statements and so all three are equivalent. ∎

The number of edges of a tree is always one less than the number of vertices, as the following induction proof shows. This proof is an excellent example of how the strong (second) version of the principle of mathematical induction is used.

Theorem 2.2. A tree with n vertices has $n - 1$ edges.

Proof. A tree with one vertex has no edges and a tree with two vertices has one edge. Let S be the set of all integers n such that a tree with n vertices has $n - 1$ edges. Suppose all integers smaller than k are in S, and let G be a tree with k vertices. Let x and y be two vertices of G such that $\{x, y\}$ is an edge. Deleting $\{x, y\}$ from the edge set of G

leaves a graph with two connected components, V_1 and V_2. Any other edge of G connects two vertices of V_1 or two vertices of V_2. Thus, using i to stand for 1 or 2, V_i together with the edges of G connecting vertices in V_i is a connected graph G_i without cycles. Thus each G_i is a tree. If V_1 has k_1 vertices and V_2 has k_2 vertices, then k_1 and k_2 are in S, so G_1 has $k_1 - 1$ edges and G_2 has $k_2 - 1$ edges. Thus, since $k = k_1 + k_2$, G has $1 + k_1 - 1 + k_2 - 1 = k - 1$ edges. Thus, k is in S. By the principle of mathematical induction, all positive integers are in S. ∎

This theorem may be modified to give two more descriptions of a tree.

Theorem 2.3. If a graph has n vertices and $n - 1$ edges, then
(a) it is a tree if it is conected.
(b) it is a tree if it has no cycles.

Proof. The proofs that these are descriptions of a tree are similar to the proofs of the last two theorems and are asked for in the exercises. ∎

Example 2.1. Show that a saturated hydrocarbon which has k carbon atoms has $2k + 2$ hydrogen atoms.

The graph of a saturated hydrocarbon is a tree, so if it has k carbon atoms and m hydrogen atoms, it has $k + m - 1$ edges.

An exercise in the last section asked for a proof that the sum of the degrees of the vertices of a graph is twice the number of edges of the graph, so that because the graph has k vertices of degree 4 and m vertices of degree 1,

$$4k + m = 2(k + m - 1),$$

or

$$4k - 2k = m - 2,$$

so that

$$m = 2k + 2.$$ ∎

This example shows the value of the next theorem.

Theorem 2.4. The sum of the degrees of the vertices of a multigraph is twice the number of edges.

Proof. Exercise 9 of Section 1. ∎

Corollary 2.5. A tree has at least two vertices of degree 1.

An important chemical question is whether there are two saturated hydrocarbon compounds with the same number of carbon atoms (and therefore the same number of hydrogen atoms) but different chemical properties. At first we might think that since they would have the same number of hydrogen atoms, two saturated hydrocarbons with the same number of carbons would have the same properties. Figure 2.4 shows pictures of butane and isobutane, two saturated hydrocarbons with four carbon atoms. As you might expect, butane and isobutane are *not* exactly the same in terms of chemical properties. Different saturated hydrocarbons with the same number of carbon atoms are called isomers of one another. What, though, do we mean when we say two saturated hydrocarbons are different? From the discussion of butane and isobutane we would expect them to be different unless they have essentially the same graphs. The question, "When are two graphs the same?" leads us to the notion of isomorphism that is taken up in Section 4. At that time, we will at least be able to give a precise meaning to the question, "How many different isomers of a saturated hydrocarbon with k carbon atoms are there?" Tools to answer the question have been developed but are somewhat too complex to present here.

Figure 2.4 Isobutane and butane.

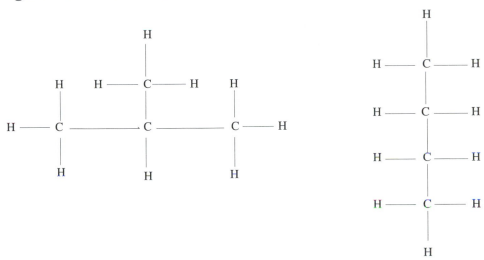

Spanning Trees

Trees play an important role in graph theory for many other reasons. Suppose we have a graph whose vertices represent cities and whose edges represent video communications links. A television program originating in one city is to be transmitted to all other cities along links of the net. The links used in the process define an *edge-subgraph* of the original graph, a new graph with the same vertex set but with an edge set that is a subset of the original one. There is no point in sending the program to a city in two different ways. Thus every pair of vertices should be connected by a unique path in the new graph; that is, the new graph should be a tree.

If $G = (V, E)$, a **spanning tree** of G is a tree whose vertex set is V and whose edge set is a subset of E. In this terminology, the links of the communications network should be the edges of a spanning tree of the graph of all potential links. A *spanning forest* of G is a forest whose vertex set is V, such that two vertices connected by a path in G are also connected by a path in the forest. All graphs have spanning forests (this is an exercise), but only connected graphs have spanning trees.

Theorem 2.6. A connected (multi)graph has a spanning tree.

Proof. Let $G = (V, E)$ be connected. If G has no cycles, it is its own spanning tree. Let S be the set of all integers n such that a graph with n cycles has a spanning tree. Zero is in S. Assume now that all integers less than k are in S and that G has $k > 0$ cycles. Let $\{x, y\}$ be an edge in a cycle of G. Then $G' = (V, E - \{\{x, y\}\})$ is connected (because there is a path from x to y not including $\{x, y\}$). G' has $k - 1$ or fewer cycles. Thus G' has a spanning tree, which is automatically a spanning tree of G. Thus k is in S, and so by version 2 of the principle of mathematical induction, S contains all positive integers. ∎

The proof of the theorem tells us that to find a spanning tree, we locate a cycle and break it; then we iterate the process. As we shall see, there are better methods. If we are going to use a spanning tree as a network for broadcasting television programs, we will have one vertex that is the source of all (or most of) the shows—the network headquarters. Thus, we might try to build our tree by working out from the source one vertex and edge at a time. This is the essence of the following algorithm, which we refer to as *algorithm Spantree*. The algorithm is applied to a graph $G = (V, E)$ with a selected vertex x_0.

(1) Let $x_0 = x$, $V' = \{x_0\}$, and $E' = \emptyset$.
(2) Repeat the following step until there is no vertex y of the type described.
   If there is a vertex x in V' and a vertex y not in V' such that $\{x, y\}$ is in E, adjoin y to the set V' and adjoin $\{x, y\}$ to the set E'.

At each stage of the algorithm, the graph (V', E') is connected (we could prove this by induction if it were not clear). According to Theorem 2.3, a connected graph with n vertices and $n - 1$ edges is a tree. Since (V', E') has one more vertex than edge at each stage (again we could prove this by induction), we have proved our next theorem.

Theorem 2.7. The graph (V', E') produced by algorithm Spantree when it is applied to a vertex x_0 in a graph $G = (V, E)$ is a tree. If $V' = V$ after the algorithm is complete, then the graph (V', E') is a spanning tree of G.

Algorithm Spantree allows us a great deal of freedom in how we choose edges for our tree.

Example 2.2. Suppose we apply algorithm Spantree to the graph of Figure 1.1 with $x_0 = CL$. Observe that one possible sequence of edges that we could choose is (CL, DE), (DE, SA), (CL, DA), (CL, AT), (CL, AL), (CL, BO). We have drawn the tree that results as the tree of heavy edges in the first picture of Figure 2.5. Observe that another possible sequence of edges that we could choose is (CL, DE), (DE, SA), (SA, DA), (DA, AT), (AT, AL), (AL, BO). We have drawn the tree that results as the tree of heavy edges in the second picture in Figure 2.5. Notice how different the two trees look. ∎

Figure 2.5

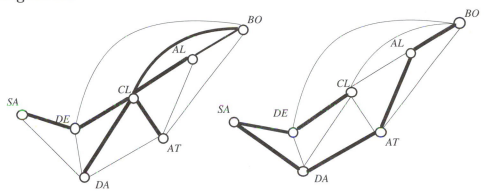

*The Number of Trees

A particularly difficult question is "How many spanning trees does a graph have?" We shall be able to answer this question later by means of some powerful techniques. If we ask the question for the complete graph with n vertices, though, it reduces to the question "How many trees are there on n vertices?" Cayley was the first person to pose and answer this question. Our solution to the problem is a modern version of Cayley's solution; it gives a sophisticated kind of generating function for trees on n vertices.

We will consider trees on the n-element vertex set $\{1, 2, \ldots, n\}$. Our generating function will involve the variables $x_1, x_2, \ldots, x_n$. We associate a monomial in these variables with each tree by using the degrees $d(1)$, $d(2),\ldots,\ d(n)$ of the vertices, namely

$$M(T) = x_1^{d(1)} x_2^{d(2)} \cdots x_n^{d(n)}.$$

Another way of constructing this monomial is

$$M(T) = \prod_{\substack{\text{edges} \\ \{u,v\} \text{ of } T}} x_u x_v.$$

The *enumerator-by-degree sequence* for trees on N is given by

$$E_N(\underline{x}) = \sum_{\substack{\text{trees} \\ T \text{ on } N}} M(T).$$

The symbol $\underline{x}$ stands for the list $(x_1, x_2, \ldots, x_n)$. The coefficient of the monomial

$$M = x_1^{d(1)} x_2^{d(2)} \cdots x_n^{d(n)}$$

in the enumerator $E_N(\underline{x})$ is the number of trees on N in which vertex 1 has degree $d(1)$, vertex 2 has degree $d(2)$, ... and vertex n has degree $d(n)$. Before considering general cases, let us examine the cases $N = \{1, 2, 3\}$ and $N = \{1, 2, 3, 4, 5\}$. The three trees on $\{1, 2, 3\}$ are shown in Figure 2.6 with their associated monomials.

Figure 2.6

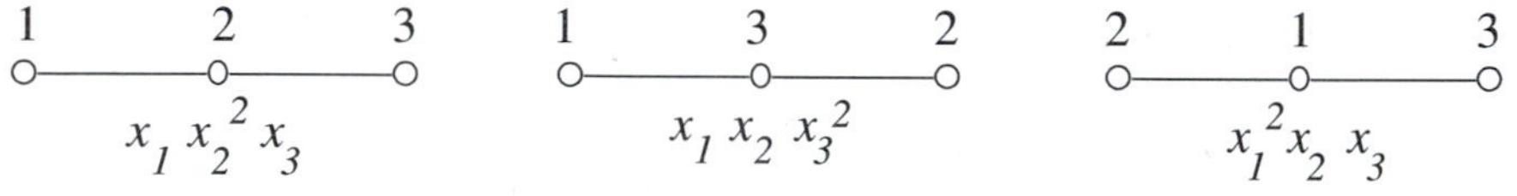

Then by adding these monomials we get

$$E_{\{1,2,3\}}(x_1, x_2, x_3) = x_1 x_2^2 x_3 + x_1 x_2 x_3^2 + x_1^2 x_2 x_3$$
$$= x_1 x_2 x_3 (x_1 + x_2 + x_3).$$

In a tree with five vertices, there will be four edges, so the sum of the degrees of the five vertices will be eight. The degree sequence will contain five positive integers that add up to eight; in other words, it will be an ordered partition of eight into five parts. The only sequences (in decreasing order) that add up to eight are $(4, 1, 1, 1, 1)$, $(3, 2, 1, 1, 1)$, and $(2, 2, 2, 1, 1)$

Figure 2.7

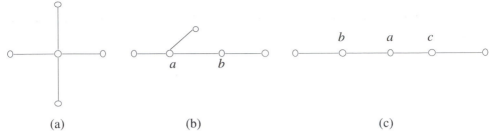

(a)   (b)   (c)

and thus, up to reordering, trees on five vertices will have one of these degree sequences.

Now there is only one way to have one vertex of degree 4 and four vertices of degree 1 in a connected graph—all the degree 1 vertices must be connected to the degree 4 vertex, as shown in Figure 2.7(a). Thus, the tree is completely determined by the choice of the degree 4 vertex.

If, as in the second sequence, we have a vertex of degree 3, say a, then three vertices must connect to a, and if all three had degree 1, there would be no way to connect the vertex of degree 2 into the tree. Thus one vertex, say b, must be the degree 2 vertex, and the third degree 1 vertex must connect to b, as shown in Figure 2.7(b).

If, as in the third sequence, we have three vertices of degree 2 and only two of degree 1, one and only one vertex, of degree 2, say a, is required to be connected to the other two degree 2 vertices, say b and c. However, since b and c each have degree 2, the degree 1 vertices must be connected to them, as shown in Figure 2.7(c).

Thus, the three trees shown in Figure 2.7 are, up to the assignment of the vertex numbers 1, 2, 3, 4, and 5, the only trees we will see on the vertex set $\{1, 2, 3, 4, 5\}$. Note that for a given assignment of degrees to vertices, there will be one tree in which vertex 1 has degree 4 and vertices 2 through 5 have degree 1, while there will be three trees in which vertex 1 has degree 3 (i.e., like a in Figure 2.7(b)), vertex 2 has degree 2, and vertices 3, 4, and 5 all have degree 1. (See Figure 2.8.)

Figure 2.8

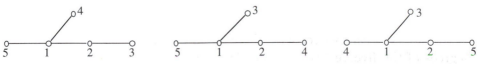

Finally, there will be six trees in which vertices 1, 2, and 3 have degree 2 (i.e., like vertices a, b, and c in Figure 2.7(c)), and vertices 4 and 5 have

Figure 2.9

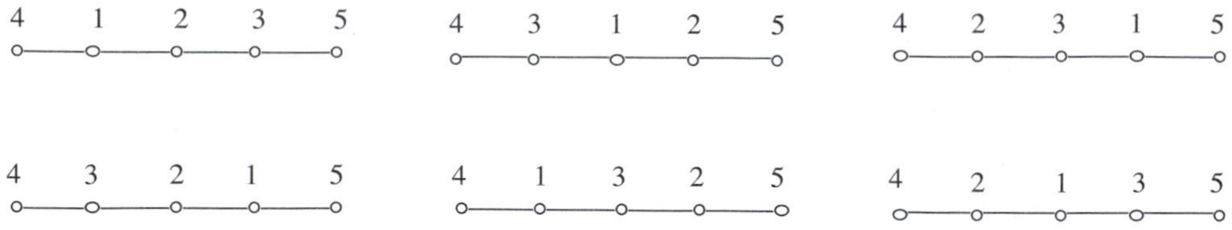

degree 1. (See Figure 2.9.)

The other assignments of degrees to vertices will give either one, three, or six trees for each degree sequence, exactly as one of the three cases we have worked out. Thus,

$$
\begin{aligned}
E_{\{1,2,3,4,5\}}&(x_1, x_2, x_3, x_4, x_5)\\
=&x_1^4 x_2 x_3 x_4 x_5 + x_1 x_2^4 x_3 x_4 x_5 + \cdots + x_1 x_2 x_3 x_4 x_5^4\\
&+ 3x_1^3 x_2^2 x_3 x_4 x_5 + 3x_1^3 x_2 x_3^2 x_4 x_5 + \cdots + 3x_1 x_2 x_3 x_4^2 x_5^3\\
&+ 6x_1^2 x_2^2 x_3^2 x_4 x_5 + 6x_1^2 x_2^2 x_3 x_4^2 x_5 + \cdots + 6x_1 x_2 x_3^2 x_4^2 x_5^2\\
=&x_1 x_2 x_3 x_4 x_5(x_1^3 + x_2^3 + \cdots + x_5^3 + 3x_1^2 x_2 + 3x_1^2 x_3 + \cdots + 3x_4 x_5^2\\
&+ 6x_1 x_2 x_3 + 6x_1 x_2 x_4 + \cdots + 6x_3 x_4 x_5).
\end{aligned}
$$

Once we factor out the $x_1 x_2 x_3 x_4 x_5$, the term in parentheses will contain each product $x_{i_1} x_{i_2} x_{i_3}$ in which i_1, i_2, and i_3 are between 1 and 5. The coefficient of x_1^3 is $\binom{3}{3,0,0}$, the coefficient of $x_1^2 x_2$ is $\binom{3}{2,1,0} = \frac{3!}{2!\cdot 1!\cdot 0!}$, and the coefficient of $x_1 x_2 x_3$ is $\binom{3}{1,1,1} = \frac{3!}{1!\cdot 1!\cdot 1!}$. In other words, by using the multinomial theorem in reverse, we see that

$$
E_{\{1,2,3,4,5\}}(x_1, x_2, x_3, x_4, x_5) = x_1 x_2 x_3 x_4 x_5(x_1 + x_2 + x_3 + x_4 + x_5)^3.
$$

This example leads us to conjecture Theorem 2.8.

Theorem 2.8. The enumerator-by-degree sequence for trees on the vertex set N is given by

$$
E_N(\underline{x}) = x_1 x_2 \cdots x_n (x_1 + x_2 + \cdots + x_n)^{n-2}.
$$

Proof. We note that a tree is completely determined by its edge set. We will show a one-to-one correspondence between edge sets of trees and monomials, that is, terms with coefficient 1, in the expansion of $(x_1 + x_2 + \cdots + x_n)^{n-2}$. It will be clear from the correspondence that the degree sequence of the corresponding tree is obtained by adding 1 to the degree of *each* x_i in this term, i.e., by multiplying the term by $x_1 x_2 x_3 \cdots x_n$. This will prove the theorem.

For a tree T, we construct its monomial as follows. First, list all edges containing a vertex of degree 1 in order of their degree 1 vertices (thus, for the first tree in Figure 2.8, we would have (3,2), (4,1), (5,1)). Now delete these vertices of degree 1 from T to get a new tree, T_1. Continue now by listing the edges containing vertices of degree 1 in T_1. (For example, for the first tree in Figure 2.8, we get the list (3,2), (4,1), (5,1), (1,2) now.) Unless T_i has only one edge, each edge listed has one vertex of degree 1, and we always list the degree 1 vertex first. We repeat the process, increasing our list each time until we have listed all the edges of T. Now the first elements of all the first $n-2$ edges and both elements of the last edge will consist of the elements of N, each appearing exactly once.

It turns out that *any* sequence (even one with repeats) of $n-2$ members of N can be the second elements of the first $n-2$ edges. Further, this sequence of $n-2$ members of N completely determines all the other entries in all the ordered pairs in the list. Note that each sequence $(k_1, k_2, \ldots, k_{n-2})$ corresponds to the term $x_{k_1} x_{k_2} \cdots x_{k_{n-2}}$ in the product $(x_1 + x_2 + \cdots + x_n)^{n-2}$; note also that the number of times x_i appears in one of these terms is one less than the number of times x_i appears in an edge. (Why?) Thus, our proof will be complete when we show why any sequence of second entries can appear from one and only one tree.

Suppose we are given a sequence of $n-2$ elements (so that our list looks like $(\ ,i_1), (\ ,i_2), \ldots, (\ ,i_{n-2}), (\ ,\ )$, including the empty ordered pair at the end so far). Note that vertices of degree 1 in T will have to be exactly those that do not appear in the sequence. At least two, and at most $n-1$, numbers do not occur; suppose there are j_1 such numbers, in increasing order, $k_1, k_2, \ldots, k_{j_1}$. Thus, our list looks like

$$(k_1, i_1), (k_2, i_2), \ldots, (k_{j_1}, i_{j_1}), (\ ,i_{j_1+1}), \ldots, (\ ,i_{n-2}), (\ ,\ ).$$

Now the edges that appear in positions $j_1 + 1$ to $n-1$ will be the edges of T_1; the numbers $k_{j_1+1}, \ldots, k_{j_2}$ not appearing in our sequence between positions $j_1 + 1$ and $n-2$, but appearing between positions 1 and j_1, will be the vertices of degree 1 in T_1, and in increasing order these will be the next j_2 numbers that occur as left-hand sides of edges. Thus, our list is now

$$(k_1, i_1), \ldots, (k_{j_1}, i_{j_1}), (k_{j_1+1}, i_{j_1+1}),$$
$$\ldots, (k_{j_2}, i_{j_2}), (\ ,i_{j_2+1}), \ldots, (\ ,i_{n-2}), (\ ,\ ).$$

In the same way, the numbers that must be the left-hand sides of each of the first $n-2$ ordered pairs in our list are uniquely determined. We never repeat a number in this list of left-hand sides, so we have $n-2$ numbers. This determines the first $n-2$ edges in the list of edges; the last edge contains the remaining two elements of N.

Now we have exactly one graph with $n-1$ edges and n vertices determined by each sequence of $n-2$ numbers from N. To prove our theorem, we must prove that this graph is a tree. First note that i_{n-2} cannot be one of the numbers we listed in the first $n-2$ left-hand sides, because the only numbers we list are those in position j_r+1 and beyond that *do not* appear in the list of right-hand sides in position j_r+1 and beyond. Thus, i_{n-2} is one member of the last and next-to-last edges in the list, so these two edges form a tree on the three vertices they include.

In general, each number in the right-hand side of an edge will appear eventually as a left-hand side of a later edge unless, perhaps, it is one of the two vertices we already know are in the last edge. Thus, if we add the edges to the edge set of a graph in reverse order from the order in which we listed them, we will always have the edge set of a connected graph, so our sequence of $n-1$ edges is the edge set of a tree. Thus, each sequence of $n-2$ numbers chosen from N corresponds to exactly one tree. ∎

Corollary 2.9. There are n^{n-2} trees on n vertices.

Proof. Set $x_1 = x_2 = \cdots = x_n = 1$ in $E_N(\underline{x})$ and you get the total number of trees on N. ∎

EXERCISES

1. We showed that all trees on three vertices have essentially the same geometric picture. In Figure 2.7 we show there are only three geometrically distinct trees on five vertices. Draw a picture for each tree on the four-vertex set $\{1, 2, 3, 4\}$. Do not draw two different pictures for the same tree, and if a tree can be drawn with a picture you have already drawn (by relabeling the vertices) do not draw it. How many pictures did you get?

2. Repeat Exercise 1 for six vertices.

3. Describe (or draw pictures of) all four-vertex forests with two connectivity classes, with three connectivity classes.

4. Show that every tree with two or more vertices has at least two vertices of degree 1.

5. How many connectivity classes are there in a forest with seven vertices and four edges?

6. Draw all pictures of saturated hydrocarbons with five carbon atoms.

7. Octane is a generic name for saturated hydrocarbons with eight carbon atoms. Draw an example of an octane molecule in which the longest path of C's has length seven. Among all imaginable octane molecules, what is the smallest possible value for the length of the longest chain of C's? Draw such a molecule.

8. Show that a connected graph on n vertices with $n - 1$ edges is a tree.

9. Show that a graph on n vertices with no cycles is a tree if it has $n - 1$ edges.

10. How many spanning trees does a cycle (closed path) on n vertices have? (Here we want the total number of spanning trees, not the number of geometrically distinct trees as in Exercise 1.)

11. Let G be the graph in Figure 2.10. Find a spanning tree of G such that no path in the tree has more than three vertices. Find a spanning tree of G with the longest possible path.

Figure 2.10

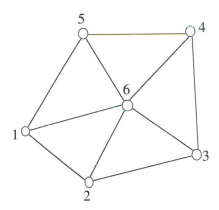

12. True or False. Any two spanning trees of a graph have an edge in common. (Prove or give a counterexample.)

13. Let T_1 and T_2 be spanning trees of a graph G and let e be an edge of T_1 not in T_2. Show that there is a spanning tree T_3 of G containing e and all but one edge from T_2.

14. Let G be a connected graph. Determine which of the following statements about G are true and which are false. (Prove or give a counterexample.)
    (a) Given any edge of G, there is a spanning tree of G containing that edge.
    (b) Given any two edges of G, there is a spanning tree of G containing them.
    (c) Given any three edges of G, there is a spanning tree of G containing them.

15. How many connected components does a forest with v vertices and e edges have? Prove that your answer is correct.

16. Show that a connected graph with $n > 2$ vertices is a cycle if and only if each pair of points may be joined by exactly two paths.

17. Use the definition of a tree rather than Theorem 2.3 to prove Theorem 2.7, which states that algorithm Spantree produces a tree (which is a spanning tree if G is connected).

18. In the previous exercise you were asked to prove algorithm Spantree works without reference to the fact that a tree on n vertices has $n - 1$ edges. Now, by using algorithm Spantree as part of your proof, show that a tree on n vertices has $n - 1$ edges.

19. Use algorithm Spantree to prove that a tree has at least one vertex of degree 1. Now use induction (removing a vertex of degree 1 in the inductive step) to show that a tree with two or more vertices has at least two vertices of degree 1.

20. By means of the technique which was used in the text to derive explicitly the enumerator-by-degree sequence for trees on $\{1, 2, 3, 4, 5\}$, derive the enumerator-by-degree sequence for trees on $\{1, 2, 3, 4\}$.

21. Show that the following three statements about a connected graph G are equivalent.
    (a) G is a tree.
    (b) Each closed walk in G uses at least one of its edges at least twice.
    (c) Each closed walk in G uses all its edges at least twice.

22. A vertex in a graph is called a cut-vertex if removing it and all edges touching it increases the number of connected components. Show that every graph with at least one edge has at least two vertices that are not cut-vertices.

23. A connected graph is called *unicyclic* if it has exactly one cycle. Prove that G is unicyclic if and only if it satisfies any two of the following properties.

(1) G is connected.

(2) G has one and only one cycle.

(3) The number of vertices of G equals the number of edges.

*24. In the proof of Theorem 2.8, there is an algorithm that produces a tree from a sequence of $n-2$ numbers between 1 and n. Write a computer program that takes a sequence as input and produces a tree as output. Modify the program to produce all trees on the set $\{1, 2, \ldots, n\}$.

Section 3 Shortest Paths and Search Trees

Rooted Trees

We observed in the previous section of this chapter that if we are going to use a spanning tree of a graph as a network for broadcasting television programs, we will have one vertex that is the source of all (or most of) the shows—the network headquarters. In certain other applications of trees we also single out a special vertex, often as a starting point of some process that the tree represents. We shall see such examples in the "search trees" we discuss in this section.

A tree in which a special vertex x has been singled out is called a **rooted tree** and the vertex x is called the **root** of the tree. Another kind of example of a rooted tree is the family tree of descendants of one person (whose descendants do not intermarry). The person whose descendants we are studying is the root for a family tree. The language of family trees has been adopted for the study of rooted trees in general. In a rooted tree with root vertex x, the vertices which lie on the (unique) path from x to a different vertex v are called the **ancestors** of v. If y is an ancestor of v, then v is called a **descendant** of y. The vertex immediately preceding v on this path is called the **parent** of v and is sometimes denoted by $p(v)$. (Notice that v has only one parent. Why?) The vertex v is called the **child** of its parent. The root, as you may have guessed, has no parent in the tree.

In Figure 3.1 we show a tree and then show the usual way of redrawing the tree when a root has been selected. Notice the way in which the tree falls naturally into levels: each vertex other than the root is on the level below its parent and the root is at the top level, usually called level 0. This is, in essence, an inductive definition of the concept of level in a rooted tree. More precisely, the root is at level zero, and the level of any other vertex is one more than the level of its parent.

Figure 3.1

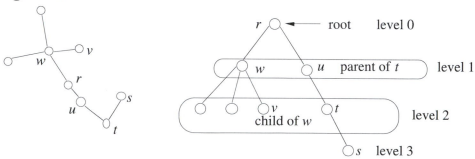

Recall that our reason for introducing rooted trees was to model the transmission of information over a network from some central source. Since there will be delay and possibility for error each time we transmit some information, we would like to choose a special kind of spanning tree—for each vertex we would like the path from the source to that vertex to require as few retransmissions as possible. The **length** of a path in a graph or multigraph is its number of edges. The **distance** between two vertices in a graph or multigraph is the length of the shortest path connecting them.

In graph-theoretic terms, we desire a rooted spanning tree such that the distance in the tree between each vertex and the root (the source of transmissions) is as small as possible. In other words, we want the distance from the root to another vertex in the tree to be the distance from the root to this other vertex in the graph. The technique of "breadth-first search" provides us with just such a tree. The technique is illustrated in Figure 3.2 and described next.

Figure 3.2   In this graph x_0 is the center x of the spanning tree; the vertex labeled x_i is the vertex added in the process of constructing the set V_i, and the edge labeled e_i is the edge added in the process of constructing the set E_i.

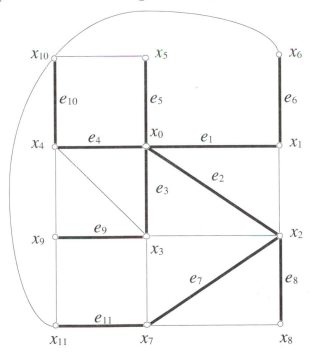

Breadth-First Search Trees

A **breadth-first search tree** centered at vertex x in a connected graph $G = (V, E)$ is a spanning tree T which can be constructed by the following rules. (The rules formalize the following description. Connect x to all vertices at distance 1 from it. Once all vertices at distance i from x have been connected into the tree, use these vertices one at a time to connect into the tree all adjacent vertices at distance $i + 1$ from x that *don't* give a cycle. Continue until you run out of vertices.)

(1) Let $x_0 = x$, $V_0 = \{x_0\}$, and $E_0 = \emptyset$.

(2) Choose a y such that $\{x_0, y\}$ is in E. Let $x_1 = y$, $V_1 = \{x_0, x_1\}$, and $E_1 = \{\{x_0, x_1\}\}$.

(3) If x_0 is adjacent to another $z \neq x_1$, let $x_2 = z$, $V_2 = V_1 \cup \{x_2\}$, and $E_2 = E_1 \cup \{\{x_0, x_2\}\}$. Otherwise if x_1 is adjacent to a $v \neq x_0$, let $x_2 = v$, $V_2 = V_1 \cup \{x_2\}$, and $E_2 = E_1 \cup \{\{x_1, x_2\}\}$.

(4) Given pairs (V_i, E_i), $i = 0, 1, 2, \ldots, k - 1$, let j be the smallest number such that x_j is adjacent to some vertex z in V but not V_k. Let $x_k = z$, $V_k = V_{k-1} \cup \{x_k\}$, and $E_k = E_{k-1} \cup \{\{x_j, x_k\}\}$.

(5) Repeat step 4 until there is no x_j of the type described.

(6) Let $T = (V_k, E_k)$, where k has the final value encountered in step 4.

Note that both step 2 and step 3 of the construction are unnecessary; they are included as a helpful mental transition from step 1 to step 4. Each is a special case of step 4. This construction gives us a list $x_0, x_1, x_2, \ldots, x_k$ of vertices as they first appear in $V_0, V_1, V_2, \ldots, V_k$. The vertex x_0 on the list is x, followed by all vertices adjacent to x_0. Next come all vertices adjacent to x_1 not already in the list, then all vertices adjacent to x_2 not already in the list, and so on. This way of numbering the vertices is called a "breadth-first numbering." The number i is called the **breadth-first number** of x_i. Notice that the algorithm we have described for breadth-first search is just a special case of the algorithm Spantree we gave in the last section. Thus we immediately have the following theorem.

Theorem 3.1. A breadth-first search tree centered at x in a connected graph G is a spanning tree.

Proof. This is a consequence of Theorem 2.7. ∎

Recall that we constructed breadth-first search trees in hopes of keeping distances in the tree as short as possible. We use $d(u, v)$ to stand for the distance from u to v. To show that our hopes are fulfilled, we need to analyze the construction process. We say that we are *adding* vertex x_k *from* vertex x_j in step 4. Note that the vertex from which we add vertex v to the tree is the parent $p(v)$ of v. We use $BFN(v)$ to stand for the breadth-first number of v.

It is intuitively clear that the vertices closer to x (with distances computed in the original graph G) have smaller breadth-first numbers, but this requires proof.

Lemma 3.2. Let d stand for the distance function in a graph G. If $d(x,v) < d(x,z)$, then $BFN(v) < BFN(z)$.

Proof. We use induction on $d(x,v)$. The statement of the lemma is clearly true when $d(x,v) = 0$, since, in that case, $x = v$. Now assume that whenever $d(x,v) \leq d$ and $d(x,v) < d(x,z)$, then $BFN(v) < BFN(z)$. Let y be a vertex with $d(x,y) = d+1$ and let z be a vertex with $d(x,z) > d+1$. Choose a shortest path from x to y, and let w be adjacent to y in this path. Observe that $d(x,w) = d$. Now for any vertex u adjacent to z, we have $d(x,u) \geq d+1$. Therefore, by the inductive hypothesis, $BFN(w) < BFN(u)$. Therefore y is added before z in step 4 of the algorithm. That is, $BFN(y) < BFN(z)$. Thus, by the principle of mathematical induction, for all values of the distance $d(x,v)$, if $d(x,v) < d(x,z)$, then $BFN(v) < BFN(z)$. ∎

Theorem 3.3. If T is a breadth-first search tree centered at the vertex x in G, then the distance from x to a vertex v in the tree is the distance from x to v in the graph G.

Proof. Our proof is by induction on the breadth-first number of v. Since v is x if $BFN(v) = 0$, the theorem is true when the breadth-first number is zero. Now assume the result holds for all v with breadth-first number less than b and let y have breadth-first number b. Let u be the last vertex on a shortest path from x to y. If $d(x,u) < d(x,p(y))$, then, by Lemma 3.2, $BFN(u) < BFN(p(y))$. But this is impossible, for $p(y)$ has the smallest breadth-first number of any vertex adjacent to y. Thus, $d(x,y) = d(x,p(y))+1$. Therefore, a shortest path from x to y through $p(y)$ has the same length as a shortest length path from x to y through u. But by the inductive hypothesis, the shortest path from x to $p(y)$ in the tree has length $d(x,p(y))$, so the shortest path from x to y in the tree has length $d(x,p(y)) + 1 = d(x,y)$. Thus, by the principle of mathematical induction, for all vertices v, the distance from x to v in the tree is $d(x,v)$. ∎

Shortest Path Spanning Trees

Theorem 3.4 tells us that the path from x to a vertex v in a breadth-first search spanning tree is a shortest path from x to v in the original graph G.

If, as we construct the tree, we record the parent of v as we add v to the tree, then we can quickly find the distance (in G) and one shortest path (in G) between any vertex v and x. We simply follow a chain of parents, counting the parents as we go, beginning with the parent of v and continuing until we reach x. The number of parents we count will be the distance. Thus, we have an efficient procedure for finding the shortest path (and its length) between two vertices—or one vertex and all others—in a graph G.

If our graph's vertices represent street corners in a city and the edges represent the streets connecting them, then breadth-first search gives us a way of finding a shortest path between any two points in the city, assuming all blocks have the same length. If the vertices of our graph represent cities and the edges represent connecting roads, then different edges could represent roadways of different length or travel time, so that breadth-first search will not necessarily give us the (geometrically) shortest path between two cities. However, if we specify the details of algorithm Spantree in a slightly different way, then we can use the resulting algorithm to solve this kind of problem. To be precise, we are given a graph G, and for each edge we are given a number called the **weight** of the edge. In this situation we say we have a *weighted graph*. We define the **weight** or **length** of a path (or walk) to be the sum of the weights of its edges. To aid our geometric intuition, we will refer to the length of a path rather than its weight. Note that the length we previously defined is just the special case where all edges have weight 1. As before, we define the **distance** between two vertices in a weighted graph to be the least length of any path between them.

For a given vertex x, in a weighted graph G, we want a tree in which the length of the path from x to each other vertex v is the length of a shortest path from x to v in G. In a sense, we adopt the strategy of breadth-first search, building up our tree one vertex at a time, always choosing the closest (to x) available vertex to add to the tree. Now, however, we measure closeness in terms of the length of a path rather than the number of edges. Once a vertex is added to the tree, its distance from x will not change as we add more vertices. We use $D(y)$ to stand for the distance (in the tree) from y to the vertex x and $w(e)$ to stand for the weight of an edge e of G. Notice that $D(y) + w(y, z)$ will be the distance from x to z in the tree if we use the edge $e = (y, z)$ as a tree edge. The following "close-first search" algorithm is in essence the algorithm called "Dijkstra's algorithm"; usually Dijkstra's name is associated with a specific version of the algorithm, discussed in the exercises, which is especially efficient. We pattern the rules for constructing the tree after those of breadth-first search, leaving out the redundant steps.

(1) Let $x_0 = x$, $D(x_0) = 0$, $V_0 = \{x_0\}$, and $E_0 = \emptyset$.

(2) Given pairs (V_i, E_i), $i = 0, 1, 2, \ldots, k$, choose an edge $\{x_j, y\}$ with $y \notin V_k$ such that $D(x_j) + w(\{x_j, y\})$ is a minimum. Let $x_{k+1} = y$, $D(x_{k+1}) = D(x_j) + w(\{x_j, y\})$, $V_{k+1} = V_k \cup \{x_{k+1}\}$, and $E_{k+1} = E_k \cup \{\{x_j, x_{k+1}\}\}$.

(3) Repeat step 2 until there is no edge $\{x_j, y\}$ of the type described.

(4) Let $T = (V_k, E_k)$, where k has the final value encountered in step 2.

Our next theorem states that our algorithm accomplishes our goals, but states it in a way that makes it straightforward to prove by induction.

Theorem 3.4. At each stage of a "close-first search," the graph (V_i, E_i) is a tree such that the distance in G from x to any vertex in V_i equals the distance in the tree from x to this vertex and the distance in G from x to any vertex of G not in the tree is at least as large as the distance in G from x to any vertex in the tree.

Proof. See Exercises. ∎

Example 3.1. In Figure 3.3 we show a weighted graph and the steps of applying "close-first search" to it. Note that $D(x_1) = 1$, $D(x_2) = 2$, $D(x_3) = 2$, $D(x_4) = 3$, and $D(x_5) = 4$. ∎

Figure 3.3

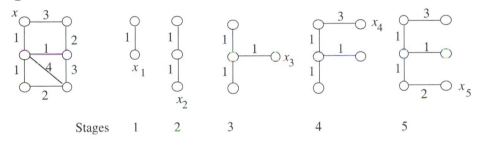

Stages   1   2   3   4   5

Bridges

We have seen how problems of designing communications networks lead us to study special kinds of spanning trees. Communications problems lead to other concepts in graph theory as well. For example, although a network whose graph is a tree has no wasted edges, it is also quite vulnerable—if any edge becomes unavailable then its endpoints can no longer communicate with each other, even indirectly. Thus, although we might use a spanning tree of our network for normal communications, we would want our network to have sufficient redundancy that removing any one edge would not disconnect the network itself. An edge of G is called a **bridge** or **isthmus** if its removal from the graph disconnects the graph. To determine if—and where—a graph is vulnerable, we must locate its bridges.

If we have a careful drawing of a graph, then we can locate all its bridges by inspection. However, if we don't already know where the bridges of a large graph are when we start drawing the graph using data of some sort, we may not draw it in such a way that the bridges are obvious. Further, if we are using a computer to analyze the graph, we cannot tell it to stare at the graph. Thus, we want a technique which may be used to search a graph for a bridge.

To visualize a strategy, think of the edges of the graph as passageways in a maze. A standard way to search for a way out of a maze is this: Put one hand—say your right one—on a wall and start walking. When you reach a dead end, you keep sliding your hand along the wall in front of you, turning around until you are facing the other way and then start walking again with your hand now opposite the wall where it was as you walked into the dead end. Whenever you have to turn a corner to keep your hand on the wall, do so. If you get back to your starting place, there is no way out of the maze (except the entrance where you presumably started). What does this have to do with finding bridges? Suppose we agree to go by exits (as if we were sliding our hand along a closed door at each exit) and keep exploring with our hand on the wall until we come back to our starting place. If one of the passageways is a bridge, then after crossing it in one direction we will have to cross it in the other direction before we get back to the starting place. Conversely, it may seem intuitively clear that if our strategy *forces* us to cross a certain passageway a second time (in the opposite direction) in order to get back to the starting place, then that passageway is a bridge. (If so, you are to be congratulated on your intuition.) In order to prove that this is the case we give a description of the search procedure we are using in precise graph-theoretic terms which don't involve "hands" and "walls." The procedure we are using may be formalized as the concept of *depth-first search*.

Depth-First Search

A ***depth-first search tree*** from the vertex x in a graph $G = (V, E)$ is constructed as follows:

(1) Let $v_0 = x$, $V_0 = \{v_0\}$, and $E_0 = \emptyset$.

(2) Given V_i and E_i, let j be the largest number such that v_j is adjacent to a vertex $y \notin V_i$. Let $v_{i+1} = y$, $V_{i+1} = V_i \cup \{v_{i+1}\}$, and $E_{i+1} = E_i \cup \{\{v_j, v_{i+1}\}\}$.

(3) Repeat step 2 until there is no j of the type described.

(4) Let $T = (V_n, E_n)$, where n is the final value of i.

A selection of a depth-first search spanning tree is shown in Figure 3.4. The edges with arrows are the edges of the tree. The arrows show the "direction" in which we are conducting our search.

Figure 3.4

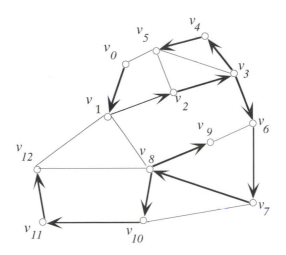

Theorem 3.5. If G is a connected graph, a depth-first search tree is a spanning tree.

Proof. This is a consequence of Theorem 2.7. ∎

Depth-First Numbering

A numbering of $v_0, v_1, v_2, \ldots, v_n$ given by depth-first search is called a ***depth-first numbering*** of G. The number i is called the ***depth-first number*** of the vertex v_i. Thus in Figure 3.4, 2 is the depth-first number of vertex v_2 and so on. The *depth-first reach* of a vertex v_i is the largest j such that vertex v_j is a descendant of vertex v_i in the tree. Thus in Figure 3.4, the depth-first reach of v_3 is 12, but the depth-first reach of v_4 is 5.

Lemma 3.6. If, in a depth-first search tree T, the depth-first reach of v_i is j, then the descendants of v_i in T are $v_{i+1}, v_{i+2} \ldots, v_j$.

Proof. By definition, v_j is a descendant of v_i. If $i < m < j$, we wish to show that v_m is a descendant of v_i in T. Suppose the contrary, i.e., that there is a smallest m such that $i < m < j$ and v_m is not a descendant of v_i in T. This means that when v_m is added to T, it is connected (in T) to a v_k with $k < i$. This means by step 2 that *no* vertices in G are adjacent to $v_i, v_{i+1}, \ldots, v_{m-1}$ except for vertices v_h with $h < i$. Thus, no vertex listed after v_m is adjacent to $v_i, v_{i+1}, \ldots, v_{m-1}$ in G, so no vertex listed after v_m is a descendant in T of v_i, contradicting the definition of j. By construction, no other vertices are descendants of v_i. ∎

Figure 3.5

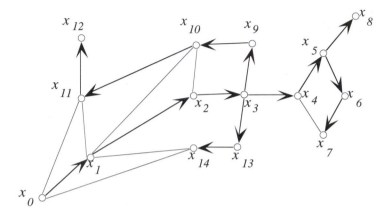

Figure 3.5 shows the result of applying depth-first search to a graph G with several bridges. The edges shown as arrows are the edges of the tree. One bridge is $\{x_3, x_4\}$ and there are no edges of G other than $\{x_3, x_4\}$ between vertices which are descendants of x_4 in T and the remainder of G. By the definition of reach and by step 2 of the search procedure, no vertices numbered greater than the reach of x_4 are descendants of x_4 in T or even adjacent in G to a descendant of x_4 in T. This suggests the following theorem.

Finding Bridges

Theorem 3.7. If T is a depth-first search tree of G, then the edge $\{v_h, v_i\}$ with $h < i$ is a bridge if and only if no descendant of v_i in T is adjacent in G to a vertex v_k with $k < i$.

Proof. By step 2 of the depth-first search algorithm and Lemma 3.6, no descendant of x_i is adjacent to a vertex whose depth-first number is greater than the reach of x_i. However, if $\{x_h, x_i\}$ is *not* a bridge, some descendant of x_i must be adjacent in G to some vertex not a descendant of x_i (otherwise, by deleting the edge $\{x_h, x_i\}$ we would separate descendants of x_i from the remainder of the graph). Thus, if $\{x_h, x_i\}$ is *not* a bridge, a descendant of x_i is connected in G to a vertex with depth-first number less than i.

If $\{x_h, x_i\}$ is a bridge, then by the description of depth-first search, x_h and vertices with depth-first number less than i are on one side of the bridge and x_i and and its descendants are on the other side. Thus, no descendant x_i is adjacent in G to a vertex with depth-first number less than i. ∎

As we can see, and the theorem implies, $\{x_{11}, x_{12}\}$ and $\{x_5, x_8\}$ are the other bridges in Figure 3.5. Theorem 3.7 gives a straightforward process for finding bridges, either by hand or by computer. However, for each tree edge, we must examine all vertices adjacent to each of the descendants of one vertex of that edge.

*An Efficient Bridge-Finding Algorithm for Computers

To use Theorem 3.7 as a basis for a more efficient algorithm for finding bridges, we introduce the vertex $L(v_i)$, which we shall call the "least vertex accessible from v_i." We *define* $L(v_i)$ to be the vertex with smallest depth-first number among all vertices adjacent in G but not in T to v_i or descendants in T of v_i. In this terminology, Theorem 3.7 states that $\{v_h, v_i\}$ is a bridge if and only if the depth-first number of $L(v_i)$ is greater than or equal to i.

The depth-first number of $L(v_i)$ can be computed as follows. Complete the depth-first search tree, assigning a depth-first number to each vertex. Now for each vertex, examine its descendants in the tree; for each, examine its neighbors, and among all of those, choose the vertex with least depth-first number. (In fact, it is possible to rewrite step 2 of the algorithm so that as you add vertices to the tree, you update information about the least vertex accessible from other vertices in the tree in such a way that when the search is done, you know all the least accessible vertices as well.)

Example 3.2. Which are the least vertices accessible from x_{10}, x_9, x_8, and x_4 in Figure 3.5? In other words, what are $L(x_{10})$, $L(x_9)$, $L(x_8)$, and $L(x_4)$? What can you conclude about bridges?
Since x_{11} is a descendant of x_{10} in T and x_0 is adjacent to x_{11} in G, $L(x_{10}) = x_0$. Since x_{10} a descendant of x_9, and $L(x_{10}) = x_0$, then $L(x_9) = x_0$. Since x_8 has no descendants and no vertices of G are adjacent to x_8 by edges not in the tree, $L(x_8) = x_8$. Since no edge of G is an arrow from x_4, x_5, x_6, x_7, or x_8 to an x with a smaller depth-first number, $L(x_4) = x_4$. Thus, $\{x_3, x_4\}$ and $\{x_5, x_8\}$ are bridges. ∎

Backtracking

Depth-first search is also known as backtracking, since we can visualize the process as follows: starting at vertex x construct a path, always going from your current vertex along an edge to a vertex not yet on your path until you have run out of such edges. When you get to a vertex which has no edges leading out—except for edges leading to another vertex already on the path—back up along your path until you find an edge that leads to a vertex not on your path and repeat the process. Keep up the repetition until you run out of vertices. It is possible to use depth-first search to determine the vertices connected to a vertex x because by Theorem 2.7, the search gives us a spanning tree of the component containing x. Thus, we can use depth-first search to find connected components of G. Other special features of graphs (e.g., bridges (as we have seen), cycles, vertices whose removal disconnects G, "bridgeless components," etc. of G) can be discovered with depth-first search as well. These kinds of applications and the terminology "depth-first-search" are largely due to Robert Tarjan. (See the Suggested Reading.) Backtracking or depth-first search is also a convenient way of listing all potential solutions to a wide variety of problems. In this way it is possible to determine whether at least one potential solution has certain desired properties. It is also possible to determine all potential solutions that have certain desired properties.

Problems that may be solved by backtracking involve lists of decisions. As an example, suppose we are searching for a way out of the maze in Figure 3.6. We are at the spot marked x, and we have a list of four choices of directions to go:

$$\{\text{up, right, down, left}\}.$$

Once we decide to try one of these directions, we start to move through the maze. Our possible tracks through the maze are shown in a graph whose vertices represent "squares" inside the maze and whose edges represent openings between the squares.

An attempted solution to the maze may be represented as a path in the graph. For example, we may decide to explore the maze by going up if possible, right if up is impossible, down if right is impossible, and left if down is impossible. To make sure we get a path and don't cycle through the graph, we agree not to return to a vertex already visited (except when backing up). Our first attempt is marked with the number 1 in the graph and leads to a dead end. Now we back up along our path until we come to a place where we can take a direction other than the one already chosen. Our next attempt is marked with 2's. Although it might seem reasonable to go left from vertex y, this would lead us back to vertex x and result in a cycle, so we do not explore this edge. On our third attempt, we get the dead-end path marked with 3's. Now we must start out again from position x. We don't want to go to vertex y, since that would put us at a vertex where all available paths failed. Thus, we explore the path marked with 4's; this takes us out of the maze.

As you see, we did not need to explore all paths leaving x. Part of this was just luck. If our exit had been at the vertex of G marked E, then

Figure 3.6

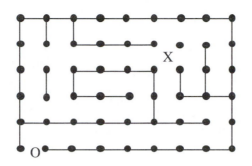

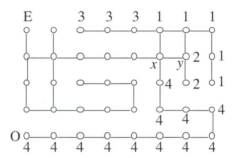

we would have had more paths to explore. Note, however, that we don't explore any path which begins $x\{x,y\}y\cdots$ because we have already seen that all paths leading out of y lead us either to dead ends or back to x. In other words, just by blindly applying the rules for depth-first search, we would never explore any paths starting $x\{x,y\}y\cdots$, so the use of depth-first search to explore the maze will save us a good deal of effort when there are many interconnected paths in the maze.

Decision Graphs

In a problem whose solution involves a sequence of decisions, we draw a *decision graph* by the following rules:

(1) Draw a vertex to represent the current situation.

(2) Draw an edge departing the initial vertex and entering a new vertex for each decision that changes the situation.

(3) For each vertex you add, determine which decisions will change the situation. For those changes that lead to a situation already associated with a vertex, connect that vertex to the associated vertex. For each change that leads to a new situation, draw an edge to a new vertex representing the new situation.

(4) Repeat step 3 as long as new situations are possible.

For example, this sequence of steps leads to the graph in Figure 3.6, which represents the situation in the maze. Here the situations are the "squares" of the maze and the decisions are always those that can be legally chosen from the set {up, right, down, left}. In Section 6 we will see how such a decision graph may be used to apply backtracking to a quite different problem.

EXERCISES

1. Describe a breadth-first search tree for a cycle with an odd number of vertices; for a cycle with an even number of vertices.

2. Find a breadth-first search tree centered at 1 for the graph of Figure 2.10 of the previous section. How many answers are possible?

3. Ignoring steps 2 and 3 (that is, considering them as special cases of later steps), find out (in terms of the number of vertices and/or edges of the graph) the maximum number of times each of the steps in the breadth-first search algorithm could be carried out when it is applied to a graph with n vertices.

4. Recall that u is an *ancestor* of v in a rooted tree centered at the vertex x in a graph G if u is on the unique path from x to v in the tree. We say an edge of G from y to z is a *cross edge* if neither y nor z is an ancestor of the other.
   (a) Show that given a breadth-first search tree of a graph G, each edge of G is either a cross edge or an edge of the tree.
   (b) Use (a) to show that the distance from x to v in the tree is the distance from x to v in G.

5. Show the results of carrying out a close-first search centered at x on the graph of Figure 3.7 by showing the tree at each stage of the construction and showing (or describing in some way) the vertices x_i as they are added to the tree.

Figure 3.7

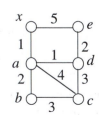

6. Prove Theorem 3.4.

7. Discuss how to use breadth-first search to find a shortest cycle in a graph.

8. Discuss how to use breadth-first search to find a shortest odd cycle in a graph.

9. Modify the close-first search algorithm by keeping track of one more list L of data with one item for each vertex. For each vertex v, let $L(v)$ be that vertex y in the tree for which $D(y) + w(\{y, v\})$ is a minimum—or let $L(v)$ be the symbol $\emptyset$ if there is no such vertex. Notice that when you add a vertex z to the tree, you can check each vertex v to see if $L(v)$ should be changed to z. Explain why this makes it easier to find which vertex to add next. Analyze the number of computational steps used in this version of the algorithm, which is Dijkstra's version.

10. Write a computer program that takes as data the adjacency table (see Exercise 26, Section 1) of a graph and produces the connected components of the graph by means of breadth-first search.

11. Write a computer program that takes as data the adjacency table of a graph and gives the distance between each two connected vertices by means of breadth-first search.

12. Compute a depth-first numbering of the vertices in the graph of Figure 3.8, starting at x.

Figure 3.8

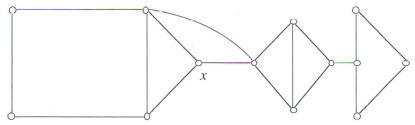

13. Compute the least vertex $L(v_i)$ for each vertex in your depth-first numbering from the previous exercise of Figure 3.8. Use this computation to illustrate Theorem 3.7.

14. Compute the least vertex $L(v_i)$ for each vertex in Figure 3.5.

15. Using the decision sequence {up, right, down, left}, search to find the vertex E in the graph of Figure 3.6.

16. Using the decision sequence {up, left, down, right}, search for a path starting at x and leading out of the maze in Figure 3.6.

17. Copy the maze in Figure 3.6 and show, by drawing the path, that if you put your hand on one wall and start moving through the maze then you will return to the entrance, having traversed each passageway at most once in each direction.

18. Using a copy of Figure 3.9, show the result of using the "hand-on-the-wall" rule to get from the entrance to the exit of the maze.

Figure 3.9

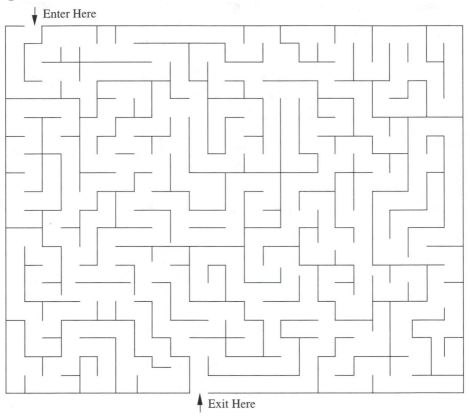

19. Cross edges are defined in Exercise 4. Show that for each depth-first search tree for a graph, there are no cross edges.

20. Given a numbered tree such as a breadth-first search tree or a depth-first search tree, found by searching a graph, there is a walk in the graph, using only tree edges, from vertex one to vertex two to vertex three and so on to vertex number n. In most cases the walk will not be a path because we may have to "back up" in the tree to get from vertex v_i to vertex v_{i+1} using only tree edges.

(a) Describe the walks associated with depth- and breadth-first search, starting at one end, on a path with n vertices.

(b) Describe the walks associated with breadth-first search and depth-first search, with vertex 1 in the middle, on a path with an odd number n of vertices.

(c) What is the maximum number of times an edge of a graph can appear in the walk associated with a depth-first search of the graph?

(d) What is the maximum number of times an edge of a graph can appear in the walk associated with a breadth-first search of a graph?

(e) Discuss the maximum possible number of edges in a walk associated with breadth-first search and one associated with depth-first search. Which kind of search represents a better way to explore a graph?

21. Draw the decision graph corresponding to the maze in Figure 3.9. (Warning: This is a time-consuming exercise!)

22. Find a path through the maze in Figure 3.9 by means of depth-first search. (Warning: This is a time-consuming exercise; it uses the previous exercise at least implicitly.)

23. Prove that the edge $\{x, y\}$ is a bridge in G if and only if every walk from x to y includes $\{x, y\}$.

24. Prove that an edge of a connected graph is a bridge if and only if it is not in any cycle.

25. The *diameter* of a graph is the maximum distance between two vertices in the graph. The *eccentricity* of a vertex v is the maximum of the distances from v to any other vertex. The *radius* of a graph is the minimum of the eccentricities of its vertices. A *central* vertex has its eccentricity equal to the radius of the graph.

(a) Show that a tree has either one or two adjacent central vertices.

(b) Show that a tree has a unique central vertex if and only if its diameter is twice its radius.

*26. An *articulation point* is a vertex whose removal disconnects the graph. Develop a method of detecting articulation points by depth-first search; apply your method to the graph in Figure 3.8.

Section 4 Isomorphism and Planarity

The Concept of Isomorphism

In Figure 4.1, you see pictures of three trees. The first tree is clearly a different structure from the other two trees, but although the second two trees have different pictures, each has one vertex of degree 3, surrounded by three vertices of degree 1. Geometrically, they appear to be essentially the same. The edge set of the second graph is $\{\{c_2, c_1\}, \{c_2, c_3\}, \{c_2, c_4\}\}$, while the edge set of the third graph is $\{\{d, a\}, \{d, b\}, \{d, c\}\}$. Note that if we rename the vertices of the second graph by changing c_2 to d, c_1 to a, c_3 to b, and c_4 to c and carry these relabelings into the edge set, then the two edge sets of the two graphs are exactly the same. As mathematical structures, then, the two graphs are essentially the same. Two mathematical objects which are essentially the same are normally said to be isomorphic (loosely translated, "isomorphic" is Greek for "same shape"). To be precise, an *isomorphism* from a multigraph $G_1 = (V_1, E_1)$ to a multigraph $G_2 = (V_2, E_2)$ is a one-to-one function f from V_1 onto V_2 such that $\{a, b\}$ is an edge in E_1 if and only if $\{f(a), f(b)\}$ is an edge in E_2 with the same multiplicity that $\{a, b\}$ has in E_1. We say G_1 is isomorphic to G_2 or that G_1 and G_2 are *isomorphic* if there is an isomorphism from G_1 to G_2. (Note that this use of language makes the assumption that the relation of isomorphism is symmetric. Is the assumption valid?)

Figure 4.1

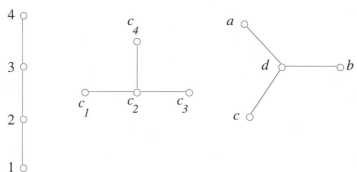

It is straightforward to prove that isomorphism between graphs is an equivalence relation, that is,

Theorem 4.1. Let S be a set of graphs or multigraphs. Then the relation "is isomorphic to" is an equivalence relation on S.

Proof. Exercise 1. ∎

Checking Whether Two Graphs Are Isomorphic

It is natural to ask how many really different graphs or really different trees, etc., there are on n vertices. We now have a concrete interpretation of this question; we are asking for the number of "isomorphism classes" of graphs for a certain set of graphs.

Figure 4.2

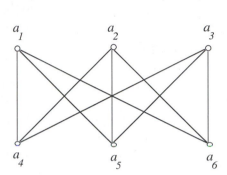
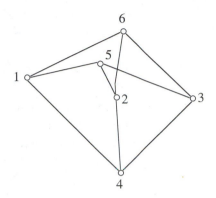

The two graphs in Figure 4.2 are isomorphic by the function $f(a_i) = i$. Once we are given the isomorphism f, it is easy to check that it has the desired properties. On the other hand, given two drawings, it may be very difficult to check whether they are drawings of isomorphic graphs. For example, is either graph in Figure 4.3 isomorphic to the graphs in Figure 4.2?

Figure 4.3

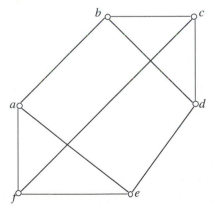
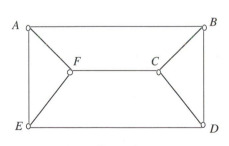

It is easy to see that the graphs in Figure 4.3 are isomorphic to each other because of the labelings; we just let the upper- and lowercase letters correspond to each other in the natural way. There is no obvious way, however, to match up the graphs in Figure 4.3 with those in Figure 4.2. *Perhaps* they are not isomorphic. Let us see how nonisomorphic graphs could be different. Each property that two isomorphic graphs must have in common corresponds to a possible way in which two nonisomorphic graphs might be different. For example, two isomorphic graphs have the same number of edges (why?). Thus if two graphs have different numbers of edges, then they cannot be isomorphic. However, all the graphs in question have nine edges. The multiset of all the degrees of vertices of one graph will be the same as the multiset of degrees of an isomorphic graph (why?). However, all the vertices of all the graphs in question have degree 3. (They are examples of so-called *regular* graphs, graphs whose vertices all have the same degree.)

One of the graphs in Figure 4.3 has been drawn without any crossings among the edges. If the graphs in Figure 4.2 and Figure 4.3 are isomorphic, then similarly we should be able to redraw the graphs in Figure 4.2 with no crossings. In this way, we would probably be able to see better what the isomorphism function is (*if* there is such a function). The second picture in Figure 4.3 divides the plane (or the flat piece of paper it is drawn on) into five regions, the two triangles and two trapezoids inside the rectangle and the region outside the rectangle. If we could draw the graph of Figure 4.2 on a plane without any crossing, the picture would also divide the plane into regions. If the graphs in question are isomorphic, there should be a redrawing of Figure 4.2 with five regions, two of them triangles, two of them quadrilaterals, and one the outside of a quadrilateral. The boundary (outside or inside, depending on the region) of each region is a cycle in the graph. From this we see that the graph of Figure 4.2 and the graph of Figure 4.3 *are not* isomorphic, because all cycles in Figure 4.2 have an even number of edges.

As this example shows, one way to show that two graphs are not isomorphic is to describe a property that two isomorphic graphs should both have or should both not have and then observe that one graph has the property and the other does not. Although this sometimes requires a good deal of insight and is not guaranteed to work, it is usually less work than the other alternative, that of writing down all the bijections between the two graphs and showing that each violates the condition of being an isomorphism. Similarly, to prove two graphs *are* isomorphic, we must give a bijection which is an isomorphism and verify that it is one. While it may require a great deal of insight and experimentation to guess the appropriate bijection to try, that may be less effort than writing out the bijections between the graphs one at a time and checking them to see if one of them satisfies the definition of an isomorphism.

Planarity

Our analysis of the graphs in Figures 4.2 and 4.3 leaves us with a thorny question: Can the graph of Figure 4.2 be drawn on a plane surface without crossings? A graph that has such a drawing is called a *planar graph*. Thus, we know the graphs of Figure 4.3 are planar, but we don't know about the graph of Figure 4.2. Euler studied this concept of planarity, especially in the case where the planar graphs are pictures in the plane of three-dimensional geometric figures called polytopes. See Figure 4.4 for the relationship between the usual picture of a cube with its faces labeled and a planar picture. The second picture in Figure 4.4 can be visualized by imagining we are looking at an elastic cube from above, with the base stretched out to allow us to see all the side faces. The one face we can't see is the one underneath the figure. We are now going to think of this figure as a planar drawing of a graph. Each face we can see defines a region inside the graph. We call this region an *inside face* of the drawing of the graph. There is also an infinite region in the plane that lies outside the graph. We call this region the *outside face* of the drawing of the graph. Notice that it has exactly the same edges on its boundary as the bottom face that lies under the figure. (Some people like to think of the bottom face of the polytope as made from infinitely stretchy material that they have punctured in the middle and stretched out around the figure so that the hole they punctured has become a circle of "points at infinity." In this way of looking at things, the outside face *is* the bottom face.) Thus there is a correspondence between faces of the polytope and the faces of the graph such that corresponding faces have the same edges and same vertices on them. Because of the relationship between graphs and polytopes, the regions of a planar drawing of a graph are also called *faces*.

Figure 4.4

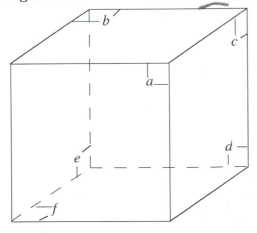

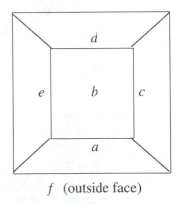

f (outside face)

Euler's Formula

For a planar drawing of a graph G, let v be the number of vertices, e be the number of edges, and f be the number of faces (including the face consisting of the region outside the drawing). Euler noticed that $v - e + f = 2$ for any graph with a planar drawing. In Figure 4.4, we get $8 - 12 + 6 = 2$, and for the planar graph in Figure 4.3, we get $6 - 9 + 5 = 2$.

Theorem 4.2. For any planar drawing (without crossings) of a connected planar graph, $v - e + f = 2$.

Proof. We prove the theorem by induction on e. Let S be the set of integers n such that Euler's formula holds for all planar graphs with $e = n$ edges. If $e = 1$, the graph has two vertices, one edge, and bounds only one region—the region outside the graph. Thus, $v - e + f = 2 - 1 + 1 = 2$. Now suppose the theorem is true for graphs with $e - 1$ edges (that is, suppose $e - 1$ is in S), and let G be a planar graph with e edges. We consider two cases. First, if G is a tree, then $f = 1$ and $e = v - 1$, so $v - e + f = 2$. Otherwise, G has at least one cycle, so that a planar drawing of G has at least two regions. Pick an edge $\{x, y\}$ of G which is between two regions in a planar drawing of G. Form G' and a planar drawing of G' by removing $\{x, y\}$ from the edge set of G and the picture of G. Then G' has $e - 1$ edges, v vertices, and, because $\{x, y\}$ separates two regions from each other, G' has $f - 1$ faces. But $e - 1$ is in S, so that $2 = v - (e - 1) + f - 1 = v - e + f$. Thus, e is in S. Then by the principle of mathematical induction, S is the set of all positive integers and the theorem is proved. ∎

At times, we may apply Euler's formula to determine the number of faces a planar drawing of a graph must have and from this see that a certain graph does not have a planar drawing. In a graph which is not a tree, a face has a cycle on its boundary and so must have at least three edges. Using this, we can come up with a test for nonplanarity that is usually easier than Euler's formula to apply in practice.

An Inequality to Check for Nonplanarity

Theorem 4.3. In a connected planar graph with three or more vertices

$$e \le 3v - 6.$$

Proof. We've noted that each face has at least three edges, and each edge separates two faces (or doesn't separate faces). Thus, the total number of ordered pairs (edge, face) with the edge touching the face is *at least* $3f$ but is *no more than* $2e$. We may express this symbolically as

$$3f \leq |\{(\text{edge}, \text{face}) | \text{the edge is on the face}\}|$$

Thus $3f \leq 2e$, but $f = 2 + e - v$, so that

$$6 + 3e - 3v \leq 2e$$

or

$$e \leq 3v - 6. \qquad \blacksquare$$

Example 4.1. The *complete graph* on five vertices K_5 is the graph with five vertices each pair of which is an edge. Show that K_5 is not planar.

Since K_5 has five vertices and $C(5; 2) = 10$ edges, we would have to have

$$10 \leq 3 \cdot 5 - 6 = 9$$

if the graph were planar, so it is not. $\blacksquare$

Notice that we were able to conclude that G was not planar in the last example because the inequality was *not* satisfied. The test does not work in the "opposite direction." That is, *even if the inequality is satisfied by a graph, the graph need not be planar.*

The graph of Figure 4.2 is called $K_{3,3}$—the "complete bipartite graph on two parts of size 3." It has six vertices and nine edges and certainly nine is less than $3 \cdot 6 - 6$. However, $K_{3,3}$ is another famous example of a nonplanar graph. We note that if it had a planar drawing, then all its faces would have to have at least *four* edges, giving us $4f \leq 2e$ rather than the less restrictive inequality $3f \leq 2e$ as in the proof of Theorem 4.3. Since the graph has nine edges, we get $4f \leq 18$ or $f \leq 4.5$; that is, $f < 5$. But Euler's formula tells us that $f = 5$ for a planar graph with six vertices and nine edges, so $K_{3,3}$ is *not planar*.

The graphs K_5 and $K_{3,3}$ are special cases of the *complete graph* K_n which has n mutually adjacent vertices and the *complete bipartite graph* $K_{m,n}$ which has an m-element set of mutually nonadjacent vertices and an n-element set of mutually nonadjacent vertices such that each vertex in one of the sets is adjacent to all vertices in the other set. These two families of graphs arise frequently as examples in graph theory.

The graphs $K_{3,3}$ and K_5 play a central role in the theory of planarity of graphs. Kuratowski showed that at least one of these graphs occurs naturally in any nonplanar graph. We say that a (multi)graph G is a *sub(multi)graph* of a graph H if the vertex set of G is a subset of the vertex set of H and the edge set of G is a subset of the edge set of H. Note that if we have a planar drawing of a graph G, then we can use it to get a planar drawing of any subgraph of G (by erasure). A subdivision of a graph G is one obtained by inserting one or more vertices of degree 2 into edges of G.

Figure 4.5

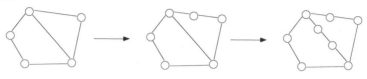

More precisely, we insert a vertex of degree 2 in the edge $\{x, y\}$ if we add a vertex z to the vertex set of G and we add $\{x, z\}$ and $\{z, y\}$ to the edge set. (See Figure 4.5.)

Kuratowski proved that

Theorem 4.4. A graph is not planar if and only if it contains a subgraph which is a subdivision to K_5 or $K_{3,3}$.

Proof. The proof is more properly part of a course in topology or graph theory and goes beyond the scope of the present book. ∎

Although Kuratowski's theorem gives us, in principle, a way to check if a graph is planar by replacing vertices of degree 2 by edges and examining subgraphs with five and six vertices, examining all $\binom{n}{6}$ six-element subsets of an n-element vertex set can be time consuming. There is a much faster algorithm (see the paper by Hopcroft and Tarjan in the suggested reading) which not only checks to see if the graph is planar but leads to a planar drawing if there is one.

Figure 4.6

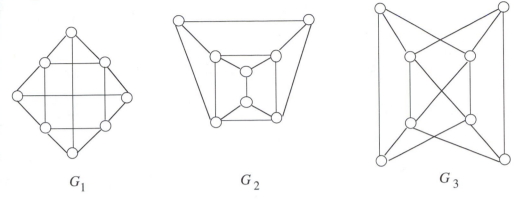

G_1 G_2 G_3

EXERCISES

1. Prove Theorem 4.1.

2. Why do two isomorphic graphs have the same number of edges?

3. Which, if any, of the graphs in Figure 4.6 are isomorphic?

4. Prove that if f is an isomorphism from G_1 to G_2, then the vertex $f(x)$ has the same degree in G_2 that x has in G_1.

5. Prove that if f is an isomorphism from G_1 to G_2 and G_1 has a cycle on four vertices, then G_2 has a cycle on four vertices as well. Is there anything special about four? What can you say about the numbers of cycles of each possible size in the two graphs?

6. Let V_1 be the positive factors of 30. Define two integers to be compatible if one is a factor of the other, and let $\{m, n\}$ be in E_1 if m and n are different but compatible. Let V_2 be the subsets of $\{x, y, z\}$ and define two sets to be *comparable* if one is a subset of the other. Let $\{S, T\}$ be in E_2 if the sets S and T are comparable but different subsets of V_2. Show that the graphs (V_1, E_1) and (V_2, E_2) are isomorphic.

7. Three houses are to be built along a line parallel to and set back 30 feet from a road. Next to the road are three utility holes, one for sewer, one for water, and one for electricity. Is it possible to connect the houses to the utility holes in such a way that no lines cross? (Lines may go behind houses.)

8. If in Exercise 7 there is a fourth hole for telephone wires, what is the minimum number of crossings if each house is connected to each hole?

9. Show that if a planar graph has no cycles of length 3, then $e \leq 2v - 4$.

10. Give as many examples as you can of trees on six vertices, no two of which are isomorphic.

11. Give as many examples as you can of forests on five vertices, no two of which are isomorphic.

12. Give as many examples as you can of graphs on four vertices with two or more nonisomorphic spanning trees.

13. How many isomorphisms are there from an n-vertex cycle to itself? An n-vertex complete graph? An n-vertex path?

14. For each combination of values of m and n, how many isomorphisms are there from the complete bipartite graph K_{mn} to itself? (This graph has a set of m mutually nonadjacent vertices and a set of n mutually nonadjacent vertices, and every vertex in one set is adjacent to every vertex in the other set.)

15. Show that a planar graph has at least one vertex of degree 5 or less.

16. The graph K_6 has 15 edges. Show that it is impossible to remove two edges to obtain a planar graph. What about three edges?

17. The Petersen graph is drawn in Figure 4.7. Show that the Petersen graph is not planar.

Figure 4.7 The Petersen graph

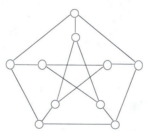

18. Show that if a planar graph has no cycles of size 5 or less, then $e \leq \frac{3}{2}v - 3$.

Figure 4.8

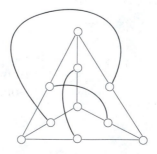

19. Is the graph in Figure 4.8 planar?

20. Is the graph in Figure 4.8 isomorphic to the Petersen graph of Figure 4.7?

21. Let the vertices of a graph correspond to the two-element subsets of a five-element set. Draw an edge between two vertices if the corresponding two-element sets have an empty intersection. Show that this graph is isomorphic to the Petersen graph in Figure 4.7. Explain why this means that there are at least 120 isomorphisms from the Petersen graph to the Petersen graph. (In fact, there are exactly 120, but this is a bit harder to explain.)

22. Show that if every vertex of a planar connected graph has degree at least 3, then there is a face with five or fewer edges on its boundary.

23. The graph $K_{1,2,3}$ has vertices $a, b_1, b_2, c_1, c_2, c_3$ such that the b's are not connected to each other and the c's are not connected to each other, but a connects to all b's and c's, and all b's connect to all c's. Is $K_{1,2,3}$ planar?

24. The graph $K_{4,4}$ is defined analogously to $K_{3,3}$. What is the minimum number of edge crossings in any drawing of $K_{4,4}$?

25. The "integer-interval graph" $I(m,n)$, defined when $m < n$, has as its vertices all closed intervals with two distinct integer endpoints between m and n (inclusive). Two vertices are connected by an edge if their intervals have at least one point in common. For what values of n is $I(0,n)$ planar?

26. Following the notation of Exercise 19, prove that $I(m,n)$ is isomorphic to $I(j,k)$ if and only if $n - m = k - j$.

27. A graph is *regular* if all its vertices have the same degree. Give one member of each isomorphism class of regular graphs on five vertices.

28. Repeat the previous exercise for six vertices.

29. An **intersection graph** is a graph whose vertices are sets and whose edges are the pairs $\{S_1, S_2\}$ of vertices such that $S_1 \cap S_2 \neq \emptyset$. Show that every graph is isomorphic to an intersection graph. (Hint: Look at the set of "neighbors" of each vertex.)

30. An **interval graph** is an intersection graph (see the previous exercise for the definition) whose vertices are intervals on the real number line. (You may assume the intervals are closed, but this is really irrelevant.) Find an example of a graph which is not isomorphic to an interval graph and explain why.

Section 5 Digraphs

Directed Graphs

Suppose we are given a map of a downtown area where all streets are one way and we are asked to find a path for a shuttle bus in which each street is traversed exactly once. This seems to be a request for an Eulerian path; but if we simply constructed an Eulerian path on the obvious graph, we might have the bus going the wrong way on some one way street! To deal with such a problem, we introduce the notion of a directed graph. A **directed graph** (**digraph** for short) consists of a set V of *vertices* and a set D of ordered pairs (called *edges*) of elements of V. A *directed multigraph* (*multidigraph* for short) consists of a set V of vertices and a multiset D of ordered pairs of vertices.

Thus, a digraph consists of a set V and some relation on V. Pictures of digraphs are frequently useful in visualizing relations. We represent a digraph using a picture in which a point is drawn for each vertex and an arrow is drawn from a to b if the ordered pair (a, b) is in the edge set. (In particular, a "loop"—an arrow from a to itself—is drawn if (a, a) is in the edge set.) We introduced such pictures in Chapter 1. The first picture in Figure 5.1 is a picture of a digraph.

Figure 5.1

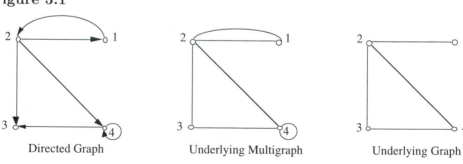

Directed Graph   Underlying Multigraph   Underlying Graph

Given a digraph, there is an obvious multigraph underlying it: the vertex set is the same as that of the digraph; for each edge (a, b) of the digraph, there is an edge $\{a, b\}$ in the multigraph. We also have an underlying graph whose edges are the two-element sets $\{a, b\}$ with (a, b) or (b, a) in the digraph edge set D. The underlying multigraph and graph are shown with the digraph in Figure 5.1. We can define connectedness, paths, walks, etc. for a digraph in terms of its underlying graph. In the case of a one way street digraph, a "walk" is allowed to traverse an edge in either direction; in cases such as a digraph of the descendants of a group of people, two people are related if they are connected in the underlying graph. Thus notions about underlying graphs are useful in analyzing directed graphs. However, they cannot be used to analyze our shuttle bus problem. Instead, we need concepts that relate to the directions as well as the connections in the digraph.

Walks and Connectivity

A *directed walk* or *strong walk* from a to b in a digraph is a sequence of vertices and edges

$$v_1(v_1, v_2)v_2 \cdots (v_{n-1}, v_n)v_n$$

with $v_1 = a$ and $v_n = b$. *Strong* or *directed paths, closed walks*, and *cycles* are defined similarly. We say a is *strongly connected* to b if there is a strong walk from a to b *and* a strong walk from b to a. It is easy to see that the relationship of being strongly connected is an equivalence relation. The equivalence classes of this relation are the "strong connectivity" classes of the directed graph.

The *indegree* of a vertex v is the number of edges in which v is the second member, i.e., the number of arrows that enter v. The *outdegree* of a vertex is the number of edges in which it is the first member. A strong Eulerian walk (which might be called an Eulerian drive) is a strong walk that includes each edge of the digraph. (A strong Eulerian walk in a multidigraph has the obvious similar definition.)

The following results are obtained by reworking the proofs of our earlier theorems about Eulerian multigraphs; thus we do not give the proofs here.

> *Theorem 5.1.* A (multi)digraph (without isolated vertices) has a closed Eulerian walk if and only if it is connected and the indegree and outdegree of each vertex are equal.

> *Theorem 5.2.* A (multi)digraph (without isolated vertices) has a nonclosed Eulerian walk if and only if it is connected, all but two vertices have equal indegree and outdegree, and of these two vertices, one has outdegree one more than its indegree and the other has outdegree one less than its indegree.

Tournament Digraphs

In a round-robin tournament, each player plays every other player exactly once. We may construct a directed graph on the set V of players by placing (x, y) in D if x beats y. The underlying graph (which is also the underlying multigraph) is complete. Thus, we define a *tournament digraph*, or simply *tournament*, as an "oriented" complete graph, that is, a directed graph on the vertex set V which contains either (x, y) or (y, x) (but not both) for every two-element subset $\{x, y\}$ of V.

In a tournament in sporting events, we wish to find a winner, or perhaps rank the players according to ability. In a three-player tournament where x beats y, y beats z, and z beats x, there is no way to decide on a winner or rank the players. A ranking of the players would be a list of all the players such that each player beats every succeeding player on the list. Although we cannot create a ranking, we can always write down a list so that each player beats the next one on the list. This corresponds to a directed path through a graph that includes every player exactly once.

Hamiltonian Paths

A path through a graph or directed graph that includes every vertex exactly once is called a **Hamiltonian path.** In general, it is very hard to find a Hamiltonian path in a graph or directed graph. In a tournament, however, it is not difficult.

> **Theorem 5.3.** If G is a tournament digraph, then it has a directed Hamiltonian path.
>
> *Proof.* Let v_1 be any vertex such that for some vertex v_2, (v_1, v_2) is an edge. This gives us two vertices and a path joining them. Suppose now we are given a directed path
>
> $$v_1(v_1, v_2)v_2 \cdots v_{i-1}(v_{i-1}, v_i)v_i.$$
>
> Let v_0 be a vertex not on this path. If there is an edge (v_0, v_1) in the graph, we can extend the original path to the path
>
> $$v_0(v_0, v_1)v_1 \cdots v_i.$$
>
> If (v_0, v_1) is not an edge, then (v_1, v_0) is an edge. If (v_j, v_0) is an edge for all j with $1 \le j \le i$, then we can extend the original path to the path
>
> $$v_1(v_1, v_2)v_2 \cdots v_i(v_i, v_0)v_0.$$
>
> If (v_j, v_0) is not an edge for all j, let k be the smallest value of j for which (v_j, v_0) is not an edge. Then (v_0, v_k) is an edge and so is (v_{k-1}, v_0). Thus, we can extend the path to
>
> $$v_1 \cdots v_{k-1}(v_{k-1}, v_0)v_0(v_0, v_k)v_k \cdots v_i.$$
>
> In this way we can continue to extend the path until it includes all the vertices of G. ∎

A tournament is said to be *transitive* if whenever (x, y) and (y, z) are edges, then so is (x, z). Note that this is equivalent to saying the relation D is transitive. If the tournament *is* transitive and $v_1, v_2, \ldots, v_n$ are the vertices listed in the order in which they appear in a Hamiltonian path, (v_i, v_j) is an edge for each $j > i$. Thus, transitive tournaments have rankings of the type we asked for earlier; in fact, they have only one such ranking.

Clearly, a tournament with a directed cycle (with more than one vertex) is not transitive.

Theorem 5.4. A tournament with no cycles is transitive.

Proof. Suppose $G = (V, D)$ is a tournament with no cycles, and let (x, y) and (y, z) be in D. Either (x, z) or (z, x) is in D, but if (z, x) is in D, then

$$x(x, y)y(y, z)z(z, x)x$$

is a nontrivial cycle. Thus G would have a cycle, so (z, x) cannot be in D. Therefore (x, z) is in D. Therefore G is transitive. ∎

Transitive Closure

In a computer program organized into related procedures, the flow of control from one procedure to another can be quite complicated. Also, data used in one procedure may or may not be passed along to a subsequent procedure when the former procedure stops and the latter starts. Other data computed in one procedure may be stored in such a way as to be accessible to some or all other procedures.

This flow of control and data gives us two digraphs associated with our program. Both have the set of procedure names of the program for their vertex set. In the "flow of control" digraph, we have an arrow from procedure P to procedure Q if P can call Q. Similarly, in the "data flow" digraph we have an arrow from P to Q if Q can access data previously accessed by P. These are special cases of communications digraphs in which vertices represent objects that can communicate with each other and arrows represent communications links along which communicators can send messages directly to other communicators. In such a situation, we are often interested in knowing who can send a message to whom along a sequence of communicators each of whom "relays" the message on. For example, if we want to modify a procedure in a way that might influence its interaction with other procedures, we need to know all the procedures that may be affected by the change. If A calls B and B calls C, a change in A may (or may not) influence C.

This leads us to the idea of the "transitive closure" of a relation or digraph. Intuitively, the transitive closure $T(R)$ of a relation R is the result of applying the transitive law (which says that if a is related to b and b is related to c, then a is related to c) again and again to the relation R. In this way, we expect to get the "smallest" transitive relation consistent with R. We can also define the transitive closure without reference to repeated applications of the transitive law. First, since a relation is just a set of ordered pairs, every relation R on a set V is a subset of at least one transitive relation, namely the set $V \times V$ of all ordered pairs of elements of V. We define the **transitive closure** $T(R)$ to be the intersection of all transitive relations containing R as a subset.

Example 5.1. The transitive relations on $\{1,2,3\}$ containg $R = \{(1,2),(2,3),(3,2)\}$ are

$$\{(1,2),(2,3),(1,3),(2,1),(3,1),(3,2),(1,1),(2,2),(3,3)\}$$
$$\{(1,2),(2,3),(1,3),(3,1),(3,2),(1,1),(2,2),(3,3)\}$$
$$\{(1,2),(2,3),(1,3),(3,1),(3,2),(2,2),(3,3)\}$$

and their intersection is $\{(1,2),(2,3),(1,3),(3,1),(3,2),(2,2),(3,3)\}$, so this the transitive closure of R. ∎

Lemma 5.5. For any relation R, $T(R)$ is transitive.

Proof. Suppose (a,b) and (b,c) are in $T(R)$. Then for each transitive relation Q containing R, (a,b) and (b,c) are in Q. Thus, (a,c) is in Q for each transitive relation Q containing R. Then (a,c) is in the intersection of all these relations Q, so (a,c) is in $T(R)$. ∎

The *transitive closure* of a digraph (V,D) is the digraph $(V,T(D))$. In our section on matrix representations of graphs, we will describe simple procedures that can be used for computing the transitive closure of a relation.

Example 5.2. The transitive closure of the directed graph in Figure 5.1 is obtained by adding the edges $(1,1)$, $(2,2)$, $(1,3)$, and $(1,4)$. ∎

Reachability

A straightforward procedure can also be based on the idea of reachability. A vertex v is **reachable** from u if there is a strong walk from u to v; v is said to be *strictly reachable* from u if there is a strong walk of length 1 or more from u to v. The only difference of reachability from strict reachability is that although x is always reachable from x, x is strictly reachable from itself only if there is a strong cycle that contains x. The *strict reachability relation* on the digraph (V,D) is the set of ordered pairs (x,y) such that y is strictly reachable from x. Notice that in Example 5.2, the edges we added to our digraph are the edges from 1 to the vertices other than 2 strictly reachable from 1 and the edge $(2,2)$ (which connects vertex 2 to the only vertex strictly reachable from vertex 2, namely itself). There are no vertices strictly reachable from vertex 3, and only vertices 3 and 4 are strictly reachable from vertex 4. Thus in this case our strict reachability relation is the edge set of our transitive closure. Our next two lemmas allow us to show that the strict reachibility relation and the transitive closure of a digraph are always identical.

Lemma 5.6. The strict reachability relation is transitive.

Proof. If there is a strong walk of length at least one from x to y and a strong walk of length at least one from y to z, then there is a strong walk of length at least one from x to z. ∎

Lemma 5.7. If T is a transitive relation containing D, then T contains the strict reachabality relation of the digraph (V, D).

Proof. If (x, y) is in the strict reachibility relation of (V, D), then there is a sequence of vertices $x_0 = x, x_1, x_2, \ldots, x_{n-1}, x_n = y$ with $n \geq 1$ such that each (x_i, x_{i+1}) is in D. Applying the transitive law to the sequence of x_i's shows that (x, y) is in T. ∎

Theorem 5.8. The strict reachability relation and the transitive closure of a digraph are identical. That is, the edge (u, v) is in the transitive closure of the digraph $G = (V, D)$ if and only if v is strictly reachable from u in G.

Proof. The theorem is an immediate consequence of Lemma 5.6 and Lemma 5.7. ∎

Modifying Breadth-First Search for Strict Reachability

In graphs or multigraphs, the ideas of reachability and connectivity are equivalent. Recall that breadth-first search is an efficient method of finding all vertices connected to a vertex x. The same sort of search can be used in directed graphs; a slight modification ensures that we are searching for vertices strictly reachable from x. We say that x is *adjacent to* y and y is *adjacent from* x if (x, y) is in D. A strict breadth-first search from x consists of the following steps:

(1) Let $x_0 = x$, $V_0 = \emptyset$, and $E_0 = \emptyset$.
(2) Given pairs (V_i, E_i), $i = 0, 1, 2, \ldots, k-1$, let j be the smallest number such that x_j is adjacent to some vertex z in V but not V_{k-1}. Let $x_k = z$, $V_k = V_{k-1} \cup x_k$, and $E_k = E_{k-1} \cup (x_j, x_k)$.
(3) Repeat step 2 until there is no x_j of the type described.

(4) Let $G' = (V_n \cup x_0, E_n)$, where n is the largest value of k encountered in step 2.

Notice that we begin the process with the empty set as the vertex set of the graph we are building. In step 4, if x_0 has not been put into the vertex set because there is no nontrivial walk from x to x, then we put it into the vertex set. The underlying graph of the final digraph we get from this process may or may not be a tree, because the final digraph we get may have a strong cycle from x to x. The vertices $x_1, x_2, \ldots, x_n$ listed (which might or might not include x_0) are all the vertices strictly reachable from x. Thus, (x, x_i) is in $T(D)$ for each i. Therefore, we can construct $T(D)$ by doing a strict breadth-first search from x for each x in V and placing all the edges we get in the process into the edge set of $T(D)$. Each application of the breadth-first search might mean examining virtually all the edges in D, and $|D|$ could be as large as $|V|^2$, so in computing the transitive closure of (V, D) by this method, we might require $|V| \cdot |V|^2 = |V|^3$ steps.

Orientable Graphs

Suppose we are given a street map of a city. Is there a way to assign directions to the streets to make them one way so it is possible to get from any one point to any other point? If not, how do we find the streets which must be two way? Can we perhaps assign directions so that we can drive around town, traversing each street exactly once? This last question is the easy one, for it asks for nothing more than an Eulerian walk. The first two questions are more like questions that could come up in planning for a city or a communications or railroad network. (Questions of *efficient* assignments of one way streets have partial answers in terms of the "diameter" of a digraph; we shall not investigate these questions in the text.)

An assignment of directions to the edges of a graph so that the resulting graph is strongly connected is called an ***orientation*** of the graph. Our first question asked how we can tell if a graph has an orientation. There is one obvious obstacle to orienting a graph. Recall that an edge of G is called a ***bridge*** or *isthmus* if its removal from the graph disconnects the graph. If a graph has a bridge $\{x, y\}$, then once we have chosen a direction—say (x, y)—for this edge, its endpoints will no longer be strongly connected. Thus a graph with a bridge has no orientation.

Graphs without Bridges Are Orientable

It may be surprising that if a graph has no bridge, then it has a strong orientation; thus the answer to the first question is "Look for bridges. If there are none, then the graph has a strong orientation." This means we can answer the second question above as well—if we let all bridges be two way streets, then the graph has a strong orientation.

Theorem 5.9. A graph without bridges has an orientation.

Proof. Construct a depth-first search tree T in G, and suppose that the list $x_0, x_1, x_2, \ldots, x_n$ is a list of the vertices of G in depth-first order. Direct T by increasing depth-first number; that is, direct each tree edge from the vertex with lower depth-first number to the vertex with higher depth-first number. Direct every other edge of G from higher to lower depth-first number. Notice that all tree-descendants of a vertex are reachable from that vertex. Then by Theorem 3.8, from each vertex (except x_0) we can reach a vertex with lower depth-first number. Therefore from each vertex in G, we can now reach x_0. But from x_0 we can reach any vertex in T. Thus, from any one vertex we can reach any other vertex. ∎

Example 5.3. Notice that in Figure 3.4 of Section 3, the edges of the tree, shown as arrows, are directed by increasing depth-first number. When we direct all other edges by decreasing depth-first number, then from v_1 through v_5 we may reach v_0 by traveling along the cycle $v_0 v_1 v_2 v_3 v_4 v_5 v_0$, and from v_9 we may reach v_6, so that we can reach v_{12} from any vertex with depth-first number between 6 and 12. Thus we may reach v_0 from any of these vertices. Therefore, as in the proof of the theorem, we may reach any vertex from any other vertex. ∎

EXERCISES

1. How many directed graphs are there with the 3-element vertex set $\{a, b, c\}$? A 10-element vertex set? (How does the presence or lack of loops affect your answer?)

2. (a) What is the maximum number of edges in a directed graph on the set $N = \{1, 2, \ldots, n\}$?
   (b) What is the number of digraphs on the set $N = \{1, 2, \ldots, n\}$?

3. How many tournaments are there on the vertex set $\{1, 2, \ldots, n\}$?

4. Show that if a tournament has a ranking $v_1, v_2, \ldots, v_n$ so that (v_i, v_j) is an edge whenever $i > j$, then the tournament is transitive.

5. How many transitive tournaments are there on the set $\{1, 2, \ldots, n\}$?

6. Which digraphs in Figure 5.2 have Eulerian walks? Closed Eulerian walks?

7. Explain why the sum of the indegrees of the vertices in a directed graph equals the sum of the outdegrees of the vertices.

Figure 5.2

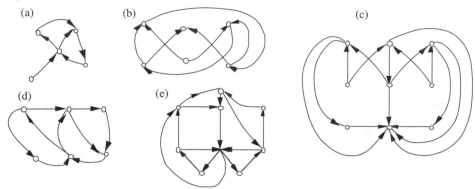

8. Around the circle in Figure 5.3 there are eight zeros and ones. By taking each of the eight intervals of three consecutive places around the circle in turn, we encounter each list of three zeros and ones exactly once. Show that 2^n zeros and ones can be arranged around a circle so that each possible list of n zeros and ones occurs as a list of n consecutive digits in the sequence. (In fact, DeBruijn showed that the number of such arrangements is $2^{2^{n-1}-n}$. This is difficult to prove.) (Hint: Start with a digraph whose vertices are lists of $n-1$ zeros and ones. Draw an arrow from vertex 1 to vertex 2 if the last $n-2$ digits of the first equal the first $n-2$ digits of the second. Now find an Eulerian path. Try to relate the Eulerian path to the list of numbers around the circle as in Figure 5.3.)

Figure 5.3

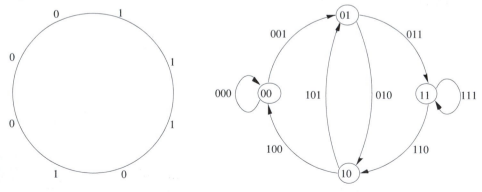

9. Prove Theorem 5.1.

10. Prove Theorem 5.2.

11. Prove that a tournament is transitive if and only if there is one and only one ranking of the vertices (that is, there is one and only one Hamiltonian path).

12. Heavy trucks are allowed only on certain roads in a city, and many of these roads are one way. A heavily laden truck must make stops at n stores. Show that if it is possible to go either from store X to store Y or from store Y to store X (but not necessarily both) for each pair of stores (without passing other stores), then there is a route which goes by each store only once and passes none by along the way.

13. (Due to C. L. Liu) n books are to be printed and bound. There is one printing machine and one binding machine. Let p_i and b_i denote the printing and binding times for the ith book. If it is known that either $p_i \leq b_j$ or $p_j \leq b_i$ for all i and j, show that there is an order in which each book may be first printed and then bound in such a way that the binding operation is kept busy from the time the first book is printed until the last book is bound.

14. Show that a tournament is transitive if and only if it has no directed cycles of length 3.

15. Show that the transitive closure of a digraph $G = (V, D)$ contains no cycles if and only if G contains no cycles.

16. Find the transitive closure of the digraphs in Figure 5.4.

Figure 5.4

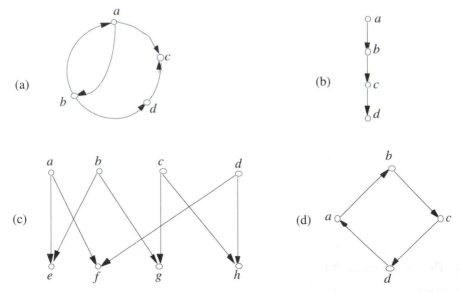

Figure 5.5

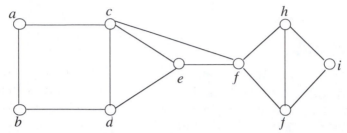

17. Find an orientation of the graph in Figure 5.5.

18. What is the transitive closure of a strongly connected graph?

19. What is the transitive closure of a directed cycle on n vertices?

20. Prove that in a tournament there is a vertex v such that for *each* x in V, either (v, x) is an edge or, for some y, (v, y) and (y, x) are edges. (This situation is described intuitively by saying that the vertex v dominates each other vertex either directly or with at most one intermediary.)

21. For each positive integer n, there is a number k such that if a digraph on n vertices has k or more edges then it has a directed cycle. What is k when $n = 4$? When $n = 10$?

22. Show that a digraph is strongly connected if and only if it has a strong closed walk that contains all the vertices.

23. Show that a finite digraph with no directed cycles has at least one vertex of outdegree zero.

24. Show that if the vertex v is reachable from the vertex u, then there is a strong path from u to v.

25. What is the maximum number of directed cycles in a strict breadth-first search graph? When will we have this number of cycles and why?

Section 6 Coloring

The Four-Color Theorem

The four-color theorem has a long and colorful history. Originally a statement about maps drawn on a plane surface or a sphere, it is now usually stated as a theorem about planar graphs. On a map, it is desirable to color different political entities which share a common boundary with different colors. As long as each entity is connected (you can move between any two points without leaving it), it is possible to get such a "proper coloring" of a planar map with four colors. This led mathematicians to ask whether this observation was in fact a theorem. Apparently the first to ask the question explicitly was Francis Guthrie, whose brother Frederick passed the question along to Augustus DeMorgan (of DeMorgan's laws) in 1852. DeMorgan's questions to others, particularly to Cayley, who discussed it in a major lecture in 1878, started to popularize the problem. About a year after Cayley's lecture, the mathematician Kempe devised what appeared to be a proof of the theorem, but 11 years after his proof, Heawood showed that though Kempe's method provided a valid proof that a planar map could be properly colored with five colors, there was an oversight that left the four-color problem unsolved. This open problem led to a great deal of new mathematics as people tried to answer the question, "Do four colors suffice?"

In the early 1970s, Apel and Haken developed a strategy, based on work begun by George David Birkhoff in 1913 and refined by Heinrich Heesch over the previous 30 to 40 years, of classifying the kinds of problems that could occur in Kempe's approach into "configurations" and showing that these configurations could be reduced to situations where the four-color problem had already been solved. They developed methods of checking for reducibility that could be programmed into a computer and set a computer to work. In 1976 they announced that any planar map can be colored with four colors; their computer work had isolated a complete set of reducible configurations. For years afterward, postage meters at the University of Illinois, their home institution, carried the legend "four colors suffice." We shall study some of the ideas that developed from the four-color problem over the years.

Chromatic Number

A proper coloring of a graph is an assignment of "colors" to the vertices in such a way that vertices connected by an edge receive different colors. More precisely, it is a function f from the vertex set V to a set C so that if $\{x, y\}$ *is an edge*, then $f(x) \neq f(y)$. The smallest size of a set C for which there is a proper coloring is called the ***chromatic number*** of G.

Theorem 6.1. If the largest degree of a vertex in a graph G is k, then the chromatic number of G is no more than $k + 1$.

Proof. Pick an arbitrary vertex, and assign it any color from a set C of size $k + 1$. Now, once j vertices are colored, pick vertex $j + 1$ and assign to it any color not already assigned to any of its neighbors. (Why is this possible?) Continue until colors are assigned to all vertices. ∎

Maps and Duals

The relation between map coloring and graphs is based on the adjacency graph of a map. On a piece of paper, we use a point to represent each region of the map and connect points representing two regions if they share a boundary line. We may draw an edge for each shared boundary line, in which case we get a multigraph. (Regions that touch only at a point are not connected by an edge in the graph.) As we shall see, a planar map yields a planar graph. A proper coloring of the graph corresponds to a map coloring in which regions that share boundary lines get different colors. If it happens that the map's regions are the faces of a drawing of a multigraph G, then the new multigraph we have constructed is called a **dual** to G. Note that any planar drawing of a graph may be regarded as a map, so that each planar drawing of G yields a dual to G.

On the other hand, coloring planar graphs is no more general a topic than coloring planar maps. To see this, we show that for each planar graph G, we can construct a map whose adjacency graph is G.

The construction of maps from graphs is intuitive and similar to the construction of graphs from maps. A detailed description of the process follows; the process is illustrated in Figure 6.1.

Figure 6.1

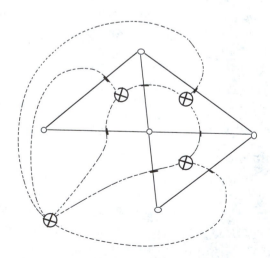

We begin with a planar graph drawn in the plane without crossings. We place a point (shown in Figure 6.1 as a circled cross) inside each face of the graph (including the outside face) and mark a point on each edge of the drawing between the endpoints. Now from the chosen point inside each face, draw line segments (shown as dashed in Figure 6.1), curved if necessary or convenient, to the marked point on each edge of the face. (Note that, since each face is a polygon, these lines can be drawn so as not to cross.) Now each vertex of the graph is surrounded by a polygon whose corners are at the previously chosen points in the faces and edges. These polygons form the regions of a planar map, and the adjacency graph of the planar map is the original graph.

Notice how we have distinguished between the points chosen inside faces and the points chosen along edges in Figure 6.1. If we think of the circled points as new vertices and the dotted lines which cross over the old edges as new edges, ignoring the marks we made on the old edges, then the new vertices and new edges give us a planar drawing of a dual to our original graph. If our original drawing had been a planar map rather than a planar graph, then essentially the same process would give us a planar drawing of the adjacency graph of the map as suggested earlier.

Note that one of the points of the original graph in Figure 6.1 corresponds to the infinite region of the dual graph. There are two ways to reconcile this idea of an infinite region with our usual idea of a map (where all the faces are finite). One is to recognize that this region is finite once the picture is drawn, because it is drawn on a finite piece of paper. The other is to realize that in addition to all the regions in the map to be colored, there is a boundary area enclosing the regions, and we wish to treat that boundary exactly as a region, i.e., color it a different color from any region it touches. In fact, drawings in a plane and on the surface of a sphere are equivalent, and on a sphere, the "outside" region of the planar drawing is just another region.

The Five-Color Theorem

We shall prove that every planar graph (and thus every planar map) can be colored with five colors. We shall translate the essence of Kempe's attempt at proving the four-color theorem into the language of graph theory because it is somewhat easier to be precise with graph-theoretic terminology. We begin with a result that says we don't have to examine all graphs.

Theorem 6.2. If each n-vertex planar graph all of whose faces are triangles has a proper coloring with $c > 1$ colors, then each n-vertex planar graph has a proper coloring with c colors.

Proof. Removing or adding a vertex of degree 1 doesn't change the fact that the graph can be colored with c colors if $c > 1$. Thus, we assume our graphs have no vertices of degree 1.

Let S be the set of all integers k such that if an n-vertex planar graph has k nontriangular faces, then it has a proper coloring with c colors. According to the hypothesis of the theorem $k = 0$ is in S. Now suppose $k \neq 0$ and $k - 1$ is in S. Let G be a graph with k nontriangular faces and let $v_1, v_2, \ldots, v_r$ be the vertices on the edges of one nontriangular face. Since we have no vertices of degree 1, the boundary of the face is a polygon. Then within that face, it is possible to draw (possibly curved) lines from v_1 to $v_3, v_4, \ldots, v_{r-1}$ so that none of the lines cross. This gives a planar drawing of the graph G' whose edge set is the edge set of G together with the edges $\{v_1, v_i\}$. G' has $k-1$ nontriangular faces and so has a proper coloring with c colors. However, the same assignment of colors provides G with a proper coloring. By version 1 of the principle of mathematical induction, S contains all positive integers; this proves the theorem. ∎

A second helpful result is an exercise in Section 4.

Theorem 6.3. A planar graph has at least one vertex of degree less than 6.

Proof. Exercise in applying $e \leq 3v - 6$. ∎

Theorem 6.4. A planar graph has a proper coloring with five or fewer colors.

Proof. Let S be the set of all integers n such that planar graphs with n vertices have proper colorings with five colors. Clearly, $n \in S$ if $n \leq 5$. Suppose now that $n > 5$ and $n - 1$ is in S. Suppose G is an n-vertex planar graph whose faces are all triangles and let x be a vertex of G of degree less than 6.

If x has degree 4 or less, let G' be the $(n-1)$-vertex graph obtained from G by deleting x and the edges including it. Then G' has a proper coloring with five colors. However, only four of these colors are used on vertices that share an edge with x in G. Thus, we may properly color G by assigning the fifth color to x and coloring the remaining vertices as in G'.

Now if x has degree 5, suppose $\{x, v_1\}$ is one of the edges containing x, and let $\{x, v_2\}$, $\{x, v_3\}$, $\{x, v_4\}$, $\{x, v_5\}$ be the remaining edges arranged in clockwise fashion around x in a (fixed) planar drawing. (See Figure 6.2.) Thus, $\{x, v_i\}$ and $\{x, v_{i+1}\}$ both lie on the boundary of a common face, as do $\{x, v_1\}$ and $\{x, v_5\}$. Since these faces are triangles, $\{v_i, v_{i+1}\}$ (for $i \leq 4$) and $\{v_1, v_5\}$ are all edges of G. Now let G' be the graph obtained from G by deleting the vertex x and the edges $\{x, v_i\}$. Since $n-1$ is in S, Theorem 6.2 tells us that G' has a proper coloring with five colors. If not all five of these colors are used up by the vertices $\{v_1, v_2, v_3, v_4, v_5\}$, then as before we may color G by assigning the unused color to x and coloring the remainder of G as G' is colored. Suppose now the five colors R, B, G, Y, and W are assigned to v_1 through v_5, respectively.

Figure 6.2

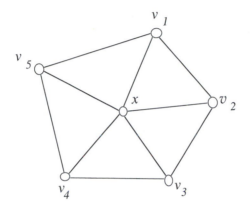

We now define two equivalence relations on the vertex set of G. They will have the property that an appropriate pair of inequivalent vertices may be colored the same color, giving a free color to use on x. We say y and z are $R-G$-equivalent if there is a path connecting y and z in G' all of whose vertices are colored R or G. Such a path is illustrated in Figure 6.3. Let V_1 be the set of all vertices $R-G$-equivalent to v_1. Then any vertex in V_1 is colored R or G, and if a vertex is in V_1, the vertices with which it shares edges *outside* V_1 are colored either B, Y, or W. Thus, if v_3 is not in V_1, we may interchange the assignments of the colors R and G to the vertices in V_1. Now we have a proper coloring of G' using only four colors on the v_i's, and so we can five-color G as before.

Now define two vertices y and z to be $B-Y$-equivalent if there is a path between them in G' all of whose vertices are colored B or Y. Define V_2 to be the $B-Y$ class containing v_2. If v_4 is not in V_2, we may interchange the colors in V_2 to obtain a proper coloring of G' in which only four colors are used on the v_i's. Thus, we can five-color G as before.

Now suppose v_3 is in V_1, so that there is a path from v_1 to v_3 consisting of vertices colored only R or G. (See Figure 6.3.) We

Figure 6.3

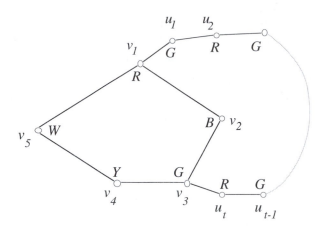

have two polygons in G' that, taken together, surround v_2, namely the pentagon $v_1 v_2 v_3 v_4 v_5$ and the closed path formed by v_1, v_2, v_3, and the other vertices of the "$R - G$ path" connecting v_1 and v_3; in Figure 6.3 we call these other vertices $u_1, u_2, \ldots, u_t$. Thus, any line segment emanating from v_2 must go either into or along an edge of one of these polygons. Since the pentagon of v's is a face of G', no edge can go inside it. Thus, any path from v_2 to v_4 must contain at least one vertex of the $R - G$ path. Then any path from v_2 to v_4 must contain a red (R) or green (G) vertex. Therefore, there is no $B - Y$ path from v_2 to v_4, i.e., v_2 and v_4 are $B - Y$-inequivalent. Thus, the graph G may be properly colored as above. ∎

Kempe's Attempted Proof

The argument here is essentially the one used by Kempe in his attempt at proving the four-color theorem. It should be clear that in the case of four colors, we would have to examine a vertex x of degree 4 and a vertex x of degree 5 separately. If there is a vertex of degree 4 it is surrounded by a quadrilateral $\{v_1, v_2, v_3, v_4\}$. If four colors are used on the quadrilateral we show we may use only three of them instead by using $R - G$ and $B - Y$ equivalence relations as above. In the degree 5 case, if all four colors appear among the vertices v_i, we can reduce the problem to the case where the W in Figure 6.3 is a B. Kempe then argued that either v_1 and v_4 are in the same $R - Y$ class or else we can interchange the R's and Y's in the class containing v_1 and extend as before to a proper coloring of G with four colors. (Similarly, we could make one $R - G$ switch unless v_1 and v_3 are in the same $R - G$ class.) In the case that v_1 and v_4 *are* connected by a path of vertices colored R or Y, and similarly v_1 and v_3 are connected by a path of vertices colored R or G (see Figure 6.4 for *one* possible picture of this situation), Kempe said we should change colors in both the $B - G$ class containing v_5 and the $B - Y$ class containing v_2. In this way we get three colors on the pentagon and can proceed as before. It is instructive to find the error here. As a hint, Kempe was fooled by a diagram equivalent to this one that was really too simplified to show everything that could happen.

Figure 6.4

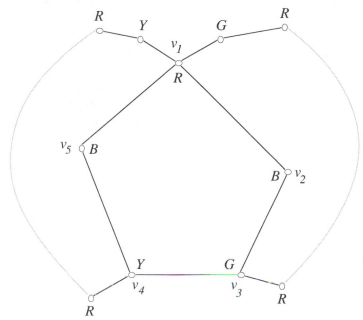

Using Backtracking to Find a Coloring

We have introduced no systematic way to find a proper coloring using a minimum number of colors or to find the chromatic number of a graph. In fact, determining whether the chromatic number of a graph is k (for any k greater than 2) is one of the problems known as NP-complete problems, a family of problems, none of which can currently be solved quickly, with the property that if any one of the problems can be solved quickly, then so can all the others. Other members of the family include determining whether a graph is Hamiltonian, determining whether a set of numbers has a subset whose members add to a given value, and determining whether one graph is isomorphic to a subgraph of another. (Curiously, at present it is not known whether determining whether two graphs are isomorphic is also in this family.) Thus, we expect that finding a coloring with a minimum possible number of colors could, for some graphs, be quite time consuming. Although this is true, there are ways to use backtracking in assigning colors to a graph, so as to explore possible colorings systematically until a proper one is found.

Recall that in backtracking, we are applying depth-first search to a decision graph, a graph that represents possible sequences of decisions we might make in trying to find a step-by-step solution of a problem. The step-by-step method we apply here is to decide on colors for vertices one at a time, constructing a sequence of "partial colorings" of our graph until we find a proper coloring or discover there is no way to make the sequence any longer without coloring adjacent vertices the same color. Our first example of a decision graph was the decision graph of a maze. The maze example is slightly misleading in that its decision graph is the same size as the maze. It is much more typical for the decision graph to be huge even if we start with what appears to be a simple or trivial problem. It is often the case that the decisions we make build up larger and larger partial solutions to a problem. In this case it is natural to take the sequences of decisions as the vertices of the decision graph and draw an edge from from one vertex to a second if the second is obtained from the first by making one more decision. Notice that we have described this edge as going from one vertex to another; that is, we have described a directed edge. Thus, in these cases we study a *decision digraph* whose vertices represent decision sequences and whose edges represent the process of adding one decision to a sequence.

Example 6.1. How many vertices are in the decision digraph for deciding, one vertex at a time, on a proper coloring of a cycle on the four vertices $\{1, 2, 3, 4\}$ with two colors, R and B?

Our initial vertex (the situation before we start coloring) represents the graph with no colors assigned. It is adjacent to eight vertices representing the assignment of one color to one vertex. There are $2 \cdot 4 = 8$ vertices representing the assignment of two colors to two adjacent vertices and another $2 \cdot 4 = 8$ vertices representing the assignment of one or two colors to two opposite vertices. There are $4 \cdot 2 = 8$ vertices representing assignments of two colors to three of the four vertices and two vertices representing the two final proper colorings. Thus, we have $1 + 8 + 16 + 8 + 2 = 35$ vertices in our decision graph. ∎

A depth-first search quickly produces both colorings but examines all 35 vertices to produce them.

Example 6.2. Using the decision rules, "choose red first" and "color the lowest-numbered available vertex first," carry out a depth-first search on the decision graph to color a four-cycle with two colors.

To describe the situation in which vertex 1 is colored red and no other vertex is colored, we use the notation $1R$. To describe the situation in which vertex 1 is colored red and vertex 2 is colored blue and no other vertex is colored, we write $1R2B$. To say compactly that the second situation follows the first, we write $1R/1R2B$. We add more situations to this list in the order in which they arise from our decision rules and backtracks. We will observe the first few colorings in the following order ($\emptyset$ here means nothing is colored).

$$\emptyset/1R/1R2B/1R2B3R/1R2B3R4B.$$

You may wish to draw the colored graphs that correspond to these sequences.

Note that this sequence of colorings has led us to a proper coloring. Even though we have found one proper coloring, we shall keep searching the decision digraph to look for others. Since we can make no further decisions at this point, we back up along the edges we have used so far to the last place where we had a choice of decisions. Since the decision "4B" was the only one we could add to the sequence $1R2B3R$, we must back up to the sequence $1R2B$. Although we made the only decision possible for vertex 3 at this point, we did not consider vertex 4 because we were taking vertices in numerical order. Since there are no new decisions we can make for vertex 3, we take the next vertex in numerical order, namely 4, and ask what decisions we may make for it. The only possibility is to color it blue since vertex 1 is red, giving us the coloring $1R2B4B$, so that our sequence of colorings is now

$$\emptyset/1R/1R2B/1R2B3R/1R2B3R4B/1R2B4B.$$

Now the only decision we could make is $3R$, which returns us to a previous vertex of the decision digraph, so there are no allowable decisions to add to this sequence. This forces us to backtrack once again to the last place where we could have made a different decision. Since we have explored all decisions we can make from $1R2B$, we now examine what decisions we may still make from the vertex $1R$. If we examine vertex 2 next, we see we could only color it blue, giving us $1R2B$, a vertex already in the decision digraph. Thus, we move on to vertex 3 of the cycle and by the "use red first" rule, we color it red, getting the sequence $1R3R$. Since the decision $2B$ leads us to a previously examined vertex in the decision digraph, we make the decision $4B$, giving us $1R3R4B$. Thus our sequence of vertices in the decision digraph has become

$$\emptyset/1R/1R2B/1R2B3R/1R2B3R4B/1R2B4B/1R3R/1R3R4B.$$

At this point the only possible decision, $2B$, leads us to a previously examined vertex of the decision graph, and so we backtrack, this time to the decision $1R/3R$, which we now change to $1R3B$, another sequence that we cannot extend without giving the same color to two adjacent vertices of the cycle. In this way we continue to get the colorings

$$\emptyset/1R/1R2B/1R2B3R/1R2B3R4B/1R2B4B/1R3R/1R3R4B/$$
$$1R3B/1R4B/1B/1B2R/1B2R3B/1B2R3B4R/1B2R4R/1B3R/$$
$$1B3B/1B3B4R/1B4R/2R/2R3B/2R3B4R/2R4R/2R4B/2B/$$
$$2B3R/2B3R4B/2B4R/2B4B/3R/3R4B/3B/3B4R/4R/4B.$$

Notice how early in the search the two proper colorings are reached. As you examine the sequence we produced, you are likely to find some ways of not making certain decisions on the ground that they would have to lead us back to vertices of the decision digraph we had already colored. One of the important topics discussed in a course in the design and analysis of algorithms is "pruning" of search trees to avoid this kind of irrelevant information. Notice also that if we want just one sequence of legal decisions rather than all the legal sequences, depth-first search should—with some exceptions—yield at least one legal sequence fairly quickly. This happens because our first priority in building a search tree is making a new decision whenever we legally can (rather than exploring all legal decisions simultaneously). Thus, we expect to get fairly long sequences of legal decisions quickly. ∎

EXERCISES

1. What is the chromatic number of the graph $K_{3,3}$? $K_{2,4}$?

2. Describe all graphs on n vertices with chromatic number n.

3. What is the chromatic number of a cycle with an even number of vertices? With an odd number of vertices?

4. What is the chromatic number of a tree?

5. A "wheel" consists of a cycle together with one more vertex connected to all vertices of the cycle. Analyze the possibilities for the chromatic number of a wheel.

6. What is the dual of a cycle?

7. Show that a wheel (see Exercise 5) is isomorphic to a graph which is its dual.

8. Find the error in Kempe's proof of the four-color theorem.

9. Prove the fact that a planar graph has a proper coloring with six colors *without* appealing to the four-color theorem or five-color theorem.

10. A college registrar must assign final exam periods to courses giving scheduled final exams in such a way that two courses having a student in common are scheduled at different times. Describe a graph whose chromatic number is the number of time slots needed.

11. A set of vertices in a graph G is called ***independent*** if no two of these vertices are joined by an edge. The *independence number* of a graph is the maximum of the sizes of its independent sets. Show that the product of the chromatic number and independence number of a graph is at least the number of vertices.

12. Show that a graph can be colored properly with two colors if and only if it contains no cycles of odd length. What does this have to do with graphs whose vertex set is a union of two independent sets (these are called *bipartite graphs*)?

13. In how many ways may we color properly the graph G with x colors if
    (a) G is an n-vertex path?
    (b) G is an n-vertex complete graph?
    (c) G is a star on n vertices, that is, a graph with one vertex connected to $n-1$ other vertices, each connected to nothing else?

14. Give an inclusion–exclusion argument to explain why the number of ways to color properly a graph G with x colors is a polynomial function of x (which, of course, may be different for different graphs). (To get started, observe that the total number of colorings is just a power of x. Now ask yourself what bad properties some of these colorings might have.) This polynomial is called the ***chromatic polynomial*** of the graph and is denoted by $\chi_G(x)$.

15. We have defined an interval graph to be a graph whose vertices are intervals of real numbers with an edge between two intervals if and only if they have a nonempty intersection. Show that the chromatic number of an interval graph is k if and only if the graph has k mutually adjacent vertices but not $k+1$ mutually adjacent vertices (in other words, if and only if the largest complete subgraph, also called the largest clique, of G has size k).

16. Show that if the longest odd cycle in a graph has n vertices, then the chromatic number of the graph is no more than $n+1$.

17. Show that $\chi_T(x)$, the number of ways to color properly a tree T on n vertices with x colors, is $x(x-1)^{n-1}$.

*18. (Brooks) Prove that a graph which is connected and not complete and whose vertices of largest degree have degree $c > 2$ can be colored properly with c colors.

19. Either prove or give a counterexample: Connected nonplanar graphs on six vertices are not four colorable.

20. In Exercise 25 of the Section 4, we introduced the integer-interval graph $I(0, n)$. For what values of n is $I(0, n)$ four colorable?

21. What is the chromatic number of the subgraph of $I(0, n)$ (see Exercise 25, Section 4) whose vertices are intervals of length 2?

22. What is the chromatic number of the subgraph of $I(0, n)$ whose vertices are intervals of length 1 or 2?

23. Show that if a planar graph has fewer than 30 edges, then it has a vertex of degree 4 or less. Prove from this that a planar graph with 30 or fewer edges has a proper coloring with four colors.

*24. A well-known and often discovered result in graph theory is that there is, for any n, a graph with no triangles and chromatic number n. (For $n = 3$, a cycle on five vertices is an example.) Find a proof of this result.

25. A graph is *n-color critical* if it has chromatic number n but the deletion of any vertex produces a graph of chromatic number $n - 1$. Give two examples of four-color critical graphs. Show that a graph of chromatic number n contains an n-color critical subgraph.

26. Draw the decision digraph described in Example 6.1.

27. Draw four copies of a cycle with the vertices 1, 2, 3, and 4 arranged in a circle and show the first four nonempty colorings of the graph described in the sequence in Example 6.2.

28. Find the number of vertices in the decision digraph for analyzing colorings of a five-vertex cycle using three colors.

29. Using the decision rules "red, then green, then blue" and "color the lowest-numbered vertex available first" write out the sequences of colorings you discover in using backtracking until you have found two different proper colorings of a cycle on $\{1, 2, 3, 4, 5\}$ using red, blue, and green.

*30. Show that the number of ways to color properly an n-vertex cycle with x colors is $(x-1)((x-1)^{n-1} - (-1)^{n-1})$.

Figure 6.5

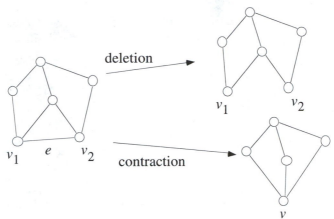

31. To delete an edge e from $G = (V, E)$ means to construct the graph $G - e = (V, E - \{e\})$. Informally, to contract an edge from G means that we remove that edge, and consider its endpoints to be the same vertex. Thus, to *contract* an edge $e = \{v_1, v_2\}$ from a graph whose vertex set does not contain a vertex named v means to construct the graph G/e whose vertex set is $V - \{v_1, v_2\} \cup \{v\}$ and whose edge set consists of E with all edges containing v_1 or v_2 removed and with each edge of the form $\{v_i, y\}$ in which $y \notin \{v_1, v_2\}$ replaced by the edge $\{v, y\}$. (Repeated edges are ignored.) (See Figure 6.5.) Show that if $\chi_G(x)$ denotes the number of ways to color G with x colors (i.e., it is the chromatic polynomial of G), then

$$\chi_G(x) = \chi_{G-e}(x) - \chi_{G/e}(x).$$

Explain how this would let us give a second proof that $\chi_G(x)$ is a polynomial in the variable X.

32. Color critical graphs were defined in Exercise 25. Show that the minimum degree of the vertices in an n-color critical graph is at least $n - 1$.

33. The symbol $\delta(G)$ is usually used to stand for the minimum of the degrees of the vertices of the graph G. Show that the chromatic number of a graph G is the maximum value that $1 + \delta(H)$ takes on for all subgraphs H of G.

34. Rather than coloring vertices of graphs we are sometimes interested in coloring edges. For example, there is a description of the Ramsey numbers we introduced at the end of Section 2 of Chapter 1 in terms

of coloring the edges of the complete graph on n vertices. Namely, the Ramsey number $R(m, n)$ is the smallest number r such that if we color the edges of a complete graph on n vertices with two colors, say A and S, then there is either a complete subgraph of size m all of whose edges are colored A or there is a complete subgraph of size n all of whose edges are colored S. Using graph-theoretic terminology, show that $R(3, 3) = 6$.

35. Restate Theorem 5.1 of Section 5 of Chapter 1 as a theorem about coloring edges of grpahs, and rewrite its proof in graph-theoretic terminology.

36. One advantage of the definition of Ramsey numbers in the previous paragraph is that it suggests natural examples of extensions of Ramsey numbers. Give a definition of $R(m, n, k)$ that involves coloring edges of a complete graph with three colors. What is $R(2, 2, 2,)$? $R(2, 2, 3)$? $R(2, 3, 3)$?

Section 7 Graphs and Matrices

Adjacency Matrix of a Graph

We assume you are familiar with the concepts of matrices, matrix addition, and matrix multiplication. If a digraph G has vertices $v_1, v_2, \ldots, v_n$, then it can be represented by an **adjacency matrix** $A(G)$ whose entry $A(G)_{i,j}$ in row i and column j is 1 if (v_i, v_j) is an edge and whose (i, j)th entry is 0 if (v_i, v_j) is not an edge. Since most computer languages allow for the storage and manipulation of matrices, an adjacency matrix provides a convenient way of storing information about a digraph in a computer. In Figure 7.1, we show a digraph and its adjacency matrix. Note that if we reassigned the symbols v_1 through v_n to the graph in a different way, we could get a different matrix. Thus, each numbering of the vertices gives a potentially different matrix as the adjacency matrix.

Figure 7.1

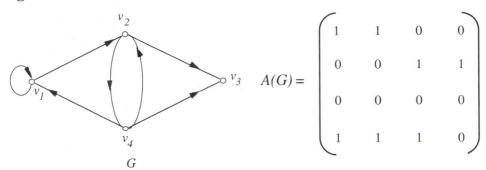

$$A(G) = \begin{pmatrix} 1 & 1 & 0 & 0 \\ 0 & 0 & 1 & 1 \\ 0 & 0 & 0 & 0 \\ 1 & 1 & 1 & 0 \end{pmatrix}$$

A multidigraph with vertices v_1 through v_n can be represented by an **adjacency matrix** whose (i, j)th entry is the number of times (v_i, v_j) appears in the edge multiset.

The **adjacency matrix** of a graph is defined similarly; we let $A(G)_{ij}$ be 1 if $\{v_i, v_j\}$ is an edge and 0 otherwise. Since saying that $\{v_i, v_j\}$ is an edge is the same as saying that $\{v_j, v_i\}$ is an edge, this means that $A(G)$ will be symmetric. The fact that a graph has no loops means $A(G)$ will have zeros along the main diagonal. For multigraphs, we let $A(G)_{ij}$ be the multiplicity of edge $\{i, j\}$. A graph and its matrix are shown in Figure 7.2.

Matrix Powers and Walks

Matrix operations help us determine certain natural properties of graphs. Recall that the *length* of a walk in a (multi)(di)graph is the number of edges in the walk.

Figure 7.2

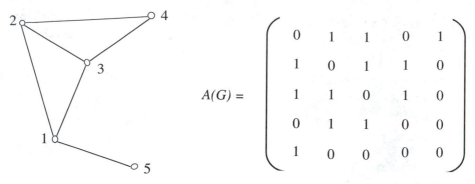

$$A(G) = \begin{pmatrix} 0 & 1 & 1 & 0 & 1 \\ 1 & 0 & 1 & 1 & 0 \\ 1 & 1 & 0 & 1 & 0 \\ 0 & 1 & 1 & 0 & 0 \\ 1 & 0 & 0 & 0 & 0 \end{pmatrix}$$

Theorem 7.1. If D is an n-vertex (multi)(di)graph with adjacency matrix A, then the (i,j)th entry of A^k is the number of (directed) walks of length k from vertex i to vertex j in D.

Proof. Clearly, A_{ij} is the number of walks of length 1 from v_i to v_j. Let S be the set of all integers m such that $(A^m)_{ij}$ is the number of (directed) walks of length m from v_i to v_j. We have seen 1 is in S; suppose all positive integers smaller than k are in S. Since

$$A^k = A^{k-1}A,$$

$$(A^k)_{ij} = \sum_{h=1}^{n}(A^{k-1})_{ih}A_{hj}.$$

However, we can partition the set of (directed) walks of length k from v_i to v_j into at most n classes; class h consists of all walks whose next-to-last vertex is v_h. By the product principle, the number of walks in class h is the number of walks of length $k-1$ from v_i to v_h times the number of walks of length 1 from v_h to v_j. This product is, by assumption,

$$(A^{k-1})_{ih}A_{hj}.$$

Thus, by the sum principle, the number of walks of length k from v_i to v_j is

$$\sum_{h=1}^{n}(A^{k-1})_{ih}A_{hj} = (A^k)_{ij},$$

and so k is in S. Thus, by version 1 of the principle of mathematical induction, S contains all positive integers and the theorem is proved. ∎

The number of (directed) *paths* from vertex i to vertex j is much more difficult to compute. For graphs, the diagonal entries of the kth power contain interesting information. For example,

Theorem 7.2. If A is the adjacency matrix of a graph, then the degree of vertex i is $(A^2)_{ii}$.

Proof.

$$(A^2)_{ii} = \sum_{j=1}^{n} A_{ij} A_{ji}$$

Since $A_{ij} A_{ji}$ is 0 unless $\{v_i, v_j\}$ is an edge, and is 1 if $\{v_i, v_j\}$ is an edge, $(A^2)_{ii}$ is the number of edges containing vertex i. ∎

Connectivity and Transitive Closure

Theorem 7.3. An n-vertex (multi)graph (or digraph) is connected (or strongly connected) if and only if every entry of $\sum_{i=0}^{n} A^i$ is nonzero.

Proof. Since any pair of connected vertices can be connected by a walk of length n or less, the theorem follows from Theorem 7.1. ∎

In fact, we need not compute each individual power and add them.

Theorem 7.4. An n-vertex (multi-)graph (or digraph) is connected (or strongly connected) if and only if every entry of $(I + A)^n$ is nonzero.

Proof. Since

$$(I + A)^n = \sum_{j=0}^{n} \binom{n}{j} I^j A^{n-j} = \sum_{k=0}^{n} \binom{n}{k} A^k$$

and since the entries of A^k are all nonnegative, the (i, j)th entry of $(I + A)^n$ will be nonzero if and only if the (i, j)th entry of some A^k is nonzero. Thus, Theorem 7.4 follows from Theorem 7.3. ∎

The matrix $(I + A)^n$ gives other information about a graph or digraph. We can define the transitive closure of a graph in a way similar to the transitive closure of a digraph. (Just regard the edge set of the graph as a symmetric relation.) Then we get

Theorem 7.5. The adjacency matrix of the transitive closure of an n-vertex (multi) (di)graph has a 1 in position i, j if and only if $(A + I)^n - I$ is nonzero in position i, j.

Proof. There is a (strong) walk from vertex i to vertex j of length k if and only if the (i, j)th position of A^k is nonzero and $\binom{n}{k} A^k$ is a summand of $(A+I)^n$. However, the term $\binom{n}{0}A^0 = I$ corresponds to walks of length 0, which are not used in the transitive closure. ∎

In a graph, two points are connected in the transitive closure if and only if they are in the same connected component. This gives a conceptually simple way to determine the connected components of a graph. The vertices in the same connected component as v_i correspond to the nonzero entries of row i of $(A+I)^n$. Thus it appears that matrix multiplication is a good way to compute connected components of a graph. However, even computing the square of an n by n matrix in the way we usually compute matrix products requires n multiplications and then adding n numbers for each of the n^2 entries of the square. Since this means that on the order of n^3 arithmetic operations are required simply to square a matrix in the usual way, it should not be surprising that other methods have been developed to use computers to determine information about graphs.

Boolean Operations

It turns out that basic computer operations are well suited for working with graphs. For example, Boolean addition, defined for just 0 and 1 by the rules

$$1 \vee 1 = 1$$
$$1 \vee 0 = 0 \vee 1 = 1$$
$$0 \vee 0 = 0$$

where the symbol $\vee$ is read as "or," is a standard computer operation. So long as we work with just 0 and 1, ordinary multiplication gives just zeros and ones also. All the usual rules like

$$a(b + c) = ab + ac$$

translate to rules like

$$a(b \vee c) = ab \vee ac$$

for multiplication and Boolean addition. In Boolean arithmetic, however, there is no subtraction concept, so even though

$$a + b = a + c \quad \text{implies} \quad b = c,$$

it is the case that

$$a \vee b = a \vee c \quad \text{need not imply} \quad b = c.$$

For example,

$$1 \vee 0 = 1 = 1 \vee 1,$$

but $0 \neq 1$. Because only multiplication and addition are used to define matrix multiplication, we can define Boolean matrix multiplication. We use $A^{(n)}$ to denote the nth power of a matrix of zeros and ones in Boolean arithmetic. The connectivity matrix of a (di)graph has a 1 in position (i, j) if vertex i and vertex j are (strongly) connected. In terms of Boolean matrix powers and multiplication, Theorems 7.3 through 7.5 may be stated as

> **Theorem 7.6.** A (multi)(di)graph is connected if and only if every entry of $(I + A)^{(n)}$ is 1; further, the transitive closure has adjacency matrix $A(I + A)^{(n-1)}$.
>
> *Proof.* Each Boolean matrix power has a 1 in each position where the ordinary powers are nonzero. Note that the second matrix is the Boolean sum of the Boolean first through nth powers of A; thus it has a one in row i and column j if and only if $(A + I)^n - I$ does. ∎

*The Matrix–Tree Theorem

More sophisticated information about a graph can be obtained from other matrix operations. The **matrix–tree theorem** lets us compute the number of spanning trees of a graph G from the adjacency matrix. We let $B(G)$ be the matrix whose (i, j)th entry is 0 if $i \neq j$ and whose (i, i)th entry is the degree of vertex v_i in G. The (i, j)th cofactor of a square matrix M is $(-1)^{i+j}$ times the determinant of the matrix obtained by deleting row i and column j of M. Thus the (2,3) cofactor of

$$\begin{pmatrix} a & b & c \\ d & e & f \\ g & h & i \end{pmatrix}$$

is $(-1)^5 (ah - bg) = bg - ah$

> **Theorem 7.7.** If G is a connected graph, then all cofactors of $B(G) - A(G)$ are equal to the number of spanning trees of G.

Theorem 7.7 may be proved as an easy (but clever) corollary of Theorem 7.8; this is left as an exercise. For a directed graph D, we let $C(D)$ be the matrix whose (i, i)th entry is the outdegree of vertex i and whose (i, j)th entry is 0 if $i \neq j$. We say a tree is directed into vertex i if its edges are directed in such a way that there is a path from any vertex to vertex i.

Theorem 7.8. The (i,i)th cofactor of the matrix $M = C(D) - A(D)$ is the number of spanning trees directed into vertex i in the connected digraph D.

Proof. (This proof may be omitted without loss of continuity.) We prove the theorem by induction on the number of edges of D. Let S be the set of all positive integers n such that if D has n edges, then the theorem holds for D. If D has one edge, say $(1,2)$, then

$$C(D) = \begin{pmatrix} 1 & 0 \\ 0 & 0 \end{pmatrix} \qquad A(D) = \begin{pmatrix} 0 & 1 \\ 0 & 0 \end{pmatrix},$$

so that

$$M = C(D) - A(D) = \begin{pmatrix} 1 & -1 \\ 0 & 0 \end{pmatrix}.$$

The $(1,1)$th cofactor of M is 0, and the $(2,2)$th cofactor of M is 1; these are the number of spanning trees directed into, respectively, vertex 1 and vertex 2. Thus 1 is in S. Now suppose all integers less than n are in S and that D is a digraph with n edges and m vertices. We divide the proof into two cases.

Case 1 is the case in which there is an edge $(1,i)$ from vertex 1 to some vertex i, so that the outdegree $od(1) > 0$. (The choice of vertex 1 is arbitrary. We could make essentially the same argument with any other vertex.) Since this edge cannot lie in any spanning tree directed into vertex 1, the number of spanning trees is the same for G and the graph G' whose edges are the edges of G other than $(1,i)$. The matrix $M' = C(G') - A(G')$ will be different from M only in the first row (and there only in column 1 and column i). Thus, M' and M have the same $(1,1)$th cofactor, which, since G' has $n-1$ edges and $n-1$ is in S, is the number of spanning trees of G' directed into vertex 1 in G'. By the preceding remarks, this is also the number of spanning trees of G directed into vertex 1.

Case 2 is the case in which the outdegree of vertex 1 is 0. This case will be further divided into three steps for clarity.

Step 1: Suppose that for each vertex i such that $(i,1)$ is an edge, vertex i has outdegree 1. Let G' be the graph obtained by choosing one such vertex, say vertex j, deleting the edge $(j,1)$, replacing each edge $(i,1)$ by the edge (i,j) instead, and deleting vertex 1. In particular, applying this construction to a spanning tree directed into vertex 1 gives a spanning tree of G' directed into vertex j. Reversing the process on a spanning tree of G' directed into vertex j gives a spanning tree of G directed into vertex 1. These correspondences are one-to-one, so the number of spanning trees of G directed into vertex 1 equals the number of spanning trees of G' directed into vertex j.

For convenience, we will assume that column i of the matrix M is the column determined by vertex i. Now the matrix M' of G' may be obtained from the matrix M of G by adding column 1 to column j and then deleting row and column 1. The matrix formed by deleting rows 1 and j and columns 1 and j from M is the same as the matrix formed by deleting row j and column j from M'. Call this common resulting matrix M^*. Now in M, row j has a -1 in position 1, a 1 in position j, and 0's elsewhere. Thus, the $(1,1)$th cofactor of M is the determinant of M^*. However, the determinant of M^* is the (j,j)th cofactor of M'; by our inductive hypothesis, this is the number of spanning trees directed into vertex j in G'. Hence the number of spanning trees of G directed into vertex 1 is the $(1,1)$th cofactor of M.

Steps 2 and 3 analyze the situation in which $(i,1)$ is an edge and i has outdegree greater than 1. We compute first the number of spanning trees that use $(i,1)$ in this situation (that computation is step 2) and then the number of spanning trees that don't use $(i,1)$ (that computation is step 3) and add these two numbers.

Step 2: We assume $(i,1)$ is an edge and that vertex i has outdegree greater than 1. If a spanning tree uses $(i,1)$, it cannot also use another edge (i,j), for the directed path from vertex j to vertex 1 would give us an (undirected) cycle in our spanning tree. Thus, the number of spanning trees using edge $(i,1)$ is equal to the number of spanning trees in the graph G'' obtained from G by deleting all edges (i,j) with $j \neq 1$. We let $M'' = C(G'') - A(G'')$; by our inductive hypothesis, we know that the number of spanning trees is the cofactor of the $(1,1)$th entry of M''. In this case, M'' differs from M only in row i; in row i and column 1 it has a -1; in row i and column i it has a $+1$ and all other entries in row i are 0.

Step 3: Now we compute the number of spanning trees that do not use $(i,1)$. This number is equal to the number of spanning trees directed into vertex 1 of the graph G^{**} obtained by deleting the edge $(i,1)$ from G. Since G^{**} has $n-1$ edges, our inductive assumption tells us that the number of spanning trees is the $(1,1)$th cofactor of the matrix $M^{**} = C(G^{**}) - A(G^{**})$.

Now let N, N'', and N^{**} represent, respectively, the matrices obtained from M, M'', and M^{**} by deleting the first row and column. Except for row i, these three matrices are identical, and row i of N is the sum of row i of N'' and row i of N^{**}. Thus, by the addition theorem for determinants, the determinant of N is the sum of the determinants of N'' and N^{**}. This sum is also the number of spanning trees directed into vertex 1. This proves that n is in S and thus, by version one of the principle of mathematical induction, proves our theorem. ∎

*The Number of Eulerian Walks in a Digraph

Knowing how many spanning trees a digraph has can help us determine how many Eulerian walks it has. Given a spanning tree directed into vertex i in an Eulerian digraph whose vertices all have equal indegree and outdegree, we may construct a directed Eulerian walk as follows. Beginning at vertex i, follow any edge (i, j_1) leading out. From vertex j_1 choose any edge—say (j_1, j_2)— not in the spanning tree. Repeat this process each time you arrive at a vertex j_k; follow an edge (j_k, j_{k+1}) leading out of vertex j_k but not in the spanning tree if there is such an edge; otherwise, follow the unique edge of the spanning tree that leaves vertex j_k. In this way, each outgoing edge from each vertex will be used, and since every vertex, except, perhaps, for vertex i and the final vertex, will have been in an even number of edges of the walk, vertex i must be the final vertex chosen. In Theorem 7.9, $od(v_i)$ stands for the outdegree of vertex i.

> **Theorem 7.9.** The number of directed Eulerian walks from vertex i to vertex i in an n-vertex digraph D each vertex of which has equal indegree and outdegree is the product of the (i, i)th cofactor of $C(D) - A(D)$ with
>
> $$od(v_i) \prod_{j=1}^{n} (od(v_j) - 1)! \, .$$

Proof. Any Eulerian walk from vertex v_i to vertex v_i yields a tree directed into vertex v_i as follows: For each vertex v_j not equal to v_i, there is a last edge (v_j, v_k) in the walk in which j appears. The vertex set V with these last edges forms a tree directed into vertex v_i. We call this tree the tree of last passage of the walk. The number of walks having a given tree of last passage may be computed by observing that each such walk can be constructed by the method described before the statement of this theorem. The first time vertex v_i is used, there will be $od(v_i)$ ways of choosing an edge leaving v_i; the first time any vertex v_j for $j \neq i$ is used, there will be $od(v_j) - 1$ edges leaving v_j to choose. On the kth use of v_i or v_j (for $j \neq i$), there are respectively $od(v_i) - k + 1$ or $od(v_j) - k$ edges leaving the vertex from which to choose. By the product principle, the number of walks with a given tree of last passage is

$$od(v_i) \prod_{j=1}^{n} (od(v_j) - 1)! .$$

One more use of the product principle proves the theorem. ∎

Unfortunately, there does not seem to be a natural parallel to Theorem 7.9 for undirected Eulerian graphs; we explore the reasons why the methods of Theorem 7.9 can't be applied to undirected graphs in the exercises.

EXERCISES

1. Draw pictures of the graphs with adjacency matrices

$$\begin{pmatrix} 0 & 1 & 1 & 0 & 0 & 1 \\ 1 & 0 & 1 & 1 & 1 & 0 \\ 1 & 1 & 0 & 1 & 0 & 1 \\ 0 & 1 & 1 & 0 & 1 & 1 \\ 0 & 1 & 0 & 1 & 0 & 0 \\ 1 & 0 & 1 & 1 & 0 & 0 \end{pmatrix} \qquad \begin{pmatrix} 0 & 1 & 0 & 0 & 0 \\ 1 & 0 & 1 & 0 & 0 \\ 0 & 1 & 0 & 1 & 0 \\ 0 & 0 & 1 & 0 & 1 \\ 0 & 0 & 0 & 1 & 0 \end{pmatrix}$$

2. Write down the adjacency matrix for the graph in Figure 7.3.

Figure 7.3

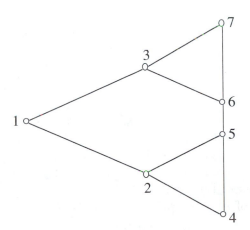

Figure 7.4

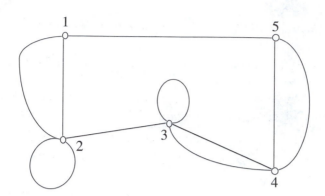

3. Write down the adjacency matrix for the multigraph in Figure 7.4.

4. Find the number of walks of length 2 and the number of walks of length 3 between vertices 1 and 5 in the graph of Figure 7.3. Do the same for paths. What about length 4?

5. Find the number of walks of length 4 between vertex 1 and vertex 4 in the multigraph of Figure 7.4.

6. Write down the adjacency matrix of the digraph in Figure 7.5.

Figure 7.5

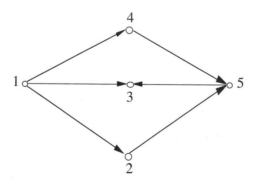

7. Find the number of walks of length 3 and 4 from vertex 1 to vertex 3 in the digraph of Figure 7.5.

8. (a) Under what circumstances will the (i, i)th entry of $A(G)^3$ be nonzero for a graph G? (That is, what property must a graph G have in this case?)

   (b) How is the number of triangles in a graph G related to the trace (sum of the main diagonal entries) of $A(G)^3$?

9. Show that a tournament D is transitive if and only if the trace (see Exercise 8) of $A(D)^3$ is zero.

10. (a) Compute the number of spanning trees directed into vertex 3 of the digraph of Figure 7.5.
    (b) Compute the number of spanning trees directed into any other vertex of the digraph of Figure 7.5.

11. (a) Find the adjacency matrix of the digraph shown in Figure 7.6.
    (b) Find the number of four-edge walks from vertex 1 to vertex 3 in Figure 7.6.

12. (a) Compute the number of Eulerian walks starting and ending at vertex 1 in Figure 7.6.
    (b) Compute the number of Eulerian walks starting and ending at vertex 3 in Figure 7.6.

Figure 7.6

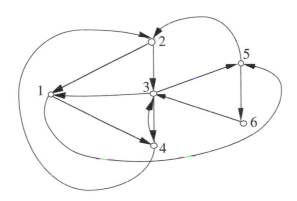

13. Discuss why we need at most $2\log_2(n)$ matrix multiplications to apply Theorems 7.4 and 7.5. What about Theorem 7.3?

*14. Explain why any spanning tree directed into vertex 3 in Figure 7.6 must contain the edge $(6,3)$.

*15. Explain why in an Eulerian digraph there is at least one spanning tree directed into each vertex.

*16. Given a graph G, its double DG is a digraph that has edges (i,j) and (j,i) for each edge (i,j) of G.
    (a) Show that there is a one-to-one correspondence between the spanning trees of G and the spanning trees of DG directed into vertex i.
    (b) Show that the diagonal cofactors of $B(G) - A(G)$ (the (i,i)th cofactors, that is) are each equal to the number of spanning trees of G.

(c) Show that the rows and columns of $B(G) - A(G)$ all add up to zero. It is a theorem of linear algebra that all cofactors of a matrix with zero row and column sums are equal. Thus you have proved Theorem 7.7.

*17. An Eulerian walk starting and ending at vertex i provides two orientations that turn G into a digraph. There is a tree directed into vertex i associated with such a walk. Also, *any* spanning tree of G (as a graph, not a digraph) can be oriented in exactly one way to become a spanning tree directed into vertex i. Why can't these facts be used as in Theorem 7.9 to compute the number of Eulerian walks in G?

18. Find the Boolean squares and fourth powers and eighth powers of the matrices of Exercise 1.

19. The Boolean sum $R \vee S$ of two row vectors R and S is the vector whose components are the Boolean sums of the components of the two vectors. Use A_{i-} to stand for row i of the matrix A. The following algorithm, called Warshall's algorithm (but attributed to Kleene by Reingold, Nievergelt, and Deo), is very effective for computing the transitive closure because the $\vee$ operation is especially easy to implement electronically. Show that this algorithm computes the transitive closure of a digraph with $n \times n$ adjacency matrix A and stores the matrix of the transitive closure in the matrix A.

> For $j = 1$ to n.
>
> For $i = 1$ to n.
>
> If $A_{ij} = 1$, then let $A_{i-} = A_{i-} \vee A_{j-}$.

20. Apply Warshall's algorithm (see the previous exercise) to compute the adjacency matrix of the transitive closure of the digraph in Figure 7.5

21. When we have a weighted graph or digraph (i.e., one with numerical weights on the edges), we form the weighted adjacency matrix A by letting $A_{ij} = 0$ unless there is an edge from vertex i to vertex j, in which case we let A_{ij} be the weight of this edge. We interpret weights on edges in a weighted graph or digraph as distances between adjacent vertices. Then the weighted adjacency matrix A has the distance from vertex i to vertex j as the entry A_{ij} if vertex i and j are adjacent and has zero as the entry otherwise. It would be nice to have instead a matrix in which the entry A_{ij} is the distance, that is, the length of a shortest path, from vertex i to vertex j, assuming they are connected. Show that if we set $A_{ii} = 0$ and replace each other zero in the matrix A by an ∞ (or simply by a number far larger than any distance under

consideration), and if there is no cycle along which the sum of the weights is negative, then the following algorithm, usually called Floyd's algorithm, computes a matrix in which the A_{ij} entry is the length of a shortest path from vertex i to vertex j or is ∞ if no such shortest path exists.

> For $k = 1$ to n.
>
> For $i = 1$ to n.
>
> For $j = 1$ to n.
>
> If $A_{ik} + A_{kj} < A_{ij}$, then let $A_{ij} = A_{ik} + A_{kj}$.

Hint: When will we know the shortest path from vertex i to vertex j using vertex 1? The shortest path using vertices 1 and 2?

22. Show the result of applying Floyd's algorithm of the previous exercise to the weighted graph of Figure 3.7 in the exercises of Section 3 of this chapter. Since there are 216 possible combinations of i, j, and k, it is unreasonable to show the matrix A which arises for each step. Thus show the six values of A after running through all the i and j values with k equal to x, a, b, c, d, e.

23. Give an example of a cycle on four vertices with weights on its edges so that Floyd's algorithm does not give a matrix in which A_{ij} is the total weight of a path of minimum total weight between vertices i and j.

24. In the theorems of this section about nth powers of matrices, the n may often be replaced by $n - 1$, but under certain conditions it may not. Discuss this. (Hint: There is a difference between graphs and digraphs.)

25. This is an exercise for people familiar with linear algebra. Suppose that the diameter of the graph G is d; that is, the largest distance between two vertices is d. Let A be the adjacency matrix of G, and let $x_0, x_1, \ldots, x_d$ be a shortest path between two vertices of distance d from each other. Explain why $A_{0j}^i = 0$ if $i < j$. Is $A_{0j}^j = 0$? What does this say about the possibility of finding numbers c_0 through c_d such that

$$\sum_{i=0}^{d} c_i A^i = 0?$$

What does this say about the dimension of the vector space generated by the powers of A? What does this say about the number of distinct eigenvalues of A?

26. Let G be a graph, and let $w_{ij}(k)$ be the number of walks of length k from vertex i to vertex j. Now, let $W_{ij}(x) = \sum_{k=0}^{\infty} w_{ij}(k) x^k$, the

generating function for the number of walks from vertex i to vertex j. This gives us a matrix W of generating functions. Show that if A is the adjacency matrix (and I is the identity matrix and x is a variable), then $W = (I - xA)^{-1}$.

Suggested Reading

Biggs, N. L., 1993. *Algebraic Graph Theory.*   Cambridge: Cambridge University Press.

Biggs, N. L., E. K. Lloyd, and R. J. Wilson. 1976. *Graph Theory 1736–1936.* New York: Oxford University Press.

Bondy, Adrian, and U. S. R. Murty. 1976. *Graph Theory and Its Applications* New York: Elsevier Science Publishing.

Chartrand, Gary, and Linda Lesniak. 1986. *Graphs and Digraphs.* 2d ed. Belmont, CA: Wadsworth.

Harary, F. 1969. *Graph Theory.* Reading, MA: Addison-Wesley.

Harary, F., R. Z. Norman, and D. Cartwright. 1975. *Structural Models: An Introduction to the Theory of Directed Graphs.* New York: Wiley.

Hopcroft, J. E., and R. E. Tarjan. 1974. Efficient planarity testing. *Journal of the Association for Computing Machinery* 21:549–568.

Liu, C. L. 1968. *Introduction to Combinatorial Mathematics.* New York: McGraw-Hill.

Nijenhuis, Albert, and Herbert S. Wilf. 1975. *Combinatorial Algorithms.* Orlando, FL: Academic Press.

Reingold, E. M., J. Nievergelt, and N. Deo. 1977. *Combinatorial Algorithms.* Englewood Cliffs, NJ: Prentice Hall.

Roberts, F. S. 1976. *Discrete Mathematical Models.* Englewood Cliffs, NJ: Prentice Hall.

Tarjan, R. E. 1972. Depth-first search and linear graph algorithms. *SIAM Journal on Computing* 1:146–160.

Tucker, Alan. 1984 *Applied Combinatorics.* 2d ed. New York: Wiley.

Tutte, W. T. 1975. Chromials. In *Studies in Graph Theory II, MAA Studies in Mathematics*, vol. 12, ed. D. R. Fulkerson. Washington, DC: Mathematical Association of America.

West, Douglas B. 1995. *Introduction to Graph Theory.* Englewood Cliffs: Prentice Hall.

Whitney, Hassler, and W. T. Tutte. 1975. Kempe chains and the four color problem. In *Studies in Graph Theory II, MAA Studies in Mathematics*, vol 12, ed. D. R. Fulkerson. Washington, DC: Mathematical Association of America.

5
Matching and Optimization

Section 1   Matching Theory

The Idea of Matching

Suppose that a school district has begun to advertise for teachers and plans to continue advertising until it has found enough qualified applicants for all the positions. Since a number of the applicants are qualified to teach more than one subject, it may not be entirely clear when enough qualified candidates have been found to fill all the positions. It is natural to ask whether there is a systematic way of determining when there are enough candidates. The problem has a natural graph-theoretic representation. We construct a graph whose vertices are the positions and the applicants. In this graph, we draw an edge joining a position to an applicant if the applicant is qualified for the position. If we can find a set of edges such that no two of them have a vertex in common and if every position is touched by one of these edges, then we can use the edges to match positions to people who qualify for them.

A *matching* M of size m in a graph G is a set of m edges no two of which have a vertex in common. A vertex is said to be *matched* (to another vertex) by M if it lies in an edge of M. We defined a *bipartite graph* G with parts V_1 and V_2 to be a graph whose vertex set is the union of the two disjoint sets V_1 and V_2 and whose edges all connect a vertex in V_1 with a vertex in V_2. The graph for the school district's problem is bipartite. A *complete matching* of V_1 into V_2 is a matching of G with $|V_1|$ edges. A complete matching of positions to candidates is what the school district wants. Examples of bipartite graphs, first with and then without a complete matching, are shown in Figure 1.1.

If there were no complete matching of positions to candidates, the school district would still want a matching as large as possible or a matching

Figure 1.1

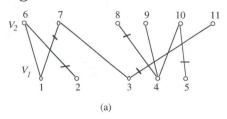 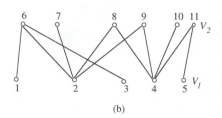

(a)   (b)

that fills a certain critical set of positions. Thus it will be useful to have methods to build matchings of maximum size and tests to determine either the size of a maximum-sized matching or whether one part of a bipartite graph has a complete matching into the other part. Our approach to these problems follows that of Berge, which parallels work of Norman and Rabin (all occurring in the mid- to late 1950s). Although our emphasis is on bipartite graphs, for which some results (which we describe as they arise) were known in the 1930s, Berge's fundamental theorems hold for arbitrary graphs as well. We begin with a result describing the interaction of two matchings. When we say that edges of a walk are alternately in M_1 and M_2, we mean that if e_1 and e_2 are edges with a vertex in common, then one is in M_1 and the other is in M_2. The following seemingly technical lemma will have surprising applications. It is illustrated in Figure 1.2

> **Lemma 1.1.** Let M_1 and M_2 be matchings of the graph $G = (V, E)$, and let $E' = M_1 \cup M_2 - (M_1 \cap M_2) = (M_1 - M_2) \cup (M_2 - M_1)$. Then each connected component of $G' = (V, E')$ is one of the following three types:
>
> (1) A single vertex.
> (2) A cycle with an even number of edges whose edges are alternately in M_1 and M_2.
> (3) A path whose edges are alternately in M_1 and M_2 and whose two end vertices are each matched by one of M_1 or M_2 but not both.

Proof. A connected component could be of type 1. We now assume we have a connected component which is not a single vertex. Since a vertex is in at most one edge of M_1 and at most one edge of M_2, it has degree at most 2. Thus, the sum of the degrees of a connected component with n vertices is at most $2n$. Since the component is connected, it has at least $n - 1$ edges, so the sum of the degrees of its vertices is either $2(n - 1)$ or $2n$. (Why can't it be $2n - 1$?)

If the sum of the degrees is $2n$, then each vertex has degree 2, so the component must be a cycle. Since a vertex cannot be in two edges

Figure 1.2 The dark solid lines indicate one matching in the graph of Figure 1.2(a) and the dark dashed lines represent a second matching. Figure 1.2(b) shows these two matchings with the vertices drawn in different places to make the two kinds of connected components described in Lemma 1.1 easier to see.

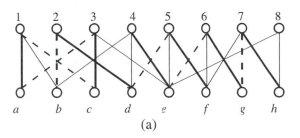

(a)

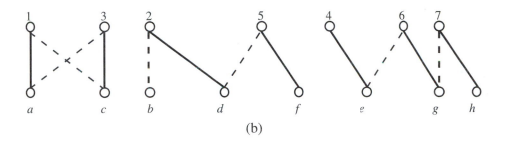

(b)

of an M_i, the edges must be alternately in M_1 and M_2. In particular, we have an even number of edges.

If the sum of the degrees is $2n - 2$, the connected component is a tree, but since each vertex has degree 2 or less, it must be a path. Since the end vertices have degree 1, each of them is in an edge of M_1 or an edge of M_2 but not both. As in case 2, the edges must be alternately in M_1 and M_2. ■

From Lemma 1.1 we can derive a criterion for a maximum-sized matching in a graph. We shall call this criterion the "Berge criterion" in recognition of Berge's proof that it characterizes maximum-sized matchings in graphs in general. The criterion was known earlier for bipartite graphs as a result of the Hungarian mathematicians König and Egerváry, who each published variations of the result in 1931. In their honor the general method we are developing was named the "Hungarian method" by Harold Kuhn when he described the method in English in 1955. The criterion uses Petersen's idea of an alternating path for a matching, an idea which goes back to the late nineteenth century, when Petersen introduced it for special matchings (known as 1-factors) in special kinds of graphs. If M is a matching, a path $v_0 e_1 v_1 e_2 \cdots e_n v_n$ is an **alternating path** for M if whenever e_i is in M, e_{i+1} is not and whenever e_i is not in M, e_{i+1} is in M. Our first step in developing the criterion shows how alternating paths may be used to enlarge matchings into bigger matchings.

Making a Bigger Matching

> **Theorem 1.2.** Let M be a matching in a graph G and let P be an alternating path with edge set E' beginning and ending at unmatched vertices. Let $M' = M \cap E'$. Then
>
> $$(M - M') \cup (E' - M'),$$
>
> which is the same set as
>
> $$(M - E') \cup (E' - M)$$
>
> is a matching with one more edge than M.
>
> *Proof.* Every other edge of P is in M. However, P begins and ends with edges not in M, so there is a number k such that P has k edges in M and $k + 1$ edges not in M. Now the first and last vertices of P are unmatched and all other vertices in P are matched by M', so no edge in $M - M'$ contains any vertex in P. Thus, the edges of $M - M'$ have no vertices in common with the edges of $E' - M'$. Further, since P is an alternating path for M, every other edge of the path is in M and thus the set of edges of the path not in M, namely $E' - M'$, is the complementary set of every other edge of the path. Therefore the edges of $E' - M'$ have no vertices in common. Thus,
>
> $$(M - M') \cup (E' - M')$$
>
> is a matching and by the sum principle, it has $m - k + k + 1 = m + 1$ edges. Elementary set theory shows that
>
> $$(M - M') \cup (E' - M') = (M - E') \cup (E' - M). \qquad \blacksquare$$

The Berge criterion for a maximum-sized matching essentially states that when Theorem 1.2 does not apply, we cannot enlarge the matching.

> **Theorem 1.3.** (Berge) Suppose G is a graph and M is a matching. Then M is a matching of maximum size (among all matchings) if and only if there is no alternating path connecting two unmatched vertices.
>
> *Proof.* If there is such an alternating path, then M is not maximum-sized by Theorem 1.2.

Now suppose there is no such alternating path and let N be a maximum-sized matching. We show that M and N have the same size by applying Lemma 1.1 to prove that $M - N$ and $N - M$ have the same size. (Since $M = (M - N) \cup (M \cap N)$ and $N = (N - M) \cup (M \cap N)$, this proves M and N have the same size.) Let G' be the graph on V whose edge set is $(M - N) \cup (N - M) = (M \cup N) - (M \cap N)$. We show that each connected component of G' has the same number of edges of M as of N. Type 1 components have no edges. Type 2 components have an even number of edges alternating between M and N, so exactly half the edges of a type 2 component are in M and the other half are in N.

Since N is maximum-sized, by Theorem 1.2, a type 3 component of G' cannot both begin and end with an edge of M (since it would begin and end with vertices unmatched by N). By the hypothesis of this theorem, a type 3 component of G' cannot both begin and end with an edge of N. Therefore, exactly half the edges of a type 3 component are in M and exactly half are in N.

Since $M - N$ and $N - M$ have the same number of edges in each connected component of G', $M - N$ and $N - M$ have the same size. Therefore, M is of maximum size. ∎

By Theorem 1.3, to produce a maximum-sized matching we need only repeat the procedure described in Theorem 1.2 until it is no longer possible to find an alternating path between two unmatched vertices. To turn this into an explicit procedure for finding maximum-sized matchings, we need only develop an explicit procedure for searching for alternating paths. In a small graph it is usually possible to find the desired path by inspection. However if we want a good method that could be programmed for a computer, inspection won't suffice. In bipartite graphs it is possible to apply small variations of search techniques we have already developed in order to find alternating paths. It is here that the theory of matchings for bipartite graphs becomes easier than the theory for graphs in general.

A Procedure for Finding Alternating Paths in Bipartite Graphs

In our study of graphs we described search procedures, two of which are depth-first search and breadth-first search, that allowed us to construct spanning trees for a graph. Either of these procedures can be modified and then used to search for an alternating path. Since breadth-first search yields minimal paths, we will describe our modification in terms of breadth-first search, but the same modification may be applied to depth-first search and other standard "tree search" techniques. An *alternating search tree* centered at the vertex x in a graph G with matching M is a tree containing x such that if edges e_1 and e_2 of the tree share a vertex v, then one of the two edges

e_i is in M and the other is not. To construct an alternating breadth-first search tree, we modify the procedure of Chapter 4, Section 3 as follows.

Begin at an unmatched vertex x for the matching M, and use this as the x in step 1. This means that we let $V_0 = \{x\}$, we let $E_0 = \emptyset$, and we let $x = x_0$. Recall that steps 2 and 3 are superfluous. Suppose in step 4 we have constructed the tree (V_k, E_k) and are examining a vertex x_j adjacent to a z in V but not in V_k. Then either $j = 0$ or there is an x_i such that $\{x_i, x_j\}$ is in E_j. If $j = 0$, we proceed with step 4 without modification. If $\{x_i, x_j\}$ is in M, then no matter what z we examine, $\{x_j, z\}$ will not be in M. Thus, we proceed with step 4 without modification. If, however, $\{x_i, x_j\}$ is *not* in M, then we examine *only* vertices z (note that there is actually at most one such vertex) such that $\{x_j, z\}$ *is* in M.

With this modification, the breadth-first search procedure will produce an alternating search tree; however, it need not produce a spanning tree. If, in a bipartite graph, there are some unmatched vertices connected to x by alternating paths, this procedure will give us at least one such alternating path of minimum length to an unmatched vertex. (If G is not bipartite, this search procedure does always not suffice. You are asked to show this in Exercise 16.)

In particular, we get our alternating path (if there is one) by constructing an alternating breadth-first search tree starting at an unmatched vertex. If we find an unmatched vertex in this tree, we stop. If we do not find an unmatched vertex, we repeat the process starting at another unmatched vertex. Once we find an unmatched vertex in our search tree and stop, we work backward to find the unique path in the tree from x to the unmatched vertex.

Constructing Bigger Matchings

Example 1.1. To see how to enlarge a matching, note that in Figure 1.1(a) we can see by inspection that $M = \{\{1, 6\}, \{3, 7\}, \{4, 8\}, \{5, 10\}\}$ is a matching. Enlarge it.

The only unmatched vertex in V_1 is vertex 2. By inspection, we see that $\{2, 6, 1, 7, 3, 11\}$ is the vertex set of an alternating path.

If we prefer, we may form the alternating breadth-first search tree centered at vertex 2. It turns out to be the path with vertex set $\{2, 6, 1, 7, 3, 11\}$.

Vertex 11 is unmatched, so we have our alternating path. Along this path, the edges $\{1, 6\}$ and $\{3, 7\}$ are in the matching, while edges $\{2, 6\}$, $\{1, 7\}$, and $\{3, 11\}$ are not. Throwing out $\{1, 6\}$ and $\{3, 7\}$ from M and adding in $\{2, 6\}$, $\{1, 7\}$, and $\{3, 11\}$ gives the matching shown by means of edges marked with a slash in Figure 1.1. ∎

Example 1.2. By inspection $\{1,6\}$, $\{2,7\}$, $\{4,9\}$, and $\{5,11\}$ is a matching in Figure 1.1(b). Show it is maximum-sized.

Since 3 is the only unmatched vertex in V_1, we look for alternating paths beginning at vertex 3. By inspection, we see there is no such path.

Rather than use inspection, we may construct an alternating search tree centered at 3. The tree consists of the vertices $\{3,6,1\}$ and the edges $\{3,6\}$ and $\{1,6\}$, because $\{2,6\}$ is not in the matching. There are no unmatched vertices other than 3 in this tree.

Thus, there is no alternating path from 3 to an unmatched vertex, so by the Berge criterion, this matching is maximum-sized. (A larger matching would have to match a currently unmatched vertex in V_1 and a currently unmatched vertex in V_2. Therefore, we need not examine unmatched vertices in V_2.) ∎

Testing for Maximum-Sized Matchings by Means of Vertex Covers

In bipartite graphs, there are two well-known tests for determining the size of a maximum matching or the possible existence of a complete matching. These tests do not lead to an algorithm as efficient as the Berge criterion, but they do provide simple techniques that in many situations can be applied visually.

Our first method, which uses the idea of a vertex cover, is usually attributed to König. A **vertex cover** C of the edges of a graph G is a set of vertices such that each edge of G contains at least one vertex in C. If M is a matching and C is a vertex cover, then $|M| \le |C|$ because each edge of M will have to contain a different element of C. In Figure 1.3(a) the vertices marked 1, 2, 3, and 4 cover all the edges. Thus, we know there is no complete matching, because any matching has size four or less. There is an obvious matching of size 4 in Figure 1.3(a), namely the vertical edges. On the other hand, Figure 1.3(b) has a complete matching (can you find one?) and has a vertex covering of size 5 (the bottom row of vertices, for example).

Figure 1.3

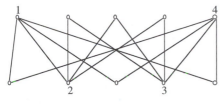

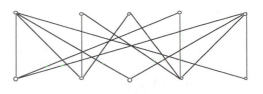

(a)   (b)

In fact, in a bipartite graph, the size of a maximum-sized matching and the size of a minimum-sized vertex cover are always equal.

Theorem 1.4. In a bipartite graph, the size of a minimum-sized vertex cover of the edges and the size of a maximum-sized matching are equal.

Proof. We use the Berge criterion to show that given a maximum-sized matching, we can construct a vertex cover of the same size. Since we already have seen that $|M| \leq |C|$ for any matching M and cover C, this will prove the theorem. Suppose G is a bipartite graph with parts V_1 and V_2, and let M be a maximum-sized matching of G. Let U_1 be the set of *unmatched* vertices in V_1. If U_1 is empty, then $|M| = |V_1|$ and we may let $C = V_1$. Otherwise, denote by A the set of vertices connected by alternating paths to vertices in U_1, let $A_2 = V_2 \cap A$, and let $A_1 = V_1 \cap A$. That is,

$$A_2 = \{v \in V_2 \mid \text{an alternating path connects } v \text{ to a vertex in } U_1\}$$

and

$$A_1 = \{v \in V_1 \mid \text{an alternating path connects } v \text{ to a vertex in } U_1\}.$$

Note that $V_1 \subseteq A_1$ since a path with no edges is trivially alternating. The set
$$C = A_2 \cup (V_1 - A_1)$$
is a vertex cover. To see why, note that if an edge contains a member of A_1, it is covered by a member of A_2, while any other edge is covered by a vertex in $V_1 - A_1$. Each member of C is matched, since members of $V_1 - U_1$ (and thus its subset $V_1 - A_1$) must be matched and if a member of A_2 were not matched, we would contradict the Berge criterion. Each member of A_2 is matched to something in A_1 (because otherwise we could increase the size of the matching); therefore no two members of C lie in the same edge of M. Thus $|C| \leq |M|$, so $|C| = |M|$. ∎

Example 1.3. Find a minimum-sized vertex cover of the graph in Figure 1.1(b). Why does this tell us the matching found in Example 1.2 has maximum size?

The set $\{6, 2, 4, 5\}$ covers all the edges. Since we have a matching of size 4, this must be a minimum-sized cover. There is no matching of size more than four since we have a four-vertex cover of the edges; therefore the matching of size four we found in Example 1.2 is a maximum-sized matching for that graph. ∎

Hall's "Marriage" Theorem

In our next test for whether there is a complete matching of V_1 into V_2 we use the idea of the set $R(X)$ of "relatives" or "neighbors" of a set X of vertices. For a subset X of the vertex set of a graph with edge set E, we let

$$R(X) = \{y \mid \{x, y\} \in E \text{ for some } x \in X\}.$$

For example, in Figure 1.1(a) $R(\{6, 7\}) = \{1, 2, 3\}$.

It is clear that if there is a complete matching, then each subset X of V_1 has as relatives at least the $|X|$ elements matched to X. In other words, to have a complete matching of V_1 into V_2, it is necessary that $|X| \leq |R(X)|$ for all $X \subseteq V_1$. In fact, it is also sufficient to check this one condition.

> **Theorem 1.5.** (König–Hall) A bipartite graph on two sets V_1 and V_2 has a complete matching from V_1 to V_2 if and only if $|X| \leq |R(X)|$ for each subset X of V_1.
>
> *Proof.* Assume $|X| \leq |R(X)|$ for all $X \subseteq V_1$, since we've already seen this holds if there is a complete matching. Since each edge relates something in V_1 to something in V_2, V_1 is a vertex cover. We show V_1 is a minimum-sized vertex cover. Suppose T is a minimum-sized vertex cover of the edges, and $T \neq V_1$. Suppose $S_1 = T \cap V_1$ and let $S_2 = V_1 - S_1$. By assumption, $|R(S_2)| \geq |S_2|$. However, $R(S_2) \subseteq T \cap V_2$ because edges not covered by vertices in V_1 must be covered by vertices in V_2. Thus, $|S_2| \leq |T \cap V_2|$, so $|V_1| = |S_1| + |S_2| \leq |S_1| + |T \cap V_2| = |T|$. Thus, V_1 is a minimum-sized vertex cover of the edges, and by Theorem 1.4 there is a matching of size $|V_1|$ which must be a complete matching from V_1 to V_2. ∎

Example 1.4. Find a set of vertices X such that $|R(X)| < |X|$ in Figure 1.1(b). Why does this tell us that there is no complete matching from V_1 into V_2 for this graph (a fact we already know)?

By inspection of the graph $R(\{1, 3\}) = \{6\}$. Thus, by the König–Hall theorem, there is no matching of V_1 into V_2 that matches all vertices of V_1 to vertices of V_2. ∎

The idea of a matching sometimes occurs in disguised forms in combinatorics. Our next example shows the form in which Hall originally discovered the König–Hall theorem in 1935.

Example 1.5. A dean is appointing a student activities committee which is to contain a representative of each student organization on campus. To avoid any conflict of interest, the dean has decided that a student can represent only one organization even if that student is a member of several organizations. Thus, the question is, "Can the dean find a system of distinct representatives, one from each organization?"

If we draw a graph in which the vertices represent students and organizations and in which a line represents membership, then a matching from organizations to students will give us a system of distinct representatives, and a system of distinct representatives would give a matching. Thus, by Theorem 1.5, the dean can find the desired system if and only if for each k, every k organizations have at least k members among them. ∎

Given a family $(S_1, S_2, \ldots, S_k)$ of sets, we call a list $(a_1, a_2, \ldots, a_k)$ a *system of distinct representatives (SDR)* if the a_i's are all different and if $a_i \in S_i$ for each i. Then using the techniques of Example 1.5, we can prove "Hall's marriage theorem":

> ***Theorem 1.6.*** A finite family **S** of sets has a system of distinct representatives if and only if for each i, the union of any i sets in **S** has at least i elements.
>
> *Proof.* Essentially given in Example 1.5. ∎

Term Rank and Line Covers of Matrices

The basic results of matching theory can be profitably reformulated in terms of matrices. By a *line* of a matrix, we mean a row or column. By a *line cover* of a matrix, we mean a set of lines containing all the nonzero entries of the matrix. An $n \times n$ matrix has n^2 positions, each containing an entry of the matrix. The ***term rank*** of a matrix is the maximum number of positions, no two in the same row or column, all containing nonzero entries. A theorem of König, which is also referred to as the König–Egerváry theorem (Egerváry's contribution is a more detailed result), relates term rank and line covers in the way matchings relate to vertex covers in bipartite graphs.

> ***Theorem 1.7.*** The term rank t of a matrix M is equal to the minimum number m of lines that contain all the nonzero entries of the matrix.
>
> *Proof.* We construct a bipartite graph G whose vertices are the lines of M. We put $\{L_1, L_2\}$ in the edge set of G if L_1 and L_2 are different and have a nonzero entry in common. Thus, our graph must be bipartite because, for L_1 and L_2 to be connected, one must be a row and the other must be a column. The term rank of M may be interpreted as the maximum number of edges of G no two of which have a vertex in common, i.e., as the maximum size of a matching. By Theorem 1.4, this is the size of a minimum vertex cover of G—and this is interpreted as a minimum-sized set of lines such that each nonzero entry in M lies

in one of these lines. Thus, the term rank is the minimum number of lines in a line cover. ∎

Permutation Matrices and the Birkhoff–von Neumann Theorem

As an example of the use of Theorem 1.7, we develop another celebrated result known as the Birkhoff–von Neumann theorem. A **permutation matrix** is a matrix of zeros and ones with exactly one nonzero element in each row and column. Multiplying a vector by such a matrix simply permutes (interchanges) the entries of the vector. We begin with a special case of the Birkhoff–von Neumann theorem for integral matrices.

Theorem 1.8. If M is an $m \times m$ matrix of nonnegative integers such that all the elements of each row and column add up to k, then M is a sum of k permutation matrices.

Proof. To prove the theorem by induction, let S be the set of all integers n such that if the row and column sums are n, then the matrix is the sum of n permutation matrices. Then $n = 1$ is in S because if the row and column sums are all 1 in a nonnegative integral matrix, then the matrix is a permutation matrix. Now suppose $k - 1$ is in S.

Now let M be an $m \times m$ matrix with all row and column sums equal to k. If all nonzero entries of M could be covered by r lines, the sum of all these nonzero entries would be no more than rk. However, the sum of all the entries in the matrix is mk, and $mk \leq rk$ implies $m \leq r$. The rows form an m-line cover, so by Theorem 1.7, there are m positions, no two in the same row or column, with nonzero entries. Let P be the zero-one matrix with 1's in exactly these positions. Then P is a permutation matrix.

Returning to the original M, note that $M - P$ is a nonnegative integral matrix whose row and column sums are all $k - 1$. Thus, since $k - 1$ is in S, $M - P$ is a sum of $k - 1$ permutation matrices, so M is a sum of k permutation matrices. Thus, by the first version of the principle of mathematical induction, the set S is the set of all positive integers. ∎

Theorem 1.9. (Birkhoff-Von Neumann) If M is a matrix of nonnegative real numbers such that each row and column adds up to 1, then $M = c_1 P_1 + c_2 P_2 + \cdots + c_k P_k$ where the c_i's are nonnegative numbers that add up to 1 and the P_i's are permutation matrices.

Proof. Outlined in the Exercises. ∎

*Finding Alternating Paths in Nonbipartite Graphs

Edmonds pointed out in 1963 how standard search techniques for alternating paths can fail in nonbipartite graphs and developed a technique without these difficulties in his article listed in the Suggested Reading. The discussion that follows is based on Edmonds' ideas. We will present his ideas as a further modification of the standard search techniques. However, since the notation we have been using becomes cumbersome in this context, we describe the search in terms of giving labels to vertices. We will still be building a tree, one vertex and edge at a time, but rather than speaking of adding vertex z and an edge $\{x_j, z\}$ to the tree, we will speak of assigning the label x_j to z; thus, labeling a vertex corresponds to adding it, and the edge joining it to its label, to a tree. The procedure we describe for assigning labels gives the user some freedom. By exercising that freedom appropriately, the user may convert the procedure to a variant of depth-first search, a variant of breadth-first search, or to some other kind of search technique.

The graph in Figure 1.4 shows how blindly applying one of our previous search techniques could block us from finding an alternating path.

Figure 1.4

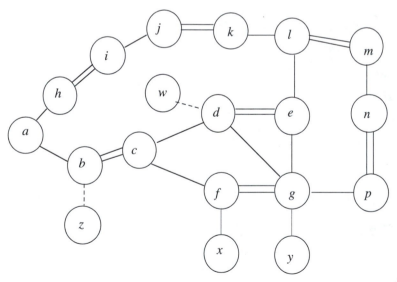

The vertices w, x, y, and z are possible unmatched vertices to which we might wish to connect the unmatched vertex a via an alternating path. The doubled edges are the matching edges. Suppose that our search leads us to vertex b and then to vertex c. If we then add vertices d and e to our alternating path and w is the only vertex of w, x, y, and z present, we have blocked ourselves from finding the useful alternating path $a\,b\,c\,f\,g\,e\,d\,w$.

Similarly, if we send the path through vertices f and g and x is the only unmatched vertex, then we have blocked ourselves. Worse still, if vertex z is the only unmatched vertex other than a present and we begin to construct an alternating path $a\,b\,c\ldots$, then we have blocked the long path

$$a\,h\,i\,j\,k\,l\,m\,n\,p\,g\,f\,c\,b\,z$$

which would be useful. Furthermore, entering the cycle $c\,d\,e\,g\,f$ by the long path won't solve our problems, for once we did so, we would block ourselves from the vertex y, which might be our unmatched vertex.

The source of our difficulty is the fact that we are searching along odd cycles, say of size $2k+1$ (k is two in the shorter troublesome cycle of Figure 1.4 and k is six in the longer one). Our cycles have k matching and $k+1$ nonmatching edges, forcing two nonmatching edges to be adjacent ($\{c,d\}$ and $\{c,f\}$ in the five-vertex cycle). This pair of nonmatching edges allows us to enter the cycle in two different directions at that point. In the language introduced by Edmonds to describe this situation, we call a cycle of length $2k+1$ with k matching edges a **blossom** if there is an alternating path from our initial unmatched vertex (a in Figure 1.4) which enters the cycle through the vertex contained in two unmatched edges. The blossom and this path are called a **flower**; the path is called the **stem** of the flower, and the vertex where the stem enters the blossom is called the *base* of the blossom. The initial vertex of the search is called the *root* of the flower. Figure 1.4 has the "obvious" flower $a\,b\,c\,d\,e\,g\,f$; it also has a flower with the "trivial" stem a, namely

$$a\,b\,c\,f\,g\,p\,n\,m\,l\,k\,j\,i\,h.$$

Unfortunately, when we enter a blossom along its stem as we search for an alternating path, we do not know which way to turn. Edmonds recognized that, once we know the edge along which we want to leave the blossom, then we can back up to find the unique path through the blossom from the base through this edge. Further, if we do not enter the blossom through its base, there will be only one direction in which an alternating path can go once it enters the blossom. Edmonds gave an elegant and simple technique for determining how we wish to leave the blossom. Once we recognize that our search has led us into a blossom through its base, we replace the entire blossom by a single vertex—connected to all vertices adjacent to something in the blossom. We continue our search in this modified graph. If we find an alternating path that happens to include the "megavertex" by which we replaced the blossom, the vertex to which we go when we leave the megavertex is also the vertex to which we wish to go when we leave the blossom. At least one vertex on the blossom is adjacent to this one by a nonmatching edge, and by backing up to the base from this vertex, beginning with a matching edge of the blossom, we can determine the alternating path we should use to substitute for the megavertex.

Now we must solve the problem of recognizing when we have entered a blossom through its base. Recall that our search process will build up a tree, one vertex and one edge at a time. From time to time, as we consider edges leaving one vertex, we discover that an edge we are considering connects to another vertex we have already added to the tree. Suppose this occurs when we have added all but one edge of a blossom and we are examining an endpoint of that edge. As you see in Figure 1.4, no matter which blossom edge we are examining, either both paths (in the blossom) from the base to its endpoints end with matching edges or else both paths end with nonmatching edges (and, in this second case, the edge we are examining is a matching edge). Perhaps surprisingly, whenever we discover we are examining an edge with this property, we have entered a blossom from its base.

Our search procedure will guarantee that each path from the root to any other vertex in the search tree is an alternating path. Thus, every other edge in such a path is a matching edge. We say a vertex in the search tree is *even* if the alternating path to it in the tree ends in a matching edge and is *odd* if the path to it in the tree ends in a nonmatching edge. In this language, we discover we have entered a blossom by its base when we find an edge whose endpoints are already in the tree and are both even or both odd. In such a situation we say that the endpoints of the edge have the same *parity*. Now suppose that in our tree we happen to find two vertices u and v with the same parity which are adjacent in G. The two paths from our root to u and v begin at the common root vertex and may have more vertices in common. However, if they take different edges at a vertex b, then b must be the last vertex they have in common, for otherwise our tree would have a cycle. Further, since b can touch only one matching edge, both paths must leave b on nonmatching edges. These two edges are the first edges of two paths which still end in adjacent vertices of the same parity. Thus, if these paths end with matching edges, they and the edge $\{u, v\}$ form a blossom. Similarly, if these paths end with nonmatching edges and $\{u, v\}$ is a matching edge, then they and the edge $\{u, v\}$ form a blossom.

Our observations have led us to the point where we can describe a search procedure to apply labels to vertices and "shrink" blossoms in the graph in order to find alternating paths or determine that there are no alternating paths. Our procedure begins with a graph G having a matching M and an unmatched vertex v_0 (which is a in Figure 1.4). We label all vertices adjacent to v_0 with the label v_0 and we use $L(x)$ to refer to the label on x when we need to. This is illustrated in the first picture of Figure 1.5. If the number of vertices we have labeled is i, then we use V_i to stand for the set consisting of v_0 and these adjacent vertices. We now define a labeled vertex x to be *odd* if $\{x, L(x)\}$ is a nonmatching edge, and we define v_0 and any other labeled vertex to be *even*. We say the edges $\{x, L(x)\}$ are *tested edges*. Now we repeat the following steps until all possible edges have been tested.

Figure 1.5

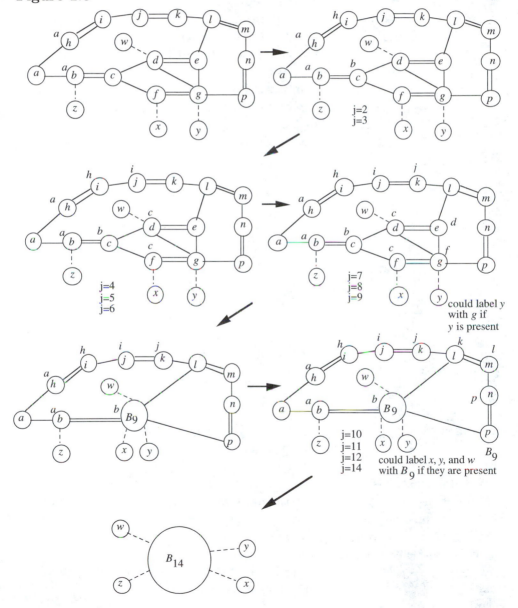

1. Given an integer j and a set V_j of labeled vertices we choose a vertex x in V_j and an untested edge $\{x, y\}$.

2. (Case 1) If y is not in V_j and labeling y with x would give x and y the opposite parity, we let $V_{j+1} = V_j \cup \{y\}$, we label y with the label x, that is, we let $x = L(y)$, and replace j by $j + 1$. We mark $\{x, y\}$ as tested.

   (Case 2) If y is in V_j and x and y have opposite parity, then mark the edge $\{x, y\}$ as tested.

   (Case 3) If y is in V_j and $\{x, L(x)\}$, $\{y, L(y)\}$, and $\{x, y\}$ are all nonmatching edges, then mark $\{x, y\}$ as tested.

   (Case 4) If y is in V_j and x and y have the same parity, then we back up along the path x, $L(x)$, $L(L(x)), \ldots$ and the path y, $L(y)$, $L(L(y)), \ldots$ until we find the first (that is, closest to x and y) vertex b they have in common. (There are various ways to carry this out; we shall not specify the precise technique here.) We replace all vertices on the paths from x and y to this common vertex by a single vertex B_j. For each labeled vertex x whose label was on one of these paths, we change the label $L(x)$ to B_j; for each edge $\{v, z\}$ such that v is not on one of these paths but z is, we change the edge to $\{v, B_j\}$. We remove the vertices on these paths from V_j, let $V_{j+1} = V_j \cup \{B_j\}$, we label B_j with $L(b)$ (if it exists) and let $L(B_j) = L(b)$, we mark $\{B_j, L(B_j)\}$ as tested if it exists, and we replace j by $j + 1$. (Note that the edge $\{b, L(b)\}$, if it exists, will be a matching edge; we let the edge $\{B_j, L(B_j)\}$ replace it in the matching but we let other matching edges along the paths from b to x and y disappear, as do any other edges joining vertices along these paths.) In any case, we say the parity of B_j is even.

This procedure is illustrated in Figure 1.5, where the five-vertex blossom is replaced by the megavertex B_9 and the larger blossom is replaced by the megavertex B_{14}. The set V_{15} now becomes B_{14}, together with any of w, x, and y that happen to be in the graph and w, x, and y have their labels changed to B_{14} if they happen to be in the graph. Finally if z happens to be in the graph it gets the label B_{14}. Notice that B_{14} has no label. How can we explain this? We say that a vertex v of G is a member of B_j if it is one of the vertices we removed when we created B_j, or if it is a member of a vertex B_i we removed when we created B_j. In this terminology, we can see that B_j is not labeled if and only if v_0 is in B_j. Since B_{14} contains a, it is unlabeled. At any given stage of the process, there can be at most one unlabeled megavertex in our graph, and this unlabeled megavertex, if it exists, must contain v_0 (which is a in our example).

It is now possible to show that once all edges of the graph (as modified in step 2, case 4) are marked as tested, there is an alternating path from v_0 to an unmatched vertex t in G if and only if t is in the tree whose vertices are the labeled vertices and whose edges are the sets $\{v, L(v)\}$. (This is true even in v_0 happpens to have been part of a blossom and so has been replaced by a megavertex as in Figure 1.5.) If v_0 is in the tree, then the path from an unlabeled vertex t to v_0 is $t L(t) L(L(t))\ldots$. If v_0 is not in the tree, there is an unlabeled megavertex, and we begin with the path $t L(t) L(L(t))\ldots$ from t to the unlabeled megavertex (which contains v_0). To obtain from this path the path from v_0 to t in G, we

1. Take the blossoms B_j in the path, in the order from the largest j to the smallest.
2. Note for which x in our path we have $B_j = L(x)$.
3. Find, in the cycle replaced by B_j, a vertex y adjacent to x (this adjacency will automatically be by a nonmatching edge).
4. Determine, by working backward from y using labels, the path beginning with a matching edge (in the cycle) from y to the base b of the cycle, and
5. Replace B_j by this alternating path.

Figure 1.6 The curved line starting at a and going to z shows which vertices are on the path between a and z at each stage.

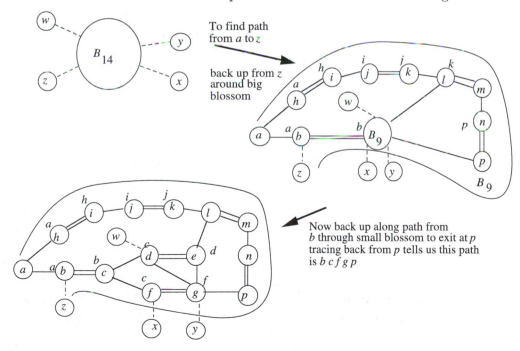

To find path from a to z

back up from z around big blossom

Now back up along path from b through small blossom to exit at p tracing back from p tells us this path is $b\,c\,f\,g\,p$

This procedure is illustrated in Figure 1.6 for the path from $v_0 = a$ to the unmatched vertex z. Our initial path is simply zB_{14}. We start with the blossom B_{14}. We note that z has the label B_{14}. We find in the cycle that B_{14} replaced the vertex b adjacent to z. Now working backward from b we find the path, starting with a matching edge, marked with the curved line joining a and z in the figure. We then repeat our procedure with the blossom B_9, and get the path shown in the final part of the figure by the curved line joining a with z.

EXERCISES

1. Explain why, if at a school dance, for each group of k boys there are at least k girls who like at least one boy in the group, then every boy can find a girl to dance with. (Assume that a girl will dance with the boys she likes and no others! For reasons related to this example, Hall's matching theorem is sometimes called the marriage theorem.)

2. Apply the Berge criterion to show that the vertical edges in Figure 1.3(b) do not form a maximum-sized matching.

3. Find a complete matching for Figure 1.3(b).

Figure 1.7

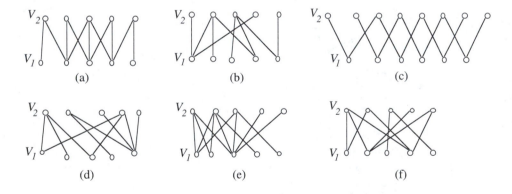

4. Which of the bipartite graphs in Figure 1.7 have complete matchings? Find matchings for those that do. (Inspection should suffice for finding matchings.)

5. Find minimum-sized vertex covers and maximum-sized matchings for the graphs in Figure 1.7 that lack complete matchings.

6. Find a system of distinct representatives for the family of sets $\{1, 2, 4, 8\}$, $\{2, 3, 5, 7\}$, $\{1, 3, 8, 6\}$, $\{2, 4, 6, 8\}$, $\{1, 3, 7, 4\}$.

7. Determine whether each family that follows has a system of distinct representatives. Find such a system if the family has one.
   (a) $\{1,2,7\}$, $\{1,4,8\}$, $\{2,4,6\}$, $\{1,6,8\}$, $\{2,3,5\}$
   (b) $\{1,7,9,4\}$, $\{2,3,5,8\}$, $\{3,8,2\}$, $\{4,6,7\}$, $\{5,6,3,8,2\}$
   (c) $\{1,2,4\}$, $\{1,4,5\}$, $\{3,4,6\}$, $\{2,5,8\}$, $\{4,5,8\}$, $\{2,4,8\}$, $\{2,4,5\}$
   (d) $\{1,3,6,9\}$, $\{3,9,7\}$, $\{1,8,7\}$, $\{3,6,8\}$, $\{6,7,8\}$, $\{7,8,9\}$, $\{1,5,6\}$, $\{3,4,5\}$,$\{1,3,7,9\}$
   (e) $\{1,5,6\}$, $\{6,8,9\}$, $\{3,5,9\}$, $\{1,7,9\}$, $\{1,6,7,9\}$, $\{2,6,8\}$, $\{2,5,7\}$, $\{2,3,6,7,8\}$

8. A Latin rectangle is an $m \times n$ matrix with $m \leq n$ whose rows are each a permutation of $1, 2, \ldots, n$ and whose columns are lists of m distinct integers. By using an SDR for the sets of numbers not used in each column, show that if $m < n$, an $m \times n$ Latin rectangle may be extended to an $(m+1) \times n$ Latin rectangle.

9. Show that Theorem 1.5 may be derived as a consequence of Theorem 1.6.

10. Derive Theorem 1.5 as a consequence of Theorem 1.7.

Figure 1.8

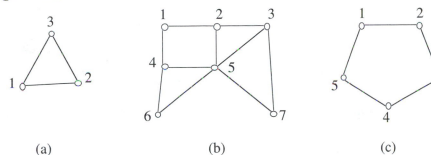

(a)   (b)   (c)

11. Find a minimum-sized vertex cover and a maximum-sized matching in the graphs of Figure 1.8.

12. What are the sizes of the minimum vertex covers and maximum matchings in
    (a) A cycle on an even number n of vertices?
    (b) A cycle on an odd number n of vertices?
    (c) A complete bipartite graph?
    (d) A complete graph on n vertices?

13. In Figure 1.1(b), construct an alternating breadth-first search tree centered at vertex 3 using the matching $\{\{2,6\}, \{4,9\}, \{5,11\}\}$. Find

Figure 1.9

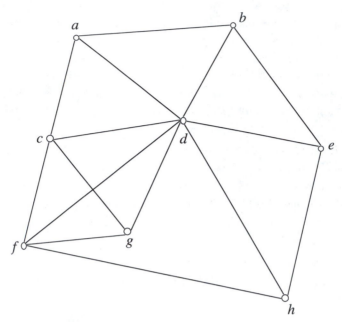

an alternating path you can use to make a larger matching. Find the larger matching.

14. In the graph of Figure 1.9, the set of edges $\{a, c\}$, $\{d, f\}$, $\{b, e\}$ is clearly a matching.
    (a) Construct an alternating breadth-first search tree centered at each unmatched vertex.
    (b) Determine whether or not there is a larger matching and find one if it exists.

Figure 1.10

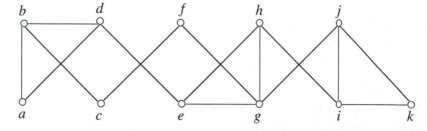

15. Beginning with the three vertical edges in Figure 1.10, find a maximum-sized matching using alternating paths. (You may find the alternating paths by trial and error or by constructing alternating search trees.)

16. Show that if there is an alternating path from an unmatched vertex x to an unmatched vertex y for a matching M in a bipartite graph, then an alternating breadth-first search tree centered at x contains an alternating path from x to an unmatched vertex. Why does this mean that repeating the process of alternating breadth-first search for alternating paths (and using the paths to make the matching bigger) will lead to a maximum-sized matching?

*17. Prove the Birkhoff–von Neumann theorem by induction on the number of nonzero entries using the following outline. Argue as to why there must be a set of m nonzero entries, one in each row and column. Pick the smallest of these m nonzero entries in M, and subtracting a multiple of a permutation matrix, subtract that smallest entry from all m positions. This gives a matrix with one more 0.

18. Prove or disprove: There is a matching from V_1 to V_2 in a bipartite graph if, for some fixed number k, there are k or more edges containing each vertex of V_1 and no more than k edges containing each vertex of V_2.

19. Suppose P and Q are partitions of N into m parts such that the union of any k classes of P is not a subset of the union of any $k - 1$ classes of Q. Prove that there is a system of common representatives of P and Q, i.e., a list of elements a_i that is an SDR for both P and Q.

20. Show that the number of $r \times n$ Latin rectangles is at least

$$ n!(n-1)! \cdots (n-r+1)!. $$

What does this say about the number of Latin squares? (See Exercise 8.)

21. (a) Suppose G is a bipartite graph. Suppose that M is a matching of G and x is an unmatched vertex that has no alternating path to any other unmatched vertex. Suppose M' is the matching we get from M by augmenting by a single alternating path. Show that there is no alternating path for M' from x to an unmatched vertex.

    (b) In this part of the exercise we will make an estimate of how time consuming it is to find a matching in a bipartite graph with v vertices and e edges by repeatedly carring out breadth-first searches and finding alternating paths. When we examine an edge in an alternating breadth-first search, one of its vertices is in the tree, so we are checking to see if its other vertex is not. At the same time we can check to see if the other vertex is unmatched. Note that a breadth-first search does not examine any edge twice. When we construct a path from a leaf of a tree to the root, we examine the parent of each nonroot vertex in the path at most once. Let us call each of the examinations we described a step. Show that the number of steps needed to find a matching of maximum size is at most ve.

22. Give an example of a matching of a nonbipartite graph in which an alternating breadth-first search tree centered at an unmatched vertex x fails to find an alternating path from x to an unmatched vertex, even though such a path exists in the graph.

*23. In Figure 1.5, the order in which the labels was assigned was breadth-first order, and since vertex k was labeled before vertex d, vertex l got k as its label rather than getting B_9. Nothing in our description of the labeling would prevent us from using the label B_9 on l instead. In fact, the result of labeling l in this way is quite interesting. Show how the remainder of the labeling and shrinking goes if we label vertex l with B_9 as the next step after B_9 is formed.

*24. In Figure 1.5, the order in which the labels was assigned was breadth-first order. Show how the process would proceed in Figure 1.5 if we use depth-first order, labeling vertex b first. (When you have a choice, make it however you wish. There are a number of possible outcomes.)

*25. In Figure 1.5, the order in which the labels was assigned was breadth-first order. Show how the process would proceed in Figure 1.5 if we use depth-first order, labeling vertex h first. (If you have a choice, make it however you wish.)

*26. Show that if, in the labeling and shrinking process described in this section, we always proceed in depth-first order, then when we discover that we have entered a blossom through its base, one of the endpoints of the edge we are examining will be the base.

*27. Show that if there is an alternating path from an unmatched vertex x to an unmatched vertex y which goes through the base of a certain blossom whose root is at x, then after that blossom is replaced by a single vertex as in the labeling and shrinking procedure, there is still an alternating path in the resulting graph from x to y. If there is an alternating path from x to y through the smaller graph, does this mean there is one in the larger graph as well?

*28. Show that if the labeling and shrinking algorithm shrinks a blossom B of a flower rooted at x in a graph G, and there is an alternating path from x to an unmatched vertex y which does not include the base of the blossom, then there is a (perhaps rather different) alternating path from x to y in the resulting graph. You may assume that B is the only blossom shrunk. As in the previous exercise, if there is an alternating path from x to y in the smaller graph, does this mean there is one in the larger graph as well?

*29. Explain how the preceding two exercises let you prove that the labeling and blossom shrinking procedure suffices to discover an alternating path from the unmatched vertex x to an unmatched vertex y if there is such a path.

30. Write a computer program that finds alternating search trees in bipartite graphs. Expand it to a program that constructs maximum-sized matchings.

Section 2   The Greedy Algorithm

The SDR Problem with Representatives That Cost Money

We began our study of matching with a problem that has a natural interpretation in terms of either systems of distinct representatives or matchings. Recall that we have a school district trying to fill positions. For each position, the school board has a set of candidates qualified to fill it, and the school board wants a system of distinct representatives that matches distinct candidates with positions for which they are qualified. We can turn any such SDR problem into a matching problem in a bipartite graph. We construct the bipartite graph whose parts are the set of positions and the set of candidates; an edge from a candidate to a position means the candidate qualifies for the position. A maximum-sized matching from the candidates to the positions gives us a set of distinct representatives which fills as many positions as possible. If all the positions are matched with candidates, then we have a system of distinct representatives. (Notice that the definition of a system of distinct representatives requires that we have a representative for each set. Thus a matching gives a set of distinct representatives, but only a complete matching gives a system of distinct representatives.) We can use the matching algorithm of the last section to find such a system of distinct representatives.

In light of financial constraints, however, the school board might not find this solution particularly attractive. Each candidate may have some minimum salary he or she will accept and the school board may wish to fill as many positions as possible at the lowest possible total cost. In other words, among all ways of choosing the set of people to fill the jobs, we wish to choose the set whose total salary is the smallest. Does this mean we must examine all sets of representatives of maximum size to choose the cheapest set? No, fortunately, there is a much less time-consuming method.

The Greedy Method

The so-called greedy method of choosing a set of representatives is as follows. First, choose the cheapest representative possible. Next, choose the cheapest representative such that these two representatives can fill two positions. Once you have a certain set of k representatives for k positions, choose as your next representative the cheapest possible representative so that this representative and the k previously chosen ones can fill $k+1$ positions among them. It turns out that if we keep this process up until it is no longer possible to find a new representative to add, then we will end up with a set of representatives which is both as large as possible *and* as cheap as possible among all maximum-sized sets of distinct representatives.

Notice that the method concentrates on the choice of the representatives and not on the assignment of the representatives to positions. For this reason we introduce new terminology to emphasize our concentration on representatives alone. In particular, given a bipartite graph with parts V_1 and V_2 we say that a subset I of V_1 is ***independent*** *for matchings of* V_1 *into* V_2 if there is a matching which matches all the elements of I to elements of V_2. In the same way, given a family **F** of subsets of a set X, we say that a subset I of X is an ***independent*** *set of representatives for* **F** if the elements of I may be matched with sets in **F** so that the elements of I (in some order) form a system of distinct representatives for the sets with which they are matched.

We want to build up maximum-sized independent sets by continually adding the cheapest element possible to the set we already have. How do we know that we will in fact have a *maximum-sized* independent set when we can no longer find representatives to add and the process stops? In other words, when we can no longer add representatives to the particular independent set we have, how do we know there isn't some other independent set that is larger? Theorem 2.1 assures us that we will get a maximum-sized independent set. This version of the theorem and the phrase "greedy algorithm" are due to Edmonds.

> ***Theorem 2.1.*** The independent sets for matchings of V_1 into V_2 in bipartite graphs and the independent sets of representatives for a family **F** of subsets of a set X satisfy the rule:
>
> ***Expansion Rule***. If I and J are independent sets and $|I| < |J|$, then there is an element x of $J - I$ such that $I \cup \{x\}$ is independent.
>
> *Proof.* We prove the theorem for independent subsets for matchings of V_1 into V_2 in a bipartite graph G. Because of the relationship between matchings and systems of distinct representatives, the same result will hold for independent sets for a family **F** of subsets of a set X.
>
> Suppose M_1 is the matching of I into V_2 and M_2 is the matching of J into V_2. As in Lemma 1.1, let G' be the graph on $V_1 \cup V_2$ with edge set
>
> $$E' = (M_1 \cup M_2) - (M_1 \cap M_2) = (M_1 - M_2) \cup (M_2 - M_1).$$
>
> From Lemma 1.1 we know that each connected component of G' is one of three types. Since $|M_2| > |M_1|$, at least one of these connected components must be a type 3 component with more edges from M_2 than from M_1. Thus, there is an alternating path (for both M_1 and M_2) whose first and last edges are in M_2. Each vertex of this path

touched by an M_1 edge is touched by an M_2 edge. Also, one V_1 vertex x (and one V_2 vertex) touched by an M_2 edge is not touched by an M_1 edge. Let E' be the edge set of the alternating path. By Theorem 1.2, the set of edges

$$(M_1 - E') \cup (E' - M_1) = M'$$

is a matching with one more edge than M_1. Thus M' is a matching of $I \cup \{x\}$ into V_2. Thus $I \cup \{x\}$ is independent and by our choice of x, we know that $x \in J - I$. ∎

To use our greedy method to build up a set of independent representatives, we need some independent set of representatives to start with. The subset rule that follows simply formalizes something we already know about independent sets and says we can start with the empty set.

Subset Rule. Every subset of an independent set is independent. In particular, the empty set is independent.

In our problem, we have a cost C_x associated with each element x of our set X of candidates, and we wish to find a set of distinct representatives such that the sum of the costs of these representatives is a minimum. In the language we have introduced, we can give the greedy method the following precise description.

The Greedy Algorithm
By the greedy algorithm, we mean the following:
Step 1. Let $I = \emptyset$.
    Repeat steps 2 through 4 until X becomes empty.
Step 2. From the set X, pick an element x with minimum cost.
Step 3. If $I \cup \{x\}$ is independent, replace I by $I \cup \{x\}$.
Step 4. Delete x from X.

The greedy algorithm works to select a maximum-sized independent set of least cost because of the expansion and subset properties. To see why, we study these properties in isolation by introducing the concept of a matroid.

Matroids Make the Greedy Algorithm Work
A **matroid** on a set X is a collection **C** of subsets of X (called independent subsets of X) satisfying the subset rule and expansion rule that follow.

Subset Rule. Every subset of an independent set is independent. In particular, the empty set is independent.

Expansion Rule. If I and J are independent sets and $|I| < |J|$, then there is an element x of $J - I$ such that $I \cup \{x\}$ is independent.

The expansion rule is the same one we stated previously for independent sets for a matching or a family of sets. (Note that we are no longer assuming we have a family **F** and are concentrating instead on the collection **C** of independent sets.) Right now the only examples we have of matroids come from matchings of a bipartite graph or independent sets of representatives of a family of sets.

In the case of a bipartite graph with parts A and B we can take as our set X the set A, and define a subset I of A to be in our collection **C** if and only if there is a matching from the vertices of X to a subset of the vertices of B. In the case of a family **F** of subsets of a set Y, we can take our set X to be Y itself, and define a subset I of Y to be independent if it can be used as a system of distinct representatives for a collection of sets in our family.

The word matroid comes from the word matrix. Given a matrix M, we can use the set of columns of M as our set X. We say a set of columns is *independent* if it is empty or if we can row-reduce M so that the corresponding set of columns in the row reduced matrix has exactly one 1 in each column, and the 1's in different columns (of this set) are in different rows. The same row reduction that demonstrates a set of columns is independent demonstrates that any nonempty subset of that set of columns is independent. Thus our independent sets of columns satisfy the subset rule. It is not so straightforward to verify that our independent sets of columns satisfy the expansion rule (although they do). Since we do not need to use this example of matroids, we will not work out the details here. However the reader who has studied linear algebra should be able to recognize the expansion rule as an application of the idea of independence of vectors, and the reader who has a solid knowledge of row reduction should be able to work out the expansion property.

Given a matroid, a set of maximum size in the collection **C** is called a **basis**. In particular, two different bases (the plural of basis) must have this same (maximum) size. An **additive cost function** f is simply a numerical function whose domain consists of all subsets of X, whose value on the empty set is zero and whose value on $\{x_1, x_2, \ldots, x_k\}$ is $\sum_{i=1}^{k} f(\{x_i\})$. In other words, f is defined on subsets by summing its values on single elements. We write $f(x_i)$ in place of $f(\{x_i\})$ when this makes our notation easier to read.

Theorem 2.1 may be reformulated to state that the independent subsets for matchings of a bipartite graph (with $X = V_1$) and the independent sets of representatives of a family **F** of subsets of X are the independent sets of a matroid on X.

Theorem 2.2. Suppose we are given a matroid on X and an additive cost function on the subsets of X. The greedy algorithm selects a basis of minimum cost (a basis such that the cost function evaluated on this basis is no more than the cost function evaluated on any other basis).

Proof. One consequence of the expansion rule is that the greedy algorithm will be able to continue adding to I until it reaches a maximum-sized independent set. Thus, it selects a basis. Suppose it selects the basis B, and A is another basis. List the elements of B as $(b_1, b_2, \ldots, b_k)$ and the elements of A as $(a_1, a_2, \ldots, a_k)$ in order of increasing cost, i.e., so that if $i < j$, then the cost of b_i is less than or equal to the cost of b_j and the cost of a_i is less than or equal to the cost of a_j. By step 2 of the greedy algorithm, $\text{cost}(b_1) \leq \text{cost}(a_1)$. If $\text{cost}(b_i) \leq \text{cost}(a_i)$ for all i, then the cost of B is no more than that of A, so assume that for some i, but for no previous i, $\text{cost}(a_i) < \text{cost}(b_i)$. Then the two sets

$$\{a_1, a_2, \ldots, a_i\}$$

and

$$\{b_1, b_2, \ldots, b_{i-1}\}$$

are both independent. Then, by the expansion property, for some a_j with $j \leq i$,

$$\{b_1, b_2, \ldots, b_{i-1}, a_j\}$$

is independent. Since $\text{cost}(a_j) \leq \text{cost}(a_i) < \text{cost}(b_i)$, a_j would have been selected by the greedy algorithm. Thus, the cost of b_i is no more than the cost of a_i and therefore B is a minimum-cost basis. ∎

The concept of a basis is especially useful in our only current example of a matroid, the matroid for a family of sets. Suppose our family **F** of subsets of X has k members. If we get a k-element set when we find a basis B for the matroid of sets independent relative to **F**, then B will be a complete set of independent representatives. Thus, we will know our family has an SDR. If B has fewer than k elements, then there is no independent set of representatives with k elements and so there is no SDR.

Frequently, we will know the size of a basis in our application; in this case, we can change the *repeat* instruction of the greedy algorithm to

Repeat steps 2 through 4 until I becomes a basis.

How Much Time Does the Algorithm Take?

The algorithm can be easily programmed for a computer once we have a representation of the elements of X in a form appropriate for our computer language and once we have a test for independence. When we study algorithms, we are faced with the practical question, "How many steps will our algorithm use?" Conceivably, even with the new repeat instruction, we would have to examine each element of X in steps 2 and 3 before getting a maximum-sized I. Further, we will need to pick the lowest cost element of X each time, and so we will either have to arrange the elements of X in increasing order of cost before starting or search for the lowest cost x each time. Finally, we won't know how many steps are involved until we know how many steps it takes to test $I \cup \{x\}$ for independence. This will vary from application to application.

Consider, for example, the application to the independent sets of representatives of a family of sets. From Theorems 1.6 and 2.1, it is clear that a set $I \cup \{x\}$ will be independent if and only if each j-element subset lies in at least j sets in the family **F**. Further, since we already know that I is independent, we need only check subsets that contain x. This is a straightforward way to proceed, but since our eventual SDR (if there is one) contains k elements, we will be making 2^k checks, one for each subset of the set of distinct representatives.

Since the time needed to use the algorithm is proportional (more or less) to the number of steps, and we have 2^k checks to make, we say this implementation of the algorithm is exponential in k. Suppose we also search for the lowest cost x each time; then potentially we must examine

$$n + (n-1) + (n-2) + \cdots + 1 = \frac{n^2 + n}{2}$$

possibilities for x. Thus, we say the algorithm is quadratic in n. Since an algorithm which requires 2^k steps would be quite time consuming, we would naturally like to improve on the time needed for the independence test.

By thinking in terms of matchings instead of representatives of a family of sets, we can develop a considerably improved implementation of the greedy algorithm for this application. Recall that we say a subset I of V_1 in a bipartite graph with parts V_1 and V_2 is independent if and only if there is a matching of I into V_2. Suppose we not only keep track of I as we apply the greedy algorithm but also, at each step, record a matching $M(I)$ of I into V_2.

We can modify the proof of Theorem 1.2 to show that given an element x not matched by $M(I)$, there is a matching of $I \cup \{x\}$ into V_2 if and only if there is an alternating path joining x to another vertex unmatched by $M(I)$. We can use inspection to find such a path or we can search for such a path by constructing an alternating search tree centered at x. From this alternating path, we write down the matching $M(I \cup \{x\})$. Thus, if we keep track of $M(I)$ as well as I, the number of steps required to check $I \cup \{x\}$ for independence is no more than the number of edges of the graph. Thus, the number of steps required by the greedy algorithm is no more than the number of edges times the number of vertices in the graph. The number of edges is less than n^2 if n is the number of vertices in the graph, and so the number of steps involved in this version of the algorithm is approximately n^3. This is certainly superior to an algorithm that is exponential in k. Often we can find a matching by examining a picture; thus in most cases when we apply the greedy algorithm to construct a set of independent representatives by hand, we will not actually use a search tree.

The Greedy Algorithm and Minimum-Cost Independent Sets

Example 2.1. Apply the greedy algorithm to find a minimum-cost independent subset of maximum size in the graph of Figure 2.1.

Figure 2.1

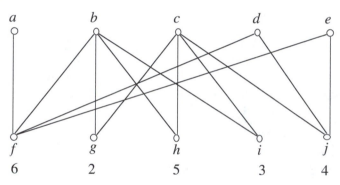

The costs of vertices in V_1 are 6 for f, 2 for g, 5 for h, 3 for i, and 4 for j. We begin with $I = \{g\}$, $M(I) = \{\{g, b\}\}$. Note we could let $M(I)$ be $\{\{g, c\}\}$ just as well. The next cheapest member of V_1 is i, so we let $I = \{g, i\}$, $M(I) = \{\{g, b\}, \{i, c\}\}$. Next we wish to add j to I; since we can match j with e, we let $I = \{g, i, j\}$ and $M(I) = \{\{g, b\}, \{i, c\}, \{j, e\}\}$. Now we ask if $I \cup \{h\}$ is independent. Since h can be matched only with b and c and they are already used, we will have to change our matching or else we won't be able to use h. You may wish to check that there is no alternating path connecting h with an unmatched vertex. In this case, however, it is simpler to note that the three vertices g, h, and i are connected to just the two vertices b and c, so by Theorem 1.5, the König–Hall theorem, there is no way to match these to three distinct vertices in V_2. (In other words, $|R(\{g, h, i\})| < |\{g, h, i\}|$.) Thus, $I \cup \{h\}$ is not independent.

Now $I \cup \{f\}$ is independent because $\{f, a\}$ is an edge without vertices in common with edges of $M(I)$. Thus, we may let $I = \{g, i, j, f\}$ and $M(I) = \{\{g, b\}, \{i, c\}, \{j, e\}, \{f, a\}\}$. Then I is the basis chosen by the greedy algorithm for the "matching matroid" and costs $2 + 3 + 4 + 6 = 15$. Note that there are many matchings of I into V_2; there is nothing special about the particular choices we wrote down as we went along. ∎

The matroid we use in studying independent sets for a matching is called a "matching matroid" or a "transversal matroid." The concept of a matroid was introduced by Hassler Whitney in 1935. He gave it this name because in the way that he studied matroids, it was quite similar to studying families of columns of a matrix. The transversal matroid was introduced by Edmonds and Fulkerson in 1965. The matroid itself consists of the set V_1 and the independent subsets of V_1. In Example 2.1 we saw several independent sets of our matroid but not all of them. For example, $\{f, h, i, j\}$ is independent as well. Can you find any other independent sets of size 4?

The Forest Matroid of a Graph

The greedy algorithm has many applications aside from the determination of lowest-cost sets of distinct representatives. That is one reason for stating it in terms of matroids; to find whether we can use the algorithm in another situation, we analyze the situation to see if it has a matroid interpretation. In our discussion of spanning trees for graphs, we introduced the problem of determining a minimum-cost spanning tree. For this problem we have a communications network connecting various cities. We know the cost of using the communication line from city i to city j (if there is such a line) and we wish to construct a system without redundant links that links together all the cities at minimum cost. Such a system is a spanning tree such that the sum of the costs of its edges has minimum value among all such sums for all spanning trees.

Based on our solution to the problem of a set of representatives of lowest cost, it seems natural to try to build our spanning tree one edge at a time. If we always choose the cheapest edge, we might first take the Concord–Boston edge, then the Miami–Atlanta edge, etc. This kind of set of edges won't look like a tree while we are building it, but so long as we keep building bigger and bigger forests of edges, eventually the pieces will join together to form a tree. A set F of edges in a graph $G = (V, E)$ is called a *forest of* G if $F \subseteq E$ and (V, F) is a forest.

The cost of each edge is given, and the cost of a set of edges is the sum of the costs of its elements. Thus, we have an additive cost function, as defined before Theorem 2.2. To apply Theorem 2.2, we need a matroid.

Theorem 2.3. The edge sets of forests of a graph $G = (V, E)$ form the independent sets of a matroid on E.

Proof. If (V, F) has no cycles, then (V, F') has no cycles for any subset F' of F, and so the forests satisfy the subset rule.

Recall that a tree is a connected graph with k vertices and $k - 1$ edges. Thus, a forest on v vertices with c connected components will consist of c trees and thus will have $v - c$ edges. Suppose F' and F are forests with r edges and s edges and suppose further that $r < s$. If no edges of F can be added to F' to give an independent set, then adding any edge of F to F' gives a cycle. In particular, each edge of F must connect two points in the same connected component of (V, F'). Thus, each connected component of (V, F) is a subset of a connected component of (V, F'). Then (V, F) has no more edges than (V, F'), so that $r \geq s$, a contradiction. Therefore, the forests of G satisfy the expansion rule, so the collection of edge sets of forests of G is a collection of independent sets of a matroid on E. ∎

Minimum-Cost Spanning Trees

Since a maximum-sized forest in a connected graph must be a tree (if it had two connected components, you could join them without adding any cycles), we get

Theorem 2.4. The greedy algorithm applied to cost-weighted edges of a connected graph produces a minimum-cost spanning tree.

Example 2.2. The greedy algorithm applied to the graph in Figure 2.2 yields the following sequence of results.

Figure 2.2

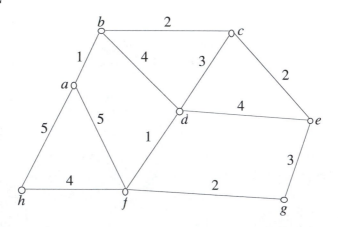

$$\{\{a,b\}\};\quad \{\{a,b\},\{d,f\}\};\quad \{\{a,b\},\{d,f\},\{f,g\}\};$$
$$\{\{a,b\},\{d,f\},\{f,g\},\{b,c\}\};\quad \{\{a,b\},\{d,f\},\{f,g\},\{b,c\},\{c,e\}\};$$
$$\{\{a,b\},\{d,f\},\{f,g\},\{b,c\},\{c,e\},\{c,d\}\};$$
$$\{e,g\}: \text{REJECT (gives cycle)};\quad \{d,e\}: \text{REJECT (gives cycle)};$$
$$\{\{a,b\},\{d,f\},\{f,g\},\{b,c\},\{c,e\},\{c,d\},\{h,f\}\}.$$

The second variant of the greedy algorithm stops with this set of seven edges since a spanning tree of a graph on eight vertices can have only seven edges. The first variant would have continued testing and rejecting edges until all edges had been tested. ∎

As applied in Example 2.2, the greedy algorithm is known as Kruskal's algorithm, published by Kruskal in 1956, for constructing a minimum-cost spanning tree. Kruskal, though, designed an efficient scheme of deciding whether to accept or reject edges. He kept track of the connected component partition of the graph whose edge set is the independent set I of edges and rejected an edge whenever it joined two vertices already connected in that connected component partition. The data structures which allow us to keep track of connected component partitions efficiently are a fascinating part of a course in the design and analysis of algorithms. Notice that this is the first example we have of an algorithm which constructs a spanning tree but is not a special case of algorithm Spantree of Section 2 of Chapter 4.

There is also a variation of Spantree which produces a minimum-cost spanning tree by constructing a spanning tree, one edge at a time, so that at each stage the edge set we have is the edge set of a tree. This algorithm, published by Prim in 1957, is the subject of Exercise 16 of this section. In order to analyze the efficiency of the algorithm, we would have to specify how the test for independence would work. We can test a set $S \cup \{e\}$ of edges for "cyclicity" by the following method. As S grows, make a connectivity table for S by placing a 1 in position i, j of a matrix if v_i and v_j are connected by a path in S. Before adding e to S, check the table to see if e connects two vertices already connected in S, and if this is the case, reject it. Otherwise, to update the table use the fact that e connects two vertices v_j and v_k and thus anything previously connected to v_j is now connected to v_k and vice versa. Now the table is ready to test another e.

Alternatively, we can keep a list in which we put a 1 in position i if v_i is already in the tree. Then to test edge $\{v_i, v_j\}$ we check to see if there are already 1's in positions i and j. If not, we may add this edge. Is one of these two methods preferable?

The concept of a matroid is quite important in modern combinatorial mathematics. There are many examples of matroids other than the two given here; for applications, the matching matroids and forest matroids are among the most important. Not every application of the greedy principle gives rise to a matroid. For example, in Section 3 of Chapter 4, we used Dijkstra's "close first search" idea to construct a spanning tree in which each path from the root to another vertex had the smallest weight possible. This algorithm does not work if the weights on a cycle add up to a negative number; by Exercise 32, this means the trees it produces are not the bases of a matroid.

EXERCISES

1. Is $\{a, b, c\}$ an independent set of representatives for the family

$$(\{a, b\}, \{a, b, c\}, \{a, c\}, \{a\})?$$

2. Is $\{a, b, c\}$ an independent set of representatives for $(\{b\}, \{a\}, \{a, b, c\})$?

3. Is $\{a, b, c\}$ an independent set of representatives for the family of sets $(\{a, b, c\}, \{a, b\}, \{a, e\})$?

4. Can a set of six edges be independent in the complete graph on six vertices?

5. Does $\{1, 2, 5\}$, $\{2, 3, 4\}$, $\{1, 5, 8\}$, $\{2, 5, 6\}$, $\{1, 3, 7\}$, $\{1, 2, 6\}$, $\{1, 2, 8\}$, $\{1, 6, 8\}$, $\{2, 5, 8\}$ have a system of distinct representatives?

6. Find a maximum-sized independent set of representatives for the family of Exercise 5.

7. Let i be the cost of using i to represent a set in Exercise 5. Find a minimum-cost basis of representatives for the family in Exercise 5.

8. Prove that if two subsets of the set X of elements of a matroid are independent, but neither is a proper subset of an independent set, then they have the same size.

9. Find a maximum-cost spanning tree in the graph of Figure 2.2.

10. Find all independent sets of size 4 in the set V_1 of Figure 2.1.

11. The graph in Figure 2.3 shows the minimum acceptable hourly wage of applicants for various construction jobs and lists which applicants are qualified for (and willing to accept) which jobs. Find a minimum-cost set of job applicants who can fill each job.

12. In how many ways may you match your set of applicants in your solution of Exercise 11 to the jobs (considering the two rough carpenter jobs to be different)?

Figure 2.3

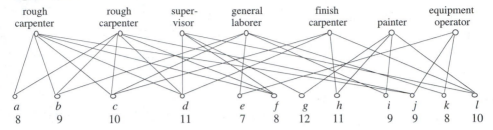

13. Find a maximum-cost set of job applicants to fill each job of Exercise 11. (To see why this problem might make sense, think of the numbers as representing years of experience; then we are getting the most experienced crew.)

14. Find a minimum-cost spanning tree in the graph of Figure 2.4.

Figure 2.4

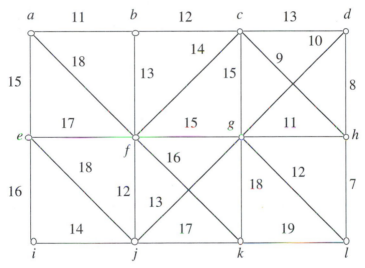

15. Find a maximum-cost spanning tree for the graph of Figure 2.4.

16. Explain why the following algorithm finds minimum-cost spanning trees. Choose a minimum-cost edge from among those which include an already chosen vertex but which do *not* include two vertices connected by edges you have already chosen. Continue until no further choices are possible.

17. Let X consist of all points in ordinary three-dimensional space. Define independence as follows. A set $I \subseteq X$ is independent if it is empty, has

one element, two elements, three elements which are not on the same line, or has four elements which are not in the same plane. Show that X together with these independent sets forms a matroid.

18. Rewrite the proof of Theorem 1.2 to show that, given an element x not matched by $M(I)$, there is a matching of $I \cup \{x\}$ into V_2 if and only if there is an alternating path joining x to another vertex unmatched by $M(I)$.

19. (a) Let I be the collection of subsets of $\{1, 2, 3, 4, 5, 6, 7, 8\}$ with fewer than four elements. Show that I is the set of independent sets of a matroid.
    (b) Let X be a set of size n. Define a subset of X to be independent if it has size less than $n/2$. Show that I is the set of independent sets of a matroid.

20. (For those who have had linear algebra.) Show that the independent sets of vectors in a vector space form the independent sets of a matroid.

21. (For those who have had linear algebra.) Suppose that X is the set of one-dimensional subspaces of a vector space and let a set I of these subspaces be independent if the dimension of the subspace sum of I is $|I|$. Explain why this gives a matroid.

22. Define a subset of $\{1, 2, 3, 4, 5, 6, 7, 8, 9, 10\}$ to be independent if its contains no more than two elements of $\{1, 2, 3, 4\}$ and no more than three elements of $\{5, 6, 7, 8, 9, 10\}$. Explain why this defines a matroid.

23. After Example 2.2 we gave a test for independence of a set of edges of a graph. Analyze the number of steps required by that test.

24. Show that if B_1 and B_2 are bases for a matroid on X and $x \in B_1$, then there is a y in B_2 such that $(B_1 - \{x\}) \cup \{y\}$ is a basis for the matroid as well.

25. Let X be a set and let **B** be a nonempty family of subsets of X with the properties that
    (1) If $B \in$ **B**, then no proper subset of B is in **B**.
    (2) If B_1 and B_2 are in B and $x \in B_1$, then there is a y in B_2 such that $(B_1 - \{x\}) \cup \{y\} \in$ **B**.
    Define a set I to be independent if I is a subset of B for some member B of **B**. Show that the set **I** of independent sets of X is the family of independent sets of a matroid and that the members of **B** are the bases of this matroid.

26. Let X be a set and let r be a function defined on the subsets of X satisfying the rules
    (1) $r(\emptyset) = 0$.

(2) For any $Y \subseteq X$ and any $x \in X$, $r(Y \cup \{x\}) = r(Y)$ or $r(Y \cup \{x\}) = r(Y) + 1$.

(3) If $r(Y \cup \{x\}) = r(Y \cup \{y\}) = r(Y)$, then $r(Y \cup \{x\} \cup \{y\}) = r(Y)$.

Define a set I to be independent if $|Y| = r(Y)$. Show that the set $\mathbf{I}$ of independent sets is the family of independent sets of a matroid on X.

27. A subset of a matroid which is not independent is called dependent. A minimal (nonempty) dependent subset of the set X of a matroid is called a **circuit**. In the forest matroid of a graph, what sets of edges are the circuits?

28. Circuits of a matroid are defined in the preceding problem. Show that circuits of a matroid have the properties that
    (1) No proper subset of a circuit is a circuit.
    (2) If C_1 and C_2 are circuits and $x \in C_1 \cap C_2$, then there is a circuit C_3 which is a subset of the set $C_1 \cup C_2 - \{x\}$.
    (2′) If C_1 and C_2 are circuits, if $x \in C_1 \cap C_2$, and if $y \in C_1 - C_2$, then there is a circuit C_3 containing y which is a subset of the set $C_1 \cup C_2 - \{x\}$.

29. Condition (2) in the preceding exercise is a consequence of condition (2′). Show that if we have a set X and a family $\mathbf{C}$ of subsets of X called circuits and satisfying rules (1) and (2′) in the preceding exercise and if we define a set to be independent if it does not have a subset which is a circuit, then the family $\mathbf{I}$ of independent sets is the family of independent sets of a matroid on X.

30. Write a program that takes for its data an adjacency matrix of a graph and a cost function and produces the adjacency matrix of a minimum-cost spanning tree (if such exists).

31. Write a computer program that takes as data the incidence matrix of a family of subsets of a set and a cost function and produces a set of distinct representatives of minimum cost if one exists. Modify the program so that it produces a minimum-cost system of distinct representatives. (The incidence matrix of a family $\mathbf{F}$ has rows corresponding to the elements of X, columns corresponding to the sets of $\mathbf{F}$, and a 1 in row i, column j if $x_i \in F_j$.)

*32. Show that a family $\mathbf{I}$ of subsets of a finite set X satisfying the subset rule is the family of independent sets of a matroid only if, for *each* way of assigning weights to X, the greedy algorithm selects a basis of minimum weight.

Section 3 Network Flows

Transportation Networks

A **network** is a directed graph (or multigraph) which has numbers (which we shall call capacities) assigned to its edges. Such a network might represent a communications network in which the capacities represent the number of messages that can simultaneously be sent from vertex i to vertex j. A network might represent transportation routes (highways, railroads, etc.) between cities; the numbers on the edges might represent the capacity of the routes in terms of the number of vehicles per hour. At times (for example, in the case in which our transportation routes were all two-way streets), we also consider networks based on graphs rather than digraphs.

A typical problem in such a network occurs when we have a *source* of material or messages destined for some destination (called a *sink* in technical terms) and we want to know how much of this material can be sent from the source to the destination in some standard unit of time. For this reason, we call a network with two chosen vertices named the **source** and **sink** (or destination) a *transportation* network. In many practical situations, edges touching the source only leave the source and edges touching the sink only enter the sink (because the material enters the network at the source and leaves the network at the sink). However, we need not assume this is always the case. In a transportation network, the numbers assigned to the edges are called **capacities**; $c(e)$ or $c(x, y)$ denotes the capacity of the edge $e = (x, y)$.

The Concept of Flow

We think of material flowing along a transportation network from the source to the sink; thus, what enters a vertex in a network must leave it—unless the vertex is the source or sink. For this reason, we define a **flow** in a network N with edge set E, capacity function c, source s, and sink t to be a function f defined on the edges of N that satisfies the conditions

(1) $0 \le f(e) \le c(e)$ for each edge e of N.
(2) For each vertex $v \notin \{s, t\}$

$$\sum_{\substack{(x,v): \\ (x,v) \in E}} f(x, v) - \sum_{\substack{(v,y): \\ (v,y) \in E}} f(v, y) = 0.$$

The interpretation of (1) is clear; in words, condition 2 says that the flow into vertex v equals the flow out of vertex v. Condition 2 is called the **balance condition**. Because of our motivating problem, it is natural to ask how

328

much material is flowing from the source to the sink. It turns out that the two expressions

$$F_s = \sum_{v:(s,v)\in E} f(s,v) - \sum_{v:(v,s)\in E} f(v,s)$$

and

$$F_t = \sum_{v:(v,t)\in E} f(v,t) - \sum_{v:(t,v)\in E} f(t,v)$$

are equal. The first difference is normally interpreted as the amount of material flowing from the source and the second is interpreted as the amount of material arriving at the sink. The common value of these two expressions is called the **value** of the flow and is interpreted as the amount of material flowing from the source to the sink.

Theorem 3.1. In a network, the flow F_s leaving the source and the flow F_t arriving at the sink are equal.

Proof. The flow entering a vertex a vertex v is

$$\sum_{\substack{(x,v): \\ (x,v)\in E}} f(x,v).$$

The flow leaving a vertex v is

$$\sum_{\substack{(v,y): \\ (v,y)\in E}} f(v,y)).$$

The expression

$$\sum_{v:v\in V} \left(\sum_{\substack{(x,v): \\ (x,v)\in E}} f(x,v) - \sum_{\substack{(v,y): \\ (v,y)\in E}} f(v,y) \right)$$

has the value $F_s - F_t$ because each term representing the flow into a vertex (other than F_s and F_t) cancels out with the term representing the flow out of that vertex because of condition 2 in the definition of a flow. However, if we distribute the terms of the sum, we obtain

$$F_s - F_t = \sum_{v:\ v\in V}\sum_{x:(x,v)\in E} f(x,v) - \sum_{v\in V}\sum_{y:(v,y)\in E} f(v,y)$$

$$= \sum_{\substack{(x,v): \\ (x,v)\in E}} f(x,v) - \sum_{\substack{(v,y): \\ (v,y)\in E}} f(v,y)$$

$$= 0$$

since both sums add up all the $f(x,y)$ for all edges (x,y) in E. Thus, $F_s - F_t = 0$ or $F_s = F_t$. ∎

A natural question to ask for a given network is, "How big can we make the flow—i.e., what is the maximum value of the flow?" Here is a way to visualize a natural limit on the size of the flow. Given a drawing of a network, imagine cutting the drawing with a pair of scissors without cutting through any vertex so that the source ends up on one piece of paper and the sink on the other. In part (a) of Figure 3.1 one piece of paper is represented by a white rectangle and one is represented by a dotted rectangle. If you further imagine a flow going along the edges of the network before it is cut, material will still be flowing out of the edges directed away from the source after the cut and in that case will leave the network when it reaches an edge we have cut through. Intuitively, it seems that the amount going from the "source half" to the "sink half" is at least as much as the flow into the sink itself. This flow can be no more than the sum of the capacities of the edges leading from the "source half" to the "sink half," and so the value of the flow itself should be no more than the sum of these capacities. This intuition holds true, and we now make these ideas sufficiently precise to show why it holds true.

Figure 3.1

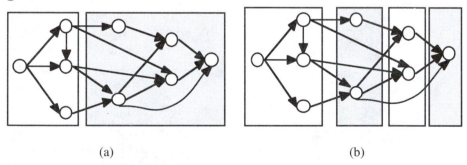

(a)   (b)

Cuts in Networks

A *cut* of a network with vertex set V is a partition of V into two (disjoint) sets S and T such that the source is in S (which is called the "source set") and the sink is in T (which is called the "sink set"). In part (a) Figure 3.1, the source set is within the white rectangle and the sink set is within the dotted rectnagle. However our definition of a cut is broad enough to allow for partitions that don't correspond to cutting through the network once with a pair of scissors. In part (b) of Figure 3.1 the vertices in the two white rectangles are the source set of a cut, and the vertices in the two dotted rectangles are the sink set of the same cut. While allowing this level of generality in the definition may seem strange at first, it doesn't hurt anything, and it makes the computations easier than they would otherwise be.

The **capacity** of the cut $\{S, T\}$, denoted by $c(S, T)$, is the sum of the capacities of the edges leading from a vertex in the source set to a vertex in the sink set. As our intuition suggests, given a flow on our network, the sum of the flow values along edges from the source set S of a cut $\mathbf{C}$ to the sink set T of $\mathbf{C}$ minus the flow in the reverse direction is the common value F of F_s and F_t.

Theorem 3.2. If $\mathbf{C}$ is a cut with source set S and sink set T, and F is the value of a flow f on the network, then

$$F = \sum_{\substack{(x,y)\in E: \\ x\in S, y\in T}} f(x, y) - \sum_{\substack{(y,x)\in E: \\ x\in S, y\in T}} f(y, x).$$

Proof. We construct a new network and flow by replacing all vertices in T with a single vertex w and construct an edge between w and a member of S (of the same capacity, in the same direction, and with the same flow) for each edge between a member of T and a member of S. (This may give a multidigraph, but that is allowable in a network.) Then w is the sink in the new network, s is the source in the new network, and F_s is unchanged, so $F_w = F$. However, F_w is simply the difference of the two sums in the statement of the theorem, so this difference must be F. ∎

Theorem 3.3. For every flow f with value F, and every cut $\{S, T\}$ of the network, $F \leq c(S, T)$.

Proof. Since $f(x, y) \leq c(x, y)$ for each edge (x, y) leading from S to T, the first sum in the expression for F in Theorem 3.2 is no more than $c(S, T)$. Subtracting the second term can only make the expression smaller. But then $F \leq c(S, T)$. ∎

Theorem 3.3 means that if for some cut, the value F of the flow f is the capacity $c(S, T)$, then the flow is as large as it can conceivably be. Thus, the maximum flow value is no larger than the minimum of the capacities of all cuts. In fact, in each network there is a maximum flow whose value is equal to the minimum of the capacities of all cuts in the network. This result is called the "max-flow min-cut theorem."

The max-flow min-cut theorem will follow from the fact that any flow which has a value less than the minimum cut capacity can be modified to give a larger value. Thus, given a flow in one network, we may ask whether and how we can improve it to get more flow from the source to the sink. Intuitively, there are two things we could do to make more material move toward the sink. If an edge is not being used to capacity, we could try to send more material through it; if an edge is working against us by sending material back toward the source, we could try to reduce the flow along this edge and redirect it in a more practical direction. These two ideas are combined in the concept of a *flow-augmenting path*. To give this idea a concrete definition, we assume there is already a flow f through our network.

Flow-Augmenting Paths

An *undirected path* (or *semipath*) connecting s and t is an alternating sequence of vertices and edges that form a path from s to t in the underlying graph. Thus an undirected path is an alternating sequence $s = v_0 e_1 v_1 e_2 v_2 \cdots v_{n-1} e_n v_n = t$ of vertices and edges such that $e_i = (v_{i-1}, v_i)$ or else $e_i = (v_i, v_{i-1})$. For example, in Figure 3.2

$$s(s,r)r(r,v)v(v,x)x(x,y)y(z,y)z(w,z)w(w,u)u(u,t)$$

is an undirected path. If $e_i = (v_{i-1}, v_i)$, we say e_i is a **forward edge** in the path; if $e_i = (v_i, v_{i-1})$, we say e_i is a **reverse edge**. In Figure 3.2, the edges $e_3 = (v, x)$ and $e_4 = (x, y)$ are forward edges but the edges $e_5 = (z, y)$ and $e_6 = (w, z)$ are reverse edges on the path we just described. (Note that as we move along the path we go across the two reverse edges in a direction opposite from the way their arrows point.) If we think of an undirected path as a walk along one-way streets, the reverse edges are the one-way streets that we walk along in the wrong way. Edges of the graph not in our undirected path are shown as dashed lines. They don't get either the name forward or reverse.

Figure 3.2

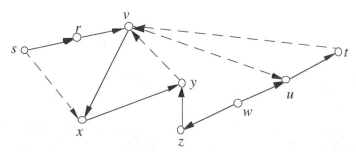

Now imagine a network whose vertices represent shopping centers and whose edges represent one-way streets connecting the shopping centers. Imagine an undirected path through the network that has some forward edges and some reverse edges, and suppose cars are flowing through our network according to the rules of a network flow. By rule 2, this means that the number of cars entering a shopping center equals the number of cars leaving it, except, perhaps, at the shopping centers marked s and t. Now suppose we increase the number of cars along each forward edge of our path by ten, and decrease the number of cars along each reverse edge of our path by ten. Then f_s will be ten more cars than before, and f_t will be ten more cars than before. If a vertex has a forward edge entering it and a forward edge leaving it on our path, then our new f still satisfy condition 2 for a flow at this vertex. (Ten more cars enter it, but ten more cars leave it.) If a vertex has a forward edge entering it and a reverse edge leaving it, then since cars must move in the direction of the arrows, cars enter the shopping center along both the roads represented by these edges. We increase the number of cars entering it on the forward edge by ten but we reduce the number of cars entering it on the reverse edge by ten. Thus this vertex still satisfies condition 2 for a flow. Similarly, the other vertices on the path will still satisfy condition 2 for a flow. This idea is the basis of the following definition of a flow-augmenting path.

We say an undirected path from s to t is a **_flow-augmenting path_** of **_worth_** $w > 0$ if adding w to $f(e_i)$ for each forward edge e_i and subtracting w from $f(e_i)$ for each reverse edge e_i yields a new flow f'. (Note that by definition the new function f' must be a flow for the path to be called flow-augmenting.)

> **_Theorem 3.4._** If f is a flow of value v in a network and if the flow f' is obtained from f by adding w to $f(e)$ for each forward edge e of a flow-augmenting path and subtracting w from $f(e)$ for each reverse edge e of this flow-augmenting path, then f' is a flow of value $v + w$.
>
> **Proof.** By assumption, f' is a flow. Note that $F_s' = F_s + w$. Thus, the flow f' has value $v + w$. ∎

The Labeling Algorithm for Finding Flow-Augmenting Paths

Our theorem tells us that the problem of improving a flow may be solved by finding a flow-augmenting path. In fact, we shall see that as big a flow as possible may be constructed by starting with the all-zero flow f_0, augmenting it with a flow-augmenting path to get a flow f_1, and repeating this process until we find a flow f_n that has no flow-augmenting path. For these reasons, we develop an algorithm to find a flow-augmenting path, if one exists, or to show that none exists otherwise. Our algorithm is a variant, due to Edmonds and Karp, of the original algorithm for finding flow-augmenting paths, the Ford–Fulkerson labeling algorithm. Vertices that might be useful in a flow-augmenting path are labeled in such a way that we can use the labels to determine the path. In essence, Edmonds and Karp's idea is to use a breadth-first search tree, placing edges in the tree only if they are useful for increasing the flow. We start at the source and search until we reach the sink or run out of edges. The importance of using breadth-first search is that we find flow-augmenting paths in increasing order of their number of edges. As Exercise 13 shows, had we not done so, we could accidentally repeat searches for flow-augmenting paths a ridiculous number of times by making a series of unfortunate choices. In Exercise 18, we analyze exactly how many searches would be required for finding a maximum flow by the Edmonds and Karp method; their method, announced in 1969 and published in 1972, was the first one for which the maximum number of searches needed did not potentially depend on the numerical values of the capacities and for which Theorem 3.7 could be proved in the case of nonrational capacities.

The steps of the labeling algorithm are:

(1) Assign to the vertex s the label "source." Make s the first element of a list L (for labeled).

(2) Repeat steps 3 and 4 until the list L is empty.

(3) Let u be the first element of the list L. (Thus, initially, u is s.)

    (a) For each unlabeled vertex v such that (u, v) is an edge and $f(u, v) < c(u, v)$, assign v the label (u, v) and place v at the end of the list L.

    (b) For each unlabeled vertex v such that (v, u) is an edge and $f(v, u) > 0$, assign v the label (v, u) and place v at the end of the list L.

(4) Remove u from the list L.

We will complete the process when (or before) all vertices of the network have been labeled. A vertex cannot have more than one label assigned to it, and different vertices are assigned different labels. Note that we have labeled vertices v with edges (u, v) if the flow from u to v can potentially be increased and have labeled vertices u with edges (u, v) if the flow from u to v can potentially be decreased. These potential increases and decreases will be important when we wish to choose the forward and reverse edges of a flow-augmenting path. Before we try to construct such a path, it would be nice to know if we already have a flow whose value equals the minimum of the capacities of the cuts. Suppose the sink is *not* labeled. Then the partition $\{P, Q\}$ of the set V in which P consists of labeled vertices and Q consists of unlabeled ones is a cut. By step 2a of our algorithm, $f(u, v) = c(u, v)$ for each u in P and v in Q such that (u, v) is an edge. (In words, our flow uses each edge leading from P to Q to full capacity.) By step 2b of our algorithm, $f(v, u) = 0$ for each v in Q and u in P such that (v, u) is an edge. Thus, the value of the flow equals the capacity of this cut $\{P, Q\}$. We have just proved a theorem that characterizes flows of maximum value.

> **Theorem 3.5.** A flow f has maximum possible value (i.e., has value equal to the minimum of the capacities of the cuts of the network) if, after the labeling algorithm is complete, the sink remains unlabeled.

As has already been suggested, if the sink is labeled, we may choose a flow-augmenting path. The construction works backward through the search tree from the sink to the source as follows. (Recall that t is the sink and s is the source.)

(1) Let $u_0 = t$.

Repeat the following step until $u_{i+1} = s$.

(2) Given a vertex u_i assigned the edge e as label, let u_{i+1} be the other vertex in e.

Now suppose we have vertices $u_0, \ldots, u_n$. Since they "range" from the sink to the source, these are in reverse order from the way we want to send material. Therefore, we let $v_0 = u_n$, and in general $v_i = u_{n-i}$. Let e_i be the edge e we used to choose v_{i-1} given v_i. Note that if $e_i = (v_{i-1}, v_i)$ (i.e., if e_i is a forward edge), it has been encountered in step 2a of the labeling algorithm, and if $e_i = (v_i, v_{i-1})$ (i.e., e_i is a reverse edge), it has been encountered in step 2b of the labeling algorithm. Thus, if $e_i = (v_{i-1}, v_i)$, then $f(e_i) < c(e_i)$ and if $e_i = (v_i, v_{i-1})$, then $f(e_i) > 0$. Let w_1 be the minimum value of $c(e_i) - f(e_i)$ for all forward edges $(v_{i-1}, v_i) = e_i$ and let w_2 be the minimum value of $f(e_i)$ for all reverse edges $(v_i, v_{i-1}) = e_i$. Now let w be the smaller of w_1 and w_2. This description of w is equivalent to the symbolic description of w in the statement of the following theorem.

> **Theorem 3.6.** Let
>
> $$w = \min(\min\{c(e_i) - f(e_i) \mid e_i \text{ is forward}\}, \min\{f(e_i) \mid e_i \text{ is reverse}\}).$$
>
> Then $v_0 e_1 v_1 \cdots e_n v_n$ is a flow-augmenting path of worth w.

Proof. Note that $f(e_i) + w \leq c(e_i)$ for each forward edge e_i and that $f(e_i) - w \geq 0$ for each reverse edge e_i. To show that adding w to each flow in a forward edge and subtracting w from each flow in a reverse edge gives a flow f', we must still show that for each vertex $v \notin \{s, t\}$,

$$\sum_{x:(x,v)\in E} f'(x, v) - \sum_{y:(v,y)\in E} f'(v, y) = 0.$$

However, each $f'(e)$ is equal to $f(e)+w$ or else $f(e)-w$, and f is a flow, so we need to show only that the w's cancel out. In particular, we need to check only vertices v on our flow-augmenting path, because otherwise f' equals f. Further, only along two edges e_i and e_{i+1} touching v_i do we change the value of f. Now if e_i and e_{i+1} are both forward or both reverse, one of the w's will occur in the left-hand sum and one in the right-hand sum, so they cancel out. If either e_i or e_{i+1} is forward while the other is reverse, then the two occurrences of w are in the same sum but with opposite sign, so they cancel out. Therefore, f' is a flow, and our path is a flow-augmenting path of worth w. ∎

The Max-Flow Min-Cut Theorem

Note that if the initial flow and capacities are integral, then the increase w in the value F will be an integer. Thus, we can continue using flow-augmenting paths to increase F until its value reaches a maximum. The same conclusion follows if all the capacities are rational numbers. This method will also produce a maximum flow if the capacities are arbitrary real numbers; see Exercise 18 for an outline of the proof. Thus we have the following theorem, first proved by Ford and Fulkerson in 1956, and called the **max-flow min-cut theorem**.

Theorem 3.7. The maximum value of a flow in a network is the minimum of the capacities of any of its cuts.

Example 3.1. Find a maximum flow in the network of Figure 3.3.
The capacity of each edge is written beside the edge. We repeat the process of labeling and constructing flow-augmenting paths.

Figure 3.3

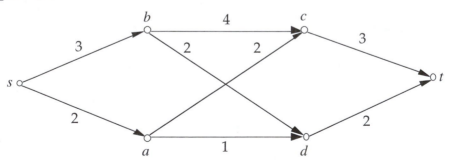

Stage 1 $f(x,y)=0$ for all x,y.
   Label s: source, b: (s,b), a: (s,a), c: (b,c), d: (b,d), t: (c,t)
   Path t (c,t) c (b,c) b (s,b) s

Find d: min(3, 4, 3) = 3
Find f′: $f'(s,b) = 3, f'(b,c) = 3, f'(c,t) = 3, f'(x,y) = 0$ for all other x, y.
Comment: We found f' by adding the value of d to $f(e)$ for each forward edge e of our path and letting $f'(e) = f(e)$ for each other edge of the network. Since there were no reverse edges, we did not subtract d from $f(e)$ for any edge e.

Stage 2 $f(s,b) = f(b,c) = f(c,t) = 3; f(x,y) = 0$ for all other x, y.
 Label s: source, a: (s,a), d: (a,d), c: (a,c), t: (d,t)
 Path t (d,t) d (a,d) a (s,a) s
 Find d: min(2, 1, 2) = 1
 Find f′: $f'(s,a) = f'(a,d) = f'(d,t) = 1$
      $f'(s,b) = f'(b,c) = f'(c,t) = 3$
      $f'(x,y) = 0$ for all other x, y.

Stage 3 $f(s,a) = f(a,d) = f(d,t) = 1, f(s,b) = f(b,c) = f(c,t) = 3$
    $f(x,y) = 0$ for all other (x, y).
 Label s: source, a: (s,a), c: (a,c), b: (b,c), d: (b,d), t: (d,t)
Comment: Note that when we are labeling from vertex c, the (potentially forward) edge from c to t is used to capacity, so we cannot label t with (c,t), but the (potentially reverse) edge from b to c has some flow in it (two units), so we label c with (b,c). This tells us that (b,c) is available to use as a reverse edge from the vertex c.
 Path t (d,t) d (b,d) b (b,c) c (a,c) a (s,a) s
Comment: Note that by following labels back from vertex s to vertex t, we have decided to use the edge (b,c). Notice that as we go from s to t along this path, we use the edge (b,c) as a reverse edge.
 Find d: min(2 − 1, 2, 3, 2, 2 − 1) = 1
 Find f′: $f'(s,a) = 2, f'(a,d) = 1, f'(a,c) = 1$
      $f'(s,b) = 3, f'(b,c) = 2, f'(b,d) = 1$
      $f'(c,t) = 3, f'(d,t) = 2$.
Comment: Note that in finding f', we subtracted 1 from $f(b,c)$ to get $f'(b,c)$ because (b,c) was a reverse edge.

Stage 4 $f(a,d) = f(a,c) = f(b,d) = 1, f(s,a) = f(b,c) = f(d,t) = 2,$
    $f(s,b) = f(c,t) = 3$
 Label s: source, no more labeling is possible.

  The value of the flow shown in stage 4 is 5; the cut with $S = \{s\}$ and $T = \{a, b, c, d, t\}$ has cut capacity 5. ∎

*More Efficient Algorithms

It may seem wasteful to label the entire network with the labeling algorithm and then use the labels only to find one flow-augmenting path. In an article which appeared in 1970, the Russian mathematician Dinic found a way to take advantage of the labeling by creating what we call a **layered** network. For this purpose, we say that the source is in layer zero by itself. We then say inductively that if u is in layer i and v is labeled with the forward edge (u, v) or the reverse edge (v, u), then v is in layer $i + 1$. To construct the layered network, we draw an edge of capacity $c(e) - f(e)$ for each edge e which goes from layer i to layer $i+1$ for some i and is not used to capacity by the current flow and an edge of capacity $f(e)$ for each edge e that goes from layer $i + 1$ to layer i and has nonzero flow in the current flow. Paths from the source to the sink in this network will correspond to flow-augmenting paths of the same length in the original network.

There are several ways people use the layered network to improve the flow. Here is a particularly straightforward one. If the sink is in layer k, then choose a vertex in layer $k - 1$ connected to it by an edge of the layered graph; continue the process as follows. From a vertex in layer i, choose a vertex in layer $i - 1$ connected to it by an edge of the layered graph. Continue until the process reaches the source. This gives one flow-augmenting path, so we augment by it. Now the idea is that all potential augmenting paths of length k still have their edges in the layered network, so that we want to repeat the process. However, augmenting by our first path will saturate—fill to capacity—some edges, effectively blocking the process we used to construct the first path at any vertex connected to the previous layer only by saturated edges.

Thus, given a layered graph and a flow-augmenting path, we now remove from the layered network all so-called blocking vertices and edges as we augment by that path. A vertex in layer i becomes **blocking** when all edges from it to layer $i - 1$ are either saturated or blocking, and an edge with an endpoint in layer j becomes **blocking** if it is saturated or it leads from a blocking vertex in layer $j - 1$. As soon as we saturate an edge, we locate all blocking edges and vertices caused by this saturation and remove them from the network. So long as we do not remove the sink, we repeat the process we used to find the first augmenting path, augment by that path, removing blocking vertices and edges each time we saturate an edge, and repeat the process until we have removed the sink. At this point we may show that there are no flow-augmenting paths of length k. We repeat the labeling and layering process and begin augmentation once again.

Example 3.2. In Figure 3.3, with the zero flow, layer 0 is $\{s\}$, layer 1 is $\{a, b\}$, layer 2 is $\{c, d\}$, and layer 3 is $\{t\}$. We choose the vertex c connected to t, then, say, b connected to c, and then s connected to b, giving us the augmenting path $sbct$. Now s cannot be blocking for no edges enter it. The vertex a is not blocking, since edge (s, a) is not saturated. However, the worth of our augmenting path is three, so that edge $\{s, b\}$ is saturated and b is blocking. Therefore, edges (b, c) and (b, d) are blocking. However, neither vertex c nor d is blocked since they are entered by nonsaturated edges. Similarly, vertex t is not blocked. Thus, our new graph has vertex set $\{s, a, c, d, t\}$ and edge set $\{(s, a), (a, c), (a, d), (d, t)\}$. In this modified graph we find the augmenting path $sadt$ and the vertex t becomes blocking. Notice that our flow now has value 4 and is the flow in stage 2 of Example 3.1. When we relabel, our labeling will be the labeling of stage 3 of Example 3.1. With this labeling, layer 0 is $\{s\}$, layer 1 is $\{a\}$, layer 2 is $\{c\}$, layer 3 is $\{b\}$, layer 4 is $\{1\}$, and layer 5 is $\{t\}$. Note how the layering presents us with the single flow-augmenting path of stage 3 of Example 3.1. Since we do not label the sink in our next stage of labeling, we have achieved a maximum flow. ∎

How does this make the process more efficient? Since the layer of the sink becomes larger each time we repeat the labeling and layering process, the number of repeats of these processes is no more than the maximum possible length of a path, the number of vertices. The process itself consists of repeating the construction of an augmenting path and removal of the resulting blocking vertices and edges. Note that in our example this gave us two flow-augmenting paths from the first labeling and layering step, saving one labeling step. In a large example, the savings can be considerably more spectacular. The number of repeats of this construction is no more than the number m of edges (since each construction saturates at least one edge), the number of removals is no more than the number of vertices plus edges, and the number of vertices examined in each construction of an alternating path is no more than the total number n of vertices. Thus, the number of steps required for this technique is no more than a multiple of $n^2 m$ (which could be about n^4), whereas the number of steps in our first algorithm can only be said to be no more than $m^2 n$ (which could be about n^5).

Work of Karzanov in 1974 and of Malhotra, Kumar, and Maheshwari in 1978 has shown that the total number of steps in constructing a maximum flow need be no more than some multiple of n^3. Although they are not simple modifications of the preceding algorithm, in effect these algorithms saturate a layer or a vertex each time we augment rather than an edge each time.

Figure 3.4

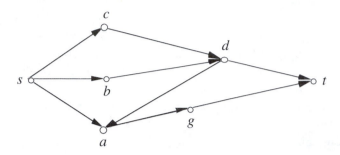

EXERCISES

Exercises 1–8 deal with the digraph in Figure 3.4.

1. If $f(e) = 1$ for all edges e of Figure 3.4, is f a flow? (Don't be concerned about capacities.)

2. Suppose $f(s,c) = f(s,b) = f(s,a) = 1$ and $f(d,a) = 0$. How should $f(e)$ be defined for other edges to achieve a flow? (Don't be concerned about capacities.)

3. What is the value of the flow of Exercise 2?

4. If all the edges of Figure 3.4 have capacity 2, locate at least seven cuts of capacity 6.

5. Using the flow of Exercise 2 and capacities of Exercise 4, apply the labeling algorithm to Figure 3.4. Find a flow-augmenting path.

6. Find a maximum flow in Figure 3.4 using the capacities of Exercise 4.

7. If all the edges of Figure 3.4 have capacity 1, what is the maximum value of a flow? Find a flow with this value.

8. Suppose $c(d,t) = c(g,t) = 3$ and all other edges have capacity 2 in Figure 3.4. Find a maximal flow and a cut whose capacity is the value of the flow.

9. In Figure 3.5, suppose all edges touching s have capacity 3, all edges touching g have capacity 2, and $c(d,t) = 6$, $c(a,c) = c(a,d) = 8$, $c(d,b) = 4$ and $c(b,a) = 2$. Determine a maximum flow.

10. In Figure 3.5 with the capacities as in Exercise 9, given a flow with $f(s,a) = 3$, $f(s,b) = 3$, $f(c,s) = 3$, $f(a,c) = 3$, $f(b,g) = 2$, $f(g,d) = 2$, $f(d,t) = 3$, $f(b,a) = 1$, and $f(a,d) = 1$, find the flow in all the remaining edges. Apply the labeling algorithm to construct a maximal flow from f by means of flow-augmenting paths.

11. Is there a way to assign capacities to the edges of Figure 3.5 such that all cuts have the same capacity and the network has a flow that is positive?

Figure 3.5

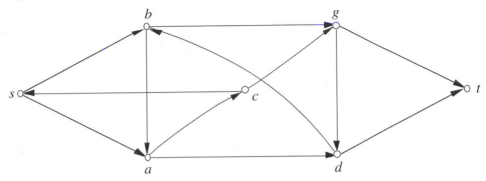

12. Redirect the edge (d, b) to be (b, d) instead in Figure 3.5. Let $c(a, c) = c(s, b) = 4$, $c(b, a) = c(b, g) = 1$, $c(d, t) = c(c, s) = c(s, a) = 3$, $c(c, g) = c(g, t) = 5$, and $c(b, d) = c(a, d) = c(g, d) = 2$. Find a maximum flow.

13. In the network of Example 3.1 (Figure 3.3), change the capacities of all the edges but (a, c) and (b, d) to one million. Change the capacities of (a, c) and (b, d) to one. Now instead of finding flow-augmenting paths in breadth-first order, suppose we always choose flow-augmenting paths which use diagonal edges, sometimes forward and sometimes backward. How many augmentations by flow-augmenting paths will be necessary before we achieve maximum flow? This example illustrates part of the importance of using breadth-first search while labeling!

14. Explain why the labeling algorithm assigns a label to each vertex at most once and labels different vertices differently.

15. Show that if all the capacities of the edges are integers, then there is a maximum flow such that the flow along each edge is an integer. Give an example which shows that there may also be maximum flows in which the flow along some edges is not an integer.

16. Suppose that for a given network we have a capacity function for each vertex other than the source and sink. That is, suppose there is a number $c(v)$ associated with each vertex v and that $f(v)$, which is the common value of the flow into v and the flow out of v, is no more than $c(v)$. Describe how to transform the problem of computing a maximum flow which obeys both the edge capacities and the vertex capacities into an ordinary network flow problem. What is the analog of a minimum cut in the network (before the problem is transformed into an ordinary network flow problem)?

17. A flow is said to be maximal if any flow-augmenting path has a reverse edge. Show that there is no relationship between the value of a maximal flow and the value of a maximum flow by showing how to construct, for any number $N \geq 1$ you choose, a network in which the value of one maximal flow is 1 but the value of a maximum flow is N.

18. This exercise shows that the algorithm we gave for constructing a maximum flow works for all capacities. Suppose that $P_1, P_2, P_3, \ldots$ denotes a sequence of flow-augmenting paths constructed by our labeling algorithm, beginning with the zero flow.

   (a) Show that if the length of P_i is more than the length of P_j, then $i > j$.

   (b) Show that if P_i and P_j have the same length, then any edge used by both paths is either a forward edge in both paths or a reverse edge in both paths.

   (c) Show that the total number of flow-augmenting paths which have a given length is no more than the total number of edges of the network.

   (d) Explain why the total number of augmenting paths found in the entire process is no more than nm, where n is the number of vertices and m is the number of edges of the network.

   (e) Explain why the total number of steps used in finding a maximum flow is no more than some multiple of nm^2.

*19. (a) Using the network of Figure 3.5 with the capacities of Exercise 9, what layered network do we get when we begin with the zero flow?

   (b) Explain why, no matter how we choose our augmenting paths in the network of part (a), it is always possible to choose two augmenting paths before we relabel and relayer.

   (c) Use the layering procedure to find a maximum flow for the network of part (a).

*20. In this exercise we see how to implement a faster algorithm for finding flows through layered networks. This is basically the Malhotra, Kumar, and Maheshwari modification of Karzanov's algorithm. Suppose you have a flow through a network with n vertices and form the associated layered network. Define the capacity of a vertex to be the minimum of the sum of capacities of its incoming edges and the sum of the capacities of its outgoing edges.

   (a) Suppose x is a vertex of minimum capacity, say of capacity C. Explain why it is possible to send C units of flow from the source to x. (Hint: If you try to work backward from x to the sink and find edges with sufficient capacity to send C units to x by splitting the C units among various edges, will you ever run into trouble?)

(b) Explain why it is possible to send C units of flow from the vertex x of part (a) to the sink.

(c) Explain how to augment the flow in the layered network which we had at the beginning of the problem so that no additional flow is possible through x in that layered network. (The use of the word augmentation here does *not* mean you must find augmenting *paths*.)

(d) Explain how to modify the layered network so that you can repeat the process with the vertex which has smallest capacity relative to this new flow.

(e) A flow in a layered network is called *blocking* if it has no flow-augmenting paths whose length (number of edges) is the number of layers of the network minus one. Describe an algorithm which produces a blocking flow in the layered network by using n or fewer augmentations.

(f) Explain how to produce a maximum flow in a network in a number of steps that is no more than a multiple of n^3.

*21. In this exercise we see how to implement another fast algorithm (Dinic's algorithm) for finding flows through layered networks. Suppose you have a flow through a network with n vertices and form the associated layered network.

(a) Using depth-first search, search in the layered network for a path from the source to the sink, removing all edges touching a vertex from which we "back up" in the process of the search. Explain why we may repeat this process until we find a set of flow-augmenting paths which produces a blocking (see part (e) of the previous exercise) flow.

(b) Explain why, if we augment by each flow-augmenting path in part (a) after we find it, then the number of times we may repeat the depth-first search is at most the number m of edges.

(c) An edge is examined in a depth-first search in part (a) each time it appears in a flow-augmenting path and also when it is deleted, but in no other cases. Explain why the total number of examinations of edges in repeating the process of part (a) until we obtain a blocking flow is no more than nm.

(d) Explain why combining the process of part (a) with relabeling and relayering gives a maximum flow in a number of steps that is no more than some multiple of n^2m.

22. Write a computer program which produces a maximum flow in a network given the $n \times n$ matrix C given by

$$C_{ij} = \begin{cases} 0 & \text{if there is no edge from } i \text{ to } j \\ c(e) & \text{if } e \text{ is the edge from } i \text{ to } j. \end{cases}$$

Assume vertex 1 is the source and vertex n is the sink.

23. It might appear that in an undirected network, the question of forward and reverse edges would be irrelevant since the edges have no directions to start with. Work out the changes needed to find a maximum flow from a source to a sink in an undirected network, that is, a graph (multigraph) with capacities assigned to the edges and a chosen source and sink. Is it still necessary to consider the concept of a reverse edge?

Section 4   Flows, Connectivity, and Matching

Connectivity and Menger's Theorem

The concept of maximum flow in a network has applications in a wide variety of mathematical problems that do not seem to be transportation problems. The first application we will discuss is the concept of n-connectivity in a graph. We say that u is connected to v in a digraph if there is a directed path from u to v. Often there will be many paths from u to v. We say that u is ***n-edge-connected*** to v if there is a set of n (directed) paths from u to v such that no two of these paths have any edges in common. (Also, u is n-edge-connected to itself.) A set of edges is called a u—v cutset (or u—v edge cutset) if deleting these edges from the edge set of the digraph yields a digraph in which u is not connected to v. It is clear that if u is n-edge-connected to v, then a u—v cutset will have to have at least n edges. (Why?)

The idea of a u—v cutset seems analogous to the idea of a cut in a network with source u and sink v. To be precise, removing the cutset edges from the edge set D of the digraph leaves a digraph with at least two strongly connected components, one of which contains u and the other v. Thus, a u—v cutset gives a cut for any network whose digraph has edge set D, source u, and sink v. Now given a digraph (V, D), we define a network by letting each edge have capacity 1. Then if u and v are n-edge-connected, we can use the n "edge-disjoint" paths to send n units of flow from u to v. The max-flow min-cut theorem then tells us that if a cut has capacity n, then n is the maximum flow that can be achieved. From such a flow we shall see that we get n edge-disjoint paths. This leads to the following theorem, which is one form of a result known as Menger's theorem.

> ***Theorem 4.1.*** If a minimum-sized u—v cutset in a digraph (V, D) has n edges, then u is n-edge-connected to v.
>
> *Proof.* As we just described, we define a network on (V, D) by letting each edge have capacity 1. Then if a minimum cut in the network has capacity n, there are n edges whose removal disconnects u from v. Suppose a set S of edges is a minimal u—v cutset, so that if all but one edge in S is removed from D, there is a path from u to v, but if all the edges in S are removed from D, there is no path from u to v. Suppose U is the set of all vertices x such that there is a path from u to x in $(V, D - S)$, and let $\mathbf{C}$ be the partition $\{U, V - U\}$. Then $\mathbf{C}$ is a cut in the network, and the edges in S are the only edges that lead from an element of U to an element of $V - U$, so the capacity of $\mathbf{C}$ is the size of S. Thus, if a minimum-sized u—v cutset has n edges, a minimum cut of the network will have capacity n.

345

From the max-flow min-cut theorem, it follows that there is a flow of value n and that all its edge values are integers. We now prove by induction that this means there are n edge-disjoint paths from u to v. Let S be the set of all n such that if there is an integral flow of value n, then there are n edge-disjoint paths from u to v. Clearly, 1 is in S; assume $k-1$ is in S and that the flow has value k. There is a path from u to v each edge of which has flow and capacity 1. Thus, the remaining $k-1$ units of flow go entirely through edges not in this path, so that deleting the edges of the path gives a flow of value $k-1$ which, since $k-1$ is in S, yields $k-1$ edge-disjoint paths. These paths, together with the deleted path, form a set of k paths no two of which have an edge in common. Thus, by version 1 of the principle of mathematical induction, S contains all the positive integers and the theorem is proved. ∎

The original form of Menger's theorem uses the idea of a u—v vertex cutset. It is a bit more difficult to understand the idea of a vertex cutset than the idea of an edge cutset for the following reason. When we remove an edge e from a digraph (V, D), we simply remove e from the set D. However, when we remove a vertex v from a digraph (V, D), we cannot simply remove it from V, for then D might have ordered pairs containing v, so that D would not consist of ordered pairs chosen from the new set V. Thus when we say that we remove a vertex v from a digraph (V, D), we mean that remove v from V, *and* we remove every edge that has v for an endpoint from D.

A u—v *vertex cutset* is a set S of vertices of (V, D) such that if S is removed from (V, D), then u and v are in the resulting digraph, but there is no path from u to v. The vertex form of Menger's theorem says that if a minimum u—v vertex cutset has n vertices, then there is a set of n paths from u to v such that no two of the paths have a vertex (other than u or v) in common. We say u and v are **n-vertex-connected** when there are n such paths. Notice that this form of Menger's theorem tells us the maximum value of n such that u and v are n-vertex-connected. There is a way to derive this theorem from Theorem 4.1; namely, for each vertex x of (V, D) other than u and v, replace x by two vertices x_1 and x_2 such that (using $u_1 = u_2 = u$ and $v_1 = v_2 = v$)

(1) (x_1, x_2) is an edge.
(2) (w_2, x_1) is an edge if and only if (w, x) is an edge.
(3) (x_2, w_1) is an edge if and only if (x, w) is an edge.

In words, we replace each vertex x (except u and v) by two vertices x_1 and x_2. Then we reattach edges coming into x so that they come into x_1 and edges going out of x so that they go out of x_2. By thus "splitting" the vertices, we are replacing each vertex cutset by an edge cutset of the same size; applying Theorem 4.1 to this new digraph provides a set of paths having none of the new edges in common; this set of paths may be translated to a set of paths in the original graph with no vertices in common. (This fact may seem clear; technically, though, it requires proof. To prove it, note that if two paths translate to paths with a vertex x in common, you can show that these paths have the edge (x_1, x_2) in common. Another technical point you need to deal with is that there is no edge cut with fewer edges than the one that comes from the vertex cutset.)

There are also two versions of Menger's theorem (a vertex form and an edge form) for ordinary (undirected) graphs. The definition of an edge cutset in a graph is the same as the definition of an edge cutset in a digraph. One version of Menger's theorem for graphs is the following.

> **Theorem 4.2.** If a minimum-sized u—v edge cutset in a graph (V, E) has n edges, then there are n paths from u to v such that no two of these paths have an edge in common.
>
> *Proof.* From the graph (V, E), construct a digraph (V, D) with (x, y) and (y, x) in D if and only if $\{x, y\}$ is in E. Then a u—v edge cutset of size n in (V, E) gives a u—v edge cutset of size $2n$ in (V, D). However, we may delete half of these edges, after which we have no directed path from u to v. (Why?) Thus, we have a u—v edge cutset in (V, D) of size n (but there is no smaller cutset), so there are n directed paths from u to v in (V, D) such that no two of them have an edge in common.
>
> Now suppose one path uses the edge (x, y) and another path the edge (y, x). Thus (as shown in Figure 4.1 geometrically), one path is
>
> $$u = x_0 e_1 x_1 \cdots x_{k-1}(x_{k-1}, x)x(x, y)y(y, x_{k+2})x_{k+2} \cdots v$$
>
> and the other is
>
> $$u = y_0 e_1' y_1 \cdots y_{j-1}(y_{j-1}, y)y(y, x)x(x, y_{j+2})y_{j+2} \cdots v.$$

We can replace these two paths with paths that use neither (x, y) nor (y, x), namely

$$u = x_0 e_1 x_1 \cdots x_{k-1}(x_{k-1}, x)x(x, y_{j+2})y_{j+2} \cdots v$$

and

$$u = y_0 e_1' y_1 \cdots y_{j-1}(y_{j-1}, y)y(y, x_{k+2})x_{k+2} \cdots v.$$

Repeating this process (using induction), we can eliminate all simultaneous occurrences of (x, y) and (y, x) among the paths. Thus, the corresponding paths in the original graph will have no edges in common as desired. ∎

Figure 4.1

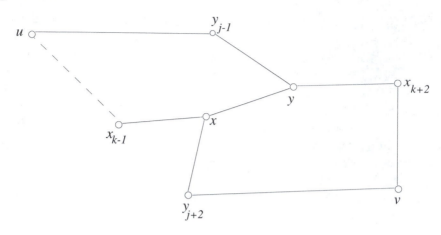

There are a vertex form and an edge form of Menger's theorem for multigraphs and multidigraphs as well; we leave the exploration of these theorems to you.

A graph, digraph, multigraph, or multidigraph is called **n-edge-connected** (or **n-vertex-connected**) if each pair of vertices is n-edge-connected (or each pair of vertices is n-vertex-connected.)

Flows, Matchings, and Systems of Distinct Representatives

In our discussion of matchings and systems of distinct representatives, we outlined algorithms for finding matchings, SDRs, minimum-cost matchings, or minimum-cost SDRs. In contrast, our flow algorithm was given in a step-by-step process suitable for writing a computer program. The reason for waiting until flows to give such details is that, with the exception of Edmonds' algorithm for matchings in nonbipartite graphs, all the algorithms we outlined are special cases of network-flow algorithms. In particular, alternating paths are special cases of flow-augmenting paths. We shall now show how flow algorithms may be applied to solve all our bipartite matching and SDR problems.

Recall that a *matching* from V_1 to V_2 in a bipartite graph on the vertex set $V_1 \cup V_2$ is a set of edges no two of which have an endpoint in common; a matching is called *complete* if each vertex of V_1 is included in it, and a matching is called a *maximum* (but not necessarily complete) matching if no other matching has more edges. A subset X of V_1 is *independent* if there is a matching of X into V_2 (or in other words, if there is a complete matching from X to V_2 in the bipartite subgraph whose vertex set is $X \cup V_2$). We use network flows to find matchings and test whether a matching of an independent set I into V_2 can be modified to give a matching of the set $I \cup \{x\}$ into V_2.

Given $G = (V_1 \cup V_2, E)$, we define a network N with vertex set $V_1 \cup V_2 \cup \{s\} \cup \{t\}$ (adding two new vertices s and t) by letting (s, x) be an edge for all x in V_1, (y, t) be an edge for all y in V_2, and (x, y) be an edge for all $\{x, y\}$ in E with $x \in V_1$ and $y \in V_2$. We assign capacity 1 to all the edges. Then a flow f whose value on each edge is either 0 or 1 will correspond to a matching; the edges in the matching are the edges of E whose directed counterparts have flow 1.

Example 4.1. Find a matching from the set $\{1, 2, 3\}$ into $X = \{1, 3\}$, $Y = \{2, 3\}$, $Z = \{1, 2\}$.

The network corresponding to this problem is shown in Figure 4.2. The process of constructing a maximal flow is outlined as follows.

Stage 1 $f(e) = 0$ for all e.

    Label $1: (s, 1), 2: (s, 2), 3: (s, 3), X: (1, X), Z: (1, Z), Y: (2, Y), t: (X, t)$

    Path $t, X, 1, s$

    Flow $f'(s, 1) = f'(1, X) = f'(X, t) = 1; \ f'(e) = 0$ for all other e.

Stage 2 $f(s, 1) = f(1, X) = f(X, t) = 1; \ f(e) = 0$ for all other e.

    Label $2: (s, 2), 3: (s, 3), Y: (2, Y), Z: (2, Z) X: (3, X), t: (Y, t), 1: (1, X)$

    Path $t, Y, 2, s$

    Flow $f'(s, 1) = f'(1, X) = f'(X, t) = f'(s, 2) = f'(2, Y) = f'(Y, t) = 1;$
             $f'(e) = 0$ for all other e.

Stage 3 $f(s, 1) = f(1, X) = f(X, t) = f(s, 2) = f(2, Y) = f(Y, t) = 1;$
        $f(e) = 0$ all other e.

    Label $3: (s, 3), X: (3, X), Y: (3, Y), 1: (1, X), 2: (2, Y), Z: (1, Z), t: (Z, t)$

    Path $t, Z, 1, X, 3, s$

    Flow $f'(s, 1) = f'(s, 3) = f'(s, 2) = f'(1, Z) = f'(3, X) = f'(2, Y) =$
             $f'(X, t) = f'(Y, t) = f'(Z, t) = 1; \ f(e) = 0$ for all other edges e.

Figure 4.2

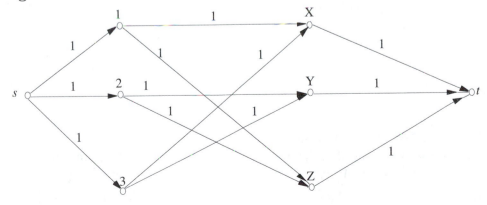

From this flow we see that one matching is $(1, Z)$ $(2, Y)$ $(3, X)$. ∎

Minimum-Cost SDRs

Now suppose we want to use the greedy algorithm to build a minimum-cost SDR for the sets X and Z of Example 4.1, given the cost function (denoted by upper-ase C) $C(1) = 10$, $C(2) = 8$, and $C(3) = 6$. The first element we would choose for our minimum-cost independent set of representatives would be 3. Vertex 3 did not appear in an edge with nonzero flow until stage 3 of the preceding flow construction process, so that process does not correspond to the greedy algorithm. Let us analyze how else we might proceed in order to develop an algorithm that will choose a matching which uses a minimum-cost set of representatives. We know that $S \subseteq \{1, 2, 3\}$ is independent if and only if it has a partial matching into $\{X, Z\}$, and we know $\{3\}$ has such a matching. We could find one such matching by observation or by applying the flow algorithm to the network in Figure 4.3 (which contains our two sets X and Z, the element 3 of interest to us, the source and sink, and all edges among these vertices) to get the flow $f(s, 3) = f(3, X) = f(X, t) = 1$; $f(e) = 0$ for all other e. This gives us the matching $\{(3, X)\}$.

Figure 4.3

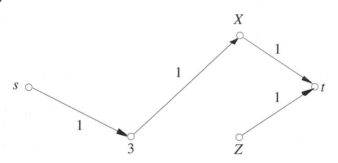

Now we ask whether we can adjoin the vertex 2 to the set $S = \{3\}$ to get a minimum-cost independent set of size 2. To find out, apply the flow algorithm to the network in Figure 4.4, starting with the flow f we just constructed. In this case, we get the flow $f(s, 2) = f(s, 3) = f(2, Z) = f(3, X) = f(X, t) = f(Z, t) = 1$. This gives us the matching $\{(2, Z), (3, X)\}$, so the set $\{2, 3\}$ is an independent set of minimum cost *and* the list $(3, 2)$ is a *system* of distinct representatives for X and Z.

In this example we could have constructed our minimum-cost SDR without the network flow algorithm; however the method shown works in all situations.

Figure 4.4

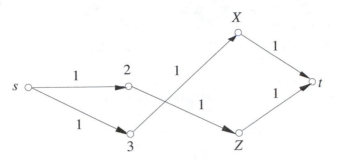

Theorem 4.3. Given a family $S_1, S_2, \ldots, S_n$ of subsets of a set X, the algorithm that follows selects a maximum-sized, minimum-cost independent set $S = \{x_1, x_2, \ldots, x_j\}$ of representatives and a matching

$$(x_1, S_{i_1}), (x_2, S_{i_2}), \ldots, (x_j, S_{i_j})$$

with $x_k \in S_{i_k}$.

Step 1. Let $M = S = \emptyset$. Let D be the digraph with vertices s, t, S_1, S_2, ..., S_n and edges (S_i, t). Let $f(e) = 0$ for each edge e.

Step 2. Choose a minimum-cost element x of $X - S$.

Step 3. Add the vertex x and the edges (s, x) and (x, S_i) for each S_i with $x \in S_i$ to D. Let N be the network with digraph D and all edge capacities 1. Let $f(e) = 0$ for all new edges e just added to D.

Step 4. Apply a flow-augmenting path algorithm to the flow f on D to increase f to a maximum flow.

Step 5. If $f(x, S_i) = 1$ for some i, add x to S. Otherwise, delete x from X.

Step 6. If $|S| = n$, let $M = \{\{x, S_i\} \mid f(x, S_i) = 1\}$ and stop.

Step 7. If $|S| = |X|$, then let $M = \{\{x, S_i\} \mid f(x, S_i) = 1\}$ and stop; otherwise return to step 2.

Proof. At each stage, we add x to S if and only if there is a maximum flow in the network with a flow of 1 out of each vertex in $S \cup \{x\}$; therefore we add x to S if and only if $S \cup \{x\}$ is independent. If the set M is nonempty when we stop, then it has the edges of a matching from the j-element independent set S into $S_{i_1}, S_{i_2}, \ldots, S_{i_j}$ with $x \in S_{i_k}$ for each $\{x, S_{i_k}\}$ in M. ∎

The matching M in Theorem 4.3 gives us a minimum-cost *system* of distinct representatives. This would mean, for example, in the case of a school board wishing to match applicants to positions, that the school board had not only a minimum-cost choice of qualified applicants but also an assignment of the chosen applicants to the positions. Note that step 4 involves constructing one flow-augmenting path and then augmenting the flow. That process requires several steps, the number of which is proportional to the number of edges. Since we carry out the process only once for each x_i (after listing them in increasing order of cost), the number of steps needed is proportional to the product of $|X|$ and the number of edges. This product is no more than $|X|^2 n$.

Minimum-Cost Matchings and Flows with Edge Costs

A subtly different kind of problem of finding minimum-cost systems of distinct representatives or minimum-cost matchings is as follows. Suppose that we have a family $S_1, S_2, \ldots, S_k$ of sets, a set $x_1, x_2, \ldots, x_m$ of representatives, and the price p_{ij} it costs to use x_i to represent S_j. We want to find a system of distinct representatives of minimum total price. This problem was first considered on its own by Philip Hall, who discovered the "marriage theorem"; however, the problem falls naturally into the class of network flow problems. Suppose we are given a network with not only a capacity $c(e)$ defined for each edge e but also a price $p(e)$ per unit of flow defined for each edge e. Thus, it costs $xp(e)$ to send x units of flow along edge e. Then the cost of a flow f on our network is the sum

$$\sum_{e \in E} f(e)p(e) = \text{cost}(f).$$

In this situation, we desire a maximum-valued flow of minimum cost—that is, a flow of maximum value whose cost is no more than that of any other flow of maximum value. To get a minimum-cost matching, we construct a network from the x_i's and S_j's as in Example 4.1. Using C_{ij} as the cost of edge (x_i, S_j) and 0 as the cost of edges including s or t, we find a maximum flow of minimum cost. If we had to construct all possible maximum flows and compute their costs to solve this problem, we would have a time-consuming chore. Fortunately this is not necessary, as the following theorem shows.

> **Theorem 4.4.** Let f be a minimum-cost flow of value V. Augmenting f by means of a minimum-cost flow-augmenting path of worth w yields a minimum-cost flow of value $V + w$.
>
> *Proof.* We assume that each $f(e)$ is rational and that among all flows of value $V + w$ the flow g is a rationally valued flow of minimum cost. We multiply by the least common multiple of all the denominators to convert f, g, w, and all capacities and prices to integers. Note now that if we can prove the theorem with this new w equal to 1, we can

iterate the process to take care of arbitrary positive integral values of w. Dividing by the least common multiple again preserves relative costs of flows and so we may assume all relevant numbers are integers and w is 1. Now consider a new network with the same vertices, edges, and prices but with $c(v_i, v_j)$ the maximum of $f(v_i, v_j)$ and $g(v_i, v_j)$. Then the capacity of the network is at least the value of g, so by labeling we may find a flow-augmenting path of worth 1 in this network. Among all such paths, pick one of minimum cost. Note that this is a flow-augmenting path of worth 1 in our original network.

An edge e used in this flow-augmenting path must be an edge with $g(e) > f(e)$. If we reduce the value of g by 1 in each of these edges, we get a flow g' of value V, the value of f. The *cost* of g is the cost of g' plus the cost of the flow-augmenting path; the cost of f is no more than the cost of g', so the cost of g is at least the cost of f plus the cost of the flow-augmenting path. ∎

The Potential Algorithm for Finding Minimum-Cost Paths

This theorem reduces the problem of finding a minimum-cost flow to that of finding a minimum-cost flow-augmenting path. Although we have already learned an algorithm to find paths of minimum cost (that is, minimum path weight), this algorithm assumes that all the prices (edge weights) are nonnegative. When we use an edge in reverse, however, its price is negative, for it costs us less to send less flow along the edge. However, it is possible to show that because we always augment with a minimum-cost augmenting path, there are no cycles of forward and reverse edges whose total cost (that is, for which the sum of the prices of the edges, with reverse edges counted as negative) is negative. In this case, it is still possible to find minimum-cost paths from a given vertex to any other vertex rather efficiently, although the number of steps taken by the algorithm on an n-vertex network is now no more than a multiple of n^3 rather than n^2.

The algorithm we use is a variant of the labeling algorithm; Fulkerson (see the Suggested Reading) called this variant a *potential* algorithm. (The idea behind the algorithm is also the idea behind Floyd's algorithm to find all distances between all pairs of vertices in a weighted graph, discussed in the exercises following Section 7 of Chapter 4.) We start by assigning either a number or the symbol ∞ as a label $P(v)$ on each vertex v. We update these numbers as we search through the network to give us our current "best estimate" of the cost of getting a unit of flow from the source to vertex v. Our algorithm is the following.

(1) For each vertex v other than the source, let $P(v) = \infty$. Let $P(s) = 0$ for the source.

Repeat step 2 until $P(u) + p(u,v) \geq P(v)$ for each edge (u,v).
(2) For each edge (u,v), if $P(u) + p(u,v) < P(v)$, replace $P(v)$ by $P(u) + p(u,v)$.

In step 2 we examine each edge of the network once. If we examine them in a random way, we might take more steps than necessary. A good order for examining edges is the order in which they appear in a breadth-first search; however, other orders might be better in certain circumstances.

Theorem 4.5. In a network with no negative cycles the potential algorithm stops in a finite number of steps, and when it stops, $P(t)$ is the cost of a minimum-cost path from s to t.

Proof. We may show that if the edges are taken in breadth-first order, then after step 2 is applied i times, the value of $P(v)$ is the minimum of costs of paths with fewer than $n-i$ more edges than a minimum-length path. Alternatively, we could show that each time step 2 is completed, there is at least one more vertex u such that $P(u)$ is the cost of a minimum-cost path form s to u. Thus, step 2 is repeated at most n times, where n is the number of vertices of the network. (The details are left as an exercise.) ∎

Theorem 4.6. A minimum-cost flow-augmenting path for a flow f may be found by subtracting $f(e)$ from $c(e)$ for each edge e, adding the edge (v,u) of capacity $f(v,u)$ and cost $-c(u,v)$ whenever $f(u,v) > 0$, applying the potential algorithm, and then tracing back from t to s along edges (u,v) satisfying $P(u) + p(u,v) = P(v)$ (starting with t and continuing until s).

Proof. Exercise 17. ∎

Finding a Maximum Flow of Minimum Cost

Example 4.2. Find a maximum flow of minimum cost in the network of Figure 4.5. The circled numbers are the prices associated with the edges shown; the uncircled numbers are the capacities.

One application of the potential algorithm assigns to each vertex v the potential $P(v)$ shown in the square near the vertex. The two boxes near vertex t show that, using edge (c,t), we may first assigned potential 5 to t, and then using edge (d,t), assign potential 4 to t. This assignment now satisfies the rule $P(v) \leq P(u) + p(u,v)$ for each edge (u,v). By working backward from t, we see that the path whose vertices are t, d, a, and s is a path of cost 4 leading to s from t, so $s(s,a)a(a,d)d(d,t)t$ is our flow-augmenting path. The worth of this path is 1, so our first flow is

Figure 4.5

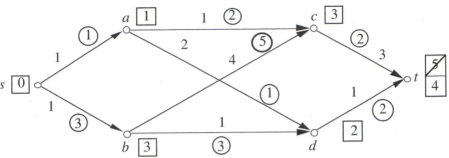

$f(s, a) = f(a, d) = f(d, t) = 1$. This flow gives us the digraph in Figure 4.6, to which we apply the potential algorithm. No capacities are shown in Figure 4.6, for once we have the new digraph the capacities are irrelevant to the potential algorithm.

Figure 4.6

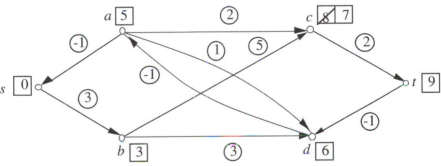

The boxes show the potentials assigned to the vertices when they are traversed in the order $s\ b\ d\ a\ c\ t$. Note that the potential assigned to c drops from 8 to 7 in the process. Working back from t along the edges and vertices with $P(v) = P(u) + p(u, v)$ gives us the sequence $t\ c\ a\ d\ b\ s$. Therefore, our minimum-cost flow-augmenting path, which we choose by examining our sequence of vertices in Figure 4.5 rather than Figure 4.6, is

$$s(s, b)b(b, d)d(a, d)a(a, c)c(c, t)t.$$

The edge (a, d) is a reverse edge on this path. This illustrates a subtle point. When we apply the potential algorithm to an auxiliary digraph (such as the digraph in Figure 4.6) and obtain an edge *not* in our network, the edge of our network pointing in the opposite direction is a reverse edge in the flow-augmenting path. Our flow-augmenting path has worth 1, so our new flow is

$$f(s, a) = f(a, c) = f(c, t) = f(s, b) = f(b, d) = f(d, t) = 1$$

and $f(e) = 0$ for all other e. This flow has value 2 and the cut capacity of the network is 2, so we have a maximum flow of minimum cost. The cost of the flow we have found is $1 + 2 + 2 + 3 + 3 + 2 = 13$. ∎

EXERCISES

1. Find the largest n such that a is n-edge-connected to b in the digraph of Figure 4.7.

Figure 4.7

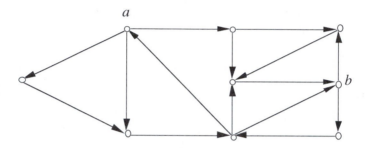

2. Use a network-flow algorithm to find a maximum-sized set of edge-disjoint paths from a to b in the digraph of Figure 4.7.

3. Find a minimum-sized a—b cutset in the digraph of Figure 4.7.

4. Find the largest n such that a is n-vertex-connected to b in the digraph of Figure 4.7.

5. Find a maximum-sized set of vertex-disjoint paths from a to b in the digraph of Figure 4.7.

6. Change the directed edges to undirected edges in Figure 4.7 and determine the largest m and n such that a is m-edge-connected and n-vertex-connected to b.

7. Change the directed edges to undirected edges and find a maximum-sized edge-disjoint set of paths in Figure 4.7 connecting a to b.

8. Find an SDR for the family $\{1, 3, 5\}$, $\{2, 3, 5\}$, $\{7, 5, 3\}$, $\{6, 2, 5\}$, $\{2, 3, 7\}$ or explain why none exists. (Use network flows.)

9. Let the cost of i be i and find a minimum-cost SDR for the family of Exercise 7, or explain why none exists.

10. Do Exercise 11 of Section 2 (finding an assignment of people to jobs).

11. Prove that the relation "a is n-edge-connected to b" is an equivalence relation.

12. Write out the details of the proof of the vertex form of Menger's theorem for directed graphs.

13. State and prove the vertex form of Menger's theorem for graphs.

14. State and prove the edge form of Menger's theorem for multigraphs.

15. Write a computer program that finds a minimum-cost SDR for a family of sets.

16. Write out a complete proof of Theorem 4.5.

17. Prove Theorem 4.6.

18. In Figure 4.5, change the capacity of edge (s, a) to 2 and then find a maximum-sized flow of minimum cost. What is the cost of the flow?

19. In Figure 4.5, change all capacities to 1 and the costs of edges touching s or t to 0. Why may this network be interpreted as the network of a minimum-cost matching problem? Find the minimum-cost matching using network flows.

20. In Figure 4.5, change the capacities of edges (s, b) and (s, a) to 2 and find a maximum-sized flow of minimum cost. What is the cost of the flow?

21. In the graph of Figure 2.2 in Section 2 of this chapter, make use of the potential algorithm to find a minimum-cost path from h to c. (Assume all edges are directed in both directions and have the same cost in each direction.)

22. (a) In the network of Figure 4.5, use a variant of the potential algorithm (which you must somehow change slightly) to find a maximum-cost path from s to t. The circled numbers are the prices.
    (b) Find a flow of maximum cost by using the potential algorithm to find flow-augmenting paths.

23. What must you know about a network in order to use (a slight variant of) the potential algorithm in order to find a maximum-length path between a vertex and each other vertex? Why were you able to find a longest path in the previous problem?

24. In Exercise 8, let the cost of representing set S_j by integer i be i if i is the minimum element of S_j and otherwise let it be i plus the minimum number in S_j. Find a minimum-cost SDR.

25. In Exercise 11 of Section 2, let the jobs be numbered $10 = $ rough carpenter, $9 = $ general laborer, $10 = $ painter, $8 = $ equipment operator, $11 = $ finish carpenter, and $11 = $ supervisor. Let the cost of assigning person i to job j be the maximum of the number just assigned to the job and the number beneath person i. (Government jobs, for example,

sometimes have the requirement that the minimum pay for a job must be the "prevailing wage rate.") Find a minimum-cost assignment of people to jobs.

26. In Example 3.1 (Figure 3.2), let the costs be $p(s, a) = c(a, d) = 3$, $p(s, b) = p(b, d) = 4$, and $p(e) = 2$ for all other e. Find a minimum-cost flow of maximum value.

27. Either give an example of a digraph with prices on the edges for which the potential algorithm does not stop after all edges have been examined once (in breadth-first ordering) or explain why there is none.

28. Write a computer program to find minimum-cost paths from one vertex to each other vertex in a digraph with a price function on the edges, assuming that there are no cycles whose total price is negative.

29. Write a computer program to find maximum matchings of minimum cost.

Suggested Reading

Berge, Claude. 1973. *Graphs and Hypergraphs,* ch. 7. Amsterdam: North Holland.

Edmonds, Jack 1963.   Paths, Trees and Flowers. *Canadian Journal of Math* 17:449–467.

Even, Shimon. 1979. *Graph Algorithms.* Rockville, MD: Computer Science Press.

Fulkerson, D. Ray, Ed. 1975. *Studies in Graph Theory I.* MAA Studies in Mathematics, vol. 11. Washington DC: Mathematical Association of America.

Lawler, Eugene L. 1971. *Combinatorial Optimization: Networks and Matroids.* New York: Holt, Reinhart, & Winston.

Welsh, D. J. A. 1976. *Matroid Theory.* Orlando, FL: Academic Press.

Wilf, Herbert S. 1986. *Algorithms and Complexity.* Englewood Cliffs, NJ: Prentice Hall.

6

Combinatorial Designs

Section 1 Latin Squares and Graeco–Latin Squares

How Latin Squares Are Used

Suppose we have five varieties of gasoline formulated in different ways, each designed to provide high mileage per gallon. We wish to test the different varieties against each other so as to choose an optimal formulation. Each variety is to be tested in a 400-mile drive. It would be nice to use the same car for each test, but then we would have to make the tests on different days and face such variable extraneous factors as wind, temperature variation, humidity, traffic, etc., any one of which might affect gasoline mileage as much as the gasoline variety itself. Thus, we use five different cars so that they may all drive the same course in the same day. Now we have the problem that different cars have different gasoline mileage characteristics! Suppose, however, instead of running the experiment just once, we run it five times in such a way that each car uses each variety of gasoline exactly once. If a variety of gasoline is truly superior, we should be able to notice that fact because the car using that gasoline on a given day will have improved relative performance (relative to the other cars on other days). Correspondingly, if one of the cars is a consistently superior performer, we will be able to notice that. Figure 1.1 is a possible schedule of cars and varieties of gasoline.

Figure 1.1

Gasoline Variety	Day 1	2	3	4	5
1	A	B	C	D	E
2	B	C	D	E	A
3	C	D	E	A	B
4	D	E	A	B	C
5	E	A	B	C	D

The cars are designated by capital letters A through E and the gasoline formulations are numbered 1 to 5. The array of letters in Figure 1.1 is called a Latin square. A **Latin square** is a square matrix of symbols such that each symbol occurs in each row exactly once and in each column exactly once. It is clear from the figure that given n symbols we can always construct an $n \times n$ Latin square, also called a Latin square of *side n*.

The Latin square in Figure 1.1 is said to be in "standard form." What does this mean? Notice that we could permute the columns in any of the $n!$ ways we could choose and we would still get a Latin square. Similarly, we could permute the rows in any way that we choose and still get a Latin square. What is special about the square in Figure 1.1 is that both row 1 and column 1 appear in their natural (alphabetical) order. In any different Latin squares we get by applying one of the $(n!)^2$ pairs of permutations, they would not be in their natural order. (Do all these $(n!)^2$ rearrangements give squares different from the one we started with?) Thus, when there is a natural ordering in which to write the symbols in a Latin square, we say a Latin square with these symbols is a *standard Latin square* if both the first row and the first column are in their natural order. In the use of a Latin square for an experimental design, the experimenter will usually rearrange the rows and the columns according to randomly chosen permutations before using the design.

Randomization for Statistical Purposes

The purpose of rearranging the rows and columns before using the design is to reduce the probability that some unrecognized feature of the experiment would bias the results. As an example, suppose the letters in Figure 1.1 still represent five cars, but now we are trying to decide what concentration of each of two gasoline additives we should use in order to best enhance the gasoline mileage. Suppose we use 1, 2, 3, 4, or 5 parts per million of each additive in an otherwise standardized blend of gasoline and use each combination in a car for 400 miles. Since there are 25 combinations and we have just five cars for testing, we will still need to run tests on 5 different days to test all the combinations. On the first day we test, we might use 1 part per million (ppm, for short) of additive one in all the cars, and use 1 ppm of additive 2 in car A, 2 ppm of additive two in car B, 3 ppm of additive two in car C, 4 ppm of additive two in car D, and 5 ppm of additive two in car E, driving each car 400 miles. On the second day, for the second 400 miles, we might use 2 parts per million of additive one and 1, 2, 3, 4, and 5 parts per million of additive two in the cars in the order they are shown in row 2, and so on. This gives us the table shown in Figure 1.2.

Thus, as before, with the five varieties of gasoline, if one level of additive one is best, we should be able to determine that. For example, if each car does better with 3 ppm of additive one than with any other level (in other words, each car has its best performance in row 3), then then we would conclude this is the best level for additive 1. By interchanging the roles of rows and columns, if one level of additive two is best, we should be able to determine that. We might be able to conclude even more. Suppose the levels that we determined to be best were 1 part per million for additive one and 4 parts per million for additive two. Then the pair (1,4) was tested in car D, so if this pair had the best results in all five tests made in car D and the best results in all the tests on day 4, we would be reasonably convinced we had found the best combination of additives to use.

Figure 1.2

Additive 1 Level	Additive 2 Level				
	1 ppm	2 ppm	3 ppm	4 ppm	5 ppm
1 ppm	A	B	C	D	E
2 ppm	B	C	D	E	A
3 ppm	C	D	E	A	B
4 ppm	D	E	A	B	C
5 ppm	E	A	B	C	D

Of course, we might not be lucky enough to get such clear experimental results as we just described. In that case, although statistical reasoning would help us estimate the optimal levels of either additive, a statistical analysis of this experiment might not be able to help us determine the combination of levels of the two additives that is best.

Just as before, it would appear that if one car was a consistently good performer, we would notice that as well. However, notice that car E is always tested with 6 parts per million of additives. If the additives have reasonably similar effects, then car E will appear to have virtually the same gas mileage in all five tests, while the other cars will appear to be erratic. However, the "erratic" and "consistent" behavior will be a result of the total concentration of additives in the various test runs, not a result of an inherent property of the cars. Thus, our experiment is biased in favor of showing consistency for car E. Other systematic factors we might not see might introduce other kinds of bias. We can show that if we averaged our results over all squares we could get by rearranging rows and columns, then the biases would average out. By instead choosing a random rearrangement, we are hoping that we have a high probability of choosing a design with low bias. We can show that if we repeat the experiment with even a relatively small number of randomly rearranged squares, then averaging our results will make the probability that our results have low bias be near one. We could also increase our probability of picking out the pair of levels of the additives that is optimal, if there is such a pair.

Orthogonal Latin Squares

Let us turn now to a different kind of problem that might arise with either experiment we have described. Since the five different cars are to be driven at the same time, they must have different drivers. However, the gasoline mileage that a car gets depends on who is driving it. If we could arrange things so that each driver drives each car exactly once and each driver uses each variety of gasoline exactly once, then perhaps we could sort out which effects are due to the drivers and which are due to the gasoline. Figure 1.3 is an arrangement in which the lowercase letters stand for the drivers.

Figure 1.3

$$
\begin{array}{ccccc}
Aa & Bb & Cc & Dd & Ee \\
Bc & Cd & De & Ea & Ab \\
Ce & Da & Eb & Ac & Bd \\
Db & Ec & Ad & Be & Ca \\
Ed & Ae & Ba & Cb & Dc
\end{array}
$$

This arrangement is called a *Graeco–Latin square* or a pair of *orthogonal Latin squares*. More precisely, a pair of matrices S and s of the same dimensions are **orthogonal** (or *combinatorially orthogonal*) if all ordered pairs (S_{ij}, s_{ij}) are different. Notice that this definition applies to matrices of any shape. In our example, there are 25 ordered pairs and 25 positions, so orthogonality is equivalent to saying that each ordered pair occurs exactly once in the positions of the superimposed squares. Sometimes it is useful to have a notation for the matrix of all ordered pairs (S_{ij}, s_{ij}); we use (S, s) to stand for this matrix so that $(S, s)_{ij} = (S_{ij}, s_{ij})$. The statement that S and s are orthogonal is the same as the statement that the entries of (S, s) are all different. With this matrix we can say exactly what we mean by a Graeco–Latin square; we call (S, s) a **Graeco–Latin square** if S and s are Latin squares of the same size and all the entries of (S, s) are different. (The use of the phrase Graeco–Latin square becomes obvious if you substitute Greek letters for the upper- or lowercase Latin letters we used!)

What if we had four, or even six, varieties of gasoline? This leads us to ask, "Is there a pair of orthogonal n by n Latin squares for each value of n? Is there a pattern in Figure 1.3 that can be duplicated for other values of n besides 5?" In Figure 1.3, we can obtain the second row of uppercase letters from the first by shifting all entries of row 1 (except the first entry) to the left and moving the first entry to the end of the row. We get succeeding rows by repeating this process. If you imagine erasing the uppercase letters in Figure 1.3, you will see that for the lowercase letters, we get the second row by shifting everything but the first two entries *two* places to the left, moving the first two to the end, and then repeating the process to get the remaining rows. Is this a general method of constructing orthogonal squares? If we tried it with $n = 6$, we would not get a Latin square for the lowercase letters because the fourth row would be the same as the first. We shall soon show that whenever the construction method gives two Latin squares, they are orthogonal. It is easy to show that there is no pair of orthogonal Latin squares of side 2. (Can you explain why?)

Euler's 36-Officers Problem

Euler observed that no orthogonal pair of 2×2 squares existed and found he was able to construct examples of $n \times n$ orthogonal squares for n up to 5 but had trouble with 6. He proposed the following problem.

> Thirty-six officers of six ranks and from six different regiments are to march in a square formation of size 6×6. Each row and each column of the formation is to contain one and only one officer of each rank and one and only one officer from each regiment. Is such a formation possible?

The officers are the 36 places of a Graeco–Latin square in which the symbols are the six names of the regiments and the six ranks. In 1782, Euler conjectured that such an arrangement was not possible. Further, he conjectured that so long as $n = 4k + 2$ for some integer k, then no pair of $n \times n$ orthogonal Latin squares exists. Around 1900, Tarry showed that there is no pair of 6×6 orthogonal Latin squares by systematically checking all possible constructions.

Euler's more general conjecture remained unsettled until 1959, when Bose, Shrikhande, and Parker gave a general method that allowed them to construct an orthogonal pair of 22×22 Latin squares. Then in 1960 they gave a general method that allowed them to construct a pair of 10×10 orthogonal Latin squares. By combining the techniques and showing how to build new orthogonal Latin squares from old ones, they were able to complete the proof that there is a pair of orthogonal $n \times n$ Latin squares if n is different from 2 or 6. We shall not prove this, but after we show that the construction used in Figure 1.3 can be generalized to yield many pairs of orthogonal squares, we will describe one general technique which constructs a 10×10 Graeco–Latin square from a 3×3 Graeco–Latin square. Our constructions use the concept of congruence modulo an integer n; we now review this concept.

Congruence Modulo an Integer n

We define an equivalence relation on the integers by saying a is **congruent** to b **modulo** n, written $a \equiv b \pmod{n}$, if and only if n is a factor of $a - b$.

Let us check that this relation is an equivalence relation. Since n is a factor of $a - a = 0$, our relation is reflexive. Since n is a factor of $b - a = -(a - b)$ whenever it is a factor of $a - b$, our relation is symmetric. Now suppose $a \equiv b \pmod{n}$ and $b \equiv c \pmod{n}$, so $a - b = kn$ and $b - c = hn$. Then

$$a - c = a - b + b - c = kn + hn = (k + h)n,$$

so our relation is transitive. Thus, it is an equivalence relation. The equivalence class containing an integer i will also contain $i + n$ and $i - n$; in fact, it will contain $i + kn$ for each (positive or negative) integer k. Further, if $x \equiv i \pmod{n}$, then for some k, $x - i = kn$ since n is a factor of $x - i$; this gives $x = i + kn$. Thus, the n equivalence classes,

$$\{0 + kn \mid k \text{ is an integer}\}, \{1 + kn \mid k \text{ is an integer}\},$$

$$\ldots, \{n - 1 + kn \mid k \text{ is an integer}\}$$

are the complete set of equivalence classes for congruence modulo n. We will use the symbol $\underset{\sim}{i}$ to stand for the equivalence class containing i.

We will define arithmetic operations on equivalence classes so that we can talk about $\underset{\sim}{i} + \underset{\sim}{j}$ and $\underset{\sim}{i} \cdot \underset{\sim}{j}$. For this purpose, we prove the following theorem.

Theorem 1.1. If $i \equiv i_1 \pmod{n}$ and $j \equiv j_1 \pmod{n}$, then

$$i + j \equiv i_1 + j_1 \pmod{n}$$

and

$$ij \equiv i_1 j_1 \pmod{n}.$$

Proof. Note that $i_1 = i + hn$ and $j_1 = j + kn$. Then

$$i + j - (i_1 + j_1) = -(h + k)n$$

so that $i + j \equiv i_1 + j_1 \pmod{n}$. Also,

$$ij - i_1 j_1 = ij - (i + hn)(j + kn)$$
$$= ij - (ij + ikn + jhn + hkn^2)$$
$$= -(ik + jh + khn)n$$

so that $ij \equiv i_1 j_1 \pmod{n}$ ∎

By Theorem 1.1, we see that whenever i is congruent to i_1 and j is congruent to j_1, then $i + j$ lies in the same class as $i_1 + j_1$ and ij lies in the same class as $i_1 j_1$. Thus, if we define $\underline{i} + \underline{j}$ to be $\underline{i + j}$, then we will have

$$\underline{i} + \underline{j} = \underline{i_1 + j_1} \text{ whenever } \underline{i} = \underline{i_1} \text{ and } \underline{j} = \underline{j_1}$$

and

$$\underline{ij} = \underline{i_1 j_1} \text{ whenever } \underline{i} = \underline{i_1} \text{ and } \underline{j} = \underline{j_1}.$$

Thus, defining $\underline{i} + \underline{j}$ to be $\underline{i + j}$ and $\underline{i} \cdot \underline{j}$ to be $\underline{ij}$, respectively, makes sense since the definition doesn't depend on which member of $\underline{i}$ and $\underline{j}$ we choose—that is, we could just as well have defined them to be $\underline{i_1 + j_1}$ and $\underline{i_1 j_1}$ and we would have gotten the same sets.

It is convenient to visualize $\underline{i}$ as being represented by the smallest nonnegative integer in the set $\underline{i}$. We don't introduce a new notation though, but we call the smallest nonnegative element of $\underline{i}$ the *residue* of i (mod n). When it is clear from context that we are discussing residues rather than integers, it is traditional to use i to stand for the residue of i (mod n). We shall follow this tradition once the difference between integers and their residues is clear. It is also traditional to use i in place of $\underline{i}$ when it should be clear that one is discussing congruence classes rather than integers.

Example 1.1. When n is 5, the classes are $\underline{0}, \underline{1}, \underline{2}, \underline{3}$, and $\underline{4}$; we often let 0 represent $\underline{0}$. Since $\underline{0} = \underline{5}$, we are letting 0 represent $\underline{5}$ as well, but we don't let 5 represent $\underline{5}$. Rather than writing $\underline{4} + \underline{3} = \underline{7}$, it is customary to write $\underline{4} + \underline{3} = \underline{2}$ or $4 + 3 \equiv 2 \pmod{5}$. Note that $\underline{4} \cdot \underline{3} = \underline{2}$ as well; that is, $4 \cdot 3 \equiv 2 \pmod{5}$. We call 2 the residue of 7 mod 5. It is also the residue of 12 or $-3 \pmod{5}$. ∎

Using Arithmetic Modulo n to Construct Latin Squares

We state the relationship between arithmetic modulo n and Latin squares as a theorem.

Theorem 1.2. If the rows of a matrix are labeled as rows 0 through $n - 1$ and the entries of row i are the residues of the numbers from i to $n + i - 1$ in the order of the numbers, then this matrix is a Latin square.

Proof. Each of the numbers from i to $n - 1$ appears as a residue in row i—but the numbers from 0 to $i - 1$ are the residues of n through $n + i - 1$, and so each residue appears in each row exactly once. Number the columns from left to right as 0, 1, through $n - 1$. Then column k

consists of the residues of $k + i$ for $i = 0, 1, \ldots, n - 1$ and thus, as with the rows, contains each residue exactly once. ∎

Now it is clear that we could have used the residues of 5 to construct the array in Figure 1.1, the array in Figure 1.2, and the uppercase array in Figure 1.3. Of course, the lowercase letters in Figure 1.3 could be replaced by symbols denoting residues too. In fact, if we replace the lowercase letters by residues modulo 5 in the natural way, we get the array in Figure 1.4 in place of the lowercase letters. Note that we get the second row by adding 2 to everything in the first row and taking residues modulo 5.

Figure 1.4

$$\begin{array}{ccccc} 0 & 1 & 2 & 3 & 4 \\ 2 & 3 & 4 & 0 & 1 \\ 4 & 0 & 1 & 2 & 3 \\ 1 & 2 & 3 & 4 & 0 \\ 3 & 4 & 0 & 1 & 2 \end{array}$$

In fact, we get each row by adding 2 to everything in the preceding row and taking residues modulo 5. If we think of this as a matrix B whose rows and columns are indexed by the numbers 0 through 4 (rather than 1 through 5, as we would normally index them), we can write

$$B_{i+1,j} \equiv B_{ij} + 2 \pmod 5$$

and since row 0 is given by

$$B_{0j} = j,$$

we get

$$B_{1j} \equiv B_{0j} + 2 \equiv 2 + j \pmod 5,$$

and in general,

$$B_{ij} \equiv B_{0j} + 2i \equiv 2i + j \pmod 5.$$

From this, it is clear that the residues of $2i, 2i+1, \ldots, 2i+5-1$ all appear in row i, and so all residues mod 5 appear in row i. Since the residues of 0, 2, 4, 6, and 8 are 0, 2, 4, 1, and 3, the first column contains distinct elements. Column j is derived by adding j to column 0, so every other column contains distinct elements because adding a fixed j to distinct elements gives distinct residues. (To prove this statement, you would assume that $\underline{i} + \underline{j} = \underline{i} + \underline{k}$ and add $-\underline{i}$ to both sides.) Therefore, B is a Latin square.

The Latin square of Theorem 1.2 is given by $A_{ij} \equiv i + j \pmod n$. Let us consider the general matrix B we get by replacing the integers between

0 and 4 by the integers between 0 and $n-1$ and otherwise carrying out the construction of B just as before. This will give us $B_{ij} \equiv 2i + j \pmod n$. To check that A and B are orthogonal, we must show that each pair of residues appears once and only once in the form of the pair

$$(A_{ij}, B_{ij}).$$

Thus, we ask for solutions to the symbolic equation

$$(A_{ij}, B_{ij}) = (A_{kh}, B_{kh})$$

which reduces to

$$i + j \equiv k + h \pmod n$$

$$2i + j \equiv 2k + h \pmod n$$

by setting the two components of the pairs equal. Now applying Theorem 1.1, and subtracting the first equation from the second, we get

$$2i + j - (i + j) \equiv 2k + h - (k + h) \pmod n$$

$$i \equiv k \pmod n.$$

Now subtract this equation from

$$i + j \equiv k + h \pmod n$$

to get

$$j \equiv h \pmod n.$$

Thus, two pairs are equal only if they are in the same position in the paired arrays. Notice that this computation did not use the actual value of n; however, we have seen that if n were 6, then the matrix B would not be a Latin square because rows 1 and 4 would be identical. (Thus, the fact that the computation applies to $n = 6$ is irrelevant.) In fact, for any even n, the matrix B would not be a Latin square for the same reason. However, the preceding methods yield a general theorem.

>   ***Theorem 1.3.*** If n is odd and if the matrices (indexed by 0 through $n-1$) A and B have $A_{ij} \equiv i + j \pmod n$ and $B_{ij} \equiv 2i + j \pmod n$, then A and B are orthogonal Latin squares, that is, the matrix (A, B) is a Graeco–Latin square.

Orthogonality and Arithmetic Modulo n

By the same sort of computations, we can prove a theorem which will give us more than one pair of orthogonal squares for many values of n.

Theorem 1.4. Suppose that r and s are numbers such that for $i = 0$ to $n - 1$, the products $r \cdot i$ are all distinct modulo n and the products $s \cdot i$ are all distinct modulo n. If $r - s$ has no factors (other than 1) in common with n, then the arrays A and B of residues (mod n) with $A_{ij} \equiv ri + j$ and $B_{ij} \equiv si + j$ (mod n) are orthogonal Latin squares, that is, (A, B) is a Graeco–Latin square.

Proof. As above, if

$$(A_{ij}, B_{ij}) = (A_{kh}, B_{kh}),$$

then

$$ri + j \equiv rk + h \quad (\text{mod } n)$$
$$si + j \equiv sk + h \quad (\text{mod } n)$$

so that subtraction gives

$$(r - s)i \equiv (r - s)k \quad (\text{mod } n)$$

which is the same as

$$(r - s)(i - k) \equiv 0 \quad (\text{mod } n).$$

But then n is a factor of $(r - s)(i - k)$, and since n has no factors in common with $r - s$, either it must have all its factors in common with $i - k$ or else $i - k$ must be 0. Since i and k are both less than n, the only possibility is that $i - k$ is 0, so that $i = k$. Then by substituting into one of the first two equations, we get $j = h$ as well. ∎

Corollary 1.5. If n is a power of a prime p, then there are $p - 1$ $n \times n$ Latin squares all orthogonal to each other.

This corollary tells us that we could, for example, also consider the effects of tires on gas mileage at the same time we are testing the five formulations of gasoline with the five different drivers in the five different cars—if we were willing to change tires from car to car following each run! In fact, if n is a power p^m of a prime, there are $p^m - 1$ mutually orthogonal Latin squares. This follows from the theory of vector spaces over finite fields and thus is beyond the scope of this book.

Compositions of Orthogonal Latin Squares

Now that we have shown that there are p by p orthogonal Latin squares for primes $p > 2$, we can demonstrate the existence of many other Graeco–Latin squares by building up new ones from old ones. The most standard method of building new Latin squares from old is what we call the *composition* of Latin squares. For this purpose we go back to thinking of the rows of an $m \times n$ matrix matrix as rows 1 through n, not rows 0 through $n - 1$.

We have defined a Latin square to be a special kind of square matrix of symbols; to define the composition $A \otimes B$ of an $n \times n$ Latin square A with an $m \times m$ Latin square B, we must say what the symbols of $A \otimes B$ are and specify what symbol appears in each possible row and column. Thus, we will simply be describing a new matrix. For this reason, we will describe the composition of two arbitrary matrices A and B. For each symbol a of A and each symbol b of B the ordered pair $[a, b]$ is a possible symbol of $A \otimes B$. (There are two reasons why we are using square brackets instead of parentheses around our ordered pairs. One will become clear in a few paragraphs, when we talk about removing parentheses, and the other when we begin to discuss orthogonality.)

Now, in the composition of A and B we will want to pair up each entry of A with each entry of B in a systematic way. When A has just one entry, there is a natural way to do this. Using $\otimes$ to stand for composition, we write

$$(a) \otimes \begin{pmatrix} b_{11} & b_{12} & \cdots & b_{1n} \\ \vdots & \vdots & & \vdots \\ b_{m1} & bm2 & \cdots & b_{mn} \end{pmatrix} = \begin{pmatrix} [a, b_{11}] & [a, b_{12}] & \cdots & [a, b_{1n}] \\ \vdots & \vdots & & \vdots \\ [a, b_{m1}] & [a, bm2] & \cdots & [a, b_{mn}] \end{pmatrix}$$

Now if A has more than one entry, we could first pair up each entry of B with the first entry in row 1, then pair up each entry of B with the second entry in row 1, and so on. One way to capture this idea in symbols would be to say that

$$\begin{pmatrix} a_{11} & a_{12} & \cdots & a_{1,s} \\ \vdots & \vdots & & \vdots \\ a_{r1} & a_{r2} & \cdots & a_{rs} \end{pmatrix} \otimes \begin{pmatrix} b_{11} & b_{12} & \cdots & b_{1n} \\ \vdots & \vdots & & \vdots \\ b_{m1} & bm2 & \cdots & b_{mn} \end{pmatrix}$$
$$= \begin{pmatrix} (a_{11}) \otimes B & (a_{12}) \otimes B & \cdots & (a_{1,s}) \otimes B \\ \vdots & \vdots & & \vdots \\ (a_{r1}) \otimes B & (a_{r2}) \otimes B & \cdots & (a_{rs}) \otimes B \end{pmatrix}$$

There is a slight problem with this definition, as the following example shows.

Example 1.2. Find the composition of $\begin{pmatrix} 1 & 2 \\ 3 & 4 \end{pmatrix}$ and $\begin{pmatrix} a & b & c \\ d & e & f \end{pmatrix}$.

We write

$$
\begin{pmatrix} 1 & 2 \\ 3 & 4 \end{pmatrix} \otimes \begin{pmatrix} a & b & c \\ d & e & f \end{pmatrix} = \begin{pmatrix} 1 \otimes \begin{pmatrix} a & b & c \\ d & e & f \end{pmatrix} & 2 \otimes \begin{pmatrix} a & b & c \\ d & e & f \end{pmatrix} \\ 3 \otimes \begin{pmatrix} a & b & c \\ d & e & f \end{pmatrix} & 4 \otimes \begin{pmatrix} a & b & c \\ d & e & f \end{pmatrix} \end{pmatrix}
$$

$$
= \begin{pmatrix} \begin{pmatrix} [1,a] & [1,b] & [1,c] \\ [1,d] & [1,e] & [1,f] \end{pmatrix} & \begin{pmatrix} [2,a] & [2,b] & [2,c] \\ [2,d] & [2,e] & [2,f] \end{pmatrix} \\ \begin{pmatrix} [3,a] & [3,b] & [3,c] \\ [3,d] & [3,e] & [3,f] \end{pmatrix} & \begin{pmatrix} [4,a] & [4,b] & [4,c] \\ [4,d] & [4,e] & [4,f] \end{pmatrix} \end{pmatrix}
$$

∎

Thus with the way we have defined the composition, we do not get a matrix containing symbols such as $[1, a]$, but rather a matrix of matrices. However, we can change our definition slightly to say that the **composition** $A \otimes B$ is the matrix we get from

$$
\begin{pmatrix} (a_{11}) \otimes B & (a_{12}) \otimes B & \cdots & (a_{1,s}) \otimes B \\ \vdots & \vdots & & \vdots \\ (a_{r1}) \otimes B & (a_{r2}) \otimes B & \cdots & (a_{rs}) \otimes B \end{pmatrix}
$$

by removing the innermost parentheses. Now let us try the example again.

Example 1.3. Find the composition of $\begin{pmatrix} 1 & 2 \\ 3 & 4 \end{pmatrix}$ and $\begin{pmatrix} a & b & c \\ d & e & f \end{pmatrix}$. We remove the innermost parentheses from

$$
\begin{pmatrix} \begin{pmatrix} [1,a] & [1,b] & [1,c] \\ [1,d] & [1,e] & [1,f] \end{pmatrix} & \begin{pmatrix} [2,a] & [2,b] & [2,c] \\ [2,d] & [2,e] & [2,f] \end{pmatrix} \\ \begin{pmatrix} [3,a] & [3,b] & [3,c] \\ [3,d] & [3,e] & [3,f] \end{pmatrix} & \begin{pmatrix} [4,a] & [4,b] & [4,c] \\ [4,d] & [4,e] & [4,f] \end{pmatrix} \end{pmatrix}
$$

and get

$$
\begin{pmatrix} [1,a] & [1,b] & [1,c] & [2,a] & [2,b] & [2,c] \\ [1,d] & [1,e] & [1,f] & [2,d] & [2,e] & [2,f] \\ [3,a] & [3,b] & [3,c] & [4,a] & [4,b] & [4,c] \\ [3,d] & [3,e] & [3,f] & [4,d] & [4,e] & [4,f] \end{pmatrix} .
$$

∎

The reader who finds this rather informal definition of composition unsettling will be glad to know there is another way to describe the same matrix. If we start with an $r \times s$ matrix A and an $m \times n$ matrix B, we can describe the composition as follows. The ordered pair $[A_{ij}, B_{i'j'}]$ will be the entry in row $(i-1)m+i'$ and column $(j-1)n+j'$ of the composition $A \otimes B$. Notice that each integer between 1 and rm may be written in exactly one way as $(i-1)m + i'$ with $1 \leq i \leq r$ and $1 \leq i' \leq m$ (and similarly with sn

and $(j-1)n+j')$, so that we assign exactly one ordered pair to each position of $A \otimes B$. Symbolically, we write

$$(A \otimes B)_{(i-1)m+i',(j-1)n+j'} = [A_{ij}, B_{i'j'}].$$

Let us return to our previous example.

Example 1.4. Find the composition of $\begin{pmatrix} 1 & 2 \\ 3 & 4 \end{pmatrix}$ and $\begin{pmatrix} a & b & c \\ d & e & f \end{pmatrix}$.

The entry $[4,f]$ appears in the last row and column; our rule says it belongs in row $(2-1) \cdot 2 + 2$ and column $(2-1) \cdot 3 + 3$, that is, row 4 and column 6. Thus our composition has four rows and six columns. Of course, in row 1 and column 1 we have the entry $[1, a]$. As a typical example, in row $1 \cdot 1 + 1 = 2$ and column $3 \cdot 1 + 2 = 5$ we have $[A_{12}, B_{13}] = [2, e]$. Following our rules in this way gives us

$$\begin{pmatrix} [1,a] & [1,b] & [1,c] & [2,a] & [2,b] & [2,c] \\ [1,d] & [1,e] & [1,f] & [2,d] & [2,e] & [2,f] \\ [3,a] & [3,b] & [3,c] & [4,a] & [4,b] & [4,c] \\ [3,d] & [3,e] & [3,f] & [4,d] & [4,e] & [4,f] \end{pmatrix}.$$ ∎

Now we are prepared to apply composition to Latin squares.

Theorem 1.6. If A is an $n \times n$ Latin square and B is an $m \times m$ Latin square, then $A \otimes B$ is an $mn \times mn$ Latin square.

Proof. We know that $A \otimes B$ is an $mn \times mn$ matrix. To show we have a Latin square, we have to show each row is a permutation of the same set and each column is a permutation of this same set. Notice that a given row contains all pairs we get from pairing each element of a fixed row of A with each element of a fixed row of B. If we let S be the set of symbols in a row of A (remember, all rows use the same set of symbols) and we let T be the set of symbols in a row of B, then what we have just said is that each row of $A \otimes B$ contains all the elements of the set $S \times T$ of ordered pairs. Since this set has mn elements and a row of $A \otimes B$ has mn elements, each row must be a permutation of $S \times T$. The same analysis applies to each column. Thus $A \otimes B$ is a Latin square. ∎

The composition of Latin squares is especially important because it extends to Graeco–Latin squares; that is, the composition of two Graeco–Latin squares is, in a slightly unorthodox way, another Graeco–Latin square. Our next result explains this.

Theorem 1.7. Suppose that A and A' are orthogonal Latin squares and B and B' are orthogonal Latin squares. Then the Latin squares $A \otimes B$ and $A' \otimes B'$ are orthogonal Latin squares.

Proof. We must show that $(A \otimes B, A' \otimes B')$ is a Graeco–Latin square, that is, that its elements are distinct. The elements of $A \otimes B$ are ordered pairs $[a, b]$ and those of $A' \otimes B'$ are ordered pairs $[a', b']$. Thus, the entries of $(A \otimes B, A' \otimes B')$ are ordered pairs *of ordered pairs* of the form $([a, b], [a', b'])$. In essence, these are four-tuples (a, b, a', b') chosen, of course, according to certain rules. In our introductory discussion of the theorem, we said that we would show that the composition of Graeco–Latin squares is, essentially, a Graeco–Latin square. The squares (A, A') and (B, B') are Graeco–Latin squares and their composition is $(A, A') \otimes (B, B')$, so it consists of ordered pairs of the form $[(a, a'), (b, b')]$. In essence these pairs of pairs are four-tuples (a, a', b, b'), chosen, except for the order in which they are written down, according to the same rules as the four-tuples (a, b, a', b') mentioned previously. Thus, the first matrix of four-tuples—our pairs of pairs—consists of distinct entries if and only if the second matrix of four-tuples—or pairs of pairs—does. However, all the members of (A, A') are different and all the members of (B, B') are different. Each position in $(A, A') \otimes (B, B')$ is computed from the entries in a different pair of positions in (A, A') and (B, B'). Therefore, the entries of $(A, A') \otimes (B, B')$ are all different. Therefore, so are the entries of $(A \otimes B, A' \otimes B')$, so $(A \otimes B, A' \otimes B')$ is a Graeco–Latin square. ∎

From the proof you see what we mean by saying we must view a composition of Graeco–Latin squares in an unorthodox way for it to be a Graeco–Latin square; we must view it as $(A \otimes B, A' \otimes B')$ rather than $(A, A') \otimes (B, B')$. That is why we didn't actually use the word composition in the statement of our theorem.

Example 1.5. Is there a 25 by 25 Graeco–Latin square?

Yes, there is; consider the pairs S and s of 5×5 orthogonal Latin squares of Figure 1.3 and the pair A and B of orthogonal Latin squares of Theorem 1.3 with $n = 5$. By Theorem 1.7, $S \otimes A$ and $s \otimes B$ are 25 by 25 orthogonal Latin squares. (We could use S and s a second time in place of A and B.) ∎

Orthogonal Arrays and Latin Squares

The idea of orthogonality for arbitrary matrices allows us to describe Latin squares in another useful way.

Example 1.6. Show that the following three 1×25 matrices are orthogonal.

$$R = (1\ 1\ 1\ 1\ 1\ 2\ 2\ 2\ 2\ 2\ 3\ 3\ 3\ 3\ 3\ 4\ 4\ 4\ 4\ 4\ 5\ 5\ 5\ 5\ 5)$$
$$C = (1\ 2\ 3\ 4\ 5\ 1\ 2\ 3\ 4\ 5\ 1\ 2\ 3\ 4\ 5\ 1\ 2\ 3\ 4\ 5\ 1\ 2\ 3\ 4\ 5)$$
$$S' = (A\ B\ C\ D\ E\ B\ C\ D\ E\ A\ C\ D\ E\ A\ B\ D\ E\ A\ B\ C\ E\ A\ B\ C\ D)$$

Notice that the last matrix consists of the five rows of the Latin square S of Figure 1.1. In the matrix R, we see ones above row one of S, twos above row two of S, and so on. In the matrix C, we see ones above the entries of column one of S, twos above the entries of column two of S, and so on. Thus, we see i in row R above j in row C above x in row S' if and only if $S_{ij} = x$.

We must show that R and C are orthogonal, C and S' are orthogonal, and R and S' are orthogonal. All five ordered pairs $(1, j)$ appear with 1 over j in the first five places of R and C. The ordered pairs $(2, j)$ appear in the second five places of R and C, and the ordered pairs (i, j) appear in the ith five places of R and C. Thus, there are 25 different pairs of elements appearing as R_{1k} and C_{1k}, so R and C are orthogonal.

To see that R and S' are orthogonal, note that since each element in the set $\{A, B, C, D, E\}$ appears in row i of S, each ordered pair (i, X) with $X \in \{A, B, C, D, E\}$ appears as (R_{1k}, S'_{1k}) for some k, so these pairs are distinct from one another. Thus, R and S' are orthogonal, and in a similar way C and S' are orthogonal. ∎

We have seen how the fact that we began with a Latin square S made the three arrays orthogonal. In fact, if we had some other row matrix T' orthogonal to R and C, then the 5×5 matrix T whose ith row consists of the ith five elements of T' would again be a Latin square. The orthogonality of T' with R means that each element of row i of T will be different and the orthogonality of T' with C means that each element of column j of T will be different. By using the techniques we have just discussed, we can prove the following theorem.

Theorem 1.8. Let L be an n by n matrix whose entries are chosen from an n-element set. Let L' be the row matrix formed by listing first row one of L, then row two of L, and so on. Let R be the row matrix consisting of a block of n ones, followed by a block of n twos, and so on. Let C is the row matrix consisting of the list $1, 2, \ldots, n$ repeated n times. Then L is a Latin square if and only if R, C, and L' are orthogonal.

Proof. The proof is essentially the analysis in the preceding example. ∎

We call L' the **row representation** of the Latin square L. Our theorem gives us a way to see how a pair of orthogonal Latin squares generalizes the idea of a single Latin square. Notice that two Latin squares S and T will be orthogonal if and only if their row representations S' and T' are orthogonal. Interpreting this observation carefully gives us our next theorem.

Theorem 1.9. If the matrices R, C, S', and T' are defined in the same way as R, C, and L' in the previous theorem, then the associated square matrices S and T are orthogonal Latin squares if and only if the matrices R, C, S', and T' are orthogonal.

The theorems appear to be rather specialized since the two special row matrices R and C are so special. In fact, the only relevant hypothesis is that the four row matrices are mutually orthogonal. To explain this point, we need the idea of "lexicographic"—or dictionary—ordering. Just as we say that the word *able* comes before the word *ace*, which comes before *bat*, we say

$$\begin{pmatrix} 1 \\ 1 \\ 2 \end{pmatrix} \quad \text{comes before} \quad \begin{pmatrix} 1 \\ 2 \\ 1 \end{pmatrix} \quad \text{comes before} \quad \begin{pmatrix} 2 \\ 1 \\ 3 \end{pmatrix}.$$

We say that the column vector of numbers

$$\begin{pmatrix} a_1 \\ a_2 \\ \vdots \\ a_k \end{pmatrix} \quad \text{comes before} \quad \begin{pmatrix} b_1 \\ b_2 \\ \vdots \\ b_k \end{pmatrix}$$

in **lexicographic order** if one of the following statements holds.

$$a_1 < b_1 \quad \text{or}$$
$$a_1 = b_1 \text{ and } a_2 < b_2 \quad \text{or}$$
$$a_1 = b_1 \text{ and } a_2 = b_2 \text{ and } a_3 < b_3 \quad \text{or}$$
$$\vdots$$
$$a_1 = b_1 \text{ and } a_2 = b_2 \cdots \text{ and } a_{k-1} = b_{k-1}, \text{ and } a_k < b_k.$$

Thus, just as all words beginning with a come before all words beginning with b, all column matrices starting with 1 come before all column matrices starting with 2. Further, among all column matrices with first entry 1, all those with second entry 1 come before all those with second entry 2 or more.

The idea of lexicographic order lets us explain why it is the orthogonality that is essential in the last two theorems.

Theorem 1.10. Suppose M is a $4 \times n^2$ matrix with entries in $\{1, 2, \ldots, n\}$. If each pair of rows of M is orthogonal, then by arranging the columns of M in lexicographic order we obtain a matrix whose last two rows are the row representations of two orthogonal Latin squares.

Proof. Since row 1 and 2 are orthogonal, each pair of numbers between 1 and n appears once and only once as the first two entries of a column. This means that, after the rearrangement of M, row 1 will be R and row 2 will be C. ∎

The assumption in the previous theorem that the entries of M are integers is not restrictive, because we can rename the entries of a row of M without changing the orthogonality; thus, if we have a matrix whose entries are not integers, we may convert it to a matrix whose entries are the integers from 1 to n.

*The Construction of a 10 by 10 Graeco–Latin Square

Now we apply Theorem 1.10 to show how an m by m Graeco–Latin square can be used to create a $3m + 1$ by $3m + 1$ Graeco–Latin square. Thus, a 3×3 Graeco–Latin square gives a 10×10 Graeco–Latin square and a 7×7 Graeco–Latin square gives a 22×22 Graeco–Latin square.

Theorem 1.11. If there is a pair of orthogonal $m \times m$ Latin squares, then there is a pair of orthogonal $3m + 1 \times 3m + 1$ Latin squares.

Proof. Suppose M is the four-row matrix corresponding to the $m \times m$ Graeco–Latin square. Relabel the entries of M so that *each* row (including the first two) has entries in the set $\{2m+1, 2m+2, \ldots, 3m\}$. Now we construct several useful row matrices of length m. We let

$$c_i = (i, i, \ldots, i)$$
$$d_i = (i+1, i+2, \ldots, i+m)$$
$$e_i = (i-1, i-2, \ldots, i-m)$$
$$f = f_i = (2m+1, 2m+2, \ldots, 3m),$$

for $i = 0, 1, 2, \ldots, 2m$, interpreting the entries of the first three row matrices as remainders (residues) modulo $2m + 1$. Notice that all the vectors f_i are equal.

Now we let

$$C = (c_0, c_1, \ldots, c_{2m})$$
$$D = (d_0, d_1, \ldots, d_{2m})$$
$$E = (e_0, e_1, \ldots, e_{2m})$$
$$F = (f_0, f_1, \ldots, f_{2m}).$$

Thus, C, D, E, and F are each row vectors with $m(2m+1)$ entries. Let R_1, R_2, R_3, R_4 be the four rows of M. Let N be the matrix described schematically as

$$N = \begin{pmatrix} C & D & E & F & e_m & m & d_m & R_1 \\ D & C & F & E & e_m & m & d_m & R_2 \\ E & F & C & D & e_m & m & d_m & R_3 \\ F & E & D & C & e_m & m & d_m & R_4 \end{pmatrix}.$$

The number of entries in a row of N is

$$4(m(2m+1)) + 2m + 1 + m^2$$

$$= 9m^2 + 6m + 1 = (3m+1)^2.$$

Now examine the first two rows of N. Where a c_i is over a d_i we get ordered pairs of the form $(i, i+j)$, and where a d_i is over a c_i we get ordered pairs of the form $(i, i-j)$. This gives us all ordered pairs of residues (i, k) with $0 \leq k \leq 2m$ and $i \neq k$. When an e_i is over an f_i we get pairs of the form $(i-k, 2m+k)$; for a given k we get all $2m$ pairs of the form $(i', 2m+k)$ because as i varies from 0 to $2m$, so do the residues $i' = i - k$. The pairs (i, i) with $0 \leq i \leq 2m$ come from the d_m over a d_m, the e_m over an e_m, or the m over m. Where R_1 is over R_2 we get ordered pairs (i, k) with i and k between $2m+1$ and $3m$. An analysis of any two rows of N is similar, so each pair of rows of N is orthogonal. Thus, when we rearrange the columns of N into lexicographic order, the last two rows of N are the row representations of two orthogonal Latin squares. ∎

The theorem gives us 10 by 10 and 22 by 22 Graeco–Latin squares; however, it does not give us 14 by 14 Graeco–Latin squares. By using similar construction techniques, Bose, Shrikhande, and Parker were able to construct enough building blocks to construct $n \times n$ Graeco–Latin squares for any integer $n > 6$.

EXERCISES

1. Write down all the 4×4 Latin squares with first row and first column consisting of 1, 2, 3, and 4 in order. Trial and error is the most appropriate method.

2. Explain why there is no pair of orthogonal 2×2 Latin squares.

3. Find a 3×3 Graeco–Latin square.

4. Find a pair of 7×7 orthogonal Latin squares.

5. Give an example using the integers mod 6 to show that the rule "if $\underline{ab} = \underline{0}$, then $\underline{a} = \underline{0}$ or $\underline{b} = \underline{0}$" doesn't always hold.

6. Prove that $\underline{a} + (\underline{b} + \underline{c}) = (\underline{a} + \underline{b}) + \underline{c}$.

7. Prove that $\underline{a}(\underline{b} + \underline{c}) = \underline{ab} + \underline{ac}$.

8. (a) Show that if n is a prime number and a is a fixed number between 1 and $n - 1$, then the products $\underline{a} \cdot \underline{1}, \underline{a} \cdot \underline{2}, \ldots, \underline{a} \cdot (\underline{n-1})$ are all distinct.
   (b) Show that if n is a prime, then for each class $\underline{a}$, there is a class $\underline{b}$ such that $\underline{ab} = \underline{1}$. (Hint: Use (a) and the pigeonhole principle.)

9. Using the integers mod 15, find a class $\underline{a}$ for which there is no class $\underline{b}$ with $\underline{a} \cdot \underline{b} = \underline{1}$.

10. Prove Corollary 1.5.

11. Describe how you could construct four mutually orthogonal 35 by 35 Latin squares. (We say squares are mutually orthogonal if each one is orthogonal to each other.)

12. Find three 5×5 Latin squares all orthogonal to one another.

13. Find a pair of 4×4 orthogonal Latin squares.

14. Construct a 12 by 12 Graeco–Latin square.

*15. Explain why there is an orthogonal pair of n by n Latin squares for every n congruent to 1 (mod 12).

16. Explain why if the smallest prime dividing n is p, then there are at least $p - 1$ mutually orthogonal n by n Latin squares.

17. Find three 4×4 Latin squares all orthogonal to each other.

18. Is there a Latin square orthogonal to

$$
\begin{array}{cccc}
1 & 2 & 3 & 4 \\
2 & 4 & 1 & 3 \\
3 & 1 & 4 & 2 \\
4 & 3 & 2 & 1
\end{array} \text{?}
$$

19. Under what conditions on n does the cancellation law "$\underline{a} \cdot \underline{b} = \underline{a} \cdot \underline{c}$ implies $\underline{b} = \underline{c}$" hold for all $\underline{a} \neq \underline{b}$?

20. A Latin square orthogonal to its transpose is called self-orthogonal.
    (a) Prove that no 3×3 Latin square is self-orthogonal.
    (b) Find a 4×4 self-orthogonal Latin square.

21. Show that it is impossible to find k mutually orthogonal $k \times k$ Latin squares.

22. Prove Theorem 1.9, including the observation about S and T in the paragraph immediately preceding the statement of Theorem 1.9.

23. A Latin rectangle is an $m \times n$ matrix whose rows are permutations of $1, 2 \ldots, n$ and whose columns each contain m distinct elements. Using the theory of systems of distinct representatives, show that it is always possible to keep adding rows to a Latin rectangle until you get a Latin square.

24. Extend the 3 by 8 Latin rectangles

$$
\begin{array}{cccccccc}
1 & 2 & 3 & 4 & 5 & 6 & 7 & 8 \\
2 & 1 & 4 & 3 & 6 & 5 & 8 & 7 \\
3 & 4 & 1 & 2 & 7 & 8 & 5 & 6
\end{array}
\quad \text{and} \quad
\begin{array}{cccccccc}
1 & 3 & 2 & 4 & 5 & 7 & 6 & 8 \\
4 & 2 & 3 & 1 & 6 & 8 & 5 & 7 \\
7 & 5 & 8 & 6 & 1 & 3 & 2 & 4
\end{array}
$$

to a pair of orthogonal Latin squares.

25. In the Exercises we have shown that there are Graeco–Latin squares of order 4 and 8. Show that if

$$ n = 2^{k_2} 3^{k_3} 5^{k_5} \cdots p^{k_p} $$

and k_2 is not 1, then there is a Graeco–Latin square of order n.

26. Show that the number of different Latin squares that arise by permuting the rows and columns of an $n \times n$ Latin square is $n!(n-1)!$.

27. Write a backtracking program to list $n \times n$ Latin squares whose first row and first column each consist of the symbols 1 through n in order.

28. (a) Using inclusion and exclusion, compute the number of 2×6 Latin rectangles whose first row consists of the symbols 1 to 6 in their usual order.
    *(b) Do the same for 3×6 Latin rectangles whose first row is 1 through 6 in order.

29. Write a backtracking program that searches for one $n \times n$ Latin square orthogonal to a given $n \times n$ Latin square. Experiment with squares of size 5, 6, and 7.

30. Write a program to confirm Tarry's result that there is no pair of orthogonal 6×6 Latin squares. (Warning: There are 9408 6×6 Latin squares with first row and column in increasing order.)

Section 2 Block Designs

How Block Designs Are Used

Suppose now we are interested in determining the effect of seven motor oil formulations on gas mileage. Each variety of motor oil will have to be tested over a significant distance, say 10,000 miles. A car will have to be broken in for 5,000 or 10,000 miles before we begin testing, and if a car goes farther than, say 40,000 miles, its age will have some effect on gasoline consumption. Thus, we should not run our cars too long, or else we will get bad data. In addition, it would take some time to drive more than 40,000 miles in a car, especially if the driving is to be done to simulate normal driving habits. For these reasons, we decide to test only three motor oils per car. If two motor oils are never compared against each other in the same car, we will have reason to be nervous about any conclusions concerning performance differences among these motor oils. Thus, for each pair of motor oils, we want to be sure there is one car in which both of them have been used (at different times, of course). We will arrange the seven motor oils into a collection of overlapping blocks of size 3, each block to be tested in a given car. Each block is tested in exactly one car, so if each pair gets into exactly one block together, then each pair gets compared by exactly one car. How many cars do we need? Since we have $C(7;2) = 21$ pairs to be compared and each car is used to make $C(3;2) = 3$ comparisons, we need seven cars. This means our arrangements of motor oils into blocks requires seven blocks. Does such an arrangement exist? We shall see that it does.

The arrangement of sets we have described for the motor oil experiment is called a "balanced incomplete block design." We shall explain this terminology now. A *design* (also called a *hypergraph*) on a set V of v objects called *varieties* (or, in hypergraph terminology, *vertices*) is a multiset of subsets of V. (If the subsets happen to be two-element subsets, then the design is a multigraph with vertex set V. This explains the "hypergraph" terminology.) The varieties are also called points in geometric contexts, plots in agricultural contexts, and treatments in certain statistical contexts. The subsets of V are called *blocks*; thus, in the multigraph case, the blocks are the edges of the multigraph. A design is called *complete* if each subset consists of all of V. (This is not consistent with hypergraph terminology.) When we use Latin squares as in Section 1 for experimental purposes, each row of the Latin square may be regarded as a block of a complete design; the arrangement of objects within rows and columns guarantees that each column may be thought of as a block of a complete design. (Alternatively, each row or column may be thought of as a complete design with one block if that fits a statistician's needs.) A design in which at least one block is not all of V is called *incomplete*.

A design is called **k-uniform** (this terminology arose from hypergraphs) or sometimes *proper* if each block has the same size k. A design is called a *(pairwise) balanced* design of *index* λ if each pair of distinct varieties appears together in exactly λ blocks of the design. Notice that a design can be balanced without being uniform. For example,

$$\{1,2\}\ \{1,3\}\ \{1,4\}\ \{2,3,4\} \tag{2.1}$$

is a balanced incomplete design on a four-vertex set with $\lambda = 1$, while

$$\{1,2\}\ \{1,3\}\ \{1,4\}\ \{2,3\}\ \{2,4\}\ \{3,4\} \tag{2.2}$$

is a uniform balanced incomplete design, also called a "balanced incomplete block design," with $k = 2$ and $\lambda = 1$.

A design is called **linked** if each two blocks have exactly μ elements in common (for some constant μ called the *link number*). Notice that design (2.1) is linked with link number $\mu = 1$ while design (2.2) is not linked. It will turn out that the design we have asked about for motor oils will be linked.

The **replication number** r_i of an element i of a design is the number of blocks in which it appears; thus, in design (2.1), $r_1 = 3$ while $r_2 = r_3 = r_4 = 2$. In design (2.2) all varieties have replication number 3. A design is called a **regular** design if each variety has the same replication number, that is, if all the r_i's have a common value r. We can prove that a uniform balanced incomplete design is regular. A design is called a **block design** if it is both uniform and regular. A design with v varieties arranged into b blocks of size k in such a way that each vertex appears in r blocks and each pair of vertices appears together in λ blocks is called a (b, v, r, k, λ) design. If $k < v$, a (b, v, r, k, λ) design is a **balanced incomplete block design**. Balanced incomplete block design is often abbreviated BIBD. The central theme of this section and the next is balanced incomplete block designs.

The Latin square designs we used in Section 1 were complete $(5, 5, 5, 5, 5)$ designs; the motor oil design we just asked for is a $(7, 7, r, 3, 1)$ design. In fact, we haven't demanded that each oil be tested in exactly the same number r of cars, but this occurs nonetheless. Our first theorem tells us how to compute r when we know b, k, and v.

Basic Relationships among the Parameters

Theorem 2.1. In a uniform regular design $bk = vr$.

Proof. Since there are b blocks each with k elements, the number of ordered pairs consisting of a block and an element of that block is bk. However, since each element of V must occur in r blocks, there must be

vr ordered pairs consisting of a vertex and a block containing it. Since bk and vr are the number of elements in the same set of ordered pairs, they are equal. ∎

Theorem 2.2. In a (b, v, r, k, λ) design, $r(k - 1) = \lambda(v - 1)$.

Proof. For each x in V, both sides of this equation count the number of ordered pairs consisting of a block in which the vertex x lies and one other element of that block. ∎

Theorem 2.3. In a balanced incomplete uniform design, each variety appears in the same number of blocks.

Proof. Exercise. ∎

Given any three of the parameters b, v, r, and k of a block design, we can compute the fourth, and given any three of the five parameters of a (b, v, r, k, λ) design, we can use them to compute the other parameters. For this reason a (b, v, r, k, λ) design is often called a (v, k, λ) design. The choice of which list of parameters you give usually depends on whether you want to emphasize b and r or not.

Example 2.1. In the motor oil example we had seven varieties of oil in V; further, we wanted three varieties per block and wanted each pair of varieties to occur together in a single block. Thus, $v = 7$, $k = 3$, and $\lambda = 1$. We computed b already, but will compute it here differently. Since $r(3 - 1) = 1(7 - 1)$, $2r = 6$, so $r = 3$ and since $bk = vr$, $b \cdot 3 = 7 \cdot 3$ and thus $b = 7$. ∎

Theorem 2.4 provides another example of how the basic equations are used.

Theorem 2.4. In a BIBD, $\lambda < r$.

Proof. We know that in an incomplete design, $k < v$ because the blocks are proper subsets. Since $\lambda(v - 1) = r(k - 1)$, we conclude that $\lambda/r = \frac{(k-1)}{(v-1)} < 1$ so $\lambda < r$. ∎

The Incidence Matrix of a Design

Each design has an ***incidence matrix*** N whose rows are indexed (labeled) by the elements of V and whose columns are indexed by the blocks. (Think of the vertices and blocks as being numbered.) The entry N_{ij} is the number of times variety i appears in block j. For any design, N_{ij} is either 0 or 1 because each block is a set. The matrix NN^t (where N^t stands for the transpose of N) is a matrix with v rows and v columns. Its form gives us considerable insight into the parameters of the design. In Theorem 2.5,

I stands for the $v \times v$ identity matrix and J stands for the $v \times v$ matrix consisting of 1's exclusively.

Theorem 2.5. If N is the incidence matrix of a BIBD on a v-element set, then

$$NN^t = (r - \lambda)I + \lambda J.$$

Conversely, if the incidence matrix N of a uniform design satisfies this equation, then the design is a BIBD.

Proof. The ij entry of NN^t is formed by taking row i of N and column j of N^t (in other words, row i and row j of N), multiplying corresponding entries, and adding up the results. However, row i and row j will both be 1 in position k if and only if vertex i and vertex j are in block k. Thus the sum, if $i \neq j$, is the number of blocks in which i and j occur together, namely λ, and if $i = j$ the sum is the number of blocks variety i lies in, namely r. Thus, $(NN^t)_{ij}$ is λ if $i \neq j$ and r if $i = j$, which are precisely the i, j entries of $(r - \lambda)I + \lambda J$. The proof of the converse is an exercise. ∎

An important application of the concept of an incidence matrix is in the proof of Fisher's inequality, proved by Fisher in 1940 without the use of the incidence matrix and by Bose, who used the incidence matrix, in 1949.

Theorem 2.6. In a balanced incomplete block design, $b \geq v$.

Proof. Note that adding one or more columns of 0's to a matrix A does not change the value of AA^t. For example,

$$\begin{pmatrix} a & b & 0 & 0 \\ c & d & 0 & 0 \end{pmatrix} \begin{pmatrix} a & c \\ b & d \\ 0 & 0 \\ 0 & 0 \end{pmatrix} = \begin{pmatrix} a^2 + b^2 & ac + bd \\ ca + db & c^2 + d^2 \end{pmatrix} = \begin{pmatrix} a & b \\ c & d \end{pmatrix} \begin{pmatrix} a & c \\ b & d \end{pmatrix}.$$

Assume that $b < v$. Then the $v \times b$ matrix N has more rows (v) than columns (b). We can change N into a square matrix M by adding $v - b$ columns of 0's to N. Then

$$NN^t = MM^t.$$

But since M is a square matrix with a column of 0's, $\det M = 0$. Therefore $\det MM^t = 0$, so $\det NN^t = 0$.

Now we show that the determinant of NN^t is nonzero. This contradiction will show that the assumption that $b < v$ is invalid. Recall the form of NN^t given in Theorem 2.5. Subtract column 1 from all the other columns. This gives a matrix A whose first column is the same as the first column of NN^t, whose first row is $\lambda - r$ in positions 2 through v, whose diagonal entries are $r - \lambda$ in positions 2 through v, and whose other entries are zero. Now add rows 2 through v of A to row 1. This gives a matrix which is 0 above the main diagonal and nonzero in each diagonal position. Thus, because the determinant of a triangular matrix is the product of its diagonal entries, it has a nonzero determinant. Therefore, it is impossible for b to be less than v, so that $v \leq b$. ∎

An Example of a BIBD

In general, the construction of BIBDs with specific parameters can be quite difficult and require the use of sophisticated mathematical techniques—or be impossible even though the parameters satisfy all the conditions we have shown are necessary. However, in the case with which we began, we can make a fairly straightforward construction.

Example 2.2. Construct a $(7, 7, 3, 3, 1)$ BIBD.

We will use the symbols 0, 1, 2, 3, 4, 5, 6 for our seven varieties. We may assume $\{0, 1, 2\}$ is one of our three-element sets. Now 0 appears in only three sets and must appear with 3, 4, 5, and 6, so we may assume two other sets are $\{0, 3, 4\}$ and $\{0, 5, 6\}$. (If the symbols 3, 4, 5, and 6 occur differently, we can rename them.) Now 1 occurs in two other sets but has already occurred with 0 and 2, so in these sets it must occur with 3, 4, 5, and 6. However, 3 has already appeared with 4, so when it appears with 1, only 5 or 6 can occur with it. Assume 5 occurs with 1 and 3. (If not, we have the option of interchanging the labels of 5 and 6.) Then $\{1, 3, 5\}$ and $\{1, 4, 6\}$ must be the two sets containing 1. There are two sets left to construct. Now, 0 and 1 have been used three times and so only 2, 3, 4, 5, and 6 may be used. Similarly, 3, 4, 5, and 6 have been used twice, but 2 has been used once, so both sets contain 2. Now 4 and 5 have not appeared together, but 3,4 and 5,6 have, so one of the sets with 2 must be $\{2, 4, 5\}$. Then the other set with 2 must be $\{2, 3, 6\}$. Thus, our blocks are the sets

$$\{0, 1, 2\}, \ \{0, 3, 4\}, \ \{0, 5, 6\}, \ \{1, 3, 5\}, \ \{1, 4, 6\}, \ \{2, 4, 5\}, \ \{2, 3, 6\}. \quad ∎$$

Isomorphism of Designs

Note that, except for our labeling of the elements of V, there really is no choice as to how we might construct the design. Two designs on V and V'

are *isomorphic* if there is a one-to-one function f from V onto to V' such that for each block B of design 1,

$$f(B) = \{f(x) \mid x \in B\}$$

is a block of design 2 and conversely each block of design 2 is $f(B)$ for some block B of design 1. The function f is called an *isomorphism*. Thus, since we had no choice except for labeling in the construction, any two $(7, 7, 3, 3, 1)$ designs are isomorphic.

The Dual of a Design

Any matrix M of zeros and ones is the incidence matrix of some design. If the matrix has n rows, we may associate the members of an n-element set $V = \{x_1, x_2, \ldots, x_n\}$ with the rows of M. Then column j of M determines a block B_j as follows:

$$x_i \in B_j \text{ if and only if } M_{ij} = 1.$$

Further, if we begin with the incidence matrix N of a design with b blocks and v varieties, then by transposing this matrix to get N^t we get the incidence matrix of a design with v *blocks* and b *varieties*. The design whose incidence matrix is N^t is called the *dual* of the original design. We denote the dual of a design D by D^t. If D is a regular design, so that each x in V appears in r blocks of V, then N has r entries in each row. Thus, N^t will have r entries in each column, so each block of D^t will have r entries. Thus, the dual of a regular design is a uniform design (and the dual of a uniform design is a regular design). To say that any two varieties appear together in λ blocks means that any two rows have matching ones in exactly λ positions; from this we may conclude that the dual of a balanced design of index λ is a linked design with link number λ and vice versa. By Fisher's inequality, the dual of a block design can be a block design only if $b = v$; in fact, if $b = v$ the dual of a block design will always be a block design. (Can you see why?)

Example 2.3. Find the dual of the design of Example 2.2.

First we construct the incidence matrix N, taking the blocks in the order they are given in Example 2.2. This gives us the matrix N.

$$N = \begin{pmatrix} 1 & 1 & 1 & 0 & 0 & 0 & 0 \\ 1 & 0 & 0 & 1 & 1 & 0 & 0 \\ 1 & 0 & 0 & 0 & 0 & 1 & 1 \\ 0 & 1 & 0 & 1 & 0 & 0 & 1 \\ 0 & 1 & 0 & 0 & 1 & 1 & 0 \\ 0 & 0 & 1 & 1 & 0 & 1 & 0 \\ 0 & 0 & 1 & 0 & 1 & 0 & 1 \end{pmatrix}.$$

Then we write

$$N^t = \begin{pmatrix} 1 & 1 & 1 & 0 & 0 & 0 & 0 \\ 1 & 0 & 0 & 1 & 1 & 0 & 0 \\ 1 & 0 & 0 & 0 & 0 & 1 & 1 \\ 0 & 1 & 0 & 1 & 0 & 1 & 0 \\ 0 & 1 & 0 & 0 & 1 & 0 & 1 \\ 0 & 0 & 1 & 0 & 1 & 1 & 0 \\ 0 & 0 & 1 & 1 & 0 & 0 & 1 \end{pmatrix}.$$

We now read off the blocks from the columns as

$$\{0,1,2\}, \ \{0,3,4\}, \ \{0,5,6\}, \ \{1,3,6\}, \ \{1,4,5\}, \ \{2,3,5\}, \ \{2,4,6\}. \quad \blacksquare$$

Notice that N^t is not identical to N, but they do have the same first three rows and the same first three columns. Recall that in Example 2.2 we made the assumption that 1, 3, and 5 appear together (see column 4 of N). In N^t, 1, 3, and 6 appear together instead (see column 4 of N'). Thus, N^t shows us the other way in which we could complete the first three blocks given into a design. Notice, however, that this design is isomorphic to the (7,3,1) BIBD design which we constructed in Example 2.2.

Symmetric Designs

In Examples 2.2 and 2.3, we had $v = b$ rather than $v < b$; a design is called **symmetric** if $v = b$. Since $bk = vr$ in a regular design, a symmetric regular design also has $k = r$. Sometimes, in older books, only a symmetric design is called a (v, k, λ) design; however, if you are ever confused by terminology, you can compute b from v, k, and λ and find out whether or not the design is symmetric. In Exercise 6, you will be asked to show that $\det N N^t$ is $(r - \lambda)^{v-1} \cdot rk$; for a symmetric design, this is $(r - \lambda)^{v-1} \cdot r^2$. Also, for a symmetric design, N and N^t are square, and $\det N = \det N^t$ gives us

$$\det(N)^2 = \det(N)\det(N^t) = \det(NN^t) = (r - \lambda)^{v-1} \cdot r^2.$$

Since $\det(N)$ is an integer, the left-hand side is a square of an integer, so the right-hand side must be also. Since r^2 is a square, the same goes for $(r-\lambda)^{v-1}$. If v is odd, this will happen automatically; if v is even, then $r - \lambda$ must be a square integer. This yields the easier part of our next theorem, sometimes called the Bruck–Ryser or Bruck–Ryser–Chowla theorem; the part we proved was also discovered by Schutzenberger and by Shrikhande at about the same time.

Theorem 2.7. For a symmetric BIBD with v even, the number $r - \lambda$ is a square. For a symmetric BIBD with v odd, the equation

$$x^2 = (r - \lambda)y^2 + (-1)^{(v-1)/2}\lambda z^2$$

has a solution in integers x, y, and z, not all zero.

Proof. The proof of the theorem for even v has already been given; for odd v it is beyond the scope of this book. See one of the books by Ryser, Hall, Street and Street, or Wallis in the Suggested Reading. ∎

Example 2.4. Is there a $(46, 10, 2)$ design?

Note that this would be a $(46, 46, 10, 10, 2)$ design. The parameters satisfy the equations $bk = vr$ and $r(k-1) = \lambda(v-1)$. However, $r - \lambda = 10 - 2 = 8$, and 8 is not a square. ∎

Theorem 2.8. If D is a symmetric BIBD, then the dual design D^t is a symmetric BIBD as well.

Proof. We know that if N is the incidence matrix of D, then

$$NN^t = (r - \lambda)I + \lambda J.$$

Further, $JN = kJ$ and $NJ = rJ$. Since $k = r$ in a symmetric design, $NJ = JN$. Also, since $\det(N^tN) \neq 0$, N is an invertible matrix, and so

$$N^t = N^{-1}(r - \lambda)I + N^{-1}\lambda J.$$

This gives us

$$\begin{aligned} N^tN &= N^{-1}(r - \lambda)IN + N^{-1}\lambda JN \\ &= (r - \lambda)I + \lambda N^{-1}JN \\ &= (r - \lambda)I + \lambda JN^{-1}N \\ &= (r - \lambda)I + \lambda J. \end{aligned}$$

Thus, by Theorem 2.5, D^t is a BIBD; D^t is symmetric because $b = v$. ∎

Corollary 2.9. A symmetric BIBD is linked.

Proof. The dual of a balanced design is linked; since the dual of the dual of a design is itself, every symmetric BIBD is the dual of a balanced design. ∎

In the proof that the dual of a symmetric BIBD is a BIBD, the symmetry was essential since we used the invertibility of the incidence matrix and only square matrices have inverses. There are two other constructions of BIBDs from a symmetric design D which use the symmetry in another way. If we have a symmetric BIBD with blocks $B_1, B_2, \ldots, B_v$, then $B_v \cap B_i$ has size λ and $B_i - B_v$ has size $k - \lambda$. Thus, the sets $B_v \cap B_i$ form a uniform design on the vertex set B_v. This is called the **derived design** of D on B_v. Similarly, the sets $B_i - B_v$ with $i < v$ form a design on $V - B_v$. This design is called the **residual design** of D on $V - B_v$.

Theorem 2.10. If D is a symmetric BIBD, then the derived design and residual designs are BIBDs.

Proof. The sets $B_i \cap B_v$ form an incomplete design for otherwise at least one B_i would contain B_v, so the link number would be k, which implies that D is complete, and this contradicts the fact that D is incomplete. The design is balanced because if x and y appear together in B_v, they also appear together in $\lambda - 1$ other blocks B_i; therefore each x and y in B_v appear in $\lambda - 1$ blocks of the derived design. We leave the proof for residual designs to the Exercises. ∎

Example 2.5. Write out the blocks of the derived design and the residual design for the (13,9,6) symmetric BIBD D whose blocks are given in the following. (The techniques by which this design was constructed will be discussed in the next section.)

$\{0,1,3,5,6,9,10,11,12\}$ $\{1,2,4,6,7,10,11,12,0\}$ $\{2,3,5,7,8,11,12,0,1\}$
$\{3,4,6,8,9,12,0,1,2\}$ $\{4,5,7,9,10,0,1,2,3\}$ $\{5,6,8,10,11,1,2,3,4\}$
$\{6,7,9,11,12,2,3,4,5\}$ $\{7,8,10,12,0,3,4,5,6\}$ $\{8,9,11,0,1,4,5,6,7\}$
$\{9,10,12,1,2,5,6,7,8\}$ $\{10,11,0,2,3,6,7,8,9\}$ $\{11,12,1,3,4,7,8,9,10\}$
$\{12,0,2,4,5,8,9,10,11\}$

The block B_v is the last one; its intersections with the other blocks are listed next.

$\{0,5,9,10,11,12\}$ $\{0,2,4,10,11,12\}$ $\{0,2,5,8,11,12\}$ $\{0,2,4,8,9,12\}$
$\{0,2,4,5,9,10\}$ $\{2,4,5,8,10,11\}$ $\{2,4,5,9,11,12\}$ $\{0,4,5,8,10,12\}$
$\{0,4,5,8,9,11\}$ $\{2,5,8,9,10,12\}$ $\{0,2,8,9,10,11\}$ $\{4,8,9,10,11,12\}$

Thus, the derived design is a $(12,9,8,6,5)$ design—one we would not expect to find by guesswork.

The residual design is found by removing each block of the derived design from the corresponding block of the original design, giving us

$$\{\{1,3,6\},\ \{1,6,7\},\ \{1,3,7\},\ \{1,3,6\},\ \{1,3,7\},\ \{1,3,6\}$$
$$\{3,6,7\},\ \{3,6,7\},\ \{1,6,7\},\ \{1,6,7\},\ \{3,6,7\},\ \{1,3,7\}\}.$$

This is a $(12,4,9,3,6)$ design. Notice that different blocks have different multiplicities; we would have been unlikely to construct this design by trial and error! ∎

The Necessary Conditions Need Not Be Sufficient

It is tempting to assume that if the parameters b, v, r, k, and λ satisfy all the tests we have devised, then there is a design with these parameters.

However, the sequence $(21, 15, 7, 5, 2)$ passes our tests and there is no design with these parameters. (See the book by Hall in the Suggested Reading for further explanation.) When we need to know if there *is* a design with certain parameters that pass our tests, there is no substitute for constructing such a design. Generally, the construction process is one of intelligent guesswork supplemented by good luck as in Example 2.2. In the next section, we give a few general construction methods that are sometimes helpful. There are tables that compactly describe designs that have given parameters or say that no design with certain parameters exists; for example, the book by Hall in the Suggested Reading has such a table.

EXERCISES

1. What are k and λ in a $(4, 4, 3, k, \lambda)$ design?

2. What are k and r in a $(7, 7, r, k, 2)$ design?

3. Show that if $\lambda = r$ in a balanced block design, then the design is complete.

4. Discuss the following sentence; in particular, in what sense is it true or false? "Normalized $n \times n$ Latin squares are pairs of symmetric complete block designs."

5. Show that $C(v; 2) \cdot \lambda = C(k; 2) \cdot b$ in any block design. (Hint: The most interesting proof comes from observing that $C(v; 2) \cdot \lambda$ is the total number of pairs of vertices that appear among the blocks of the design.) Write out the resulting equation in explicit form.

6. Show that the determinant of NN^t is $(r - \lambda)^{v-1} \cdot rk$.

7. Show that the inverse of NN^t is

$$\frac{1}{r - \lambda}\left(I - \frac{\lambda}{r + (v - 1)\lambda}J\right).$$

8. Write out the matrices N and NN^t for Example 2.2.

9. (a) Show that if N is the incidence matrix of a BIBD, then $NJ = rJ$.
   (b) What is JN for a BIBD?

10. Explain why the dual of the dual of a design is the original design.

11. Are there (b, v, r, k, λ) designs with the following parameters?
    (a) $b = 8$, $v = 6$, $r = 5$, $k = 3$, $\lambda = 2$
    (b) $b = 12$, $v = 8$, $r = 6$, $k = 4$, $\lambda = 2$

12. How many $(4, 4, 3, 3, 2)$ designs are there? Find them.

13. Find a $(7, 7, 4, 4, 2)$ design.

14. Find the residual design of the design of Exercise 13.

15. Find the derived design of the design of Exercise 13.

16. Are all $(7, 7, 4, 4, 2)$ designs isomorphic?

17. Find all $(6, 4, 3, 2, \lambda)$ designs.

18. Find a $(14, 7, 6, 3, 2)$ design.

*19. Find a symmetric $(11, 5, 2)$ design.

*20. Find the derived and residual designs of the design of the previous exercise.

21. Find a $(10, 6, 5, 3, 2)$ design.

22. Prove Theorem 2.3, that a balanced uniform design is regular.

23. True or false (prove or give a counterexample): A balanced regular design is uniform.

24. Show that given v and k, there is a design with

$$b = C(v; k), \quad r = C(v - 1; k - 1), \quad \lambda = C(v - 2; k - 2).$$

25. True or false: In a BIBD, if $k = 3$, then either λ is even or v is odd. (Prove or give a counterexample.)

26. Show that the residual design of a symmetric BIBD is a BIBD. Point out explicitly how you are using the fact that the design is symmetric.

27. Prove the converse of Theorem 2.5.

28. Is it possible to distribute 22 brands of detergent to different households so that each household tests seven brands over a 7-month period and each pair of brands is tested by two different households?

29. (a) Why is Exercise 28 made significantly harder if 22 is replaced by 16 and seven is replaced by six?
    *(b) Solve this more difficult problem.

*30. Show that if M_1 and M_2 are the following matrices, then there are three more matrices whose first column is the first column of M_2 such that the rows of these five matrices form the blocks of a $(16, 20, 5, 4, 1)$ design.

$$M_1 = \begin{pmatrix} 1 & 2 & 3 & 4 \\ 5 & 6 & 7 & 8 \\ 9 & 10 & 11 & 12 \\ 13 & 14 & 15 & 16 \end{pmatrix} \qquad M_2 = \begin{pmatrix} 1 & 5 & 9 & 13 \\ 2 & 6 & 10 & 14 \\ 3 & 7 & 11 & 15 \\ 4 & 8 & 12 & 16 \end{pmatrix}$$

(Hint: Put three orthogonal Latin squares with first row $ABCD$ down on M_1, in which case A, B, C, and D each land on a four-element set.)

31. Discuss the potential use of backtracking for finding BIBDs. Write a backtracking program to search for designs and apply it to various *small* parameter values. Discuss how many steps a backtracking search for a symmetric $(7, 3, 1)$ design would take in comparison with checking all families of seven three-element sets chosen from a seven-element set.

Section 3 Construction and Resolvability of Designs

A Problem That Requires a Big Design

The director of the Lebanon Softball League recently asked the following question. "We have fourteen teams—probably, but there may be a couple dropouts. We can use the municipal field at night just long enough to play three games before the lights go out. We would like to have three teams show up each time and play three games; that way, one team can do the umpiring, etc. while the other two play. Is there any way to schedule the season of about 12 weeks so that each team plays each other team twice and each team comes out just once a week?"

Although the statement of the problem wasn't exactly a problem of design, it is fairly clear that the question deals with the existence of some sort of $(b, v, r, 3, 2)$ design where v would probably be 14 but might be less. In the case $v = 14$, we can write

$$b \cdot 3 = 14 \cdot r$$

and

$$r(2) = 2(13),$$

so $r = 13$ and thus $b \cdot 3 = 14 \cdot 13$. However, since 3 is not a factor of either 13 or 14, there is no solution. If the softball league had a dropout or two or consolidated some teams, then v would be 13 or 12. Then we would have either

$$b \cdot 3 = 13 \cdot r$$
$$r \cdot 2 = 2 \cdot 12$$

so that

$$r = 12, b = 52$$

or

$$b \cdot 3 = 12 \cdot r$$
$$r \cdot 2 = 2 \cdot 11$$

so that

$$r = 11, b = 44.$$

The next question to ask is whether there *are* any $(52, 13, 12, 3, 2)$ designs or any $(44, 12, 11, 3, 2)$ designs.

Cyclic Designs

Since we have as yet given no systematic methods for the construction of designs, our only method is that of trial and error.

Just as the arithmetic of integers modulo n helped us in the construction of Latin squares, here it provides us with a powerful tool for the construction of block designs.

A balanced block design is said to be *cyclic mod n* with *base* **B** if

(1) The elements of V are equivalence classes mod n.

(2) **B** is a set of blocks of the design.

(3) Every block C in the design may be obtained in exactly one way (exactly as often as it appears in the design) by adding some $\underline{m}$ (which is determined by C) mod n to each element of a block B in **B**.

If $B \in \mathbf{B}$ is the set $\{\underline{b}_1, \underline{b}_2, ..., \underline{b}_k\}$, then we denote the set

$$C = \{\underline{b}_1 + \underline{m}, \underline{b}_2 + \underline{m}, ..., \underline{b}_k + \underline{m}\}$$

by $B + \underline{m}$. We call $B + \underline{m}$ the *translate* of B by $\underline{m}$.

Example 3.1. If $V = \{\underline{0}, \underline{1}, \underline{2}, \underline{3}, \underline{4}\}$ is the set of equivalence classes mod 5, and $\mathbf{B} = \{\{\underline{0}, \underline{1}\}, \{\underline{2}, \underline{4}\}\}$, then **B** is a base for the design

$$\{\underline{0}, \underline{1}\}, \{\underline{1}, \underline{2}\}, \{\underline{2}, \underline{3}\}, \{\underline{3}, \underline{4}\}, \{\underline{4}, \underline{0}\}$$

$$\{\underline{2}, \underline{4}\}, \{\underline{3}, \underline{0}\}, \{\underline{4}, \underline{1}\}, \{\underline{0}, \underline{2}\}, \{\underline{1}, \underline{3}\}.$$

Note that the first row of sets can be written

$$\{\underline{0} + \underline{0}, \underline{1} + \underline{0}\}, \{\underline{0} + \underline{1}, \underline{1} + \underline{1}\}, \{\underline{0} + \underline{2}, \underline{1} + \underline{2}\}, \{\underline{0} + \underline{3}, \underline{1} + \underline{3}\}, \{\underline{0} + \underline{4}, \underline{1} + \underline{4}\},$$

or

$$\{\underline{0}, \underline{1}\} + \underline{0}, \{\underline{0}, \underline{1}\} + \underline{1}, \{\underline{0}, \underline{1}\} + \underline{2}, \{\underline{0}, \underline{1}\} + \underline{3}, \{\underline{0}, \underline{1}\} + \underline{4}.$$

The second row is obtained in the same way from the set $\{\underline{2}, \underline{4}\}$. Straightforward checking shows this is a BIBD with parameters $(10, 5, 4, 2, 1)$. ∎

Notice that one pair of **B** consists of adjacent elements and one pair of **B** consists of two elements 2 units from each other. In particular, the differences

$$\underline{1} - \underline{0} = \underline{1}, \ \underline{0} - \underline{1} = \underline{4}, \ \underline{4} - \underline{2} = \underline{2}, \ \underline{2} - \underline{4} = \underline{3}$$

include all the nonzero integers modulo 5 exactly once. When we add $\underline{1}$ to each element of $\{\underline{0}, \underline{1}\}$, we get another pair 1 unit apart. Continuing in this way, we get *all* pairs 1 unit apart. We also get the pair $\{\underline{4}, \underline{0}\}$, which is also a pair 1 unit apart if we arrange $\underline{0}, \underline{1}, \underline{2}, \underline{3}, \underline{4}$, around a circle in natural order. It is this circular visualization of distance that underlies the use of the word cyclic. In the same way, we get every pair 2 units apart by repeatedly adding 1 to the pair $\{2, \underline{4}\}$. By shifting our base blocks around the circle one place at a time, we ensure that every point appears the same number of times as every other point. The fact that each difference appears once means that as we move around the circle, each pair of numbers with difference $\underline{1}$ appears once, each pair with difference $\underline{2}$ appears once, and so on.

In Section 1 of this chapter we pointed out that it is traditional to use i rather than $\underline{i}$ to stand for the congruence class of i when it is clear from context that we are discussing congruence classes. Since the elements of a cyclic block design are congruence classes, it should be clear from context that we are working with congruence classes, so that for the remainder of our discussion of cyclic designs, we will dispense with underlining.

In general, suppose we have a base of blocks for our design. We visualize them as sets on a circle labeled by the integers modulo n. Imagine "picking up" a base block and shifting each element one unit to the right around the circle. This corresponds to adding 1 modulo n to each element of the block. Thus, part 3 in the definition of a cyclic design says geometrically that we obtain all blocks by shifting the base blocks repeatedly around the circle. From this, we can guess how to arrange a bunch of sets around a circle in such a way as to be a base for a design. Namely, if a pair $\{a, b\}$ is in some block (that is, is a subset of some block), then since its block can be shifted into one of the base blocks, we will find that for some m, $\{a + m, b + m\}$ is in a base block. We can capture this property without mentioning m by observing that whenever $\{a, b\}$ is in a block, then $a - b = a + m - (b + m)$ is a difference between two elements of a base block. If $\{a, b\}$ appears in two different blocks, then $a - b$ will appear twice as a difference between two elements in the base blocks. Our analysis suggests the following theorem.

> **Theorem 3.1.** A set **B** of k-element sets of equivalence classes of integers mod n is a base for a uniform balanced block design which is cyclic mod n if and only if each nonzero integer mod n occurs the same number of times as a difference $a - b$ of distinct elements chosen from the same set in **B**.
>
> *Proof.* Since we know the translates of the sets in **B** will form a uniform block design, we need only show that the design is balanced if and only if each nonzero integer mod n occurs the same number of times as a difference $a - b$ of distinct elements chosen from the same set in **B**.
>
> For this purpose, suppose that x and y are members of V, and $B \in \mathbf{B}$. Then $\{x, y\} \subseteq B + i$ if and only if there is a pair $\{a, b\} \in B$ such that $x = a + i$ and $y = b + i$. Given x, y, and i, there can be only one such pair in B. Thus for each i such that $\{x, y\} \subseteq B + i$, there is a unique pair $\{a_i, b_i\} \subseteq B$ such that $x - y = a_i - b_i$ and $x - a_i = i$.

Now suppose we have an a and b in B such that $x - y = a - b$. Then we also have that $x - a = y - b$, and if we let $i = x - a$, we have that $x = a + i$ and $y = b + i$, so that $\{x, y\} \subseteq B + i$. There is exactly one i such that $x = a + i$ and $y = b + i$ because i must be equal to the common value of $x - a$ and $y - b$. We will denote this common value of $x - a$ and $y - b$ by $i_{(a,b)}$. Then what we just said may be rephrased as follows. For each pair (a, b) of elements of B with $a - b = x - y$ there is a unique $i_{(a,b)}$ such that $\{x, y\} \subseteq B + i_{(a,b)}$.

The function that sends i to (a_i, b_i) is therefore a bijection and its inverse is the function that sends (a, b) to $i_{(a,b)}$. Therefore there are exactly λ values of i such that $\{x, y\} \subseteq B + i$ for some $B \in \mathbf{B}$ if and only if there are exactly λ triples (a, b, B) such that $B \in \mathbf{B}$, $\{a, b\} \subseteq B$, and $a - b = x - y$. Thus every $\{x, y\}$ is a subset of exactly λ blocks if and only if every equivalence class mod p that appears as an $x - y$, also appears exactly λ times as a difference $a - b$ with $\{a, b\} \subseteq B \in \mathbf{B}$. Since every equivalence class will occur as a difference $x - y$, this proves the theorem. ∎

Example 3.2. The sets $\{0, 1, 3\}$, $\{2, 6, 7\}$, $\{5, 10, 12\}$, and $\{4, 8, 11\}$ form a base for a cyclic block design on the integers mod 13 with $\lambda = 2$. To see this, we form the differences $0 - 1$, $1 - 0$, $0 - 3$, $3 - 0$, $1 - 3$, $3 - 1$ for the first set in our proposed base and the differences $2 - 6$, $6 - 2$, $2 - 7, \ldots$ and so on for the other sets in our base. Thus, we get four sets of differences, one for each set in the base. The sets we get are $\{\pm 1, \pm 2, \pm 3\}$, $\{\pm 4, \pm 5, \pm 1\}$, $\{\pm 5, \pm 2, \pm 7\}$, and $\{\pm 4, \pm 3, \pm 7\}$. These sets include each nonzero equivalence class mod 13 exactly twice since $6 = -7$, $8 = -5$, etc. This design is a $(52, 13, 12, 3, 2)$ design that *almost* meets the criteria we have for our softball league problem. If we use the blocks given in week 0 and add i (mod 13) to the elements in the base blocks to get the blocks of teams who play in week i, then each week all teams but one play once (each has 1 week off) and each team plays each other team exactly twice in a 13-week period. ∎

Example 3.3. The (13,9,6) design of Example 2.5 in this chapter is cyclic mod 13, as is evident from the definition of cyclic designs. However, it is not immediately obvious from studying the 13 sets that they do form a balanced design. How do we know that they do?

Note that the nonzero differences mod 13 of distinct members of the first block are $1 - 0 = 1$, $0 - 1 = 12$, $3 - 0 = 3$, $0 - 3 = 10$, 5, 8, 6, 7, 9, 4, 10, 3, 11, 2, 12, 1, 2, 11, 4, 9, 5, 8, 8, 5, 9, 4, 10, 3, 11, 2, 2, 11, 3, 10, 6, 7, 7, 6, 8, 5, 9, 4, 1, 12, 4, 9, 5, 8, 6, 7, 1, 12, 2, 11, 3, 10,

1, 12, 2, 11, 1, 12. Thus each equivalence class mod 13 appears six times as a difference, and by our theorem, the design is balanced. ∎

Resolvable Designs

Our solution to the softball problem did not have each team play *exactly* once each week. A solution that did would be called resolvable. A (b, v, r, k, λ) design is **resolvable** if the blocks may be grouped into s sets each of which is a partition of V. (Note, we partition the blocks into sets of blocks each set of which is a partition of V into parts of size k.) Each of the partitions of V is called a **parallel class** of blocks in analogy to parallel lines in plane geometry.

> **Theorem 3.2.** If a design is resolvable, then $\frac{b}{r}$ and $\frac{v}{k}$ are the same integer.
>
> *Proof.* Since V is a union of some number m of disjoint sets of size k, $v = km$. Thus, $\frac{v}{k}$ is an integer. But since $bk = vr$, we get by division $\frac{v}{k} = \frac{b}{r}$. ∎

Example 3.4. There is no perfect solution to the softball problem with 13 teams because if each team were to play in a week, the blocks for each week's play would be a partition of V and our design would be resolvable. However, $\frac{13}{3}$ is not an integer. (Note that since 13 is not a multiple of 3, we see right away that no schedule could have each team play exactly once a week with three games a night!) This is exactly what it means to say that there is no resolvable design. ∎

Example 3.5. Find a resolvable design for a softball league with nine teams.

Since we know we are to construct a resolvable design, we may choose our first three blocks to be any partition of a nine-element set into three parts. We choose

$$\{1, 2, 3\}, \ \{4, 5, 6\}, \ \{7, 8, 9\}.$$

Since $b \cdot 3 = 9 \cdot r$, $b = 3r$. Since $1 \cdot (9 - 1) = r \cdot (3 - 1)$, $r = 4$ and thus $b = 12$. Therefore, we must find three more parallel classes of three blocks each. Since 1 must be with each of 4, 5, 6, 7, 8, and 9 once in a block and since when 4 is with 1, the numbers 5 and 6 cannot be, we make the following choices:

$$\{1, 2, 3\}, \ \{4, 5, 6\}, \ \{7, 8, 9\}$$
$$\{1, 4, 7\}$$
$$\{1, 5, 8\}$$
$$\{1, 6, 9\}.$$

Since 2 and 3 must be in different blocks and not in a block with 1, we may further fill in the array as follows.

$$\{1,2,3\}, \ \{4,5,6\}, \ \{7,8,9\}$$
$$\{1,4,7\}, \ \{2, \quad \}, \ \{3, \quad \}$$
$$\{1,5,8\}, \ \{2, \quad \}, \ \{3, \quad \}$$
$$\{1,6,9\}, \ \{2, \quad \}, \ \{3, \quad \}.$$

Now there are four elements to be placed in each of the four missing places of each row, and by experimentation we discover that we may fill in the places as follows.

$$\{1,2,3\}, \ \{4,5,6\}, \ \{7,8,9\}$$
$$\{1,4,7\}, \ \{2,5,9\}, \ \{3,6,8\}$$
$$\{1,5,8\}, \ \{2,6,7\}, \ \{3,4,9\}$$
$$\{1,6,9\}, \ \{2,4,8\}, \ \{3,5,7\}.$$ ∎

∞-Cyclic Designs

We need a slightly more complicated idea to deal with a 12-team schedule. Suppose we introduce a new element ∞ to the integers mod n and define

$$\infty + \underline{m} = \infty - \underline{m} = \infty$$
$$\underline{m} - \infty = -\infty - \underline{m} = -\infty$$

for all $\underline{m}$.

We say a balanced design is ∞-**cyclic** mod n if V consists of ∞ and the equivalence classes of integers mod n and if it has a base **B** satisfying conditions 2 and 3 of the definition of a cyclic design. Essentially the same proof that we used for Theorem 3.1 shows that

> *Theorem 3.3.* If **B** is a family of k-element subsets of the equivalence classes of integers mod n and ∞, then **B** is a *base* for an ∞-cyclic design if and only if ∞, $-\infty$, and each residue mod n occur the same number of times as a difference of distinct elements of the same set in **B**.

Example 3.6. The sets $\{\underline{0}, \underline{1}, \underline{3}\}, \{\underline{4}, \underline{5}, \underline{9}\}, \{\underline{2}, \underline{6}, \underline{8}\}$, and $\{\infty, \underline{7}, \underline{10}\}$ form a base for a $(44, 12, 11, 3, 2)$ ∞-cyclic design mod 11. This is the kind of design we need for our softball problem with 12 teams. Further, we can use the four base blocks for the schedule for the first week, because each team would then go to the field once that week. After we add 1 (mod 11) to each entry in each block, the teams are still all represented. For each succeeding week, we again add 1 to each entry of a block; thus, we have a perfect solution to the softball problem with 12 teams. ∎

The idea used to show that all teams play each week in the example may be used to prove

> **Theorem 3.4.** A cyclic or ∞-cyclic design is resolvable if it has a base which is a partition of V.

Triple Systems

A balanced design with $k = 3$ such as those we have constructed is called a *triple system*. A balanced design with $k = 3$ *and* $\lambda = 1$ is called a **Steiner triple system** (although these designs were first studied by Kirkman). Steiner triple systems are important in their own right but are also useful as "building blocks" for triple systems with $\lambda = 2$.

> **Theorem 3.5.** If there is a triple system on v vertices, then $\lambda(v - 1)$ is even and $\lambda v(v - 1)$ is a multiple of 6.

Proof. We substitute $k = 3$ in the relations of Theorems 2.1 and 2.2 of this chapter and solve for r and b, getting

$$r = \frac{\lambda(v - 1)}{2} \quad \text{and} \quad b = \frac{\lambda v(v - 1)}{6}.$$

Since r and b must be integers, the conclusion of the theorem follows. ∎

> **Theorem 3.6.** If there is a Steiner triple system on V, then $v = 6t + 1$ or $v = 6t + 3$ for some t.

Proof. Either $v = 6t, 6t + 1, \ldots$, or $6t + 5$ for some t. Since 2 is a factor of $v - 1$, the only possibilities are $v = 6t + 1, 6t + 3, 6t + 5$. However, if $v = 6t + 5$, then $v(v - 1) = (6t + 5)(6t + 4) = 36t^2 + 54t + 20$. But 6 cannot divide this sum since it divides the first two terms but doesn't divide 20. Thus, the only possible v's have the form $6t + 1$ or $6t + 3$. ∎

Steiner and Kirkman both asked whether, for each v allowed by Theorem 3.6, there is a Steiner triple system with this v. The answer is yes, as Kirkman showed around 1850. (Steiner asked the question in 1857, unaware of Kirkman's work.)

Hanani has shown that for any v and λ which satisfy the conditions of Theorem 3.5, there is a triple system with this v and λ. The method is basically to construct many families of cyclic, ∞-cyclic, and related designs and then use methods such as the following to build new designs from old ones.

Theorem 3.7. If S_1 is a Steiner triple system on V_1 and if S_2 is a Steiner triple system on V_2, then there is a Steiner triple system on $V = V_1 \times V_2$.

Proof. As in Theorem 3.1, if we exhibit a family of k-element subsets of $V_1 \times V_2$ such that each pair of distinct elements of $V_1 \times V_2$ is in exactly one such subset, then we have exhibited a block design with $\lambda = 1$ on $V_1 \times V_2$. Consider the following subsets of $V_1 \times V_2$.

(Type 1) $\{(a_i, b) \mid \{a_1, a_2, a_3\}$ is a block of $S_1\}$ (for each $b \in V_2$ we have such a set).

(Type 2) $\{(a, b_i) \mid \{b_1, b_2, b_3\}$ is a block of $S_2\}$ (for each $a \in V_1$ we have such a set).

(Type 3) $\{(a_i, b_i) \mid \{a_1, a_2, a_3\}$ is a block of S_1 and $\{b_1, b_2, b_3\}$ is a block of $S_2\}$.

Now we want to show that each pair of elements of $V_1 \times V_2$ lies in exactly one block. We have three cases, corresponding to our three kinds of blocks. First, if a pair $\{(a_1, b_1), (a_2, b_2)\}$ of distinct elements of $V_1 \times V_2$ satisfies $b_1 = b_2$ then $a_1 \neq a_2$ and it is in one type 1 block and no blocks of type 2 or 3. Second, if a pair of distinct elements $\{(a_1, b_1), (a_2, b_2)\}$ of $V_1 \times V_2$ satisfies $a_1 = a_2$, then $b_1 \neq b_2$ and the pair lies in one type 2 block and no blocks of type 1 or 3. Third, if a pair of distinct elements $\{(a_1, b_1), (a_2, b_2)\}$ of $V_1 \times V_2$ satisfies neither the first nor the second condition, then $a_1 \neq a_2$ and $b_1 \neq b_2$, so that the pair lies in one type 3 block and no blocks of type 1 or 2. Thus, each pair lies in exactly one block of type 1, one block of type 2, or one block of type 3. ∎

Kirkman's Schoolgirl Problem

Steiner triple systems need not be resolvable. It is far more difficult to find a *resolvable* triple system than to find just any triple system. In fact, Kirkman posed but could not solve the following problem. "If once every day for a week 15 schoolgirls are to walk in five rows of three girls each, is it possible for each girl to be in a row with each other girl exactly once?" Kirkman was asking for a resolvable $(35, 15, 7, 3, 1)$ design. (Can you explain why?)

Kirkman's problem is especially intriguing since a triple system on 15 vertices with $\lambda = 1$ has

$$b \cdot 3 = r \cdot 15, \text{ so } b = 5r$$

and

$$r \cdot (2) = (14) \cdot 1, \text{ so } r = 7$$

and there are 35 blocks of size 3. Since there are 5 blocks walking together each day for a week, we will use exactly 35 blocks. A solution to Kirkman's problem is given in Figure 3.1. The notation is intended to suggest a more general kind of cyclic construction that was used to generate the solution. See Exercise 28 for details.

Figure 3.1

MONDAY			TUESDAY			WEDNESDAY			THURSDAY		
0	a^0	b^0	0	a^1	b^1	0	a^2	b^2	0	a^3	b^3
a^1	a^2	a^4	a^2	a^3	a^5	a^3	a^4	a^6	a^4	a^5	a^0
a^5	b^1	b^6	a^6	b^2	b^0	a^0	b^3	b^1	a^1	b^4	b^2
a^3	b^2	b^5	a^4	b^3	b^6	a^5	b^4	b^0	a^6	b^5	b^1
a^6	b^3	b^4	a^0	b^4	b^5	a^1	b^5	b^6	a^2	b^6	b^0

FRIDAY			SATURDAY			SUNDAY		
0	a^4	b^4	0	a^5	b^5	0	a^6	b^6
a^5	b^6	a^1	a^6	b^0	a^2	a^0	b^1	a^3
a^2	b^5	b^3	a^3	b^6	b^4	a^4	b^0	b^5
a^0	b^6	b^2	a^1	b^0	b^3	a^2	b^1	b^4
a^3	b^0	b^1	a^4	b^1	b^2	a^5	b^2	b^3

More generally, we might ask for what values of v we may have a resolvable triple system. One can show that $v \equiv 3 \pmod{6}$ is a necessary condition. By extending Hanani's construction techniques, Ray-Chaudhuri and Wilson have shown that such "Kirkman triple systems" may be constructed whenever $v \equiv 3 \pmod{6}$.

Constructing New Designs from Old

Although the cyclic designs are quite useful, constructing them "from scratch" can be quite difficult, and further it is unreasonable to expect to be able to answer every construction question with a cyclic or ∞-cyclic design. There are other construction techniques based on various ideas of abstract algebra, there are techniques based on geometry, some of which are discussed in the next section of this chapter, and there are techniques based on putting two designs together or modifying an existing design to get a new one. For example, Theorem 3.7 demonstrates one way to construct designs by the "composition" of two triple systems. The derived and residual designs of Section 2 are an example of modifying an existing design to get a new design. There are quite a few other elementary and straightforward ways to construct new designs from existing ones; we describe a number of these ways here.

Complementary Designs

One straightforward construction technique is to replace each block B of a design D on V by its complement $V - B$. This design, which we denote by $\overline{D}$, is called the *complementary design* or complement of D. It is clear that the complement of a uniform design is uniform and that the complement of a symmetric design is symmetric. Some other properties of designs are also preserved by this construction.

Theorem 3.8. If D is a regular design with b blocks and replication number r, then $\overline{D}$ is a regular design with replication number $b - r$.

Proof. Straightforward application of the definitions. ∎

Theorem 3.9. If D is a balanced regular design with b blocks and replication number r, then $\overline{D}$ is a balanced regular design with index $b - 2r + \lambda$.

Proof. Since each x is in r blocks and each pair (x, y) is in λ blocks of D, there are $2r - \lambda$ blocks containing one or the other of x or y. Thus, $b - (2r - \lambda) = b - 2r + \lambda$ blocks of the complement contain both the variety x and the variety y. ∎

Notice that if $b - 2r + \lambda$ is zero, the theorem tells us we have a balanced block design with $\lambda = 0$. Such a design is rather trivial since the blocks are either empty or singleton sets (why?). However, such a design is a balanced design according to the definition we gave.

Example 3.7. The complementary design of the symmetric design of Example 2.5 of the last section is

$$\{2, 4, 7, 8\}, \ \{3, 5, 8, 9\}, \ \{4, 6, 9, 10\}, \ \{5, 7, 10, 11\},$$
$$\{6, 8, 11, 12\}, \ \{7, 9, 12, 0\}, \ \{8, 10, 0, 1\}, \ \{9, 11, 1, 2\}, \ \{10, 12, 2, 3\},$$
$$\{11, 0, 3, 4\}, \ \{12, 1, 4, 5\}, \ \{0, 2, 5, 6\}, \ \{1, 3, 6, 7\}.$$

We see that this design has $\lambda = 1$, and we note that $13 - 2 \cdot 9 + 6 = 1$. If we were not given this design as the complement of a balanced design, how would we know it is balanced? It is clear from the blocks that this design is cyclic; however, it is not clear that it is balanced. To apply Theorem 3.1 we check the 12 differences $2 - 4 = 11$, $4 - 2 = 2$, $2 - 7 = 8$, $7 - 2 = 5$, $2 - 8 = 7$, $8 - 2 = 6$, $4 - 7 = 10$, $7 - 4 = 3$, $4 - 8 = 9$, $8 - 4 = 4$, $7 - 2 = 8$, and $7 - 2 = 5$ of equivalence classes mod 13 to see that each nonzero class appears exactly once as a difference. This is far easier than the 72 distinct differences we had to check in Example 3.3. Thus, to show that the design of Example 2.5 is a balanced design, it would have been much easier to construct this design, verify that it is balanced, and then verify that the design of Example 2.5 is its complement. ∎

We would conjecture from the example that the complement of a cyclic design will be cyclic. (See Exercise 14.) Will the complement of an ∞-cyclic design be ∞-cyclic or cyclic?

A second construction that uses complementation requires that we start with a simple uniform design (in a simple design, all blocks have multiplicity 0 or 1). If we have a design D with block size k, then clearly the set D^* of all k-element sets not in D is another simple uniform design with block size k. Again, D^* has the properties of regularity and balance when D does. We call D^* the *inverse* of D.

> **Theorem 3.10.** If D is a simple k-uniform regular design on a set of v varieties with replication number r, then D^* is a simple uniform regular design with replication number $\binom{v-1}{k-1} - r$.
>
> *Proof.* Immediate from the fact that $\binom{v-1}{k-1}$ k-element subsets contain a given element. ∎

> **Theorem 3.11.** If D is a balanced simple k-uniform design of index λ on a set of v varieties, then D^* is a balanced simple k-uniform design of index $\binom{v-2}{k-2} - \lambda$.
>
> *Proof.* Similar to the preceding proof. ∎

Example 3.8. What are the parameters of the inverse of the design in Example 3.7?

Since $v = 13$, $b = 13$, $k = 4$, $r = 4$, and $\lambda = 1$, in the inverse design we have $v^* = 13$, $b^* = \binom{13}{4} - 13 = 715 - 13 = 702$, $r^* = \binom{12}{3} - 4 = 220 - 4 = 216$, and $\lambda = \binom{11}{2} - 1 = 54$. ∎

Unions of Designs

Notice that when we put a simple design D and its inverse design D^* together we obtain the set of all subsets of size k chosen from V. Thus, putting these two balanced incomplete block designs together gives us another balanced incomplete block design. This is a special case of the next theorem, in which we define the ***design union*** of two designs.

> **Theorem 3.12.** If D_1 and D_2 are block designs on a set V, then their design union $D_1 \sqcup D_2$ in which the multiplicity of a block B is the sum of its multiplicities in D_1 and D_2 is
>
> (1) k-uniform if D_1 and D_2 are k-uniform.
> (2) Regular if D_1 and D_2 are regular.

(3) Balanced if D_1 and D_2 are balanced.

Proof. Immediate from the definitions. ∎

Example 3.9. The design union of Example 2.2 and the design of Example 2.3 (both from the previous section) is a $(7, 3, 2)$ design in which some blocks—such as $\{0, 1, 2\}$—appear twice, others—such as $\{2, 4, 5\}$—appear once, and other three-element sets do not appear at all as blocks. ∎

Product Designs

We gave a construction of a triple system on $V_1 \times V_2$ for two triple systems on V_1 and V_2. In order to construct a triple system on the product of the vertex sets we had to be careful in our choice of blocks. If, instead, we make the most natural choice of blocks of the product of the two vertex sets, we get a very elementary construction of a design. Given designs D_1 and D_2 on sets V_1 and V_2, we define the **product design** $D_1 \times D_2$ to be the design on the set of ordered pairs $V_1 \times V_2$ whose blocks are the sets $B_1 \times B_2$ with B_1 a block of D_1 and B_2 a block of D_2. Again, this construction preserves the fundamental structure of a design.

> **Theorem 3.13.** If D_1 and D_2 are designs, then the product design $D_1 \times D_2$ is
> (1) Uniform if D_1 and D_2 are uniform.
> (2) Regular if D_1 and D_2 are regular.
> (3) Balanced if D_1 and D_2 are balanced.

Proof. The proofs are straightforward applications of the definitions. ∎

The construction techniques just described will allow us to reduce some questions about existence of designs to simpler ones.

Example 3.10. Is there a $(49, 9, 1)$ design?

Since v is 7^2 and k is 3^2, we are immediately led to consider the product of a $(7, 3, 1)$ design with itself; the previous theorem tells us this is indeed a $(49, 9, 1)$ design. ∎

Composition of Designs

Our final example of a straightforward construction technique is similar to the composition of Latin squares. The idea of the construction is to replace each block of one design by another entire design. If D is a k-uniform design on V and D' is a k'-uniform design on a k-element set, then we can construct an isomorphic copy $D'(K)$ of D' on each block K of D. Then the set of *all* the blocks of all the designs $D'(K)$ will be a set of blocks of a uniform design of block size k' on V. This kind of design is called a *composition* of D with D'.

Theorem 3.14. If D is a (v, k, λ) design and D' is a (k, k', λ') design, then a composition of D with D' is a $(v, k', \lambda\lambda')$ design.

Proof. Each pair of varieties appears in λ different blocks K of size k in D. For each of these blocks K, the pair will appear in λ' blocks of D. ∎

Theorem 3.15. If D and D' are resolvable designs, then a composition of D and D' is resolvable.

Proof. Straightforward application of the definitions of resolvability and composition. ∎

By limiting the b, v, r, k, and λ considered to those that pass the two tests $bk = vr$ and $\lambda(v-1) = r(k-1)$, it is possible to make a list of all possible parameters of (relatively) small size (say for v up to 100) for which these construction techniques do not reduce the size of the problem. There are tables giving at least one known design fitting such parameters for which designs are known; a table with $r \leq 20$ is given in the back of the book by Hall in the Suggested Reading. More extensive tables are published and revised regularly in the statistical literature as new designs become known. The *CRC Handbook of Combinatorial Designs* listed in the Suggested Reading is one such source. It gives tables of symmetric designs, triple systems, etc., as well as tables of block designs. A statistician needing to design an experiment will consult one of these tables early in the design process.

*The Construction of Kirkman Triple Systems

We now describe a composition technique which has roots in a method used by E. H. Moore in 1893 to construct Steiner triple systems on $2v+1$ varieties from Steiner triple systems on v varieties. The particular construction we use is a special case of a construction of Mullin and Vanstone. This construction can be used to give a nice proof of the Ray-Chaudhuri and Wilson result that resolvable triple systems, also known as **Kirkman triple systems**, on $6t + 3$ varieties exist for each integer $t > 0$.

Theorem 3.16. If there is a $(v, k, 1)$ design and a resolvable triple system on $2k + 1$ varieties, then there is a resolvable triple system on $2v + 1$ varieties.

Proof. We describe the construction in the case that the triple system on $2k + 1$ varieties has $\lambda = 1$; we omit the details of proof that the construction gives a resolvable triple system. For the general case and the details of the proof, see the book by Wallis in the Suggested Reading.

Let T be the triple system on $2k+1$ varieties, and assume 0 is one of its varieties. Since $\lambda = 1$, $r = k$, so that 0 is in k blocks,

$$\{0, a_1, b_1\} \ \{0, a_2, b_2\} \ \cdots \ \{0, a_k, b_k\}.$$

Now if $a_i = a_j$ for $i \neq j$, the pair $\{0, a_i\}$ would appear in two different blocks, as it would if $b_i = b_j$ or $a_i = a_j$ for $i \neq j$. Since $\lambda = 1$, this is impossible, and therefore the a_is and b_is must be the remaining elements of the set U of varieties of T; that is,

$$U = \{0, a_1, b_1, a_2, b_2, \ldots, a_k, b_k\}.$$

Now let $V = \{1, 2, \ldots, v\}$ be the set of varieties of the given $(v, k, 1)$ design D. We need a $2v + 1$-element vertex set for the resolvable triple system we desire to construct. For this purpose, let

$$V' = \{0, x_1, y_1, x_2, y_2, \ldots, x_v, y_v\}.$$

From our design D we will construct a design D' with blocks of size $2k + 1$ on V'. Next we will duplicate the triple system T on every one of these blocks, giving us many isomorphic triple systems. To get our resolvable triple system T' on V', we take all the three-element blocks of all these isomorphic triple systems.

To construct D', for each block H of D, let

$$H' = \{0\} \cup \{x_i | i \in H\} \cup \{y_i | i \in H\}.$$

Then H' has $2k + 1$ elements. The sets H' form a uniform design on V'. (This design is not balanced, but fortunately that is irrelevant to the construction.) To duplicate T on H', we first choose a bijection φ from $K = \{1, 2, \ldots, k\}$ to H. Then we define a bijection F_H from V to H by

$$f_h(0) = 0$$
$$f_h(a_i) = x_{\varphi(i)}$$
$$f_h(b_i) = y_{\varphi(i)}.$$

Now let T_H be the triple system whose blocks are the set

$$\mathbf{B}_H = \{\{f_H(c), f_H(d), f_H(d)\} | \{c, d, e\} \text{ is a block of } T\}.$$

Here we are just using the bijection f_H to copy the structure of T isomorphically onto the set H'. Finally, we let $\mathbf{B}$ be the set union (not the multiset union) over all blocks H of D of the sets $\mathbf{B}_H$; that is, $\mathbf{B}$ is the set of all the blocks in one or more of the sets $\mathbf{B}_H$. We let T' be the design on V' whose blocks are the set $\mathbf{B}$. T' is the desired resolvable triple system on $2v + 1$ vertices.

It is not too difficult to show that we have constructed a Steiner triple system on V, but we omit the careful analysis required.

The resolvability of the design will have to result from the resolvability of the triple system T. Each of the blocks $\{0, a_i, b_i\}$ of T will have to be in a different parallel class of the resolution of T, and each parallel class will have to contain one of these blocks. In our triple system T' on V', 0 now appears in v blocks, the blocks of the form $\{0, x_i, y_i\}$. Thus a resolution of T' will have to have v parallel classes, one containing each of these blocks. For each block H containing i, we have an induced triple system T_H on H' containing the block $\{0, x_i, y_i\}$. In this triple system T_h there is a parallel class $P(H, i)$ containing the block $\{0, x_i, y_i\}$. If we let

$$\mathbf{P}_i = \cup_{H:i \in H} P(H, i),$$

then the sets $\mathbf{P}_1, \mathbf{P}_2, \ldots, \mathbf{P}_v$ turn out to be the parallel classes of a resolution of the design T'. While this is not difficult to show once one masters the notation, it does require a careful analysis (which we omit) of how the blocks of T' arise from those of T and D. ∎

Example 3.11. Can we use the previous theorem to construct a solution to the Kirkman schoolgirl problem (with $v = 15$)?

Since $2 \cdot 7 + 1 = 15$, we could use a design D on 7 varieties with blocks of size k and then a resolvable triple system on $2k + 1$ varieties to apply our theorem. There are $(7, 3, 1)$ and $(7, 4, 2)$ designs on 7 varieties, so the possible nontrivial values of k are 3 and 4. If we take $k = 3$, we would need a resolvable triple system on $2 \cdot 3 + 1 = 7$ vertices, and we know there is none. However the $(7, 4, 2)$ design is not a $(v, k, 1)$ design, and so the theorem does not apply. Thus the theorem does not help us construct a solution to the Kirkman schoolgirl problem. ∎

Example 3.12. There is a $(16, 20, 5, 4, 1)$ design. Why may we conclude that there is a Kirkman triple system on 33 varieties?

Since 9 is $4 \cdot 2 + 1$ we may apply Theorem 3.16 with the $(16, 20, 5, 4, 1)$ design and the $(9, 3, 1)$ design of Example 3.5 to construct a resolvable triple system on 33 varieties. ∎

EXERCISES

1. Write out all the blocks of the design in Example 3.2.

2. Write out all the blocks of the design in Example 3.6.

3. Find all triple systems on four vertices.

4. For what values of λ is there a triple system on six vertices? (Hint: Exercise 17 in the previous section contains a useful fact.)

5. Find all resolvable triple systems on seven vertices or show there is none.

6. Show that if there is a resolvable Steiner triple system on v vertices, then $v \equiv 3 \pmod{6}$.

7. An experimenter has nine suntan lotions. In order to test them, the experimenter identifies three similarly exposed patches of skin on sunbathers. Since some conditions vary from day to day, the experimenter wishes to test all nine lotions each day the experiment runs. In addition, each lotion should be tested against each other lotion on the same sunbather exactly once. Describe what kind of design is needed and give an example of such a design.

8. Show that any two Steiner triple systems on seven vertices are isomorphic. (This problem can be solved by gathering together bits and pieces of information you have already learned.)

9. Find a cyclic symmetric $(7, 3, 1)$ design. Is this the design given in Section 2? Explain how this and the result of Exercise 8 fit together.

10. Is there a cyclic symmetric $(11, 5, 2)$ design?

11. Show that any two Steiner triple systems on nine vertices must be isomorphic.

12. On the basis of the preceding exercise, why may you conclude that a Steiner triple system on nine varieties must be resolvable?

13. Prove Theorem 3.8, which states that the complement of a regular design is regular.

14. Show that the complement of a cyclic design is cyclic.

15. Prove or give a counterexample: The inverse of a cyclic design is cyclic.

16. Determine whether or not the complement of an ∞-cyclic design is ∞-cyclic. Determine whether it is cyclic.

17. Prove Theorem 3.11, which states that if D is a balanced simple k-uniform design of index λ on a set of v varieties, then D^* is a balanced simple k-uniform design of index $\binom{v-2}{k-2} - \lambda$.

18. Are the sets $\{1, 3, 9\}$ and $\{2, 5, 6\}$ a base for a cyclic design mod 13? What would the parameters of such a design be?

19. Are the sets $\{\infty, 0, 3\}$, $\{0, 1, 4\}$, $\{0, 2, 3\}$, $\{0, 2, 7\}$ a base for an ∞-cyclic design on 10 vertices? If so, what are its parameters?

20. Find a cyclic $(13, 4, 1)$ design.

21. Describe how to construct a resolvable triple system on 27 varieties.

22. Find a cyclic $(21, 5, 1)$ design.

23. Can the construction of Theorem 3.7 be used to create a new design with $\lambda = 1$ from two designs with $\lambda = 1$ and the same k?

24. Can the construction of Theorem 3.7 be applied when $\lambda > 1$?

25. If you are given v and k, how should you decide what values of n are worth trying for a cyclic design mod n? How should you decide how many blocks to try in a base?

*26. Prove Theorem 3.3.

*27. If $v = 6t + 6$, show that the following sets form a base for a design with $\lambda = 2$ over the integers mod $6t + 5$.
$$\{\infty, 0, 3t + 2\}$$
$$\{0, i, 2t + 3 - i\} \quad i = 1, 2, \ldots, t + 1$$
$$\{0, 2i, 3t + 3 + i\} \quad i = 1, 2, \ldots, t$$

28. (a) Consider the sets $\{\infty, a, b\}$, $\{ax, ax^2, ax^4\}$, $\{ax^5, bx, bx^6\}$, $\{ax^3, bx^2, bx^5\}$, $\{ax^6, bx^3, bx^4\}$. Think of a and b as ax^0 and bx^0; use the rule $\infty x^i = \infty$ for all i. Show that adding the integers mod 7 to the *exponents* in the way we added integers mod n to elements of base blocks gives a solution to Kirkman's schoolgirl problem with these base blocks.

  *(b) There is a similar construction of a $(35, 15, 7, 3, 1)$ design using seven base blocks, constants a, b, and c, and the integers mod 5; however, there is no element that plays a special role like ∞. Find the base. Is the resulting design resolvable?

29. Show that blocks of two different parallel classes of a resolvable $(16, 4, 1)$ design must have exactly one vertex in common. Explain why this means we may assume a resolvable $(16, 4, 1)$ design has the blocks

$$\{1, 2, 3, 4\}, \quad \{5, 6, 7, 8\}, \quad \{9, 10, 11, 12\}, \quad \text{and} \quad \{13, 14, 15, 16\}$$

and also the blocks

$$\{1, 5, 9, 13\}, \quad \{2, 6, 10, 14\}, \quad \{3, 7, 11, 15\}, \quad \text{and} \quad \{4, 8, 12, 16\}.$$

Beginning with these blocks, construct a resolvable $(16, 4, 1)$ design.

30. A bridge club has 16 members. In bridge, four people are seated around a table to play a game, so the club has four tables of people playing bridge each time it meets. Can the club devise a scheme that allows its members to play bridge every week for a certain number of weeks and have played at a table with each other member exactly once? If so, how many weeks are required and how should they arrange the tables? If not, why not? What about exactly three times, once seated to the person's left, once across from the person, and once to the person's right?

31. In the proof of Theorem 3.16, show that the design we have constructed is balanced.

32. Write a computer program that uses backtracking to generate appropriate partitions of a set V and checks to see if they form a base for a cyclic resolvable block design on V.

Section 4 Affine and Projective Planes

Affine Planes

There are analogies between block designs and geometry that are quite useful in the construction of designs. For example, in plane geometry every pair of points lies in a unique line. In a balanced block design with $\lambda = 1$, every pair of points lies in a unique block. In plane geometry two lines either intersect in one point or are parallel. In a block design with $\lambda = 1$, two points can't lie in two different blocks, so two blocks intersect in one point or they do not intersect at all. Thus, taking the points in the plane as our varieties and the lines in the plane as blocks would give us an infinite analog of a block design. Perhaps by taking a finite set of points we could use the lines through these points to construct interesting finite block designs. In Figure 4.1 we show nine points in the plane.

Figure 4.1

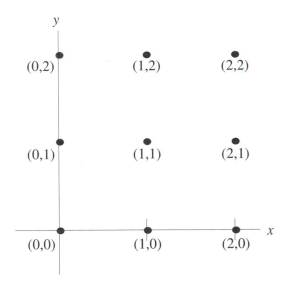

Horizontal lines and vertical lines through this array of points each determine three-element blocks. The line of slope one through $(0,1)$ and $(1,2)$ determines only a two-element block, while the line with slope one through $(0,0)$ and $(1,1)$ determines a three-element block. Although designs with varying block sizes are sometimes useful, uniform designs are usually more interesting. By interpreting the coordinates as integers modulo 3 we can make *every* "line" have exactly three points. By a line, we mean a set of points (x, y) such that for some a, b, and c, $ax + by = c$.

Example 4.1. Show that among the points (i, j) in which i and j represent integers modulo 3, there are finitely many lines in Figure 4.1 and find how many lines there are.

We consider the possible solutions $(x, y) = (i, j)$ to an equation of the form $ax + by = c$. With $a = 0$ and $b = 1$ we get $y = 0$, $y = 1$, and $y = 2$, the equations of the three horizontal "lines" in Figure 4.1. The equations $x = 0$, $x = 1$, and $x = 2$ are the equations of the three vertical "lines." Further, $y = x$ or $x - y = 0$ (which is also $x + 2y = 0$ in the integers modulo 3) describes the diagonal line through $(0, 0)$. The equation $x + y = 2$ is the equation of the opposite diagonal. So far, there have been no surprises. Consider, however, the equation $x + 2y = 1$, which may also be written as $x - y = 1$ or $y = x - 1$. The point $(1, 0)$ and the point $(2, 1)$ both satisfy this equation, as we expect by looking at the figure and thinking about the graph of the equation $y = x - 1$. So does the point $(0, 2)$ since

$$2 \equiv 0 - 1 \pmod{3}.$$

Thus, the line with equation $x + 2y = 1$ is the set $\{(1, 0), (2, 1), (0, 2)\}$. Similarly, each other line will have exactly three points. (There are three possibilities for x and then we "solve" the equation for y.) Since we have a total of nine points, we can have only a finite number of three-element subsets and thus only a finite number of lines. In fact, the lines form the blocks of a 3-uniform design. Notice that the equation $x + y = 1$ and the equation $2x + 2y = 2$ determine exactly the same line, so each line corresponds to at least two equations. The total number of equations, including $0 = 0$, $0 = 1$ and $0 = 2$ is 3^3, the number of ways to choose a, b, and c. The three equations we just mentioned with a and b both zero do not determine lines, so we have $27 - 3 = 24$ equations that represent lines. Since there are at least two equations per line, we see that there should be at most 12 lines. However, it is clear that for a fixed choice of a and b the three lines $ax + by = 0$, $ax + by = 1$, and $ax + by = 2$ must all be different. Thus, our three horizontal lines, our three vertical lines, and our three lines with equations $1x + 1y = c$ are all different. Simple algebra (done modulo 3) shows that a line with equation $1x + 1y = c$ and a line with equation $1x + 2y = c$ have exactly one point in common, so we have three more lines with equations of the form $1x + 2y = c$. Thus, we have 12 lines. ∎

Example 4.2. Show that each pair of distinct points of the form (x_1, y_1) and (x_2, y_2), with x_i and y_i integers modulo 3, determines a unique line.

As in ordinary plane geometry, if x_1 and x_2 are different, the two points (x_1, y_1) and (x_2, y_2) lie in the line

$$y - y_1 = \frac{y_2 - y_1}{x_2 - x_1}(x - x_1)$$

(just substitute (x_1, y_1) or (x_2, y_2) for (x, y) and verify the equality). If x_1 and x_2 are the same, then the corresponding equation with $y_2 - y_1$ in the denominator describes a line containing our points. From the previous example, we see that each line has an equation of the form $x = c$, $y = c$, $x + y = c$, or $x + 2y = c$. As in the previous example, simple algebra (done modulo 3) shows that lines whose equations have different forms have exactly one point in common, and lines whose equations have the same form have no points in common. Thus, two points cannot lie on two different lines. Therefore, each pair of points determines a unique line. ∎

From the last two examples, we see that the lines we have described are the blocks of a balanced incomplete $(12, 9, r, 3, 1)$ design, and we can compute that $r = 4$. This block design has exactly the parameters of our resolvable block design on nine varieties in Example 3.5. Although this coincidence of parameters could just be an accident, it is natural to ask if the geometric approach has led us to a resolvable design as well. By looking at Figure 4.1 we see that the vertical lines (those with slope 0) form a partition of the variety set and the horizontal lines form a partition of the variety set. The other lines are those with slope one or two, those with the forms $x + 2y = c$ or $x + y = c$. Below we list first the points in the vertical, then horizontal, then slope 1, then slope 2 lines. We can see that we have listed the parallel classes of a resolvable design.

$$\{(0,0), (0,1), (0,2)\}, \ \ \{(1,0), (1,1), (1,2)\}, \ \ \{(2,0), (2,1), (2,2)\},$$
$$\{(0,0), (1,0), (2,0)\}, \ \ \{(0,1), (1,1), (2,1)\}, \ \ \{(0,2), (1,2), (2,2)\},$$
$$\{(0,0), (1,1), (2,2)\}, \ \ \{(0,1), (1,2), (2,0)\}, \ \ \{(1,0), (2,1), (0,2)\},$$
$$\{(0,0), (1,2), (2,1)\}, \ \ \{(0,1), (1,0), (2,2)\}, \ \ \{(0,2), (1,1), (2,0)\}.$$

Each of our parallel classes consists of all lines of a given slope—lines we would think of as parallel in ordinary geometry. Thus, it appears we have an outstandingly intuitive tool for the construction of resolvable block designs. We would take a square array of points—thus we would have $v = n^2$, and using the integers mod n, we would use lines given by equations as our blocks and families of lines with the same slope as our parallel classes. Unfortunately, this works only when n is a prime, because to compute a slope of the form

$$\frac{y_2 - y_1}{x_2 - x_1}$$

we must be able to divide. As we saw in Section 1, this is possible in the integers modulo n only when n is a prime. (In abstract algebra we learn to construct so-called fields with p^k abstract elements; fields are algebraic structures in which we can divide, so this geometric construction using a field would give us resolvable designs whenever $v = (p^k)^2 = p^{2k}$.)

In general, the "geometry" we construct in this way using the integers modulo a prime p will have p^2 points, $p^2 + p$ lines, p points per line, every point in $p+1$ lines, every two points determining a unique line, and every pair of lines either parallel (that is, determined by equations with the same slope) or intersecting in exactly one point. This kind of geometry is called an "affine plane" or "affine plane geometry" of order p. Although it is sometimes called a Euclidean plane geometry, especially by specialists in block designs, we shall not do so because in other branches of mathematics the word Euclidean implies the existence of a special way to measure distances.

Postulates for Affine Planes

When you began your study of geometry, you studied geometry from the point of view of properties of points and lines, not from the point of view of ordered pairs. You had axioms and postulates to work with and used them to prove facts about geometric objects. Similarly, there is a second description of affine plane geometries that avoids the use of ordered pairs and the concept of slope. An **affine plane** consists of a set P of points and a family L of subsets of P called lines such that

(1) Each pair of points lies together in exactly one line.
(2) Each line has at least two points.
(3) Given a line l and a point p not on l there is one and only one line l' in L such that $p \in l'$ and $l \cap l' = \emptyset$.
(4) There are (at least) three distinct points, not all on the same line.

While there are affine planes that cannot be constructed using ordered pairs and equations, the examples of such affine planes are complex. Further, as yet such examples have p^{2k} points and so have the same parameters as affine planes that are constructed by using ordered pairs and equations (although it is necessary to use the "fields" mentioned above). The following theorems show that affine planes of this more general type lead to resolvable block designs. As they are analogous to other theorems we shall prove later, we leave their proofs as exercises. The concept of parallelism is quite useful in visualizing the proofs of these theorems. We say lines l_1 and l_2 are **parallel** if they do not intersect. Notice that postulate (3) may be reworded as

(3) Given a line l and a point p not on l there is one and only one line l' in L such that $p \in l'$ and l' is parallel to l.

Theorem 4.1. In a finite affine plane there is an integer n such that all lines have n points.

Proof. Exercise 14. ∎

The number n of the previous theorem is called the ***order*** of the affine plane; notice that the order of a plane is *not* the number of points in the plane.

Theorem 4.2. In an affine plane of order n, each point lies on $n + 1$ distinct lines.

Proof. Exercise 15. ∎

Theorem 4.3. An affine plane of order n has n^2 points.

Proof. Exercise 16. ∎

Theorem 4.4. An affine plane of order n has $n^2 + n$ lines.

Proof. Exercise 17. ∎

Theorem 4.5. In an affine plane of order n each line intersects n^2 other lines, each in one point.

Proof. Exercise 18. ∎

Theorem 4.6. In an affine plane of order n, the relation of parallel or equal to is an equivalence relation and there are $n + 1$ equivalence classes each containing n parallel lines.

Proof. Exercise 20. ∎

From these theorems it is clear that the lines of an affine plane of order n are the blocks of a resolvable $(n^2, n, 1)$ design. On the other hand, given a resolvable $(n^2, n, 1)$ design, Postulates 1, 2, and 4 are clearly satisfied. Proving that Postulate 3, the parallel postulate, is satisfied will prove our next theorem.

Theorem 4.7. The blocks of a resolvable $(n^2, n, 1)$ design with $n > 1$ are the lines of an affine plane on these n^2 points.

Proof. We must show that given a block B and a point p not in B, there is one and only one block B' containing p and not intersecting B. If we let B' be the block containing p of the parallel class of B, then B' is a block containing p and not intersecting B. Since each parallel class partitions the set of varieties, there is exactly one such block in the parallel class of B. Now suppose B^* is a block in a different parallel class. Since there are n^2 varieties and n varieties per block, there must be n blocks per parallel class. However, B^* cannot have two points in common with any block parallel to B, for then it would equal that block (since $\lambda = 1$) and yet we know that B^* is not in the parallel class of B. Thus, since each of the n varieties of B^* must be in a different one of the n blocks in B's parallel class, B^* must intersect each of these blocks, including B, in exactly one point. Therefore, B^* cannot be a block containing p and not intersecting B. Therefore, all four postulates for an affine plane are satisfied, so the varieties and blocks of the design are the points and lines of an affine plane. ∎

In fact, we may leave the word resolvable out in the previous theorem.

Theorem 4.8. Each $(n^2, n, 1)$ design D on a set V is resolvable.

Proof. Let B be a block of D. Let x be an element of V not in B. For each point a of B, there is a unique block of D containing a and x. Further, no two of these blocks can be the same because the only block containing two points of B is B itself. Since x is in $n+1$ blocks and B has n points, there is a unique block B_1 containing x but no points of B. For each y in B_1, the block B_1 is the unique block containing y but no points of B. Now let z be a point of V not in B or B_1. Then the unique block containing z but no points of B cannot contain any y in B_1 because B_1 is the unique block containing each of its y's and skipping B. We continue in this fashion, choosing blocks $B_1, B_2, \ldots, B_{n-1}$ until we run out of elements of V. Let F_1 be the family of blocks $B, B_1, \ldots, B_{n-1}$. Let B' be a block of D not in F_1. Then B' must intersect each block in F_1, because if B' skips every point in B_i, it is the unique block containing a certain point and missing B_i, so it would be in F_1.

We now repeat our process with B' to construct a family F_2 which is a partition of V. Repeating the process until we run out of blocks, we get a resolution $F_1, F_2, \ldots, F_{n+1}$ of D into $n+1$ families of n blocks each. ∎

We see now that $(n^2, n, 1)$ designs and affine planes of order n are, in essence, the same.

The Concept of a Projective Plane

In ordinary geometry, as in affine planes, two lines intersect in either one point or no points. This means that affine plane geometries will not be useful in the construction of symmetric block designs, because we have seen that symmetric designs are linked, that is, each pair of blocks intersects in exactly $\mu = \lambda$ points. There is another kind of geometry, projective geometry, that is analogous to symmetric designs. In projective geometry, there are additional "points at infinity" so that two lines which would be parallel in Euclidean geometry intersect in a point at infinity. Although these points at infinity strain the imagination for a while, they make the geometry easier.

To have a good imaginary picture of points at infinity, you might try to visualize the line at infinity which contains all these points. Try to imagine a point at the "end" of the positive x-axis, a point at the "end" of the positive y-axis, and a semicircle that starts at the "endpoint" of the x-axis, goes up through the "endpoint" of the y-axis and comes down toward the "negative end" of the x-axis without ever reaching it as in Figure 4.2.

Figure 4.2   In a projective plane, three lines we would normally think of as parallel lines of slope one all intersect at a point on the line at infinity, a line visualized as a semicircle infinitely far from the origin.

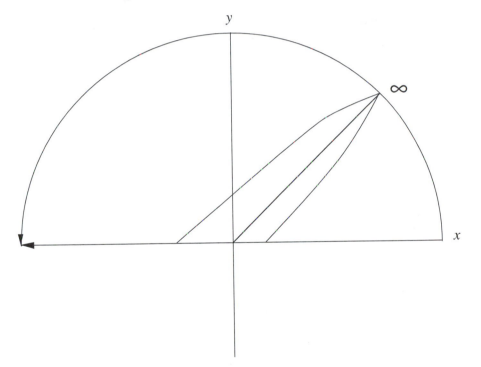

This semicircle is the line at infinity. Then all lines parallel to the x-axis intersect at the right end of the x-axis, all lines parallel to the y-axis intersect at the point at its end, and all lines with slope 1 intersect at the point on the semicircle "halfway" between the x-axis and y-axis. Each possible slope determines a whole family of normally parallel lines; in this projective plane, these lines all intersect at the point we get by "projecting" lines with that slope out to the semicircle. It is possible to describe most projective planes, as it is most affine planes, by using "coordinates." However, it is more complicated to describe projective planes than affine planes in this way, and so we deal with postulates directly in order to describe projective planes.

We define a ***projective plane*** π to be a set P of *points* and a family L of subsets of P called *lines* such that

(1) Each pair of points lies together in exactly one line.
(2) Each pair of lines intersects in exactly one point.
(3) There are (at least) four distinct points no three of which lie together in the same line.

Note that Postulate 2 can be turned into Postulate 1 and vice versa by interchanging "intersect" and "lie together," and "point" and "line." We say Postulates 1 and 2 are *dual* postulates. No dual to Postulate 3 is listed because its dual is implied by Postulates 1 through 3.

Theorem 4.9. In a projective plane, there are (at least) four distinct lines no three of which intersect in the same point.

Proof. Suppose the four points given by Postulate 3 are p_1, p_2, p_3, and p_4. Then by Postulates 1 and 3, there are four lines L_{12}, L_{23}, L_{34}, L_{41} with L_{ij} containing points p_i and p_j but not the other two p's. Suppose three of these lines intersect in some point q, say

$$L_{12} \cap L_{23} \cap L_{34} = \{q\}.$$

Then $L_{12} \cap L_{23}$ must be $\{p_2\}$ by Postulate 2. Thus, since

$$\{p_2\} \cap L_{34} = \{q\},$$

$p_2 = q$. Thus, p_2 is in L_{34} along with p_3 and p_4; however this contradicts Postulate 3. ∎

You can further show that Theorem 4.9, Postulate 1, and Postulate 2 imply Postulate 3. What this means is that whenever we use only the postulates in a proof of a theorem about points and lines and then interchange the words "point" and "line," and likewise the word "intersect" and the phrase "lie in," then this new statement can be proved by interchanging the same terms in the proof of the original theorem. This gives us the ***duality principle*** for projective planes.

Theorem 4.10. If a theorem is a consequence of Postulates 1 through 3 for projective planes, then it can be turned into another theorem by interchanging "point" with "line" and "intersect" with "lie in."

A theorem that is converted into a restatement of itself is called *self-dual*. For example, in Exercise 8 we state that "In a finite projective plane, the number of points that lie on a given line equals the number of lines that contain [intersect in] a given point." This statement is self-dual.

Projective planes on a finite set of points are symmetric balanced incomplete block designs, as we should have suspected; the proof of this fact will be smoother if we first present certain additional information about projective planes.

Basic Facts about Projective Planes

Theorem 4.11. For any two lines L_1 and L_2 in a projective plane, there is a one-to-one function from L_1 onto L_2.

Proof. By Postulate 3, there must be a point q not on L_1 or L_2, or else each of these has two points $x_1, y_1 \in L_1$ and $x_2, y_2 \in L_2$ not on the other. In this case, the line joining x_1 and x_2 and the line joining y_1 and y_2 intersect in a point q not on L_1 or L_2. For each point p on L_1, there is a line L_{pq} different from L_1 and L_2 containing p and q. Define $f(p)$ to be the unique intersection of L_{pq} and L_2. Then f is a function from L_1 to L_2. If p and r are different points of L_1, then to have $f(p) = f(r)$ would mean q and $f(p)$ lie on the same unique line that q and $f(r)$ lie on. But this line intersects L_1 in only one point and contains p and r as does L_1. Thus, if $f(p) = f(r)$, then $p = r$ because they are both points on the intersection of a certain line with L_1. Therefore f is one-to-one. If s is a point on L_2, then the line containing q and s must intersect L_1 in some point p; thus, $f(p) = s$, so f is onto. ∎

We have just shown that for each finite projective plane, any two lines have the same number of points. (We used the correspondence principle again.) Thus, there is a number n, called the *order* of the plane, such that all lines of the plane have $n + 1$ points. (The n versus $n + 1$ choice is purely technical and results in greater consistency between affine and projective planes later on.)

Theorem 4.12. Suppose we are given a point p in a projective plane and a line L not containing p. There is a one-to-one function from the set of lines containing p onto the set of points of L.

Proof. For each line L' containing p, define $f(L')$ to be the intersection of L and L'. If $f(L') = f(L'')$, then L' and L'' have two points (p and the common value of f) in common, therefore $L' = L''$. If q is on L, then p and q lie in a line L' containing p. Thus $q = f(L')$. ∎

From this theorem we can conclude that the number of points on any given line equals the number of lines containing any given point. Now we can explain the relationship between projective planes and block designs.

Projective Planes and Block Designs

Theorem 4.13. The lines of a projective plane of order n on a set V of points form an $(n^2+n+1, n^2+n+1, n+1, n+1, 1)$ balanced incomplete block design.

Proof. We have seen that each line has $k = n+1$ points, that each point appears in $r = n+1$ lines, and that each pair of points appears together in $\lambda = 1$ lines. Thus, the lines form a block design. Since $bk = vr$ and $r = k$, we get $b = v$. Because

$$r(k-1) = \lambda(v-1)$$

it follows that

$$(n+1)n = v-1,$$

or

$$v = n^2 + n + 1. \qquad \blacksquare$$

Theorem 4.14. The blocks of an $(n^2+n+1, n^2+n+1, n+1, n+1, 1)$ balanced incomplete block design with $n > 1$ on V form the lines of a projective plane on V.

Proof. We know that every pair of points lies in a unique block. Now suppose two blocks B_1 and B_2 do not intersect. Let x be a point in B_1. Then for each point y in B_2, x and y are in a unique block. Also if z is in B_2, the block containing x and y must be different from the block containing x and z because B_2 is the only block containing y and z. Thus, x would have to be in $n + 2$ blocks, one for each point of B_2 and B_1 as well. Since this is impossible, any two blocks must intersect. Now suppose we have two points x and y and let B_1 be the unique block containing them. There are more than $n + 1$ points, so there is another point z not in B_1. Let B_2 and B_3 be the blocks containing $\{z, x\}$ and $\{z, y\}$. Since $n \geq 2$, then $n^2 + n \geq 3n$, so $n^2 + n + 1$ is greater than the total number of distinct points in B_1, B_2, and B_3. Thus, there is a fourth point w not in any of these blocks. That is, there are four points, no three of which are in a block. $\blacksquare$

Example 4.3. The block design introduced in Examples 2.1 and 2.2 of this chapter is a $(7, 7, 3, 3, 1)$ design and so is a projective plane of order 2. This plane is shown schematically in Figure 4.3. Note that six of the lines are represented as line segments; the seventh is represented by the circle. $\blacksquare$

Figure 4.3

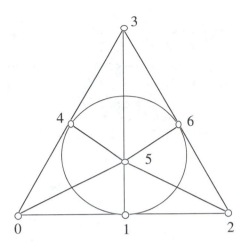

It turns out that there is at least one projective plane of order p^n for any prime number p and any positive integer n; further, no examples of projective planes with other orders are known. One of the exciting problems of twentieth century combinatorial mathematics has been whether there is a projective plane of order 10. A combination of deep mathematical work by many people and extensive computer work led to an announcement by C. W. Lam (of Concordia University in Montreal) in late 1988 that there is no projective plane of order 10.

Example 4.4. (For those who have studied vector spaces.) Let W be a three-dimensional vector space. Let P be the set of one-dimensional subspaces of W and let $\mathbf{L}$ be the set of two-dimensional subspaces of W. Then two elements of P lie in a unique member of $\mathbf{L}$. Further, since the dimension of the vector space sum of two members of $\mathbf{L}$ must be 3, the dimension of their intersection must be 1. Thus, two members of $\mathbf{L}$ intersect in a unique member of P. Three linearly independent vectors together with their sum generate four one-dimensional subspaces no three of which lie in the same two-dimensional space. Thus, P and $\mathbf{L}$ are the points and lines of a projective plane.

If you are familiar with abstract algebra, you know that any vector space is associated with a field and that any finite field has p^n elements for some prime p and some integer n. Further, there are $p^n + 1$ one-dimensional spaces contained in a given two-dimensional space. Thus, the parameters we get are the expected ones. ∎

Planes and Resolvable Designs

More examples of affine and projective planes may be obtained from sets of mutually orthogonal Latin squares. To see how this happens, we must analyze the designs associated with planes. Recall that we said one can visualize a projective plane by adding an imaginary line at infinity to an ordinary plane as in Figure 4.2. Our next theorem tells us that given an affine plane, it is possible to add an additional $n + 1$ points and one line containing them and insert the additional points into existing lines in such a way that we create a projective plane.

Theorem 4.15. Given a resolvable BIBD D on V with parameters $(n^2 + n, n^2, n + 1, n, 1)$, we may adjoin one additional block with $n + 1$ additional vertices and add these additional vertices to existing blocks to obtain a design D' with parameters $(n^2+n+1, n^2+n+1, n+1, n+1, 1)$.

Proof. Suppose D has been resolved into families $F_1, F_2, \ldots, F_{n+1}$ of blocks each family of which is a partition of the set V.

Add the $n + 1$ symbols $f_1, f_2, \ldots, f_{n+1}$ to V to get V', adjoin f_i to each block in F_i and let D' be the design on V' whose blocks are these enlarged blocks and the block $\{f_1, f_2, \ldots, f_{n+1}\}$. Then D' consists of V' together with a family of $(n + 1)$-element sets of V' such that each point in V' lies in $n + 1$ of these sets. Further, two points of V' are either both f_i's, in which case they lie in the added block (and only that one) or are both in V, in which case they are in exactly one of the blocks created by adding an f_i to a block of D, or else one of the points is in V and the other is an f_i. But each point in V is in exactly one block of F_i. Thus, in any case, these two points lie in exactly one block of D'. Therefore, we have constructed a balanced incomplete $(n^2 + n + 1, n^2 + n + 1, n + 1, n + 1, 1)$ design. ∎

Planes and Orthogonal Latin Squares

The connection between Latin squares and planes is a result of

Theorem 4.16. If there are $n - 1$ mutually orthogonal $n \times n$ Latin squares, then there is an $(n^2 + n, n^2, n + 1, n, 1)$ design.

Proof. Let V consist of the ordered pairs (i, j) with $1 \leq i \leq n$, $1 \leq j \leq n$. Suppose the n symbols in the Latin squares are $A_1, A_2, \ldots, A_n$ and let the Latin squares be given as matrices $M(1), M(2), \ldots, M(n - 1)$. Then let the following sets be blocks. (For the sake of intuition and an idea of how we might come up with these blocks, compare them to the sets of ordered pairs which formed the blocks in the design that resulted from Example 4.2.)

(Type 1) $\{(i, j) \mid i \text{ is fixed}, j = 1, 2, \ldots, n\}$ for each i
(Type 2) $\{(i, j) \mid j \text{ is fixed}, i = 1, 2, \ldots, n\}$ for each j

(Type 3) $\{(i,j) \mid M(k)_{ij} = A_h\}$ for each k and h.

Thus, we have $n^2 + n$ blocks, and since each A_h appears n times in $M(k)$, each block has size n. Each (i,j) occurs in $n+1$ blocks. Suppose we are given two ordered pairs (i,j) and (r,s). Assume they are both in two different blocks B_1 and B_2. If both B_1 and B_2 are type 3 blocks, and if B_1 comes from $M(k)$, then

$$M(k)_{ij} = M(k)_{rs} = A_t$$

for some t, and if B_2 comes from $M(h)$, then

$$M(h)_{ij} = M(h)_{rs} = A_u$$

for some u. Thus, the pair (A_t, A_u) appears twice among corresponding places in the Latin squares $M(k)$ and $M(h)$, and so $M(k)$ and $M(h)$ are not orthogonal, contradicting our assumption.

Now suppose B_1 is a type 1 block. Then $i = r$, so if B_2 is a type 3 block, then for some k, $M(k)_{ij} = M(k)_{is}$. Thus, $M(k)$ is not a Latin square, contradicting our assumption. If B_2 is a type 2 block, then $j = s$ and the two pairs are not distinct, while if B_2 is a type 1 block, then $B_2 = B_1$. All other cases for B_1 and B_2 are covered in a similar way. Thus, each pair (i,j) and (r,s) of distinct ordered pairs lies in at most one block. However, we have n pairs per block and $n^2 + n$ blocks, so among the blocks we will find

$$(n^2 + n)\binom{n}{2} = \frac{n^2(n+1)(n-1)}{2} = \frac{n^2(n^2-1)}{2} = \binom{n^2}{2}$$

two-element sets of pairs. Thus, since none of these two-element sets of ordered pairs occurs twice in different blocks, each two-element set of ordered pairs must occur exactly once among the blocks. Therefore, we have a block design. ∎

Example 4.5. From Corollary 1.5, we know that if n is a prime, there are $n - 1$ orthogonal $n \times n$ Latin squares. Applying Theorems 4.7, 4.8, 4.15 and 4.16, we see that, therefore, there are both an affine plane and a projective plane of order n for each prime number n. In particular, depending on which family of orthogonal Latin squares we choose, the affine planes that result may or may not be realizable as sets of ordered pairs, as our original examples were. ∎

Example 4.6. In Exercise 17 of Section 1 we constructed three mutually orthogonal 4×4 Latin squares. Therefore, there are an affine plane of order four and a projective plane of order four. ∎

EXERCISES

1. How many lines are there in the affine plane of ordered pairs of integers modulo 5? How many lines are in an equivalence class of parallel lines? How many of these parallel classes are there?

2. What are the five BIBD parameters of an affine plane of order 3?

3. What are the five BIBD parameters of a projective plane of order 3?

4. Explain why a block design with 16 varieties and 20 blocks of size 4 must be resolvable.

5. What are the parameters of a projective plane that arises from the design of Exercise 4?

6. Given three lines in a projective plane of order n, what is the maximum number of points you can find among these lines? What about four lines?

7. Answer the questions of the previous exercise for three lines in an affine plane of order n.

8. Explain why in a finite projective plane the number of points contained in a given line equals the number of lines containing a given point.

9. Under what circumstances will a Steiner triple system be a projective plane? Under what circumstances will a Steiner triple system be an affine plane?

10. Find a projective plane with 13 points and 4 points on a line.

11. Why would a symmetric $(21, 5, 1)$ design be a projective plane? Find one.

12. What are the parameters of the resolvable design associated with the plane of the previous exercise? Find such a resolvable design.

13. Can the blocks of a resolvable block design form the lines of a projective plane?

14. Prove Theorem 4.1.

15. Prove Theorem 4.2.

16. Prove Theorem 4.3.

17. Prove Theorem 4.4.

18. Prove Theorem 4.5.

19. In the proof of Theorem 4.16, explain how to deal with the cases where B_1 is of type 2 and B_2 is of type 3.

20. Prove Theorem 4.6.

21. Construct a resolvable design on four vertices using a single 2×2 Latin square.

22. How does Exercise 30 in Section 2 relate to affine planes?

23. Is Exercise 7 in Section 3 related to affine planes?

24. Show that the removal of a line and all its points from a projective plane yields an affine plane.

25. Construct the design and then the projective plane associated with two orthogonal 3×3 Latin squares.

26. Repeat Exercise 25 with three orthogonal 4×4 Latin squares.

Section 5 Codes and Designs

The Concept of an Error-Correcting Code

When messages are sent from one computer to another—for example, from an automatic teller machine to a bank's central computer—they are sent in a digital form, that is, as a sequence of digits. The digits are usually just 0 and 1, but they might be 0–9, or they might be the alphabet letters a–z, together with 0–9 and some punctuation. Changing even one digit in a message could have a significant impact—for example, a withdrawal of 100 dollars could become a withdrawal of 1100 dollars or a withdrawal of zero dollars. This problem has led people to design schemes of encoding the messages in such a way that if the received message contains an error the receiver will be able to determine that there has been an error and perhaps even correct the error. For example, if we encode our message in such a way that the number of nonzero digits is even, then a receiver that receives a message with an odd number of nonzero digits will know that the received message is incorrect.

There is a standard way to encode a message which is a sequence of $n-1$ zeros and ones into an n-tuple of zeros and ones with an even number of ones. If the original $n-1$ digit sequence itself has an even number of ones, we let digit n be zero, but if the original sequence has an odd number of ones, we let digit n be one. This system, called an overall parity check, can be expressed by the equation

$$x_n \equiv x_1 + x_2 \cdots + x_{n-1} \pmod 2$$

or the equivalent equation

$$x_1 + x_2 \cdots + x_{n-1} + x_n \equiv 0 \pmod 2.$$

(The equivalence is an elementary fact about arithmetic modulo 2.) This gives us a quick way to detect any single error that may occur when a message is transmitted.

How, though, could we *correct* an error in the received message? As one example, suppose our messages are 12-tuples of zeros and ones and we know that a possible message has 0, 3, 6, 9, or 12 nonzero digits. If a received message has one nonzero digit and one error has occurred, the message sent must have been the all-zeros message, because two or more errors would be required to distort a message with 3, 6, 9, or 12 digits into a sequence with one nonzero digit. This suggests a scheme for actually correcting a single error. If the received message has the wrong number of nonzero entries, experiment with changing each digit (and changing it back again afterward) until you find a single digit you can change to get a legal message. Once you find such a legal message, conclude that it was the message sent.

Hamming Distance

In a sense our decoding scheme tells us to choose the legal message "closest" to our received message in order to decode the received message. The idea of Hamming distance makes this concept more precise. The **Hamming distance** between two n-tuples is the number of coordinates in which they are different. We use $d(x, y)$ to stand for the Hamming distance between two n-tuples x and y.

Example 5.1. $d\big((1,0,0,1),(1,1,0,0)\big) = 2$, $d\big((0,0,0,0),(1,0,1,1)\big) = 3$, and $d\big((a,b,d,e,c),(a,b,c,c,e)\big) = 3$. ∎

The system of experimentally changing one digit at a time amounts to saying "Find out whether there is a legal message at a distance of one from the received message and decode to that legal message if this is the case." A slight generalization of this principle is "Find the legal message whose Hamming distance to the received sequence is smallest and decode to that message." This is called *minimum distance decoding.* It is up to the designer of the encoding scheme to determine which messages are legal. Ideally the scheme will be designed so that there is one closest legal coded message to each possible received message. It is standard to call the set of legal messages a **code** and to call the legal messages **codewords**. We think of the possible entries of the messages as an **alphabet** and refer to the n-tuples of alphabet symbols as **words** of length n.

It is natural to think of the relationship between messages to be sent and the codewords as being similar to the Morse code, which converts each letter to a few dots and dashes and then converts words to longer sequences of dots and dashes. However, it is up to the designer of the communication system to assign actual messages to codewords. For example, a telephone company might use the scheme

$(0,0,0) \leftrightarrow$ The number you have reached is not in service . . .

$(0,1,1) \leftrightarrow$ All circuits are busy now . . .

$(1,0,1) \leftrightarrow$ To call the number you have dialed, you must first dial a one.

$(1,1,0) \leftrightarrow$ We are unable to complete your call, please . . . .

Thus, in a context with a small number of possible messages, each codeword might represent an entire message. In other contexts—say banking transactions—some messages would likely require several codewords since it would be difficult to design an efficient code large enough to encompass all possible messages. Since the scheme used to relate actual messages to codewords does not influence the code's error-correcting capability, we concentrate our study on the codes themselves rather than on encoding schemes. There are two conflicting desires we might have in designing a code. First, we would like to be able to correct as many errors as possible.

Example 5.2. The two-word code consisting of

$$\{(0,0,0,0,0,0,0,0,0,0,0,0,0), (1,1,1,1,1,1,1,1,1,1,1,1,1)\}$$

will allow us to send one of two messages and correct up to six errors by choosing the codeword closest to the received word. ∎

Of course, if we want to send more than two messages, this code will not be useful. To send more messages, we need more words in our code, and as we put more words into a code we are likely to reduce the minimum distance between codewords, so the goal of correcting many errors conflicts with the goal of sending many messages. Thus, we are interested in two parameters, the number of messages we can send—the size of the code—and the error correction capability. It is clear from the examples we have seen that the ability to correct errors is related in a direct way to the distances between codewords.

> *Theorem 5.1.* If there is a positive integer e such that the minimum distance between two codewords is $d = 2e + 1$, then the system of choosing the codeword closest to a received word will allow us to determine the codeword x from the n-tuple x' if x' is obtained from x by introducing e or fewer errors.
>
> *Proof.* We need to use the fact (whose proof is an exercise) that for any three n-tuples x, y, and z,

$$d(x, z) \leq d(x, y) + d(y, z),$$

the so-called ***triangle inequality***. (The inequality gets its name since we can visualize it in ordinary geometry by thinking of x, y, and z as the vertices of a triangle.) Now we show that if a word x' differs from some codeword x in e or fewer places, then x' is closer to x than any other codeword is. For this purpose note that, for any other codeword z, $d(x, z) \geq 2e + 1$. Now since $d(x, x') \leq e$, we have the inequalities

$$2e + 1 \leq d(x, z) \leq d(x, x') + d(x', z) \leq e + d(x', z).$$

Subtracting e from the first and last terms tells us that $d(x', z) \geq e + 1$, so that x is the only codeword at distance e or less from x'. ∎

There are many systems for building codes to correct errors. These codes are designed so that we can send a reasonably large number of different messages and also correct a reasonably large number of errors. Most often, constructing these codes and showing they have the desired properties makes heavy use of linear algebra and abstract algebra. One application of the resulting codes is sending messages from satellites in space to earth. For example, the original black and white pictures from satellites circling Mars were divided by computers on the satellites into small squares, and each square was assigned to a sequence of 0's and 1's that indicated how light or dark the square was. This sequence was then encoded by an error-correcting code and transmitted to earth. While static in the transmission often changed the codeword before it was received, the error correcting capability of the code allowed engineers to recover the original picture, or at least something close to the original picture. Error correcting codes are also used to encode the digital version of music and pictures stored on compact disks and digital video disks, so that if the disks get dirty or scratched, the original sound or image can still be recovered by the machine that plays the disk.

Perfect Codes

Now suppose we wish to design a code that uses codewords of length n and corrects as many as e errors. To determine how many different messages we can send, we ask "What is the maximum possible number of codewords we can have?" We use K to stand for the *size* of the code, that is, for the number of codewords. (This is not consistent with the notation we have been using for integers, but it is the notation usually used for codes to emphasize its relationship with another number, usually denoted by k which we shall see later.) In symbols, we are asking, for a given n and e, "How large can we make K?"

The set of words at a distance e or less from a codeword x is called the *sphere* of radius e around x. To say that our minimum-distance decoding scheme allows us to correctly decode any word with e or fewer errors is to say that, for any two distinct codewords x and y, the sphere of radius e around x and the sphere of radius e around y have no words in common.

Now if our alphabet has q symbols, the number of words at distance one from x is $(q-1)n$ (we can change any one of the n positions of x to be any of the $q-1$ other symbols). Thus, the number of words in a sphere of radius one is $(q-1)n + 1$ since we must include x in the sphere. To get the number of words in a sphere of radius two, we add to this the number of words at distance two from x. This is the number, $\binom{n}{2}$, of ways to choose two places of x to change times the number $(q-1)^2$ of ways to make both these changes. Continuing this argument gives us our next theorem.

Theorem 5.2. The number of words in a sphere of radius e around a word over an alphabet with q symbols is

$$1 + \binom{n}{1}(q-1) + \cdots + \binom{n}{e}(q-1)^e.$$

Proof. Outlined above. ∎

What does the size of the sphere of radius e have to do with the number K of codewords? Since the spheres don't overlap, K times the size of a sphere is the total number of words in the spheres around codewords and so is no more than the number of words (n-tuples) possible. This gives an inequality known as the "sphere packing bound" on K, namely

$$K \sum_{i=0}^{e} \binom{n}{i}(q-1)^i \le q^n.$$

A code which made perfect use of the spheres around codewords would be one in which no words were wasted, that is, one in which

$$K \sum_{i=0}^{e} \binom{n}{i}(q-1)^i = q^n.$$

Codes that satisfy this equation are called **perfect codes**.

Since most digital computers deal with information in sequences of zeros and ones, it is natural to ask whether there are any *binary* perfect codes, that is, codes with $q = 2$. If a K-word perfect binary code of length n corrects e errors, then we have the equation

$$K\left(1 + \binom{n}{1} + \binom{n}{2} + \cdots + \binom{n}{e}\right) = 2^n.$$

We see immediately that K and the sum of binomial coefficients must be powers of two. Thus, we ask for what values of n and e the sum of binomial coefficients is a power of two. In Table 5.1 we show the result of experimenting with various values of n and e, computing the sum $\sum_{i=0}^{e} \binom{n}{i}$.

Table 5.1

$e \backslash n$	1	2	3	4	5	6	7	8	9	10	11	12	13	14	15
1	2	3	4	5	6	7	8	9	10	11	12	13	14	15	16
2	2	4	7	11	16	22	29	37	46	56	67	79	92	106	121
3	2	4	8	15	26	42	64	93	130	176	232	299	378	470	576
4	2	4	8	16	31	57	99	163	256	386	562	794	1093	1471	1941

When n is small the powers of 2 in the table are giving us information about rather trivial codes. For example, with $e = 3$ and $n = 3$ we see that

$$1 + \binom{3}{1} + \binom{3}{2} + \binom{3}{3} = 8;$$

however, the equation

$$K\left(1 + \binom{3}{1} + \binom{3}{2} + \binom{3}{3}\right) = 2^3$$

gives $K = 1$. Thus, for $e = 3$ and $n = 3$ our code would have one word. A one-word code has little relationship to the correction of errors, because, with a one-word code, only one message can be sent. Therefore, no matter which places are wrong (even all three) we will know which message has been sent! In fact, for the minimum distance between codewords to be $2e + 1$, we must have $n \geq 2e + 1$ to have a nontrivial example. The cases $e = 1$, $n = 3$, $e = 2$, $n = 5$, $e = 3$, $n = 7$, and $e = 4$, $n = 9$ all give $K = 2$, so we have a two-word perfect code consisting of the all-zero word and the all-one word, so these codes are just barely nontrivial! Thus, for $e = 1$ we see $n = 7$ and 15 could lead to a nontrivial example with more than two words. None of the other combinations of e and n can lead to perfect codes. Noting that 3, 7, and 15 have the form $2^m - 1$ leads us to check that

$$1 + \binom{2^m - 1}{1} = 1 + 2^m - 1 = 2^m,$$

so that whenever $n = 2^m - 1$ we may be able to find a perfect single-error-correcting code of length n.

How, though, do we go about trying to find such a code? By assuming that $n = 2^m - 1$ and $1 + \binom{n}{1} = 2^m$, we get $K \cdot 2^m = 2^{2^m - 1}$ so that $K = 2^{2^m - m - 1}$. Thus, we know how many codewords we must have, how long they must be, and that, in order to have a perfect single-error-correcting code, the minimum distance between them must be 3. For $n = 3$ we get $K = 2$, and since the minimum distance must be 3, we have $\{(0, 0, 0), (1, 1, 1)\}$ as the only possible code. For $n = 7$, such ad hoc methods are less likely to suffice. Thus, we investigate how we might systematically construct a code.

Linear Codes

One systematic method was introduced when we considered the code of words of length n with an even number of nonzero entries. This code is the set of all solutions of

$$x_1 + x_2 + \cdots + x_n \equiv 0 \pmod{2}$$

which we now write simply as

$$x_1 + x_2 + \cdots + x_n = 0.$$

Since the all-zeros word and a word with two ones satisfy this equation, we know that the equation does not describe our hoped-for perfect single-error-correcting code (because the code it describes contains words a distance two apart). Perhaps another equation or a system of equations would do so instead. This equation came from the instruction to compute x_n by adding x_1 through x_{n-1} modulo 2. The digit x_n is called a **parity check bit**. Note that if n is 7, we have 2^6 words in the code consisting of words satisfying the equation, one word for each way of choosing x_1 through x_6. Even if we had some other equation for computing x_n, we would still have 2^6 words in the code consisting of all words satisfying the equation. Since, for the perfect code we want, the equation

$$K(1 + 7) = 2^7$$

gives $K = 2^4$, we will not be able to describe our code by just one equation.

If we had an equation to compute x_6 from x_1 through x_5 and one equation to compute x_7 from x_1 to x_5, then we would have 2^5 words in our code, one for each choice of x_1 to x_5.

With three equations, one for x_5, one for x_6, and one for x_7, all in terms of x_1 through x_4, we would get 2^4 codewords. Our three equations will have the form

$$a_1 x_1 + a_2 x_2 + a_3 x_3 + a_4 x_4 = x_5$$
$$b_1 x_1 + b_2 x_2 + b_3 x_3 + b_4 x_4 = x_6$$
$$c_1 x_1 + c_2 x_2 + c_3 x_3 + c_4 x_4 = x_7,$$

or

$$a_1 x_1 + a_2 x_2 + a_3 x_3 + a_4 x_4 + x_5 \qquad\qquad = 0$$
$$b_1 x_1 + b_2 x_2 + b_3 x_3 + b_4 x_4 \qquad + x_6 \qquad = 0$$
$$c_1 x_1 + c_2 x_2 + c_3 x_3 + c_4 x_4 \qquad\qquad + x_7 = 0.$$

These equations can be expressed in matrix form as

$$\begin{pmatrix} a_1 & a_2 & a_3 & a_4 & 1 & 0 & 0 \\ b_1 & b_2 & b_3 & b_4 & 0 & 1 & 0 \\ c_1 & c_2 & c_3 & c_4 & 0 & 0 & 1 \end{pmatrix} \begin{pmatrix} x_1 \\ x_2 \\ x_3 \\ x_4 \\ x_5 \\ x_6 \\ x_7 \end{pmatrix} = 0.$$

So that we may think of codewords as 7-tuples written horizontally, we write this matrix equation as

$$HX^t = 0.$$

The matrix H is called the parity check matrix of the code; the code is the set of all 7-tuples satisfying the matrix equation $HX^t = 0$. If we think of the matrix H as having columns C_1 through C_7, we may write

$$(C_1 \quad C_2 \quad C_3 \quad C_4 \quad C_5 \quad C_6 \quad C_7) X^t = 0,$$

or

$$x_1 C_1 + x_2 C_2 + x_3 C_3 + x_4 C_4 + x_5 C_5 + x_6 C_6 + x_7 C_7 = 0.$$

Now one 7-tuple in the code is 0, the all-zeros 7-tuple. Thus, we want no tuples with one or two nonzero entries in the code since their Hamming distance from 0 would be one or two. The **weight** of a word is the number of nonzero entries it has; we've just said we want no words of weight one or two. How can we select the columns of H to ensure this happens? Note that if we have a word in the code of weight 1 whose only 1 is in position i, then our column form of the matrix equation becomes

$$x_i C_i = 1 C_i = 0.$$

This can't happen if column C_i is not zero. Thus, by choosing our parity check matrix so that it has no all-zero columns, we will ensure that the code has no words of weight one.

Now note that if we have a word of weight two in the code whose only ones are in positions i and j, then our column form of the matrix equation becomes

$$x_i C_i + x_j C_j = C_i + C_j = 0,$$

or, since we are carrying out our arithmetic modulo 2,

$$C_i = C_j.$$

Thus, to ensure that there are no words of weight two in the code we must choose a parity check matrix that does not have two equal columns, that is, a matrix whose columns are all different.

The Hamming Codes

It turns out that we have almost completely described our matrix. There are eight ways to choose a column of three zeros and ones, one of which is the all-zeros column. Thus, the seven columns of the matrix must be the seven different nonzero column vectors and our only choice is in what order we write the first four columns down. Thus, we settle on an order and choose

$$H = \begin{pmatrix} 1 & 1 & 1 & 0 & 1 & 0 & 0 \\ 1 & 1 & 0 & 1 & 0 & 1 & 0 \\ 1 & 0 & 1 & 1 & 0 & 0 & 1 \end{pmatrix}.$$

The set of 7-tuples X such that $HX^t = 0$ is our code. It is known as the 4-dimensional Hamming code of length 7. R. W. Hamming introduced the systematic study of linear codes in a series of papers appearing around 1950.

Given an $n - k$ by n matrix H whose last $n - k$ columns form an identity matrix, we call the set of zero–one n-tuples X such that $HX^t = 0 \pmod 2$ a **binary linear code** of **dimension k**. The matrix H is called a **parity check matrix for the code.** A binary linear code of dimension k has 2^k members, one for each choice of the first k entries of a codeword. A p-ary linear code is defined similarly with matrices whose entries are integers modulo p. (For those familiar with fields of order q, q-ary linear codes are defined similarly with matrices whose entries are chosen in a field of order q.) If M is an m by n matrix whose columns may be rearranged to make the last m columns equal to the identity, the set of n-tuples X such that $MX^t = 0$ is still called a linear code; in this book we will not need to be this general.

We have not yet demonstrated that our Hamming code of length seven is perfect, because we have not yet shown that its minimum distance is three. However, it is somewhat easier to compute the minimum distance of a linear code than an arbitrary code. Notice that if X and Y are codewords so that $HX^t = 0$ and $HY^t = 0$, then $0 = HX^t - HY^t = H(X^t - Y^t)$ so that $X - Y$ is in the code. Note that if X and Y are different in d places, then $X - Y$ is nonzero in d places (and is therefore different from 0, a codeword, in d places). This proves our next theorem.

Theorem 5.3. The minimum distance of a linear code is the minimum of the weights of nonzero words in the code.

Proof. Given before the statement of the theorem. ∎

Example 5.3. What is the minimum distance of the 4-dimensional Hamming code of length 7?

We have designed the code so that it has no words of weight one or two. However,

$$H \begin{pmatrix} 0 & 1 & 1 & 1 & 0 & 0 & 0 \end{pmatrix}^t = 0,$$

so the code has a word of weight 3. Thus by our theorem, the code has minimum distance three. ∎

The preceding example completes the proof that the binary Hamming code of length 7 is perfect. The Hamming code we constructed is one member of a family of single-error-correcting codes. The **binary Hamming code** of length $2^m - 1$ is the code whose parity check matrix has all the nonzero column vectors of length m for its columns. Clearly there are several orders in which we could write the columns down; we call the code a Hamming code regardless of the order.

Theorem 5.4. The binary Hamming code of length $2^m - 1$ has dimension $2^m - m - 1$ and minimum distance 3.

Proof. Since the columns of the parity check matrix are distinct and nonzero, there are no codewords of weight one or two. It is straightforward to find a codeword of weight 3. Since there are $2^m - 1$ columns and we have chosen to write the parity check matrix with the m columns of the identity matrix last, the code has dimension $2^m - m - 1$. ∎

Theorem 5.5. The binary Hamming code of length $2^m - 1$ is perfect.

Proof. The number of codewords is

$$2^{2^m - m - 1}$$

and so

$$K \cdot \left(1 + \binom{2^m - 1}{1}\right) = 2^{2^m - m - 1} \cdot 2^m = 2^{2^m - 1};$$

therefore, the code is perfect. ∎

There are nonbinary Hamming codes as well; their description is a natural extension of the way in which we described the binary Hamming codes. These codes are perfect as well. There are only two other perfect linear codes with more than two codewords. These fascinating codes, discovered by Marcel Golay, are described in detail in the books by Sloane and MacWilliams and by Pless in the Suggested Reading.

Constructing Designs from Codes

While the subject of perfect codes is clearly interesting in its own right, it is also interesting from the point of view of designs. For each word w in a code, the **support** of w is the set of positions in which the word is nonzero.

Example 5.4. What are the supports of the codeword $(0, 1, 1, 1, 0, 0, 0)$ and the codeword $(1, 0, 0, 0, 1, 1, 1)$ in the Hamming code of length 7?

The support of the first word is $\{2, 3, 4\}$ and the support of the second word is $\{1, 5, 6, 7\}$. ∎

Given a code, the support sets of codewords form a design on the set of coordinate positions. The codewords of a binary code may be thought of as the characteristic functions, written as n-tuples, of the blocks. Thus, in a sense the theory of binary codes is a branch of the theory of designs, and vice versa. The design that corresponds to all the codewords has several different block sizes. Frequently the words of minimum weight form an especially interesting design, occasionally a BIBD.

Example 5.5. Show that the supports of words of weight 3 in the Hamming code of length 7 form a BIBD.

We must show that for each i and j there are the same number λ of coordinates k such that $\{i, j, k\}$ is the support of a codeword. Thus, given i and j, we want to find λ words of weight 3 that are nonzero in positions i and j. Now the seven-tuple with ones in positions i and j and zeros elsewhere is a word w_{ij} of weight two and therefore not in the code. Now w_{ij} must be in one and only one sphere of radius one around a codeword w. Since w_{ij} differs from w in one position, this codeword w must have weight three, and w must be the only word of weight 3 that is nonzero in positions i and j. Thus, given i and j, there is one and only one three-element set $\{i, j, k\}$ that is the support of a codeword, so i and j appear together in $\lambda = 1$ support sets. Thus, our code is a $(7, 3, 1)$ design—which we know is a projective plane of order 2. ∎

Highly Balanced Designs

The designs associated with codes often show an even higher degree of balance.

Example 5.6. Let E be the code obtained from the Hamming code of length seven by adjoining to each word $(x_1, x_2, x_3, x_4, x_5, x_6, x_7)$ one more digit x_8 which is the sum of x_1 through x_7 (mod 2). (This gives a code of length 8 called the extended Hamming code.) Let D be the design whose blocks are the support sets of words of weight 4 in the code. Show that for each *three* elements of $\{1, 2, 3, 4, 5, 6, 7, 8\}$ there is a block of the design in which they lie together. Show that each two elements lie together in a fixed number λ_2 of blocks. What is λ_2?

We show that each set $\{i, j, k\}$ with $i < j < k$ lies in a unique block of the design. We deal with the cases $k = 8$ and $k < 8$ in different ways. Suppose $k < 8$, and let u be the word of length seven with ones in positions i, j, and k and zeros elsewhere. Then u is in a sphere of radius one around a codeword w of the perfect Hamming code of length seven. Thus, either $u = w$ or u is a distance one from w so that w has weight four. If $u = w$, then adjoining $u_8 = 1$ to u gives a word u' (in the code E) of weight 4. If

$u \neq w$, then w is a word of weight 4 and adjoining $w_8 = 0$ to w gives a word w' of weight 4 in E. But the support of u is in the support of u' or w', so $\{i, j, k\}$ is in the support set of a codeword. Now if the support of u were in the support of two different codewords of weight 4, their sum would be a word of weight 2 in the extended Hamming code, an impossibility. This shows that for each three elements of $\{1, 2, 3, 4, 5, 6, 7\}$ there is exactly one block of the design contianing these three elements.

Now in the case $k = 8$, we let w be the word of weight 3 of the Hamming code of length seven whose support contains i and j, and we let w' be the word in E obtained by adjoining $w_8 = 1$ to w so that w' has weight 4. Then $\{i, j, k\}$ is in the support of w', and as before w' is the unique codeword whose support contains $\{i, j, k\}$.

Finally, each two-element subset of $\{1, 2, 3, 4, 5, 6, 7, 8\}$ is contained in six three-element subsets, each of which is contained in a different block of the design of the code. Therefore, each two-element set is contained in six blocks of the design, giving $\lambda_2 = 6$. ∎

t-Designs

A design D is called a t-(v, k, λ) design if it consists of k-element subsets of a v-element set V such that each t-element subset of V is contained in exactly λ blocks of the design. Thus, a 2-(v, k, λ) design is a BIBD. The design associated with the extended Hamming code of length eight is both a 3-$(8, 4, 1)$ design and a 2-$(8, 4, 6)$ design. A t-design with $\lambda = 1$ is called a Steiner system of type $S(t, k, v)$ or simply an $S(t, k, v)$. The Steiner triple systems we have already studied are Steiner systems of type $S(2, 3, v)$. The 3-design associated with the extended Hamming code is an $S(3, 4, 8)$.

By using the methods we used for the Hamming code of length seven, we can prove the following.

> **Theorem 5.6.** Let C be a perfect code of minimum distance $d = 4j + 3$ and length n. Let $\hat{C}$ be the code formed from words in C by adjoining an overall parity check. Then the supports of words of C of weight d form an $S(2j + 2, d, n)$ Steiner system and the supports of words of weight $d + 1$ in $\hat{C}$ form an $S(2j + 3, d + 1, n + 1)$ Steiner system.
>
> *Proof.* Similar to the proof for the Hamming code of length 7. ∎

There are many other circumstances where the codewords of a given weight in a code must form a t-design; thus codes are a good source of highly balanced designs. The relationship between codes and designs also gives codes from designs. We can take the columns of the incidence matrix of a design as the code, or we can take the columns along with some or all the linear combinations of the columns. If we take all the linear combinations of these columns we get a linear code; otherwise we get a nonlinear code.

Theorem 5.7. The code consisting of the columns of the incidence matrix of a t-$(v, k, 1)$ design has length v, contains $\binom{v}{t}/\binom{k}{t}$ codewords and has minimum distance at least $2(k - t + 1)$.

Proof. Each t-element subset of the variety set V of the design lies in a unique k-element block of the design. Thus, the number of pairs (B, T) where B is a block and T is a t-element subset of that block is simply the number of t-element subsets of V, or $\binom{v}{t}$. But this number of pairs is also $b\binom{k}{t}$. (Why?) Therefore, $b = \binom{v}{t}/\binom{k}{t}$. We leave the proof that the minimum distance is at least $2(k - t + 1)$ as an exercise. ∎

Codes and Latin Squares

We might expect that the relationship between codes and designs would give a relationship between codes and orthogonal Latin squares. In fact, there is a direct relationship between single-error-correcting codes and Latin squares which makes it exceptionally easy to construct certain families of orthogonal Latin squares.

Theorem 5.8. There is a code of length $m + 2$, minimum distance $m + 1$, and with q^2 words over an alphabet with q symbols if and only if there are m mutually orthogonal $q \times q$ Latin squares.

Proof. Suppose we have m mutually orthogonal $q \times q$ Latin squares, $S^{(1)}, S^{(2)}, \ldots, S^{(m)}$. Suppose each square uses the symbols $1, 2, \ldots q$. Then the array

$$
\begin{array}{ccccccccc}
1 & 1 & 1 & \cdots & 1 & 2 & 2 & 2 & \cdots & q \\
1 & 2 & 3 & \cdots & q & 1 & 2 & 3 & \cdots & q \\
S_{11}^{(1)} & S_{12}^{(1)} & S_{13}^{(1)} & \cdots & S_{1q}^{(1)} & S_{21}^{(1)} & S_{22}^{(1)} & S_{23}^{(1)} & \cdots & S_{qq}^{(1)} \\
& & & \vdots & & & & & & \\
S_{11}^{(m)} & S_{12}^{(m)} & S_{13}^{(m)} & \cdots & S_{1q}^{(m)} & S_{21}^{(m)} & S_{22}^{(m)} & S_{23}^{(m)} & \cdots & S_{qq}^{(m)}
\end{array}
$$

consisting of the rows R, C, and the linear representations (as studied in Section 1) of the squares has q^2 columns. These are the (transposes of) words of the associated code.

Because the rows are orthogonal, two columns cannot be equal in two different positions, as we shall explain. First, observe that they can't be equal in row 1 and row 2 by construction, or row 1 and row i for $i > 2$ by the Latin square property. Similarly, they can't be equal in row 2 and any other row by the Latin square property. If they were equal in row i and row j, for $i, j > 2$, then we would have $S_{hk}^{(i-2)} = S_{rs}^{(i-2)}$ and $S_{hk}^{(j-2)} = S_{rs}^{(j-2)}$ for some h k, r, and s. If $h = r$ or $k = s$ this would violate the Latin square property; otherwise this would violate the orthogonality of square $i-2$ and square $j-2$. Thus, any two columns are different in at least $m + 1$ places.

For the converse, list the words of the code as columns of a matrix. Replace the q symbols by the numbers 1 through q and sort the columns lexicographically. You can use the fact that the minimum distance is $m + 1$ to show that any two rows of the resulting array are orthogonal. Thus, by Theorem 1.10 of Section 1 of this chapter, the array consists of the linear representations of m mutually orthogonal Latin squares. ∎

Example 5.7. Give another proof that there are $p - 1$ mutually orthogonal $p \times p$ Latin squares.

We consider the two $p + 1$-tuples

$$(1, 0, 1, 2, 3, \ldots, p - 1) = v$$
$$(0, 1, 1, 1, \ldots, 1) = u.$$

The set of all linear combinations—that is, sums of the form $iu + jv$—of these $(p + 1)$-tuples (mod p) has p^2 different $(p + 1)$-tuples in it. This set is our code. (It is in fact a linear code as those who have studied linear algebra should note.) The distance between two codewords is the weight of their mod p difference. Thus, the minimum distance of the code is the minimum number of nonzero elements in a vector of the form $iu + jv$.

Now if i or j is zero, $iu + jv$ has weight $q - 1$. Otherwise, the first two entries of $iu + jv$ are not zero and each other entry has the form

$$i \cdot 1 + j \cdot k = i + jk.$$

But if $i + jk \equiv 0 \pmod{p}$ then $jk \equiv p - i \pmod{p}$. However, by Exercise 8(b) of Section 1, there is an integer j' such that $j'j \equiv 1 \pmod{p}$. Therefore, $k \equiv j'(p - i) \pmod{p}$, and since k is between 0 and $p - 1$, this determines k completely. Therefore, there is at most one position of $iu + jv$ which is zero (mod p) and so $iu + jv$ has weight at least p. Thus, by applying Theorem 5.8 to this code, we obtain $q - 1$ mutually orthogonal Latin squares. (This construction may be regarded as a variation of the construction used in Section 1, so our theorem above is a generalization of the techniques used in Section 1 to construct orthogonal Latin squares.) ∎

EXERCISES

1. Find the Hamming distances between each pair of the following words: $(1, 1, 0, 0, 1, 1, 1)$, $(0, 0, 0, 0, 0, 0, 0)$, $(1, 1, 1, 0, 0, 0, 0)$, $(0, 0, 1, 0, 1, 1, 1)$.

2. Write down a parity check matrix for the Hamming code of length 15.

3. Write down the 16 codewords of the Hamming code of length seven.

4. Write down the 16 codewords of the extended Hamming code of length 8.

5. Write down a matrix H' which has a row of ones and has the property that the extended Hamming code of length eight consists of the 8-tuples X such that $H'X^t=0$. Explain why H' is not, technically speaking, a parity check matrix for the code. Perform elementary row operations on H' to convert it to a parity check matrix.

6. Table 5.1 shows that for $e = 2$ and $n = 5$, $1 + n + \binom{n}{2}$ is a power of two. Describe the very simple perfect code which corresponds to this observation.

7. Find an $n > 5$ such that $1 + 2n + 4\binom{n}{2}$ is a power of three. What kind of perfect code does this suggest that we should look for? (Marcel Golay looked for and found such a linear code.)

8. Consider the word $w = (0, 1, 0, 1, 0, 1)$.
   (a) Write down the members of the sphere of radius one around this word.
   (b) Write down the members of the sphere of radius two around this word.

9. There are seven three-element sets of columns of the parity check matrix of the Hamming code of length 7 which sum to the column of zeros. Find them and the words of the Hamming code that correspond to them.

10. (a) What is the maximum number of nonzero, length-three column vectors of integers modulo three we can find so that no one is a multiple of the other? If you use these as the columns of a parity check matrix of a linear code over the integers modulo three, what are its length, dimension, and minimum distance? How many codewords are there?
    (b) Answer the same questions for length n column vectors.

11. Find the support sets of each of the following words: $(1, 1, 0, 0, 1, 1, 1)$, $(1, 1, 1, 0, 0, 0, 0)$, $(0, 0, 1, 0, 1, 1, 1)$, $(0, 0, 0, 0, 0, 0, 0)$, $(1, 1, 1, 1, 1, 1, 1)$.

12. Write down the support sets of words of weight 3 in the Hamming code of length seven. Verify directly that they form a $(7, 3, 1)$ design.

13. Write down the support sets of words of weight 4 in the Hamming code of length 7. Do they form a BIBD? What are its parameters if they do (and why not if they don't)?

14. Find the block of the design of the extended Hamming code in which 1, 2, and 3 lie.

15. Show that a t-(v, k, λ) design is an i-(v, k, λ_i) design for each $i < t$. Compute λ_i in terms of t, v, k, and λ.

16. Prove that in a t-(v, k, λ) design,

$$b\binom{k}{t} = \lambda\binom{v}{t}$$

so that $b = \lambda\binom{v}{t}/\binom{k}{t}$.

17. Prove the triangle inequality for Hamming distance, namely that

$$d(x, z) \le d(x, y) + d(y, z).$$

18. Prove Theorem 5.6.

19. Prove the statement about minimum distance in Theorem 5.7.

20. Show that, in a binary linear code, either half the codewords have even weight or all the codewords have even weight.

21. Note that $6^5(1 + \binom{7}{1} \cdot 5) = 6^7$.
    (a) What would the parameters, including alphabet size, of a perfect code corresponding to this equation be?
    (b) Show that if there is a code with these parameters, then there is a code with 36 words, length 4, and minimum distance 3. (Hint: How many words of the original code do you get if you consider all codewords whose first three entries have some specified values?)
    (c) Explain why there is no perfect code with the parameters given in part (a).

22. (Golay) (a) Find an n such that

$$2^{78}\left(1 + \binom{n}{1} + \binom{n}{2}\right) = 2^{90}.$$

    (b) What must the length, dimension, and minimum distance of a perfect code arising from these parameters be?
    (c) Show that if there is a linear perfect code with these parameters, then its parity check matrix H is a 12 by 90 matrix with nonzero

columns such that no two, three, or four of its columns sum to zero modulo two.

(d) Show that if Y and Z are words of weight 1 or 2 and H is a matrix of the type described in part (c), then $HY^t \neq HZ^t$.

(e) Suppose that there is a matrix of the type described in part (c) and that r of its columns have an odd number of ones. How many column vectors HX^t have odd weight if X is a 90-tuple of weight one? How many column vectors HY^t have odd weight if Y is a 90-tuple of weight two?

(f) Show that if there is a perfect linear code with the parameters you gave in part (a), then

$$r(91 - r) = 2^{11}$$

and explain why this means there is no such code.

23. A code is *self-orthogonal* if for each pair of codewords w and v, thought of as row matrices, $wv^t = 0$. We say w and v are orthogonal when $wv^t = 0$. A code is self-dual if every word orthogonal to all codewords is in the code.

(a) Show that the extended Hamming code of length 8 is self-dual.

*(b) Is the extended Hamming code of length 2^m self-dual or self-orthogonal (or neither) for $m > 3$?

24. Show that if there is a Steiner system of type $S(t, k, v)$, then there is a Steiner system of type $S(t - 1, k - 1, v - 1)$.

Suggested Reading

Colbourne, Charles J. and Jeffrey H. Dinitz, eds. 1996. *The CRC Handbook of Combinatorial Designs.* Boca Raton, FL: CRC Press.

Hall, Marshall. 1978. Combinatorial Constructions. In *Studies in Combinatorics,* MAA Studies in Mathematics, vol. 17, ed. G. C. Rota. Washington DC: Mathematical Association of America.

————. 1986, *Combinatorial Theory.* 2d ed. New York: Wiley.

John, P. W. 1971. *Statistical Design and Analysis of Experiments.* New York: Macmillan.

Liu, C. L. 1968. *Introduction to Combinatorial Mathematics.* New York: McGraw-Hill.

————. 1972. *Topics in Combinatorial Mathematics.* Washington DC: Mathematical Association of America.

MacWilliams, F. J., and N. J. A. Sloane. 1977. *The Theory of Error Correcting Codes.* Amsterdam: North-Holland.

Pless, Vera. 1982. *Introduction to the Theory of Error-Correcting Codes.* New York: Wiley.

Ryser, H. J. 1963. *Combinatorial Mathematics.* Carus Mathematical Monographs, vol. 14. Washington DC: Mathematical Association of America.

Street, A. P., and D. J. Street. 1987. *Combinatorics of Experimental Design.* New York: Oxford University Press.

Street, A. P., and W. D. Wallis. 1977. *Combinatorial Theory: An Introduction.* Winnipeg: Charles Babbage Research Center, University of Winnipeg.

Wallis, W. D. 1988. *Combinatorial Designs.* New York: Marcel Dekker.

7
Ordered Sets

Section 1 Partial Orderings

What Is an Ordering?

Suppose a professor gives three examinations. Each student has a list of three grades. The professor might reasonably believe that the student with the list (g_1, g_2, g_3) is better than the student with the list (h_1, h_2, h_3) if $g_1 \geq h_1$, $g_2 \geq h_2$, $g_3 \geq h_3$, and at least one of these inequalities is strict. Thus we have a natural "ordering" relation defined on the students. However, if for some i, $g_i > h_i$ and for some j, $g_j < h_j$, then the two students with these grade lists cannot be compared. We say that the "better than" relation is an example of an *ordering* or, sometimes, a *partial* ordering.

A *strict (partial)* **ordering** relation P on a set S is a relation (set of ordered pairs) on S such that

(1) (x, x) is not in P for any x in S.
(2) If $(x, y) \in P$, and $x \neq y$, then (y, x) is not in P.
(3) If (x, y) and $(y, z) \in P$, then $(x, z) \in P$.

The reader will recognize that condition 3 says P is a **transitive** relation; condition 1 is called the **irreflexive** law and condition 2 is called the **antisymmetry** law. Thus, a strict ordering is an irreflexive, antisymmetric, transitive relation. Note that the $x \neq y$ in condition 2 is redundant in light of condition 1. In fact, as you are asked to show in Exercise 14, condition 2 is a consequence of conditions 1 and 3.

Example 1.1. The relation P on the set of *all subsets* of a set S given by $(X, Y) \in P$ if and only if $X \subset Y$ (but $X \neq Y$) is a strict ordering. We normally say that "$\subset$" is an ordering, avoiding all mention of P. Similarly, we say "$<$" is an ordering on the set of integers. ∎

Example 1.2. There is a relation "finer than" defined on the partitions of a set X. Recall that a partition of X is a collection of disjoint sets (called

442

classes) whose union is X. Some examples of partitions of $\{1, 2, 3, 4, 5\}$ are $\{\{1, 2\}, \{3, 4, 5\}\}$, $\{\{1\}, \{2, 3\}, \{4, 5\}\}$, and $\{\{1, 2\}, \{3, 5\}, \{4\}\}$. We say the partition Q is **finer than** the partition R if every class of Q is a subset of some class of R (and $Q \neq R$). Thus, $\{\{1, 2\}, \{3, 5\}, \{4\}\}$ is finer than $\{\{1, 2\}, \{3, 4, 5\}\}$ but $\{\{1\}, \{2, 3\}, \{4, 5\}\}$ is not. (Checking that the relationship described is an ordering is Exercise 14.) This ordering is usually called the *refinement ordering*. ∎

Notice that in Example 1.1 we used the parenthetical statement "(but $X \neq Y$)," and in Example 1.2 we used the parenthetical statement "(and $Q \neq R$)." We had to include these statements to ensure that the relations we were describing were irreflexive. When we think about subsets, we often think of the subset relation "$\subseteq$." That is, when we say "A is a subset of B" we mean that each element of A is an element of B. Thus, we allow B to be a subset of itself. To describe such situations we use the idea of a **reflexive ordering**. This is a relation P which is reflexive (i.e., (x, x) is in P for each x in S), antisymmetric, and transitive. Thus, "$\subseteq$" is a reflexive ordering of the subsets of a set, the ordinary "$\leq$" is a reflexive partial ordering on the integers, etc. The difference between a strict and a reflexive ordering is normally clear in context, and the word ordering may refer to either kind of ordering. A set S with an ordering P defined on it is called an **ordered set** or **partially ordered set (poset)**.

Example 1.3. The positive integers are partially ordered by a relation we call the *divides* relation. We write $m|n$ and say "m divides n" if m is a factor of n. Clearly, $m|m$ is true for every positive integer m. Therefore, the divides relation is reflexive. Also if $m|n$ and $n|m$, then $n = mk$ and $m = nj$ for some positive integers j and k. Thus, $n = njk$, so $jk = 1$. This is possible only if $j = k = 1$, so $m = n$. Thus, if $m|n$ and $m \neq n$, then $n|m$ is false. Thus, the divides relation is antisymmetric. Finally, if $m|n$ and $n|r$, then $n = mk$ and $r = nj$, so that $r = mkj$ and $m|r$. Therefore, the divides relation is transitive. Thus, the relation of "divides" is a reflexive partial order. ∎

We will write $x < y$ (mod P) or $x < y$ in P rather than $(x, y) \in P$ because the suggestive symbolism is helpful to our intuition. When it is clear from the context that P is the only ordering relation under consideration, we may write $x < y$ rather than $x < y$ (mod P). We write $x \leq y$ if $x < y$ or $x = y$. With this notation we can think of P as standing for both a strict partial ordering and the reflexive partial ordering formed by adding all pairs (x, x) to P. When we write $x < y$ (mod P), we are considering the strict ordering; when we write $x \leq y$ (mod P), we are considering the reflexive ordering. We use the symbols $>$ and $\geq$ in the expected way.

Linear Orderings

The usual ordering relation of "less than or equal to" on the integers has the property that, given any two integers m and n, either $m \leq n$ or $n \leq m$. This allows us to list a set of k integers in a line as $(m_1, m_2, ..., m_k)$ so that $m_i < m_j$ if and only if $i < j$. By analogy with this listing of integers in a line, we say that a partial ordering P of a set X is *linear* if for each two elements x and y in X, either

$$x \leq y \pmod{P} \quad \text{or} \quad y \leq x \pmod{P}.$$

(Note that a linear ordering is really the same thing as a transitive tournament, a concept we studied in connection with directed graphs.) The partial ordering of subsets by inclusion, of partitions by "finer than," and of the positive integers by "divides" are all examples of nonlinear orderings.

Example 1.4. Suppose a professor gives one test for which the students all receive different grades. Suppose further the professor decides that one student shows more knowledge than another if the first student has a higher grade than the second. Then the "shows more knowledge" relation is a linear ordering. Notice how this ordering ranks the students in a line as number 1, 2, 3, etc. As with the ordering of students in our example, any linear ordering of a k-element set "looks like" the usual ordering on the integers between 1 and k in the following sense. ∎

Theorem 1.1. If L is a linear ordering of the k-element set X, then there is a one-to-one function from the set $K = \{1, 2, ..., k\}$ onto the set X such that

$$f(i) \leq f(j) \pmod{L} \quad \text{if and only if } i \leq j.$$

Proof. We prove the theorem by induction on k, letting S be the set of all n such that the theorem is true for an n-element set. Clearly, 1 is in S; assume now $k - 1$ is in S.

We choose an element x_1 of the k-element set X. If there are elements y in X such that $x_1 < y \pmod{L}$, pick one such y and let it be x_2. Continue in this way to construct a sequence; once x_{i-1} is chosen, if there is some y with $x_{i-1} < y \pmod{L}$, then let x_i be one such y. Now the elements of the list of x's are distinct, because $x_{i+j} > x_i$.

Eventually we reach an x_j such that $x_j \not< y \pmod{L}$ for any y; since L is linear this means that $y < x_j$ for all $y \neq x_j$. Now let $X' = X - \{x_j\}$ and $K' = K - \{k\}$. Since $k - 1$ is in S, there is a function f from K' onto X' such that $f(r) < f(s) \pmod{L}$ if and only if $r < s$. As illustrated in Figure 1.1, we extend the function f to K

Figure 1.1

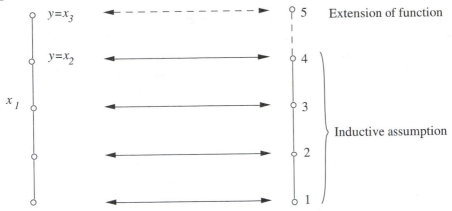

by defining $f(k) = x_j$. Since $x \leq x_j$ (mod L) for all x in X, we may conclude that $f(r) \leq f(s)$ (mod L) if and only if $r \leq s$. ∎

A function f (such as the one in Theorem 1.1) from the set X ordered by P to the set Y ordered by R is called an ***isomorphism*** if f is a one-to-one and onto function such that $f(x) \leq f(y)$ (mod R) if and only if $x \leq y$ (mod P). Thus, the preceding theorem tells us that any finite linearly ordered set is isomorphic to a set of integers with their natural ordering.

Maximal and Minimal Elements

An element x of X is a ***maximum element*** relative to the partial ordering P of X if $y \leq x$ for all y in X. When we reached the element x_j in the proof of Theorem 1.1 we had found a maximum element. Although we can find maximum elements for *linear* orderings, we cannot always do so in more general ordered sets. An element x of X is ***maximal*** relative to the ordering P if whenever $x \leq y$, in fact $x = y$. Visually, an element is maximum if it is "above everything" while an element is maximal if nothing else is "above it." Although maximum elements are also maximal, maximal elements are not necessarily maximum. *Minimum* and *minimal* elements are similarly defined.

Example 1.5. Let X be the set of integers from 1 to 12. Let P be the "divides" relation, i.e., $x \leq y$ (mod P) if and only if x is a factor of y. Then 7, 8, 9, 10, 11, and 12 are all maximal elements of this poset. On the other hand, 1 is the only minimal element and is also a minimum element. As you may suspect, when we have a minimum element we can, by definition, have only one minimum element, so we would normally say "1 is also the minimum element" rather than "1 is also a minimum element." ∎

By using much the same method we used to find the maximum element x_j in Theorem 1.1, you can prove that finite ordered sets always have maximal and minimal elements.

> ***Theorem 1.2.*** If the finite set X is ordered by P, then X has at least one maximal element and at least one minimal element.

The Diagram and Covering Graph

We can draw particularly nice graphs or digraphs to represent finite ordered sets. Given a poset X ordered by P, we use the following method, illustrated in Figure 1.2 for the ordered set of Example 1.5. Let X_1 be the set of minimal elements of X relative to P. Draw dots to represent these minimal elements in a horizontal line on a piece of paper.

Figure 1.2

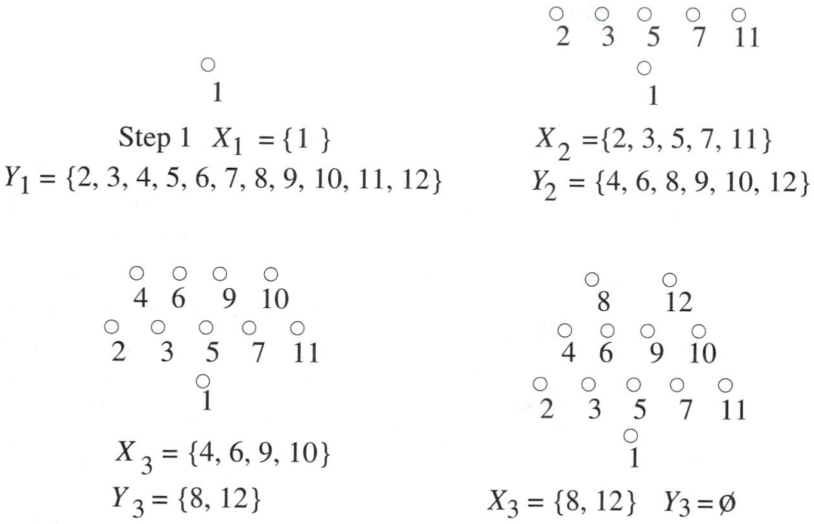
1
Step 1 X1 = {1 }
Y1 = {2, 3, 4, 5, 6, 7, 8, 9, 10, 11, 12}
2 3 5 7 11
1
X2 ={2, 3, 5, 7, 11}
Y2 = {4, 6, 8, 9, 10, 12}
4 6 9 10
2 3 5 7 11
1
X3 = {4, 6, 9, 10}
Y3 = {8, 12}
8 12
4 6 9 10
2 3 5 7 11
1
X3 = {8, 12} Y3 = ∅

Let $Y_1 = X - X_1$ and let P_1 be the partial ordering P restricted to Y_1; that is, P_1 contains exactly those ordered pairs (x, y) of P such that x and y are both in Y_1. Let X_2 be the set of minimal elements of Y_1 relative to P_1; let $Y_2 = Y_1 - X_2$. Let P_2 be the restriction of P_1 to Y_2. Draw a horizontal line of points for the elements of X_2 above the line of points for X_1. (As you see in Figure 1.2, there can be elements (such as 5, 7, and 11) that are both maximal and minimal in this restriction; for this construction we treat such elements just like any other minimal elements.) Repeat the process, constructing sets X_i, Y_i and an ordering P_i until Y_i is empty.

To draw the digraph of P, draw an arrow from the point for x to the point for y for each (x, y) in P. This digraph is shown separately in Figure 1.3.

Figure 1.3

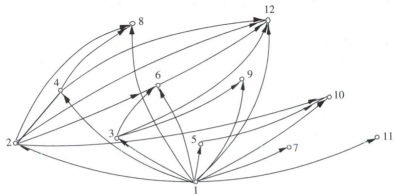

Note that all arrows point upward so the arrowheads can be removed without affecting our reading of the ordering from the picture. Some lines (for example the one from 2 to 8 in Figure 1.3) are redundant in the sense that there may also be lines connecting i (2 in our example) up to j (4 in our example) and j up to k (8 in our example). These redundant lines are required by the transitive law. On the other hand, we can remove a line from i to k whenever a sequence of intersecting lines also leads down from i to k; to interpret the resulting picture, we apply the transitive law. The picture we get by removing redundant lines (and the arrowheads) in Figure 1.3 for the integers between 1 and 12 is shown in Figure 1.4. This picture gives a much clearer view of the ordering; it is called a ***diagram*** or sometimes a *Hasse diagram* of the poset. If only redundant lines are removed and arrowheads are left in, the resulting directed graph is called a *covering digraph* of the poset.

We say that the element y ***covers*** the element x relative to the partial ordering P if $x < y$ (mod P), and whenever $x \leq z \leq y$ (mod P) then $z = x$ or $z = y$ (mod P) (i.e., there is no z such that $x < z < y$). We can draw the diagram of the poset by drawing the points in levels as above and then drawing a line from y down to x if y covers x. We do not need to draw diagrams in such a formal way; any drawing of the covering graph of the ordered set in which the vertex for x is higher than the vertex for y when $x > y$ is still called a diagram of the ordered set.

Figure 1.4

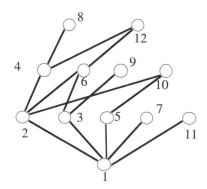

We show the covering digraph of the poset consisting of the integers $\{2, 4, 3, 12\}$ ordered by divisibility in Figure 1.5. (Note that the lines from 12 to 3 *must* skip over a level because of the way in which we decided to write down the levels. However, for the sake of understanding the ordering, we could draw the circle for 3 anywhere below the circle for 12 and still visualize the same relation; by our remark at the end of the previous paragraph, we would still call the drawing a diagram.)

Figure 1.5

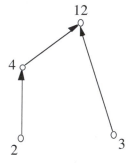

The example in Figure 1.5 is called a *restriction* or an "induced ordered subset" or "subposet" of the example in Figure 1.4. The **restriction** of the ordering P of X to Y is the set of ordered pairs in P both of whose elements are in Y. This is the same idea of restriction used to describe the method for drawing a covering graph. Note that a line in the covering graph of a restriction of P need not correspond to a line in the covering graph of P, as you can see by examining Figures 1.4 and 1.5. Can you explain this? (If not, think about how you would draw a diagram of the restriction of the divides ordering on the set of integers between 1 and 12 to the set $\{3, 12\}$.)

Ordered Sets as Transitive Closures of Digraphs

In our study of digraphs, we introduced the idea of the transitive closure of a relation or digraph. A poset is the transitive closure of its covering digraph. In fact, if $G = (V, D)$ is a digraph whose transitive closure is the poset (X, P), then every edge in the covering graph must also be an edge in D. Thus, the covering graph is the smallest digraph whose transitive closure is (X, P). This is the meaning of Theorem 1.3.

> ***Theorem 1.3.*** If a digraph $G = (V, D)$ has the poset (X, P) as a transitive closure, then every edge of the covering graph of P is an edge of G.

> *Proof.* Suppose y covers x in P, and P is the transitive closure of D. Because (x, y) is in P, we know that y is reachable from x in D. But if the path from y to x in D included another vertex z, we would have $x < z < y$ in P. Since there is no such z, (x, y) must be an edge in D. ∎

Not every digraph has a partial ordering for its transitive closure. If, however, the digraph has *no* directed cycles, then its transitive closure is a partial ordering. This can be shown by an argument using the idea of reachability. Can you see how? We will give a proof in a later section when we need to use this fact.

Trees as Ordered Sets

Recall that a *tree* is a connected graph with no cycles. Often when we use a tree for some practical purpose, one of its vertices is designated as a starting place. Such a starting place is typically called a "root." This is the kind of tree we constructed for breadth-first search, depth-first search, and close-first search spanning trees of a graph centered at a certain vertex. One reason these trees are so useful is that they give us a natural partial ordering of the vertices. A ***rooted tree*** is simply a tree with some vertex r selected and designated the root. We use (V, E, r) to stand for a rooted tree with vertex set V, edge set E, and root r. Remember that in a tree any two vertices are connected by a unique path. In a rooted tree, let us define $x \leq y \pmod{P}$ to mean that y is on the unique path connecting x to r. We call P the *ancestor relation* of the tree and we say that y is an ***ancestor*** of x and x is a ***descendant*** of y. Suppose now that P is the ancestor relation of a rooted tree. P is reflexive, because x is on the unique path connecting r to x.

To see that P is antisymmetric suppose that y is on the unique path Q connecting r to x. Then if there were a path between y and r including x, it would have to follow Q to go between y and x. Then it would have use all of Q to go from x to .. Thus it would not be a path after all.

Finally, the relation P is transitive, because if y is on the unique path from r to x and x is on the unique path from r to z, we obtain such a second path by first taking the unique path from r to x and then the unique path from x to z. Since we are in a tree, this is the unique path connecting r and z. But y is on that unique path from r to x, so y is on the path from r to z. Thus, if $z \leq x$ and $x \leq y$, we conclude that $z \leq y$.

Figure 1.5 shows an example of a rooted tree that comes from a special case of the divides ordering. The root is the vertex labeled 12. It is traditional to call x a **child** of y and to call y the **parent** of x if y covers x. Thus, 4 and 3 are the children of 12 and 2 is the child of 4. Vertex 2 and vertex 3 are said to have no children; they are called *terminal* vertices or *leaves*.

Whenever we regard a rooted tree as an ordered set, the diagram of the partial ordering is isomorphic to the original tree. The diagram makes it easier to visualize properties of the tree that are related to the natural ordering. Two more rooted trees are shown in Figure 1.6. The second tree (marked (b)) in Figure 1.6 is an example of a *binary* tree. A rooted tree is called binary if each nonterminal vertex has a unique child called a right child and/or a unique child called a left child (and no other childern). In Figure 1.6(b), b is a right child of a and j is a left child of l.

Figure 1.6

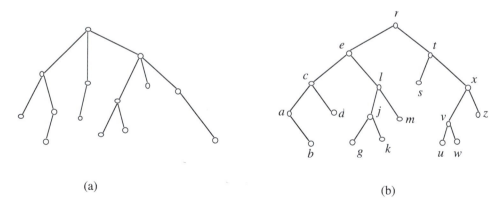

(a)

(b)

Weak Orderings

We saw in Example 1.4 how a linear ordering might arise from giving a test to a group of students all of whom get different grades. Even if the grades are not different, the relation R which contains the ordered pair (x, y) if and only if x has a lower score than y is an ordering. Since we described linear orderings by saying that each two elements are comparable, it is natural to ask for a similar description of these orderings, which are almost linear.

Notice that even though two, three, or more people with the same test score are mutually incomparable, if two people have different test scores, they can't both be incomparable with a third person. This property turns out to let us determine whether or not an ordering of students that has some incomparabilities could have come from scores on a single test. To be precise, we say the ordering P of a set X is a **weakly linear ordering**, or a **weak ordering** for short, if whenever $x < y$ in P and z is in X, then z is either over x or under y; that is, either $x < z$ or $z < y$ in P. Just as linear orderings may be visualized by assigning distinct numbers to objects, weak orderings may be visualized by assigning numbers, not necessarily distinct, to objects. Figure 1.7 shows a weak ordering with numbers assigned and Theorem 1.4 interprets the assignment in a precise way.

Figure 1.7

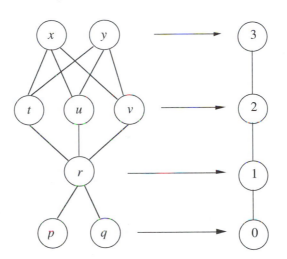

Theorem 1.4. An ordering P of a finite set X is a weak ordering if and only if there is a function f from X to a set of numbers (say, the integers) such that $x > y \pmod{P}$ if and only if $f(x) > f(y)$ (as numbers).

Proof. The proof is an inductive one similar to that for Theorem 1.1, the corresponding theorem about linear orderings. However, rather than beginning the inductive step by removing a maximum element, one begins by removing *all* maximal elements. ∎

Interval Orders

In scheduling a contest like a track meet with many teams participating, it is common to schedule some events simultaneously. The long jump might be scheduled during part of the time for which the 1000 meter race is scheduled, for example. Various events are scheduled during various intervals of time. For example, the long jump might be scheduled from 1 PM to 2 PM while the 1000 meter race might be scheduled between 1:30 PM and 3:30 PM. We would say event A comes before event B if the entire time interval for A comes before the starting time of B. In part (a) of Figure 1.8 you see a vertical number line with several (closed) intervals along it marked by brackets. In part (b) of the figure you see a diagram of the ordering of intervals that results on the set of seven intervals shown.

Figure 1.8

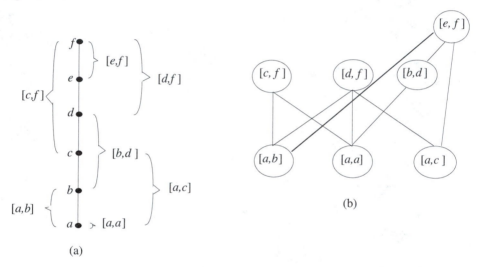

(a)

(b)

Surprisingly, it is almost as easy to test an ordering to see if it can be represented by intervals in the way we have done in Figure 1.8 as it is to test an ordering to see if it is a linear ordering or a weak ordering. We say that an ordering P is an **interval ordering** if whenever $x < y$ in P and $z < w$ in P, then either $x < w$ or $z < y$ in P. In words, P is an interval ordering if whenever we have two pairs of comparable elements, at least one of the two top elements of the pairs is over both the bottom elements.

Figure 1.9 illustrates this condition. The picture shows that whenever we have two pairs of comparable intervals, at least one of the "top" intervals is above both "bottom" intervals. Thus, in one case, looking at the pairs $x < y$ and $z < w$, the interval $y = [d, f]$ is over both the interval $x = [a, c]$ and $z = [a, b]$ and in the other case, looking at $x < y$ and $w' < z'$, the interval $w' = [e, f]$ is over both $x = [a, c]$ and $z' = [a, d]$.

It is possible to show that if P is an interval ordering on X, then there is a way to assign an interval $I(x)$ to each element x of X so that $x > y$ mod

Figure 1.9

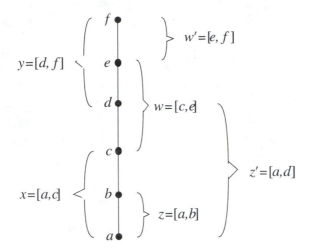

P if and only if the right endpoint of $I(y)$ is less than the left endpoint of $I(x)$. We outline the method in the Exercises.

There is a way to illustrate the definitions of linear, weak, and interval orderings with diagrams that show a natural progression. Notice that the definition of a linear ordering may be reworded to say there are not two incomparable elements. In other words, an ordering is linear if and only if no restriction of it to a two-element set has the diagram in Figure 1.10(a). The definition of a weak ordering may be reworded to say there are no elements incomparable to both members of a pair of comparable elements. In other words, an ordering is weak if and only if no restriction of it to a three-element set has the diagram in Figure 1.10(b). The definition of an interval ordering may be reworded to say that there are no two pairs of comparable elements with all elements of one pair incomparable to all elements of the other. In other words, an ordering is an interval ordering if and only if no restriction of it to a four-element set has the diagram in Figure 1.10(c).

Figure 1.10

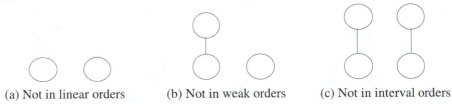

(a) Not in linear orders   (b) Not in weak orders   (c) Not in interval orders

EXERCISES

1. Let X be a set of people and let P be the relation of "is older than" on X. Show that P is a partial ordering. If no two people in X have the same age, what kind of ordering relation is P?

2. Let X be a set of individuals and let $x < y$ stand for "the weight of person x is less than the weight of person y." Show that the relation described by $<$ is a partial ordering. Is the relation reflexive or strict? If we substitute "less than or equal to" for "less than," is the resulting relation an ordering?

3. Draw the diagram for the $\subseteq$ ordering on the set of all subsets of $\{1, 2\}$ and of $\{1, 2, 3\}$.

4. Draw the diagram for the "is a factor of" partial ordering on

$$\{1, 2, 4, 6, 9, 12, 18, 36\}.$$

5. In a simplified notation for partitions, $ab|cde$ stands for the partition of $\{a, b, c, d, e\}$ into the two sets $\{a, b\}$ and $\{c, d, e\}$, while $d|bc|ae$ stands for the partition into the sets $\{d\}$, $\{b, c\}$, and $\{a, e\}$. Using this kind of notation, write down all partitions of $\{1, 2, 3\}$ and draw the diagram of the partial ordering of "is finer than" on these partitions.

6. Using the notation described in Exercise 5, draw a labeled diagram showing all partitions of $\{1, 2, 3, 4\}$ ordered by the "finer than" relation. (Hint: There are 15 such partitions.)

7. Explain why the subset relation described in Example 1.1 is a (partial) ordering.

8. Explain why the "finer than" relation described in Example 1.2 is a (partial) ordering.

9. Prove Theorem 1.4; that is, show that if P is a weak linear ordering of the k-element set X, then there is a function (not necessarily one-to-one) from X to a set $J = \{1, 2, ..., j\}$ with $j \leq k$ such that

$$f(x) < f(y) \quad \text{if and only if } x < y.$$

Explain what this means about the diagram of a weak linear order.

10. Explain how the following facts may be derived from the definitions of a maximum element and a maximal element.
    (a) If an ordered set has a maximum element, then it has only one maximum element.

(b) If an ordered set has one and only one maximal element, then that maximal element is a maximum element.

11. Prove Theorem 1.2.

12. Let a relation R be defined on the integers by

$$R = \{(x, y) | y - x \text{ is a positive even number }\}.$$

Show that R is a strict ordering.

13. Draw a diagram of the restriction of the "subset of" relation on the subsets of $\{1, 2, 3, 4\}$ to the subsets of size 1 and 3.

14. Show that condition 2 in the definition of a strict partial ordering is implied by conditions 1 and 3.

15. If X is a poset ordered by the reflexive partial ordering P, we say a pair (x, y) of elements of X is a *cover* if y covers x. Define a relation on the set C of covers of (X, P) by $(x, y) < (z, w)$ if and only if $(y, z) \in P$. Show that the relation given by $<$ is a partial ordering of C.

16. Explain how the relationship between an equivalence relation and its equivalence class partition can be used to prove that the partially ordered set of all partitions of Y with the "finer than" relation is isomorphic to the partially ordered set of all equivalence relations (considered as sets of ordered pairs) on Y ordered by the "subset" relation.

17. Draw a diagram for the rooted tree partial ordering of the vertices of Figure 1.11 obtained with vertex a as a root and for the ordering obtained with vertex b as a root.

Figure 1.11

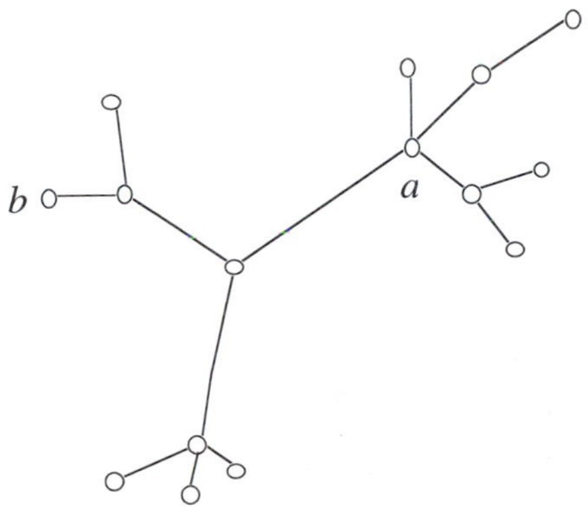

18. Why is the transitive closure of a directed graph without directed cycles a partially ordered set?

19. Give an example of an ordering P of a set X and a subset Y of X such that no line in the diagram of the restriction of P to Y is a line in the diagram of P.

20. Consider the set of functions f from a set X to a set S that is ordered by P. Let R be the relation

$$R = \{(f,g)|f(x) \le g(x) \text{ for all } x \in X\}.$$

Show that R is a reflexive ordering on this set of functions.

21. The predecessor set $\mathrm{Pred}(x)$ and the successor set $\mathrm{Suc}(x)$ of an element x in a set X ordered by a relation $<$ are defined by

$$\mathrm{Pred}(x) = \{y \mid y < x\} \quad \text{and} \quad \mathrm{Suc}(x) = \{y \mid y > x\}.$$

The predecessor and successor sets of a subset Z of X are defined by

$$\mathrm{Pred}(Z) = \{y \mid y \in X \text{ and } y < z \text{ for all } z \text{ in } Z\}$$

and

$$\mathrm{Suc}(Z) = \{y \mid y \in X \text{ and } y > z \text{ for all } z \text{ in } Z\}.$$

(a) Show that

$$\mathrm{Pred}(Z) = \bigcap_{z:z\in Z} \mathrm{Pred}(z) \quad \text{and} \quad \mathrm{Suc}(Z) = \bigcap_{z:z\in Z} \mathrm{Suc}(z).$$

(b) Describe a subset Z of X such that $\mathrm{Pred}(Z) = \emptyset$.
(c) Describe a subset Z of X such that $\mathrm{Pred}(Z) = X$.
(d) Show that if $x < y$, then $\mathrm{Pred}(x) \subseteq \mathrm{Pred}(y)$.
(e) How may $\mathrm{Pred}(Y \cup Z)$ be computed from $\mathrm{Pred}(Y)$ and $\mathrm{Pred}(Z)$? Explain why.
(f) Is there a result similar to the result of part (e) to describe $\mathrm{Pred}(Y \cap Z)$?

22. Show that if P is an interval order on the set X, then $\{\mathrm{Pred}(Y) \mid Y \subseteq X\}$ is linearly ordered by set inclusion.

23. In a linearly ordered set, we define the height of an element x, $h(x)$ to be the number of elements strictly below it. Thus, in the set $\{1,2,3\}$ with the usual ordering, 1 has height 0 and 3 has height 2. Show that

if P is an interval order on a set Y, then the function F from Y to the open intervals of real numbers given by

$$f(x) = (h(\text{Pred}(x)), h(\text{Pred}(\text{Suc}(x))))$$

is an isomorphism from the ordered set (Y, P) onto the image of f (with the usual order relation of an interval order) if no two elements of Y have exactly the same predecessors *and* successors.

24. Explain how the previous exercise may be used to show that a finite interval order has a representation using closed intervals in the real numbers.

25. Show that if a set X of *equal length* closed intervals (in the real numbers) is ordered by $[a, b] < [c, d]$ if $b < c$, then whenever x, y, and z are intervals with $x < y < z$, it follows that, for any element (interval) w of X, either $w > x$ or $w < z$.

26. Suppose that an interval ordering P of a finite set X does not contain four elements w, x, y, and z such that $x < y < z$ and w is incomparable with all three.
    (a) Suppose that no two elements have exactly the same predecessors and exactly the same successors. Define the relation R to be the one which contains (x, y) if $\text{Pred}(x)$ is a strict subset of $\text{Pred}(y)$ or $\text{Pred}(x) = \text{Pred}(y)$ and $\text{Suc}(y)$ is a strict subset of $\text{Suc}(x)$. Show that R is a linear ordering of X.
    (b) What kind of ordering arises in part (a) if two elements may have exactly the same predecessors and successors?
    *(c) Discuss the representation of ordered sets by equal length intervals of real numbers.

Section 2 Linear Extensions and Chains

The Idea of a Linear Extension

From the pictures of partially ordered sets, it seems clear that we can line up the points in a vertical straight line in such a way that if $x < y$ (mod P), then x is lower than y. We simply line up all the points on the bottom level in a covering graph in any order we choose, then line up the points on the next level above those from the bottom level, and so on.

In this way we construct a picture of a linear ordering L of the set X that is compatible with P in the sense that whenever $a \leq b$ (mod P), then $a \leq b$ (mod L) as well. Of course, unless P were linear to start with, there would be elements x and y in X with $x \leq y$ (mod L), but $x \not\leq y$ (mod P). Note that saying that $a \leq b$ (mod L) whenever $a \leq b$ (mod P) is the same as saying that (understood as sets of ordered pairs) $P \subseteq L$. In this case, we say that L is a **linear extension** of P. In computer science, linear extensions of P are sometimes called "topological sortings" of P. We have seen geometrically that each partial ordering has a linear extension. In fact, a partial ordering P has so many linear extensions that their intersection is P itself. (This will be a corollary of the next theorem.)

> **Theorem 2.1.** Suppose P is a strict ordering of a finite set X and let x and y be elements of X such that $x \not< y$ and $y \not< x$. Then P has a linear extension in which $x < y$.

Proof. We show first that given an ordering P of X and two elements x_1 and y_1 not compared by P (i.e., neither (x_1, y_1) nor (y_1, x_1) is in P), there is a *partial* ordering P_1 containing P and (x_1, y_1). Geometrically speaking, the idea is that in order to put y_1 above x_1, we must also put each element over y_1 above each element under x_1. Symbolically, we let

$$P_1 = P \cup \{(z, w) | z \leq x_1 \text{ and } y_1 \leq w \pmod{P}\}.$$

(In words, we get P_1 from P by adding all pairs (z, w) such that $z \leq x_1$ and $y_1 \leq w$, i.e., the pairs formed by a point under x_1 and a point over y_1.) Since no pair of the form (x, x) is in P_1, it is irreflexive.

Now to show that P_1 is transitive, suppose (x, y) and (y, t) are in P_1. If they are both in P, then (x, t) is in P and thus P_1. If (x, y) is not in P, then $x \leq x_1$ in P and $y_1 \leq y$ in P. Thus, $y \not\leq x_1$ in P because $y \leq x_1$ in P would imply $y_1 \leq x_1$ in P. Since y is not less than or equal to x_1 in P, (y, t) was not in the set to the right of the union in the definition of P_1. Therefore, $(y, t) \in P$. Thus, $y_1 \leq y \leq t$ in P, and so $y_1 \leq t$ in P. Thus, (x, t) *is* in the second set in the definition of P_1. Therefore, (x, t) is in P_1, which is what the transitive law requires. In a similar way, it is possible to show that if (y, t) is not in P, then (x, t) is in P_1. For this reason, P_1 is transitive. To see that P_1 is asymmetric, note that if (x, y) and (y, x) were in P_1, then by transitivity (x, x) would be in P_1—which we already know is impossible. Thus, P_1 is a partial ordering.

Now let x_1 and y_1 be the elements x and y referred to in the statement of the theorem. If the extension P_1 just constructed for x and y is not linear, we can find a pair of elements x_2 and y_2 such that $x_2 \not\leq y_2$ and $y_2 \not\leq x_2$. Using these elements, we can construct a P_2 where they are compared. Since there is only a finite number of ordered pairs of elements of X, we can repeat the process of constructing orderings P_i until P_i has no pair of elements x_{i+1} and y_{i+1} such that $x_{i+1} \not\leq y_{i+1}$ in P. Thus, P_i will be linear. This completes the proof. ∎

The theorem remains true if X is infinite; however, the proof uses set theoretic ideas that some readers will not have seen. By thinking about an ordering as a set of ordered pairs (since it *is* a relation), we obtain an interesting corollary.

Corollary 2.2. A partial ordering of a finite set X is the intersection of the collection of all its linear extensions.

Proof. The proof, which is an elementary application of Theorem 2.1, is left as an exercise. ∎

Dimension of an Ordered Set

The size of the smallest set of linear orderings whose intersection is P is called the *dimension* of P.

Example 2.1. The partitions of $\{1, 2, 3\}$ are the sets $\{\{1\}, \{2\}, \{3\}\}$, $\{\{1, 2\}, \{3\}\}$, $\{\{1, 3\}, \{2\}\}$, $\{\{1\}, \{2, 3\}\}$, and $\{\{1, 2, 3\}\}$. We use the shorthand notation $1/2/3$, $12/3$, $13/2$, $1/23$, 123 to stand for these five partitions. Three linear orderings L_1, L_2, and L_3, which are linear extensions of the refinement ordering, are given by

$$L_1 : 1/2/3 < 12/3 < 13/2 < 23/1 < 123$$
$$L_2 : 1/2/3 < 23/1 < 13/2 < 12/3 < 123$$

and
$$L_3 : 1/2/3 < 12/3 < 23/1 < 13/2 < 123.$$

Note that $L_1 \cap L_2$ is the refinement ordering on the partitions of $\{1, 2, 3\}$. However, $L_1 \cap L_3$ is not the refinement ordering because the ordered pair

(12/3, 13/2) is in $L_1 \cap L_3$, meaning that $12/3 < 13/2$ in $L_1 \cap L_3$; however, 12/3 is not a refinement of 13/2. Since, however, there are two orderings (L_1 and L_2) whose intersection is the refinement ordering and since the refinement ordering is not linear, the dimension of the refinement ordering on the partitions of $\{1, 2, 3\}$ is 2. ∎

Topological Sorting Algorithms

There are so many applications of the idea of a linear extension that the study of "topological sorting" algorithms has become an important topic in computer science. Our first geometric construction of a linear extension gives an easy algorithm for finding linear extensions.

> Find a minimal element.
> Put it on top of anything already in the linear ordering.
> Remove it from the poset.
> Repeat the process until the poset is empty.

The process of finding a minimal element could lead us through all the vertices of the poset. Thus the number of steps in the algorithm execution would be approximately equal to the number n of vertices times the number of times we find a minimal element, which is the number of vertices. Thus the number of steps in the algorithm could be a significant fraction of n^2.

There are possibly better algorithms that use a number of steps proportional to $n + e$, where e is the number of ordered pairs in the order relation. For example, we could carry out a breadth-first search of the order relation as described in Chapter 4; the breadth-first numbering defines a linear extension.

Chains in Ordered Sets

One of the ways in which companies introduce new products and build sales of old ones is by participation in national and regional "shows." A manufacturer of faucets, for example, will send people to hardware shows, plumbing shows, building shows, perhaps architectural and design shows, etc. The company must staff a show with people and supply the staff with equipment, samples, etc. A team of people and their equipment can staff two different shows only if there is time enough from the end of the first show to move the people and move and set up their equipment for the second show. This gives us a natural ordering of the shows as follows: We have a starting time and an ending time for each show. We also have numbers t_{ij} giving the time needed to move from the location of show S_i to the location of show S_j. We say S_i *precedes* S_j if the difference between the beginning of S_j and the end of S_i is at least t_{ij}.

Example 2.2. Show that the "precedes" relation is a partial ordering.

First, the precedes relation is irreflexive (the beginning of show i minus the ending of show i is negative, while t_{ij} is presumably zero or more). In a similar way the relation is antisymmetric.

To show the relation is transitive, we need to make an observation, called the **triangle inequality**, about the numbers t_{ij}. Namely, for any location k, $t_{ij} < t_{ik} + t_{kj}$. This says that the time to move from location i to location j is no more than the time needed to move from location i to j and then from j to k. Since going from i to k to j is one route from i to j, the time needed to travel from i to j is no more than the time needed to travel from i to k to j.

To prove transitivity, suppose that S_i precedes S_j and S_j precedes s_k. Then the difference between the beginning of S_j and the end of S_i is at least t_{ij}, and the time between the beginning of S_k and the end of S_j is at least t_{jk}. Thus, using D to stand for the duration of the show S_j, the time between the end of S_i and beginning of S_k is at least

$$t_{ij} + D + t_{jk}.$$

But since $t_{ij} + t_{jk} \geq t_{ik}$ by the triangle inequality, we have

$$t_{ij} + D + t_{jk} \geq t_{ik}$$

and so the time between the end of S_i and the beginning of S_k is at least t_{ij}. This shows that our relation is transitive. ∎

In order to control costs, the company will want to plan attendance at shows in a way that minimizes the number of teams needed. Notice that the same team can work at show S_i and show S_j if and only if S_i precedes S_j or S_j precedes S_i. Thus, the shows covered by a given team will be linearly ordered by the precedes relation. Further, if a set of jobs is linearly ordered by the precedes relation, then that set of shows may all be assigned to the same team. Finding a good assignment of teams to shows, then, means dividing the shows into a minimum number of these linearly ordered subsets. Note that if two shows are not compared by the precedes relation, they can't be covered by the same team, and if all the shows in a certain set are mutually incomparable, all these shows will require different teams. Thus, the number of teams needed is at least the number of shows in a maximum-sized set of mutually incomparable shows.

In an ordered set, a subset whose elements are linearly ordered (relative to each other) by the ordering is called a **chain**. (This terminology reflects the fact that a diagram of just the part of the ordered set represented by the points in question would look like a chain whose links are the lines connecting adjacent points.) One remark we made earlier about assigning people to shows may be interpreted as saying that the shows covered by a given team must be a chain.

A set of points in a partially ordered set, no two of which are compared by the ordering, is called an **antichain**. Another remark we made earlier may be regarded as saying that the number of teams needed is at least the size of a largest antichain.

In Example 1.5 (in Section 1 of this chapter), $\{1,2,4,8\}$ is a chain, as are $\{2,8\}$ and $\{1,3,9\}$. The elements 3 and 10 are incomparable, and so $\{3,10\}$ is an antichain. Two examples of larger antichains are $\{4,6,7,11\}$ and $\{4,6,7,10,11\}$.

Chain Decompositions of Posets

Our preceding analysis of assignments of teams to shows showed that if X is partially ordered by P and is a union of k sets X_i such that each X_i is a chain (i.e., each X_i is linearly ordered by P), then k is at least the size of a maximum-sized antichain of (X, P). In Example 1.5, $\{4,6,9,10,7,11\}$ is an antichain with six elements. The $k = 7$ sets $\{2,8\}$, $\{4,12\}$, $\{3,6\}$, $\{9\}$, $\{5,10\}$, $\{1,7\}$, $\{11\}$ are chains that contain among them all the elements of X. Note that the chain $\{2,8\}$ contains no element of the antichain. Each other chain in our list contains exactly one element of the antichain. Two elements of the antichain could not be in the same chain, because elements of a chain have to be comparable and elements of an antichain have to be incomparable. The sets $\{1,2,4,8\}$, $\{6,12\}$, $\{3,9\}$, $\{5,10\}$, $\{7\}$ and $\{11\}$ are $k = 6$ chains that contain all the elements of X. Notice that each element of the antichain is in one and only one of these chains. A theorem discovered by R. P. Dilworth in 1950 says that in fact such a chain decomposition using k chains will always exist. The proof we give is due to M. Perles.

Theorem 2.3. (Dilworth's theorem) If X is partially ordered by P and if the largest antichain(s) has (have) k elements, then there is a set of k chains whose union is X.

Proof. If X has one element, then X is linearly ordered by P, so X is a union of one chain, namely itself! We will use induction; let S be the set of all integers n such that if X has n elements and has a maximum-sized antichain with k elements, then X is a union of k chains. We just saw that 1 is in S and we now assume that all positive integers less than j are in S. Suppose now that X has j elements.

We consider two cases. Suppose first that the only maximum-sized antichain is the set of all maximal elements or the set of all minimal elements or that these two sets are the only maximum-sized antichains. Let x be a maximal element, y be a minimal element less than or equal to x (mod P), and let X' be the set $X - (\{x, y\})$ obtained by removing x and y from X. Let P' be the restriction of P to X', that is, the ordering obtained from P by removing ordered pairs in which x or y occurs. Any antichain of (X', P') is an antichain of (X, P) and cannot contain all the maximal elements or all the minimal elements of X. Thus, a maximum-sized antichain of (X', P') has $k - 1$ elements, so X' is a union of $k - 1$ chains because it has fewer than j elements. Since chains for P' are chains for P, X is the union of these $k - 1$ chains together with the chain $\{x, y\}$.

Now we assume that X contains an antichain A with k elements, some of which are not maximal and some of which are not minimal. Let U (for upper) consist of those elements of X greater than or equal (mod P) to some element of A and L consist of those elements less than or equal (mod P) to some element of A. Then U and L are partially ordered by the restriction $P|_U$ of P to U (which consists of pairs of P with elements only of U) and $P|_L$, the restriction of P to L. Since the sizes of U and L are integers in S, $(U, P|_U)$ and $(L, P|_L)$ are each unions of k chains because A is a k-element antichain in each. Each of these chains C_i of U has an element of A at its bottom, and this element of A is also at the top of exactly one chain D_i of L. Since each element of D_i is below that element of A, which in turn is below each element of C_i, $C_i \cup D_i$ is a chain. Thus, we have k chains whose union is X.

Therefore, by the principle of mathematical induction, S contains all positive integers and so the theorem is true. ∎

The size of a maximum-sized antichain is called the **width** of (X, P). The **height** of an element of X is the size of the largest chain of elements strictly below (and not equal to) it. The **height** of (X, P) is the largest height of any element of X and thus is one less than the size of the largest chain of P.

Example 2.3. The width of the partially ordered set of all subsets of the three-element set $\{1, 2, 3\}$ is 3; the height is also 3. A maximum-sized antichain in this ordered set is

$$\{\{1\}, \{2\}, \{3\}\}$$

and a maximum-sized chain is

$$\{\emptyset, \{1\}, \{1, 2\}, \{1, 2, 3\}\}.$$

Three chains whose union is X (the set of all subsets of our three-element set) are

$$\{\emptyset, \{1\}, \{1,2\}\}, \quad \{\{2\}, \{2,3\}, \{1,2,3\}\}, \quad \{\{3\}, \{1,3\}\}. \qquad \blacksquare$$

Example 2.4. Show that among a group of $rs + 1$ people, there is either a list of $r + 1$ people such that each person on the list is a descendant of the next person on the list or a set of $s + 1$ people no two of whom are descendants of each other.

Consider the relation "is a descendant of." If this relation is a union of m, but not fewer, chains, then a maximum-sized antichain has m elements. If the longest chain(s) has length n, then the set has no more than mn elements. Thus, we cannot have both n less than or equal to r and m less than or equal to s. A chain gives a list of descendants and an antichain is a set of people no two of whom are descendants of each other. Note we could have interchanged the roles of r and s and still have proved the corresponding result. $\blacksquare$

Finding Chain Decompositions

Our discussion of the assignment of teams to trade shows demonstrates that the practical problem is determining the actual assignments of teams to shows, that is, finding chain decompositions of partially ordered sets. If a set X with n elements is partially ordered by a relation P, then no matter what P is, there is a set of n disjoint chains whose union is P: each chain is a one-element set containing one element of X. Unless X is an antichain, though, this is not a minimum-sized chain decomposition. That is, this chain decomposition is not a chain decomposition using a minimum number of chains. We can make a smaller chain decomposition by choosing two comparable elements and joining them together to form a chain. As long as we can find a pair of comparable elements, we can join them together to make a two-element chain and reduce the number of chains in our chain decomposition. This suggests a strategy we might follow in trying to find a minimum-sized chain decomposition: start with a convenient partition of X into chains, perhaps even the trivial one above, then try to reduce the number of chains in the chain decomposition by joining chains of comparable elements together. If we interpret this strategy literally it fails to work; our following example shows how. If instead we interpret the strategy liberally, it leads us to an algorithm for finding minimum-sized chain decompositions.

Example 2.5. Given the chain decomposition

$$\{c,e\} \quad \{a,b\} \quad \{d\} \quad \{f,g\} \quad \{h,i\} \quad \{j\}$$

Figure 2.1

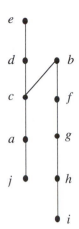

of the ordered set whose diagram is Figure 2.1, apply the procedure of joining together chains whose elements are comparable until no two chains have all their elements comparable.

In Figure 2.2 we see the initial set of chains and then the chains that result from:

> joining $\{j\}$ and $\{a, b\}$,
> joining $\{h, i\}$ and $\{f, g\}$,
> joining $\{c, e\}$ and $\{d\}$.

Figure 2.2

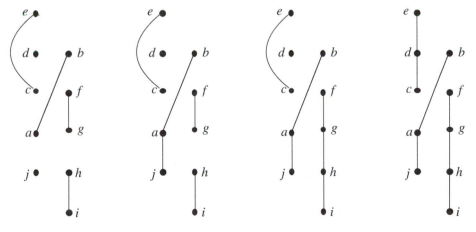

No further joining is possible after $\{c, e\}$ and $\{d\}$ have been joined because, for each pair of remaining chains, some element of one is incomparable to some element of the other. For example, $\{a, b, j\}$ and $\{c, d, e\}$ may not be joined because, as you see in Figure 2.1, b is incomparable with d (and with e as well.) ∎

The three chains in our example do not form a minimum-sized chain decomposition, for, clearly, the two chains $\{j, a, c, d, e\}$ and $\{b, f, g, h, i\}$ form a chain partition of the ordered set. Thus, our idea of joining chains together two at a time to make one chain is not sufficient; we must continue the process with a way of joining three chains to make two, and perhaps four chains to make three, and so on. In Figure 2.2 we see that the chain $\{c, d, e\}$ is over the lower part $\{a, j\}$ of the chain $\{j, a, b\}$, while the upper part $\{b\}$ of this chain is over all of chain $\{f, g, h, i\}$. Thus, by splitting the chain $\{j, a, b\}$ into two parts, we can join one part with one chain and use the second part with the other chain in order to transform the three chains into two. We split the chain $\{j, a, b\}$ by removing the edge (a, b) (a covering edge of the chain) from the diagram; next we join $\{a, j\}$ and $\{b\}$ into the other chains by adding the edges (a, c) and (f, b) of the poset to the covering diagram for the chains. In Figure 2.3 we have a similar situation which allows us to convert four chains into three.

Figure 2.3

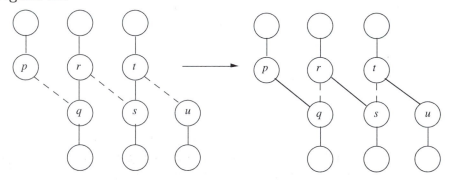

The four chains on the left-hand side of the arrow are the ones determined by the solid edges. By removing the two edges (q, r) and (s, t), we may insert the dashed edges (q, p), (s, r), and (u, t) and then get three chains. Notice the way in which the removal of the two edges (q, r) and (s, t) gives us a chain decomposition in which p, r, and t are minimum elements of their chains and q, s, and u are maximum elements of their chains. When we join p and q (or r and s or t and u) we are combining two chains whose elements are mutually comparable.

We can also visualize the transformation of Figure 2.3 into three chains sequentially. We may use the edge (q, p) to put the elements below q into a chain with the elements above p, but only at the cost of breaking the covering edge (q, r) of the second chain. Then we may use edge (s, r) to put the elements below s into a chain with the elements above r, but only at the cost of breaking the edge (s, t) of the third chain. Then we may use the edge (u, t) to put the elements below u into a chain with the elements above t—at no cost since there is nothing above u in the fourth chain.

Alternating Walks

Notice how this alternate insertion and removal of edges defines a "walk" in the figure. In fact, this is a walk (in an underlying undirected graph) in the graph-theoretic sense. The ***comparability graph*** of an ordered set has the edge $\{x, y\}$ if and only if the strict ordering has the ordered pair (x, y) or (y, x). Thus the comparability graph is the underlying graph of the digraph of the ordering (not the covering digraph or diagram). In the ordered set of Figure 2.3 the walk $p\ q\ r\ s\ t\ u$ has four important properties:

(1) Every other edge of the walk is a covering edge of a chain in the chain decomposition and is traversed upward in the walk.

(2) Each other edge of the walk is not a covering edge of a chain in the chain decomposition and is traversed downward in the walk.

(3) A vertex has at most one chain-covering and at most one non-chain-covering edge entering it on the walk and at most one of each kind of edge leaving it on the walk.

(4) The walk begins by going downward from a minimal element of a chain of the decomposition and ends by going downward to a maximal element of a chain of the decomposition.

In general, a walk in the comparability graph of an ordered set with a chain decomposition $\mathbf{C}$ is called an ***alternating*** walk for $\mathbf{C}$ if it satisfies conditions (1) through (3) and a *decreasing* alternating walk for $\mathbf{C}$, or simply a *decreasing walk* for $\mathbf{C}$, if it satisfies conditions (1) through (4). Note that a decreasing walk will have an odd number $n = 2r + 1$ of edges, because it has one more downward edge than upward edge. Our analysis of Figure 2.2 suggests that finding decreasing walks is a possible approach to finding minimum-sized chain decompositions. To show that this is indeed the case, we need a precise description of the relation we are modifying as we add and remove edges. Given a chain decomposition $\mathbf{C} = \{C_1, C_2, \ldots, C_n\}$ of the set X ordered by the relation P, we define the relation $R(\mathbf{C})$ to have the edge (x, y) if and only if y is greater than x in one of the chains C_i. We call (x, y) a *chain-covering edge* if y covers x in $R(\mathbf{C})$. This edge is an edge in the covering relation of the partial ordering $R(\mathbf{C})$ of X. (Notice that y may cover x in C_i without covering x in P; for example, e covers c in its chain in the first three diagrams of Figure 2.2; but in the ordered set of Figure 2.1 e does not cover c.) In Figure 2.2 you see the diagrams of four different covering relations for partial orderings $R(\mathbf{C})$ determined by four different chain decompositions $\mathbf{C}$.

Theorem 2.4. If $x_0 x_1 \ldots x_n$ is a decreasing walk for a chain decomposition $\mathbf{C}$ of an ordered set, then by removing the edges (x_{2i-1}, x_{2i}) from $R(\mathbf{C})$ and adding the edges (x_{2i+1}, x_{2i}) to the resulting relation we obtain the relation $R(\mathbf{C}')$ of a chain decomposition $\mathbf{C}'$ with one less chain than $\mathbf{C}$.

Proof. Removing the chain-covering edges gives a chain decomposition $\mathbf{C}^*$ in which all vertices of the walk followed by non-chain-covering edges are minimum elements of their chains and all vertices preceded by non-chain-covering edges are maximum in their chains. Now adding a non-chain-covering edge of the walk to $R(\mathbf{C}^*)$ reduces the number of chains by one (placing all elements of one chain over all elements of another in a common chain). Further, after we add a non-chain-covering edge to $R(\mathbf{C}^*)$, all vertices followed by the remaining non-chain-covering edges are still minimum in their chains and all vertices preceded by non-chain-covering edges are still maximum in their chains. Thus, we may continue adding non-chain-covering edges, reducing the number of chains each time, and obtain a chain covering $\mathbf{C}'$ such that $R(\mathbf{C}')$ is obtained from $R(\mathbf{C})$ by deleting the chain-covering edges of the walk and adding the others. Since each deletion of a chain-covering edge increased the number of chains by 1, adding each other edge of the walk decreased the number of chains by one. Because the walk had one less chain-covering edge than non-chain-covering edge, $\mathbf{C}'$ has one less chain than $\mathbf{C}$. ∎

Example 2.6. The alternating walk used to reduce from three chains to two in the last diagram of Figure 2.2 is the walk $c(a, c)a(a, b)b(f, b)f$. Notice that removing the chain-covering edge and adding the non-chain-covering edges of the walk is exactly the process we discussed intuitively following Example 2.5. ∎

Example 2.7. The last stage of Figure 2.2 represents the insertion of d into the chain $\{c, e\}$. How may this be interpreted using alternating walks?

At d we have an ordered-set edge from c. We may follow that with the chain-covering edge (c, e) and the ordered-set edge (e, d). Notice that this gives us an alternating walk from the minimum element of the chain $\{d\}$ to the maximum element of the chain $\{d\}$. Thus, we may use this alternating walk to reduce the number of chains; doing so converts $\{c, e\}$ and $\{d\}$ to $\{c, d, e\}$. ∎

Theorem 2.4 shows that a chain decomposition is *not* of minimum size if it has a decreasing alternating walk. We show now that a chain decomposition with no decreasing alternating walk is as small as possible. To show that a chain decomposition with no decreasing alternating walks has minimum size, our strategy is to find an antichain such that *each* chain of the decomposition contains a member of the antichain. As we observed before proving Dilworth's theorem, since each member of an antichain must be in a different chain, regardless of the chain decomposition, the size of a chain decomposition is at least the size of a maximum-sized antichain. Our strategy proves that the sizes of the chain decomposition and the antichain are equal, thus giving another proof of Dilworth's theorem *and* proving that a chain decomposition without alternating walks has minimum size.

To implement this strategy we show how each chain decomposition leads to an antichain that either helps us determine that a decreasing alternating walk exists or includes an element of each chain. In particular, if there is a decreasing alternating walk, our antichain will have the second-to-last element of such a walk. Thus, to test whether a decreasing alternating walk exists we need only check to see if a member of the antichain is over a maximum element of a chain. For that purpose, we want to choose the elements of the antichain to be as "high" as possible in the ordered set, consistent with keeping them on alternating walks.

Theorem 2.5. Suppose $\mathbf{C}$ is a chain decomposition of a set X ordered by P. Let A be the maximal elements of the set E of endpoints of alternating walks which begin at minimum elements of chains in $\mathbf{C}$. Then A is an antichain such that either (1) there is a decreasing alternating walk whose second-to-last member is in A or (2) every chain in $\mathbf{C}$ contains a member of A.

Proof. Since two maximal elements of E cannot be comparable, A is an antichain. Now suppose there is a chain C in $\mathbf{C}$ that contains no member of A. Note that the minimal element of C is in E. Let x be maximal among the elements of C that are in E. Since A includes all maximal elements in E, there is at least one member y of A with $x < y$. Because y is maximal in E, any edge immediately preceding it in an alternating walk is traversed upward. By conditions 1 and 2 of the definition of an alternating walk, such an edge is a chain-covering edge. Because y is not in C, the edge (x, y) is a non-chain-covering edge. Thus if there is an alternating walk $y_0 y_1 \ldots y_n y$ ending in y and not containing x, then $y_0 y_1 \ldots y_n y x$ is a decreasing walk whose second-to-last element is in A. If there is an alternating walk $y_0 y_1 \ldots y_n y$ with x in the walk, say $x = y_i$, then the edge (x, y_{i+1}) cannot be a chain-covering edge by the maximality of x Therefore either $x = y_0$ or (y_{i-1}, x) is a chain-covering edge. In either of these cases $y_0 y_1 \ldots y_n y x$ is a decreasing walk whose second-to-last element is in A. ∎

Corollary 2.6. (Dilworth's theorem) The number of chains in a minimum-sized chain decomposition of an ordered set is equal to the number of elements in a maximum-sized antichain.

Our theorem also tells us the following algorithm will yield a minimum-sized chain decomposition:

Start with a chain decomposition **C**.
Repeat the following steps until they are impossible to repeat.
(1) Find a decreasing alternating walk.
(2) Replace **C** by the chain decomposition whose chain-covering relation is obtained by removing the chain-covering edges of the walk from $R(\mathbf{C})$ and adding the non-chain-covering edges of the alternating walk to the resulting relation.

Of course, we have not shown how to find a decreasing alternating walk, so the practicality of our algorithm relies on whether there is a practical method for finding a decreasing alternating walk.

Finding Alternating Walks

Since an alternating walk need not be a path in the comparability graph (that is, since a vertex may appear twice), we should not expect to be able to use standard search techniques, such as breadth-first search, to find alternating walks. However, a simple modification of breadth-first search does work. We put labels on vertices, labeling a vertex with an edge that, it turns out, is the last edge on an alternating walk from a minimum element of a chain to that vertex. With this edge we can learn the second-to-last vertex of the walk, with its label we can learn the third-to-last vertex, and so on. Thus, if we label a maximum element of a chain, we can use the labels to write down an alternating walk to that maximum element. There will be one detail to prove, namely that if there is a decreasing alternating walk, then the labeling process *will* label a maximum element of a chain.

The labeling process is as follows:
(1) For each minimum vertex x of a chain and for each unlabeled vertex y such that (y, x) is an edge, label y with (y, x).
(2) Now, taking labels on vertices z in the order in which they have been assigned, repeat the following until no more labelings are possible:
  (a) If z is labeled by a non-chain-covering edge and (z, y) is a chain-covering edge, then label y with (z, y) unless y already is labeled with a chain-covering edge.

(b) If z is labeled by a chain-covering edge and (y, z) is a non-chain-covering edge, then label y with (y, z) unless y is already labeled with a non-chain-covering edge.

Theorem 2.7. If there is a decreasing alternating walk W for the chain decomposition **C** of an ordered set, then the labeling algorithem will label a maximum element of a chain with a non-chain-covering edge.

Proof. Suppose $W = v_0 v_1 \ldots v_n$ is an alternating walk from a minimum element of a chain of **C** to a maximum element of a chain of **C**. If v_1 is not labeled (v_0, v_1), it must be labeled (u, v_1) for some other minimum element u of a chain of **C**. Now let x be the last vertex of W labeled by the same kind of edge (chain-covering or non-chain-covering) as enters it in W. An analysis of the labeling algorithm shows that x must be a maximum element of a chain of C. We leave the details as Exercise 24. ∎

Example 2.8.   Show the result of applying the labeling algorithm to the ordered set of Figure 2.1 using the chain decomposition in the last diagram of Figure 2.2.

Vertex c is the only minimum vertex of a chain for which step 1 produces a label. Step 1 places the label (a, c) on a and the label (j, c) on b. Now since a has been labeled with a non-chain-covering edge, step 2a tells us to label a with (j, a). Now step 2b applies to vertex b, and so we give i, h, g, and f the labels shown. Since f (the maximum element of a chain) has been labeled, we can find an alternating walk from f to c. ∎

EXERCISES

1. Write down a linear extension of the eight-element ordered set consisting of all subsets of the set $\{1, 2, 3\}$ ordered by set inclusion.

2. Write down a linear extension of the ordering of Exercise 1 that contains the ordered pair $(\{1, 2\}, \{3\})$, that is, an extension that puts $\{1, 2\}$ below $\{3\}$.

3. Can the partial ordering of all subsets of a three-element set be an intersection of two linear extensions? (Hint: If a linear extension places $\{1\}$ over $\{2, 3\}$, can it place any other singleton set over its complement?)

4. What are the height and width of the set of all subsets of a four-element set?

5. Draw the six-element ordered set consisting of a, b, c, d, e, and f with $a < b$, $c < b$, $a < d < e$, and $c < f < e$. Find its width and height.

Figure 2.4

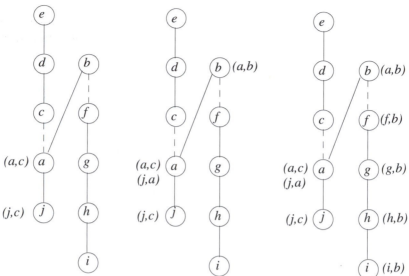

6. Write down three linear extensions of the ordered set of Exercise 5 whose intersection is the partial ordering.

7. What is the dimension of the ordered set of the previous two exercises?

8. Write down three linear extensions of the ordered set of Exercise 1 whose intersection is the partial ordering. What is the dimension of this ordered set?

9. Write down a set of chains whose union is the set of all subsets of the four-element set $\{1, 2, 3, 4\}$. Is your set as small as possible?

10. Show that the conclusion of Dilworth's theorem may be modified to state that the k chains are disjoint. (Note: It is not necessary to reprove the entire theorem.)

11. What are the height and width of the poset of all numbers between 1 and 12 ordered by "divisibility"?

12. Draw a diagram of the six-element ordered set with elements x_1, x_2, x_3, y_1, y_2, y_3 and the ordering whose only pairs correspond to $x_i < y_j$ if and only if $i \neq j$.
    (a) What are the height and width of this ordered set?
    (b) If, in a linear extension of this ordered set, x_i is above y_i, then can any other x_j be above the corresponding y_j in this linear extension?
    (c) Find three linear extensions of this ordered set whose intersection is the ordering.
    (d) Determine the dimension of this ordering.

13. There is a result similar to Dilworth's theorem that tells us the smallest number of antichains needed to partition an ordered set into antichains. State and prove this theorem.

14. Use the labeling algorithm to find an alternating walk for the chain decomposition $\{4, 8\}$, $\{2, 10, 20\}$, $\{5, 40\}$ of the set $\{2, 4, 5, 8, 10, 20, 40\}$ ordered by "is a factor of." Use this walk to give a chain decomposition with fewer chains.

15. Apply the labeling algorithm to the chain decomposition $\{2, 4, 8, 40\}$, $\{5, 10, 20\}$ of the set $\{2, 4, 8, 5, 10, 20, 40\}$ ordered by "is a factor of" and explain why there is no decreasing alternating walk.

16. Apply the construction process of Theorem 2.1 to the ordered set in Figure 2.5 using B and F as the incomparable pair. Draw a Hasse diagram for the resulting partial ordering.

Figure 2.5

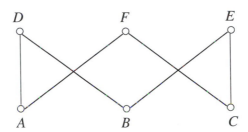

17. Show that adding an element b to (X, P) and putting (b, x) in P for all x in X does not change the dimension of P.

18. Show how the following problem (due to Dantzig and Fulkerson) is related to Dilworth's theorem. An airline company has n different routes (numbered 1 through n). A schedule set in advance gives the starting time s_i and finish time f_i for route i. Let t_{ij} be the time required to move an airplane from the destination of route i to the origin of route j. This partially orders the routes: we put (i, j) in P if $f_i + t_{ij} < s_j$. What is the minimum number of airplanes needed?

*19. A chain is *saturated* if whenever it contains elements x and y with $x < z < y$ for some z, then there is an element z' (perhaps different from z) in the chain with $x < z' < y$. A chain in an ordered set of height n is *symmetric* if whenever it contains an element of height i, it contains an element of height $n - i$ as well. Explain why the ordered set of subsets of an n-element set has a "symmetric chain decomposition," that is, a chain decomposition of minimum size into symmetric saturated chains.

*20. Sperner's theorem says that the width of the partially ordered set of all subsets of an n-element set is the largest value of $\binom{n}{k}$. Prove this theorem.

21. We say a set of partial orderings $\{P_1, P_2, \ldots, P_n\}$ of X *generates* P if P is the transitive closure of their union, $P_1 \cup P_2 \cup \cdots \cup P_n$.

    (a) Why must each cover be in one P_i in order for the P_i's to generate P?

    (b) We define a relation R on the covers of P by $(x, y) < (z, w) \pmod{R}$ if and only if $y \leq z \pmod{P}$. Show that the relation R is a strict partial ordering of X, as our notation suggests.

    (c) Show that the minimum number of chains needed to generate P is the size of a maximum antichain in the ordering of part (b).

22. Suppose that P and Q are partial orderings of X and that P is a proper subset of Q. Show that there is an ordered pair (x, y) in Q and not P such that $P \cup \{(x, y)\}$ is a partial ordering.

23. When a rooted tree is regarded as a partially ordered set, its dimension is one or two. Explain why.

24. In Theorem 2.7 we did not explain why vertex x is the maximum element of a chain of C. Explain why.

25. There is a relationship between chain decompositions and network flow, which this and the next several problems explore. We give the network without capacities in this exercise.

    (a) For each vertex x of the ordered set X, we have two vertices x_1, x_2 and a directed edge (x_1, x_2) of the network. For each edge (x, y) of the ordered set we have an edge (x_2, y_1) of the network. Add a source vertex s with an edge (s, x_1) for each x and a sink vertex t with an edge (x_2, t) for each vertex x. Remove vertices e, f, g, h, and i from the ordered set of Figure 2.1 and draw the network of the ordered set that results.

    (b) Associated with a chain in the network of an arbitrary ordered set as described in part (a), we have flow f of value 1. In particular $f(x_1, x_2) = 1$ for each x in the chain, $f(x_2, y_1) = 1$ if y covers x (in the chain, not necessarily the ordering), and $f(s, x_1) = f(y_2, t) = 1$ if x is the minimum element of the chain and y is the maximum element of the chain, and $f(s, z) = 0$ for every other vertex z. What must $f(v_1, v_2)$ be for any other edge (v_1, v_2) of the network?

    (c) A chain decomposition $\{C_1, C_2, \ldots C_n\}$ gives flows $f_1, f_2, \ldots, f_n$ according to part (b). Why is $f = f_1 + f_2 + \cdots + f_n$ a flow?

(d) The sets $\{d\}$, $\{b, j\}$, $\{a, c\}$ are a chain decomposition of the ordered set of part (a). What flow corresponds to it? What is the flow's value?

(e) The set $\{d, c\}$, $\{j, a, b\}$ is a chain decomposition of the ordered set of part (a). What flow corresponds to it? What is the flow's value?

(f) A maximum-flow problem is to maximize f subject to the capacity constraints $f(e) \leq c(e)$. For a minimum-flow problem there are certain necessities $n(e)$ associated with edges e; each $n(e)$ is a real number. We are to minimize f subject to the necessity constraints $f(e) \geq n(e)$. Which of these two kinds of problems corresponds to finding a chain decomposition? Explain why.

26. Just as a flow in a network with capacities may have a flow-augmenting path of worth w, a flow in a network with necessities may have a flow-decreasing path of worth w. (We subtract w from the flow for each forward edge of the path and add w to the flow for each reverse edge of the path.)

(a) Find a flow-decreasing path of worth 1 that converts the flow of part (d) of the previous problem to the flow of part (e).

(b) Describe how to use network-flow methods to find chain decompositions.

(c) What relationship is there between alternating walks and flow-decreasing paths? To what aspect of a flow-decreasing path does a vertex which appears on an alternating walk twice correspond?

*27. Discuss how Dilworth's theorem is related to a variation of the max-flow min-cut theorem of Chapter 5.

28. Apply the flow method to get a chain decomposition with the ordered set consisting of 2, 4, 8, 5, 10, 15, and 20 ordered by "is a factor of."

29. There is a relationship between chain decompositions of ordered sets and matchings in bipartite graphs. If X is ordered by P, construct a bipartite graph which has a vertex x_b (called the bottom of x) and a vertex x_t (called the top of x) for each x in X. There is an edge from x_t to y_b if (x, y) is an edge of the ordered set (so that $x < y$ (mod P)). This graph is the *Fulkerson graph* of the ordered set, introduced by Fulkerson in an article (in 1956) on which this problem is based.

(a) Draw the bipartite graph of the ordered set

$$\{2, 4, 8, 5, 10, 15, 20\},$$

ordered by "is a factor of."

(b) Verify that the edges $(2_t, 4_b)$, $(5_t, 10_b)$, and $(4_t, 20_b)$ are a matching in the bipartite graph in part (a) and show it in the graph.

Draw the new bipartite graph which has these edges and in addition has the edges (x_b, x_t) for each vertex x. What subsets of $\{2, 4, 8, 5, 10, 15, 20\}$ are the connected components of this graph? They correspond to chains of the ordered set. What chains do they correspond to?

(c) Find an alternating path which increases the size of the matching in the graph of part (a).

(d) Find the matching resulting from the path of part (c) and the chain decomposition which corresponds to it through the construction of part (b).

(e) Discuss the relationship between finding chain decompositions and finding maximum matchings.

30. Using the Fulkerson graph of the ordered set and the theory of matchings, prove that a chain decomposition has minimum size if and only if it has no decreasing alternating walk.

31. What theorem of matching theory corresponds to Dilworth's theorem? Derive Dilworth's theorem from this one by using the Fulkerson graph.

Section 3 Lattices

What Is a Lattice?

In the partially ordered set of all subsets of a set X with the subset ordering, there is a great deal of mathematical structure that is not typical of ordered sets in general. The union of two subsets X and Y is a set greater than or equal to both X and Y in the subset ordering. (The subsets of a three-element set are shown in Figure 3.2(b).)

In the integers from 1 to 12, ordered by "m is less than or equal to n (mod P) if and only if m is a factor of n," 3 and 4 are both less than 12, so the element 12 is greater than or equal to both of them. Further, 12 is the smallest element greater than or equal to both of them. Thus, 12 plays the role of a "union" for 3 and 4. However, for 3 and 5 there is no element that plays the role of a union, for no integer between 1 and 12 is greater than both of them in this ordering.

Figure 3.1

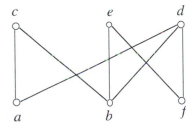

In the ordered set in Figure 3.1, the element c is "right above" both a and b, as is the element d. Thus, it is tempting to say that both c and d play the role of a union of a and b in this ordered set. On the other hand, *any* set that contains two sets X and Y in the subset ordering also contains the set $X \cup Y$, and so in Figure 3.2(b), $X \cup Y$ is the *only* set "right above" X and Y. In this sense, both c and d fail to play the role played by a union of two sets, because neither is the *only* element right above c and d.

The set $X \cup Y$ has two fundamental properties. First, it is above both X and Y in the subset ordering; we say it is an "upper bound" for X and Y. Second, given any other set above both X and Y, this other set must be above $X \cup Y$; we say $X \cup Y$ is the "least upper bound" for X and Y. In any partially ordered set, we say that u is an ***upper bound*** for x and y if u is greater than or equal to x and y. We say the upper bound u is the ***least upper bound*** of x and y if for any v with $x \leq v$ in P and $y \leq v$ in P, the element u is less than or equal to v in P. (Note that x and y could not have more than one least upper bound.) We use the suggestive notation

$$u = x \vee y$$

and also call u the **join** of x and y when u is the least upper bound of x and y: we read $x \vee y$ as "x join y."

Similarly, we say w is the **greatest lower bound** of x and y and we write

$$w = x \wedge y$$

if $w \leq xP$ in P and $w \leq yP$ in P, and whenever $v \leq x$ in P and $v \leq y$ in P, then $v \leq w$ in P as well. We also call $x \wedge y$ the **meet** of x and y, and we read $x \wedge y$ as "x meet y." Figure 3.1 shows that in general a pair of elements in a partially ordered set should not be expected to have a meet or a join.

We say a partially ordered set is a **lattice** if every pair of elements has a least upper bound (join) and a greatest lower bound (meet).

Example 3.1. The subsets of a set X, when they are ordered by set inclusion, form a lattice. ∎

Example 3.2. All the ordered sets shown in Figure 3.2 are lattices. The ordered sets in Figures 3.1, 1.4, 1.5, and 1.7 are not lattices. The ordered set shown in Figure 1.5 is called a "semilattice." Can you see why? ∎

Example 3.3. The positive integers with the usual ordering form a lattice; the join of two elements is the greater one and the meet of two elements is the lesser one. ∎

Example 3.4. The positive integers with the "is a factor of" ordering form a lattice. What are the join and meet of two integers m and n?

Since the join of m and n is the smallest number of which m and n are both factors, it is the least common multiple of m and n. Similarly, the meet of m and n is the greatest common divisor of m and n. ∎

Example 3.5. If you have had linear algebra you will be able to see that the subspaces of a vector space V form a lattice in which the meet of two subspaces is their intersection and the join of two subspaces is their subspace sum. To show this is so, we must show that the intersection is a greatest lower bound and the sum is a least upper bound. We shall consider the subspace sum here and save the intersection as an exercise. The sum $S + T$ of two subspaces is by definition the set $\{s + t | s \in S, \ t \in T\}$. Now $S = \{s + 0 | s \in S\}$ and $T = \{0 + t | t \in T\}$, so that $S \subseteq S + T$ and $T \subseteq S + T$. Thus, $S + T$ is an upper bound for S and T. Now suppose $S \subseteq U$ and $T \subseteq U$ for a subspace U of V. Then if $s \in S$ and $t \in T$, it follows that $s + t \in U$

Figure 3.2

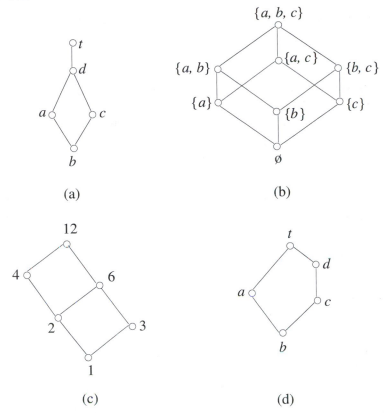

(a)

(b)

(c)

(d)

because U is a subspace. Thus, every element of $S+T$ is in U and therefore $S+T \subseteq U$. Therefore, $S+T$ is the least upper bound of the subspaces S and T. ∎

The Partition Lattice

Still another example of a lattice is the ordered set of all partitions of a set Y with the refinement ordering. The greatest lower bound of a partition $\mathbf{P} = \{P_1, P_2, ..., P_k\}$ and a partition $\mathbf{Q} = \{Q_1, Q_2, ..., Q_m\}$ is easy to describe. Since each element x of Y is in one of the P_i's and one of the Q_j's, each x is in an intersection $P_i \cap Q_j$. Either a set $P_i \cap Q_j$ is empty or else it has no element in common with any of the other sets $P_r \cap Q_s$. Thus, the nonempty sets $P_r \cap Q_s$ form the classes of a partition of Y. In any partition below both P and Q in the refinement ordering, each block is a subset of one of the P_i and a subset of one of the Q_j. Thus, each block is a subset of a $P_i \cap Q_j$. Therefore, the partition whose blocks are the nonempty intersections of P_i's and Q_j's is the greatest lower bound for $\mathbf{P}$ and $\mathbf{Q}$. In fact, in certain circumstances (which apply here) we can use meets to describe joins even if we don't know in advance that joins exist. To see the structure we need in order to use meets to describe joins, we introduce the concept of a semilattice.

An ordered set is called a ***meet-semilattice*** when any pair of elements has a greatest lower bound. Correspondingly, an ordered set is called a ***join-semilattice*** when each pair of elements has a least upper bound. Each ordered set in Figure 1.5 is a join-semilattice but not a lattice. The result of the last paragraph may be restated as follows:

Theorem 3.1. The partitions of a set form a meet-semilattice.

Now suppose X is a finite set ordered by P. It is possible to show that if any two elements of X have a greatest lower bound (mod P), and F is a subset of X, then F has a greatest lower bound. In other words, there is an element x such that $x \leq y$ for all $y \in F$, and if $z \leq y$ for all $y \in F$, then $z \leq x$. The proof for general sets F is quite analogous to the proof for the case in which F is a set with three elements, say $F = \{a, b, c\}$. In this case we let $x = (a \wedge b) \wedge c$. Then $x \leq a \wedge b$ and $x \leq c$. But since $a \wedge b$ is a lower bound for a and b, $x \leq a$ and $x \leq b$. Thus, x is less than or equal to all three elements of F. Now suppose z is less than or equal to a, b, and c. Then $z \leq a \wedge b$ (because $a \wedge b$ is the greatest lower bound for a and b) and $z \leq c$. Therefore, $z \leq (a \wedge b) \wedge c$, for the right-hand side of the inequality is the greatest lower bound of $a \wedge b$ and c. However, we have just shown that x satisfies the definition of being the greatest lower bound of F.

Lemma 3.2. If F is a nonempty finite set in a meet semilattice, then F has a greatest lower bound.

Proof. Let S be the set of all nonnegative integers n such that if F has n elements, then F has a greatest lower bound. Then 2 is in S by definition. Assume k is in S and let F have $k+1$ elements $x_1, x_2, ..., x_{k+1}$. Let $F' = \{x_1, x_2, ..., x_k\}$ and let x be the greatest lower bound of F'. Then $x \wedge x_{k+1}$ is less than or equal to x_{k+1} and x, and since x is less than or equal to all elements of F', then $x \wedge x_{k+1}$ is less than or equal to all elements of F. But if z is less than or equal to all elements of F, then it is less than or equal to all elements of F' and so is less than or equal to x (why?) and x_{k+1}. Thus, it is less than or equal to $x \wedge x_{k+1}$. Therefore, $x \wedge x_{k+1}$ satisfies the two conditions which make it a greatest lower bound of F. Thus $k+1$ is in S, and so by the principle of mathematical induction, S contains all the positive integers. This proves the theorem. ∎

We have seen that if (X, P) is a finite meet-semilattice, such as the partitions of a finite set, any nonempty subset of X has a meet. In particular, if x and y are in X and some elements of X are above both x and y, then the set of all elements above both x and y form a nonempty set F. The meet of F is above both x and y, so it is an upper bound to x and y. Any element above x and y is in F, so it is greater than or equal to the meet of F. In other words, the meet of F is the least upper bound of x and y. This observation yields our next theorem.

Theorem 3.3. If a finite meet-semilattice (X, P) has an element t such that $t \geq x$ for all $x \in X$, then it is a lattice.

Proof. Since the "top element" t is above each x and y, the set F of elements above x and y is nonempty. Thus, as before, the meet of F is a least upper bound for x and y, so every pair of elements has a least upper bound. ∎

Example 3.6. If X is the set of partitions of a set Y and P is the refinement ordering, then (X, P) is a lattice, known as the **partition lattice**. This is a direct consequence of Lemma 3.2 and Theorem 3.3 because the partition of Y with one class—Y itself—is greater than or equal to all partitions relative to the refinement ordering. ∎

There is a disadvantage to this method of showing that the poset of partitions of Y is a lattice. Namely, given two partitions $\mathbf{P}$ and $\mathbf{Q}$, we have no concrete description of the least upper bound, or join, of $\mathbf{P}$ and $\mathbf{Q}$. Let us try to find such a concrete description. Because $\mathbf{P}$ and $\mathbf{Q}$ must be finer than $\mathbf{P} \vee \mathbf{Q}$, every class of $\mathbf{P}$ must be a subset of some class of $\mathbf{P} \vee \mathbf{Q}$ and every class of $\mathbf{Q}$ must be a subset of some class of $\mathbf{P} \vee \mathbf{Q}$. Now suppose a class C of $\mathbf{P}$ and a class K of $\mathbf{Q}$ have an element x in common. Then both C and K must be subsets of the class of $\mathbf{P} \vee \mathbf{Q}$ containing x. Thus, it would appear that the classes of $\mathbf{P} \vee \mathbf{Q}$ are formed by taking unions of overlapping classes of $\mathbf{P}$ and $\mathbf{Q}$. This will be the case if we interpret "union of overlapping classes" properly. For such a proper interpretation, we must realize that while C and K may have x in common, C may have an element y in common with still another class K' of $\mathbf{Q}$, and K' may have another element z in common with a class C' of $\mathbf{P}$, and so on. Thus, we form unions until we run out of overlaps and then we have a class of $\mathbf{P} \vee \mathbf{Q}$. It was because of this somewhat complicated description of joins that we chose to use Theorem 3.3 to show that we have a lattice.

The Bond Lattice of a Graph

Example 3.7. If $G = (V, E)$ is a (multi)graph, there are some special partitions of V that give rise to an interesting lattice that will be useful later. A partition $\mathbf{P}$ of V is called a *bond* of G if, for each class C of $\mathbf{P}$, all pairs of vertices in C are connected by a walk whose vertices are all in C. Thus, for the graph in Figure 3.3, the partition $\{\{1,4\},\{2,3\}\} = 14/23$ is not a bond, because 1 and 4 are not connected by a walk in the set $\{1,4\}$. On the other hand, 12/34 is a bond because 1 and 2 are connected by edge $\{1,2\}$ and 3 and 4 are connected by the edge $\{3,4\}$. The partial ordering on bonds is the usual partial ordering on partitions. The bond lattice of G is shown next to G in Figure 3.3. Notice that the meet of 124/3 and 134/2 is *not* the partition 14/2/3 (as in the partition lattice) because 14/2/3 is *not* a bond.

The bond partitions of a graph *do* form a lattice when ordered by the usual ordering on partitions. Certainly we can check that any two bonds have a greatest lower bound and a least upper bound in the picture in Figure 3.3.

Figure 3.3

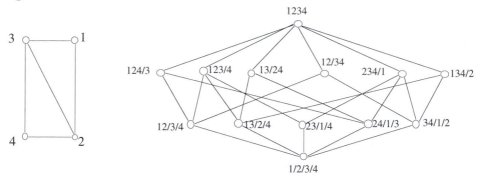

The meet of two bonds may be described rather cleverly as follows. Suppose **P** and **Q** are bonds. Let F be the set of edges of G both of whose vertices lie in the same class of $\mathbf{P} \wedge \mathbf{Q}$ in the partition lattice. Let **R** be the connected component partition of (V, F). (Note that some vertices grouped together by $\mathbf{P} \wedge \mathbf{Q}$ might not be joined by a walk whose edges are in F.) It is straightforward to show that **R** is the meet of **P** and **Q** in the bond lattice.

The join of each pair **P** and **Q** in the lattice shown in Figure 3.3 is their join in the partition lattice. It is not too difficult to use the "overlapping classes" description of the join of partitions to show that the join of two bonds in the partition lattice is their join in the bond lattice as well. ∎

The Algebraic Description of Lattices

The notation $x \vee y$ for the join of x and y and $x \wedge y$ for the meet of x and y is suggestive of the algebraic operations of $x + y$ and $x \cdot y$. The set union and intersection operations have many of the arithmetic properties of "plus" and "times." To a lesser extent, the join and meet operations share some of these properties. The most useful properties are listed in the following theorem.

Theorem 3.4. The operations $\vee$ and $\wedge$ in a lattice L satisfy the following rules:

(1) $(x \vee y) \vee z = x \vee (y \vee z)$ associative laws
     $(x \wedge y) \wedge z = x \wedge (y \wedge z)$
(2) $x \vee x = x \wedge x = x$ idempotent law
(3) $x \vee y = y \vee x$ commutative laws
     $x \wedge y = y \wedge x$
(4) $x \vee (x \wedge y) = x$ absorptive laws
     $x \wedge (x \vee y) = x$

Proof. Rules 2 and 3 follow immediately from the definition of least upper bound and greatest lower bound. The first part of rule 4 states that x is the least upper bound of x and $x \wedge y$. However, regardless of what y is, $x \wedge y \leq x$. Thus, x is greater than or equal to both x and $x \wedge y$. If z is greater than or equal to both x and $x \wedge y$, then z is greater than or equal to x. Then by definition, x is the least upper bound of x and $x \wedge y$; this proves the first absorptive law. The proof of the second is similar.

To prove the first associative law, note that if $(x \vee y) \vee z = w$, then $w \geq z$ and $w \geq x \vee y$. Then by the definition of join, $w \geq x$ and $w \geq y$. Since $w \geq y$ and $w \geq z$, it follows that $w \geq y \vee z$, and since $w \geq x$ as well, it follows that $w \geq x \vee (y \vee z)$. Thus, we get

$$(x \vee y) \vee z \geq x \vee (y \vee z).$$

A similar argument shows that

$$(x \vee y) \vee z \leq x \vee (y \vee z);$$

and by the antisymmetry of $\leq$, we get the desired equality. The second associative law is proved similarly. $\blacksquare$

It is tempting to expect the "distributive" law $x \wedge (y \vee z) = (x \wedge y) \vee (x \wedge z)$ to hold since it holds for intersections and unions of sets. (Check that it holds for sets if you don't remember why!) However, it does not hold in general for lattices, and so you must take care in making computations by using standard arithmetic rules.

Note, by the way, that $(x \wedge y) \wedge z$ is the greatest lower bound of x, y, and z, as we already showed must be the case. Thus, the associative law says that either natural interpretation of the symbol $x \wedge y \wedge z$ is the greatest lower bound of x, y, and z. A valuable observation is that rules 1–4 give us an alternate description of a lattice.

> **Theorem 3.5.** Suppose a set L has two operations $\wedge$ and $\vee$ defined on all pairs of elements in L and satisfying rules 1–4 of Theorem 3.4. Then L is a lattice with the partial ordering P defined by $x \leq y \pmod{P}$ if and only if $x = x \wedge y$.
>
> *Proof.* It is necessary to verify that the relation P is a partial ordering. This is a straightforward application of rules 1–3. Then it is necessary to verify that $x \wedge y$ is the greatest lower bound of x and y. This is also a straightforward application of rules 1–3. In order to show that $x \vee y$ is the least upper bound of x and y, we first use rule 4 to show that $x \leq y$ if and only if $y = x \vee y$; then we apply rules 1–3 again. ∎

EXERCISES

1. Let X consist of the integers from 1 to 10 and let P be the usual less than or equal to relation. Show that two elements of X have a least upper bound, namely the larger of the two numbers. Show that (X, P) is a lattice.

2. Let X consist of the positive factors of 24 (1 and 24 included), and let P be the relation "is a factor of" or "divides." Does every pair of elements have a least upper bound? If so, describe it. If not, why not?

3. Let X consist of all the positive integers between 1 and 24 (inclusive), and let P be the relation "is a factor of." Does every pair of elements have a greatest lower bound? If so, describe it; if not, why not?

4. Draw as many nonisomorphic examples as you can of five-element lattices, that is, lattices with a five-element vertex set.

5. Every positive integer can be factored in one and only one way into a product of powers of prime integers. Use this fact to show that every pair of positive integers has a least upper bound relative to the partial ordering "is a factor of." (Hint: Consider all primes that occur in the unique factorization of two integers. Multiply together all prime powers that occur in one but not both factorizations and multiply this by the higher power used for each prime appearing in both factorizations.) Notice that in proving you have a least upper bound you are proving that each pair of integers has a least common multiple.

6. (Continuation of Exercise 5.) Show that each pair of elements in the partially ordered set of Exercise 5 has a greatest lower bound. Note that you are showing that each pair of positive integers has a greatest common divisor.

7. Each equivalence relation on a set Y may be regarded as a set of ordered pairs. For two equivalence relations E and R on Y, let us say that $E \leq R$ if E is a subset of R. Show that the equivalence relations on Y form a meet-semilattice.

8. (Continuation of Exercise 7.) Show that the equivalence relations on Y form a lattice.

9. The lattice of Exercises 7 and 8 is isomorphic to another lattice studied in the text. Discover which one it is and show why the two lattices are isomorphic.

10. Find all partitions of $\{1, 2, 3, 4\}$ that are not bonds in Example 3.7 and show they are not bonds. (One such partition was described in Example 3.7.) How do you know you have found them all?

11. Let G be the graph on vertices $\{1, 2, 3, 4\}$ with edges $\{1, 2\}, \{2, 3\}, \{3, 4\}$, and $\{1, 4\}$. Find the bonds of G and draw the bond lattice of G.

12. Draw the bond lattice of the graph on vertices $\{1, 2, 3, 4\}$ with edges $\{1, 2\}, \{2, 3\}, \{1, 3\}$, and $\{1, 4\}$.

13. What is the bond lattice of the complete graph on four vertices, the graph with all possible edges connecting four vertices? (Hint: This is a lattice whose name you already know.)

14. Consider the set of all partial orderings P of a set X (thought of as sets of ordered pairs), ordered by set inclusion. Show that this ordered set is a semilattice but not a lattice.

15. Recall that a tree is a connected graph with no cycles. Give as complete a description as you can of the bond lattice of a tree on four vertices.

16. What can you say about the bond lattice of a complete graph? Of a tree?

17. (a) Show that the join of two bonds may be computed as described in Example 3.7.
    (b) Show that the meet of two bonds may be computed as described in Example 3.7.

18. Prove the second associative law in Theorem 3.4.

19. Prove the second absorptive law in Theorem 3.4.

20. We say b is the *bottom* of a poset (X, P) if $b \leq x$ for all $x \in X$ and t is the *top* if $t \geq x$ for all $x \in X$. Prove that if a lattice has a bottom and a top, then

$$x \vee b = x \quad \text{and} \quad t \wedge x = x \text{ for all } x \text{ in the lattice}$$

and
$$t \vee x = t \quad \text{and} \quad b \wedge x = b \text{ for all } x \text{ in the lattice.}$$

21. Show that the lattice of subsets of a set satisfies the distributive laws
    (a) $X \cap (Y \cup Z) = (X \cap Y) \cup (X \cap Z)$.
    (b) $X \cup (Y \cap Z) = (X \cup Y) \cap (X \cup Z)$.

22. Show that the lattice of partitions of $\{1, 2, 3\}$ satisfies neither of the distributive laws
    (a) $X \wedge (Y \vee Z) = (X \wedge Y) \vee (X \wedge Z)$.
    (b) $X \vee (Y \wedge Z) = (X \vee Y) \wedge (X \vee Z)$.

23. Show that a finite lattice has a bottom and a top (see Exercise 20).

24. (a) Show that the relation P defined in Theorem 3.5 is a partial ordering.
    (b) Show that $x \wedge y$ is the greatest lower bound of x and y in the partial ordering of Theorem 3.5.
    (c) Show that $x \vee y$ is the least upper bound of x and y in the partial ordering of Theorem 3.5.

*25. Review the concept of a matroid, which was introduced in Section 2 of Chapter 4. An element x of a matroid is said to depend on the independent set I if $I \cup \{x\}$ is *not* independent. The *flat* $F(I)$ determined by I consists of I together with all points that depend on I. Show that the flats $F(I)$ of a matroid form a lattice. (This kind of lattice is called a geometric lattice and is very important in theoretical combinatorics. The bond lattice of a graph turns out to be isomorphic to a geometric lattice, for example.)

26. A lattice is called **distributive** if for each x, y, and z in the lattice, $x \wedge (y \vee z) = (x \wedge y) \vee (x \wedge z)$. A lattice is called **modular** if for each x, y, and z in the lattice with $y \leq x$, $x \wedge (y \vee z) = y \vee (x \wedge z)$. Show that a distributive lattice is modular.

27. Distributive and modular lattices are defined in the previous exercise.
    (a) Which, if any, of the lattices in Figure 3.4 is distributive?
    (b) Which, if any, of the lattices in Figure 3.4 is modular?

28. Distributive and modular lattices are defined in Exercise 26. For what values of n is the lattice of partitions of an n-element set distributive? Modular?

29. In an ordered set we define the *closed interval* $[u, v]$ by $[u, v] = \{z \mid u \leq z \text{ and } z \leq v\}$. Show that in a modular lattice, for any two elements x and y, the mapping defined by
    $$f(t) = t \vee y$$

Figure 3.4

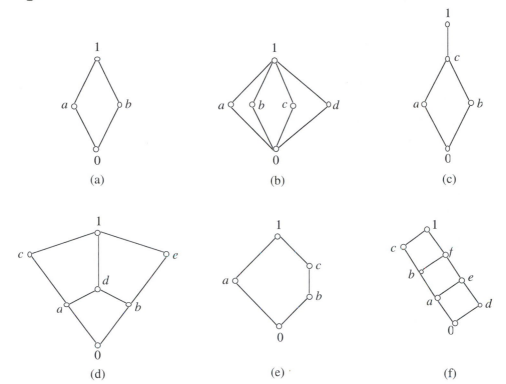

(a)   (b)   (c)

(d)   (e)   (f)

is a bijection between $[x \wedge y, x]$ and $[y, x \vee y]$. What can you conclude about the sizes of these two intervals in a finite lattice?

30. Find an example of a lattice and two elements x and y in the lattice such that the mapping f of the previous exercise is not a bijection.

31. Prove that in a modular lattice the bijection of Exercise 29 is an isomorphism.

32. If $\mathbf{P} = \{B_1, B_2, \ldots, B_k\}$ is a partition of the n-element set $\{1, 2, \ldots, n\}$, then the set $\{p, q\} \subseteq B_i$ (with $p < q$) and the set $\{r, s\} \subseteq B_j$ (with $r < s$) *cross* if $p < r < q < s$ (or $r < p < s < q$) and $i \neq j$. We say $\mathbf{P}$ is a noncrossing partition if there are no two pairs $\{p, q\}$ and $\{r, s\}$ that cross. Give an example of a crossing partition of $\{1, 2, 3, 4\}$ and a noncrossing partition of $\{1, 2, 3, 4\}$. Show that, with the usual refinement ordering, the noncrossing partitions form a lattice. Do the meet and join operations for this lattice produce the same partitions (from two noncrossing partitions) as the meet and join operations of the partition lattice?

Section 4 Boolean Algebras

The Idea of a Complement

We began our study of lattices with the lattice of subsets of a set. We soon saw that the lattice of subsets of a set "behaves" much more nicely than do other lattices. We have, for example, the distributive laws

$$A \cap (B \cup C) = (A \cap B) \cup (A \cap C)$$

and

$$A \cup (B \cap C) = (A \cup B) \cap (A \cup C).$$

We saw that these laws don't hold in all lattices; however, they are not restricted to subset lattices. For example, in the lattice of integral divisors of 12 shown in Figure 4.1, we have that

$$p \wedge (q \vee r) = (p \wedge q) \vee (p \wedge r) \tag{4.1}$$

and

$$p \vee (q \wedge r) = (p \vee q) \wedge (p \vee r). \tag{4.2}$$

Figure 4.1

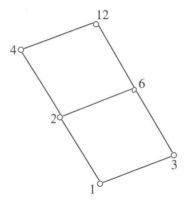

A lattice satisfying the two rules given in Equations (4.1) and (4.2) is called a ***distributive lattice***. There is, however, another property that makes subset lattices, such as the one shown in Figure 4.2, stand out from other distributive lattices, such as the one in Figure 4.1. Each set A in Figure 4.2 can be paired with one other subset A^{-1} in such a way that

$$A \cup A^{-1} = \{a, b, c\} \quad \text{and} \quad A \cap A^{-1} = \emptyset.$$

Figure 4.2

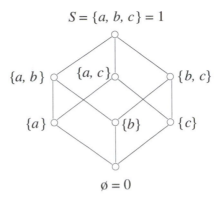

$$S = \{a, b, c\} = 1$$

For example, if $A = \{b\}$, then $A^{-1} = \{a, c\}$, and if $A = \{a, b\}$, then $A^{-1} = \{c\}$. On the other hand, in Figure 4.1 there is no element 2^{-1} such that $2 \vee 2^{-1} = 12$.

The set A^{-1} is called the complement of A in the set $S = \{a, b, c\}$. For example, $\{a\}^{-1} = \{b, c\}$ and $\{a, b\}^{-1} = \{c\}$. Thus, A^{-1} is just the set we have been denoting by $S - A$ and calling the complement of A in S all along. We have known all along that

$$(S - A) \cup A = S \quad \text{and} \quad (S - A) \cap A = \emptyset.$$

The use of the negative one exponent is designed to suggest a number of algebraic properties that will be useful to us later on.

The complementation property occurs in other areas as well, for example, in the negation of statements in elementary logic or in programming languages and in inverter circuits in circuit design. It turns out that both these important examples (which we shall discuss later) also give rise to distributive lattices. In order to give a uniform description of the applications of complementation in distributive lattices, we introduce some new terms. In this context, it has become quite standard to use 0 to stand for the bottom element of a lattice and 1 to stand for the top element of a lattice. We adopt this standard for this section.

In an (arbitrary) lattice L with bottom 0 and top 1, an element c is said to be a **complement** to a if

$$a \vee c = 1$$

and

$$a \wedge c = 0.$$

A lattice L is said to be **complemented** if every element of L has a complement.

Boolean Algebras

A **Boolean algebra** is a complemented, distributive lattice.

Example 4.1. Discuss which lattices in Figures 4.1, 4.2, and 4.3 are distributive, which are complemented, and which are Boolean algebras.

Figure 4.3

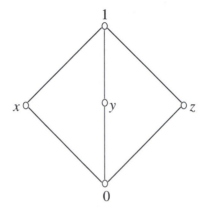

We have already observed that the lattice in Figure 4.1 is distributive and that the lattice in Figure 4.2 is distributive because it is a subset lattice. In Figure 4.3, however, $(x \vee y) \wedge z = 1 \wedge z = z$ and $x \wedge z = 0$ and $y \wedge z = 0$, so $(x \wedge z) \vee (y \wedge z) = 0$. Thus, the lattice is not distributive. We have observed that 2 has no complement in Figure 4.1; in fact, neither does 6. We have observed that Figure 4.2 represents the subsets of a set S and that for each A, the set $S - A$ is its complement. Thus Figure 4.2 is a complemented lattice. In fact, Figure 4.3 is also a picture of a complemented lattice because 0 has 1 for a complement, 1 has 0 for a complement, x has two complements y and z, y has two complements x and z, and z has two complements x and y. Among the three lattices, only the lattice in Figure 4.2 is both distributive and complemented. Thus, the lattice in Figure 4.2 is the only Boolean algebra among the three lattices. ∎

Notice that in Figure 4.3 the element x has two different complements y and z. In the lattice of subsets of $\{a, b, c\}$, the set $\{a\}$ has exactly one complement, namely $\{b, c\}$. In fact, in any Boolean algebra, each element has a unique complement.

> **Theorem 4.1.** In a Boolean algebra, each element has one and only one complement.
>
> *Proof.* The proof is an exercise in applying the distributive law to show that the assumption that an element x has two distinct complements leads to a contradiction. ∎

Boolean Algebras of Statements

Boolean algebras come up in the logical analysis of the truth and falsity of statements. What do we mean by the word statement? The one thing we assume is that a statement is either true or false. Thus "G has four edges" is a statement we could make about any graph. For some graphs it would be true, and for some it would be false, but once we know what graph we are talking about, we will know if the statement is true or false. On the other hand the phrase "G is a pretty graph" is vague, so we do not consider it to be a statement. The statement "G does not have four edges" is the opposite of the statement "G has four edges," and so it is a natural candidate to serve as its complement in the Boolean algebras we build from statements.

So that we can talk about statements in general terms, we will use symbols (typically p, q, r, ...) to stand for statements. If p stands for a statement then $\neg p$ stands for the opposite statement "it is not the case that p." Thus if p stands for "G has four edges," then $\neg p$ stands for the statement "It is not the case that G has four edges," or, more simply, "G does not have four edges." Note that $\neg p$ is true exactly when p is false. This may be described by the "truth table" in Table 4.1

Table 4.1

p	$\neg p$
T	F
F	T

Given two statements p and q, we use $p \wedge q$ to stand for the compound statement p and q. Thus if p stands for "G has four edges" and q stands for "G has five vertices," then $p \wedge q$ stands for "G has four edges and five vertices." Note that $p \wedge q$ is true exactly when p and q are both true. This is described by the truth table in Table 4.2 . For each possible combination of the truth and falsity of the constituent statements p and q, the table tells us the truth or falsity of the statement $p \wedge q$.

Table 4.2

p	q	$p \wedge q$
T	T	T
T	F	F
F	T	F
F	F	F

Given two statements p and q, we use $p \vee q$ to stand for the statement "p or q or both." Mathematicians usually read $p \vee q$ as "p or q" but in mathematical discussions interpret it as "p or q or both." Table 4.3 summarizes what we have said about how the truth and/or falsity of p and q determine that of $p \vee q$.

Table 4.3

p	q	$p \vee q$
T	T	T
T	F	T
F	T	T
F	F	F

The symbol strings we can build up from symbols standing for statements and unambiguous use of $\wedge$, $\vee$, and $\neg$ are called *symbolic statements*. If p stands for the statement "this graph has five vertices" and q stands for the statement "this graph has four edges," then it is clear that the statements $p \wedge q$ and $q \wedge p$ mean the same thing. We say that two symbolic statements are *equivalent* if they are true in exactly the same circumstances. (We will further explain the phrase "same circumstances" later on.) In the same way that we write $\frac{1}{2} = \frac{2}{4}$, even though they technically are different representations of the same number, we will write $p \wedge q = q \wedge p$ and in general write $p = q$ when p and q are equivalent symbolic statements, even though they are different representations of the same "idea."

To analyze a symbolic statement by a truth table, we first show, as in Table 4.4, all possible assignments of truth and falsity to the component propositions $p, q, \ldots$. These assignments are the "circumstances" we referred to earlier. Next, we make a column for every symbol (except for parentheses if there are any) in the expressions we are analyzing. To begin the analysis, in the column under each of the symbols standing for the original propositions, we copy the truth value of the original proposition from the left-hand columns. These columns are marked at the bottom with 1's in Table 4.4, meaning this was our first action. Next, we look for each operator ($\wedge$, $\vee$, or $\neg$) whose truth value we can determine and fill it in. These columns are marked with 2's. Now we repeat the process of looking for operators whose truth values we can fill in until all operators have truth values below them. In Table 4.4, we show this process for the two statements $(p \vee q) \wedge p$ and $(p \wedge q) \vee q$ that appear in the absorptive laws for lattices.

For the purpose of explaining step 3, we added some technically unnecessary notation to the table. We use the truth values in the two columns

Table 4.4

p	q	$(p$	$\overset{+}{\vee}$	$q)$	$\wedge$	p	•	$(p$	$\wedge$	$q)$	$\overset{+}{\vee}$	q
T	T	T	T	T	T	T		T	T	T	T	T
T	F	T	T	F	T	T		T	F	F	F	F
F	T	F	T	T	F	F		F	F	T	T	T
F	F	F	F	F	F	F		F	F	F	F	F
step		1	2	1	3	1		1	2	1	3	1

marked with a plus sign to combine the truth values of $(p\vee q)$ with those of p and to combine the truth values $p\wedge q$ with those of q. Now we note the first column marked 3, which has the truth values for the proposition $(p\vee q)\wedge p$, has exactly the same truth values as p. Thus, $(p\vee q)\wedge p$ is equivalent to p. Column 3 under $(p\wedge q)\vee q$ has the same entries as the column for q. Thus, $(p\wedge q)\vee q$ is equivalent to q. We have proved that the lattice laws $(p\vee q)\wedge p=p$ and $(p\wedge q)\vee q=q$ apply to statements.

In Table 4.5, we show the truth table analysis of $(p\vee q)\vee r$ and $p\vee(q\vee r)$. This shows that the associative lattice identity holds for statements.

Table 4.5

p	q	r	$(p$	$\overset{+}{\vee}$	$q)$	$\vee$	r	•	p	$\overset{+}{\vee}$	$(q$	$\overset{+}{\vee}$	$r)$
T	T	T	T	T	T	T	T		T	T	T	T	T
T	T	F	T	T	T	T	F		T	T	T	T	F
T	F	T	T	T	F	T	T		T	T	F	T	T
T	F	F	T	T	F	T	F		T	T	F	F	F
F	T	T	F	T	T	T	T		F	T	T	T	T
F	T	F	F	T	T	T	F		F	T	T	T	F
F	F	T	F	F	F	T	T		F	T	F	T	T
F	F	F	F	F	F	F	F		F	F	F	F	F
step			1	2	1	3	1		1	3	1	2	1

Again for convenience, we mark with a plus sign the two columns used in stage 3.

In order to show that the symbolic statements we can make with a nonempty set of statement symbols $\{p,q,\ldots\}$ form a Boolean algebra, we denote $p\wedge\neg p$ (which is interpreted as $p\wedge(\neg p)$) by 0 and denote $p\vee\neg p$ by 1. Note that 0 is thus a symbolic statement that is false in all circumstances and 1 is a symbolic statement that is true in all circumstances. With these definitions, $\neg p$ becomes the complement of the statement p. Using truth table analyses we can check the other lattice laws and the distributive laws, so we get the following theorem:

Theorem 4.2. The symbolic statements made from a nonempty set of statement symbols (with "=" interpreted as "is equivalent to") form a Boolean algebra.

Combinatorial Gate Networks

Boolean algebras provide a convenient framework for studying electrical networks built from "two-state devices." A spot on a magnetic tape can be either magnetized or not, a switch can be either on or off, a voltage in a wire can be low or high, and so on. These are all examples of two-state devices. Several standard devices can be built to combine electronically voltages $x_1, x_2, \ldots, x_n$ on n various input wires. One is an "inverter," a standard diagram of which is shown first in Figure 4.4. An inverter turns a high voltage to a low one and turns a low voltage to a high one. An "and gate," shown next in Figure 4.4, takes two voltages as input and produces a high voltage if both are high and otherwise produces a low voltage. An "or gate," shown third in Figure 4.4, takes two voltages as input, produces a low voltage if both are low, and otherwise produces a high voltage.

Figure 4.4

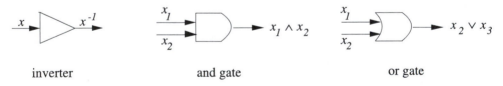

inverter   and gate   or gate

Table 4.6 shows the voltages produced by each gate using the convention 0 for low voltage and 1 for high voltage.

Table 4.6

Inputs		Devices		
x_1	x_2	x_1-inverter	And gate	Or gate
1	1	0	1	1
1	0	0	0	1
0	1	1	0	1
0	0	1	0	0

Because the 0's and 1's in this table mirror exactly the F's and T's in the tables for $\neg$, $\wedge$, and $\vee$, we may use $x_1 \vee x_2$ to stand for the result of feeding voltages x_1 and x_2 into an "or" gate, $x_1 \wedge x_2$ to stand for the result of feeding voltages x_1 and x_2 into an "and" gate, and x_1^{-1} (read as x_1 inverted) for the result of feeding x_1 into an "inverter" gate. Then any expression such as $(x_1 \vee x_2) \wedge (x_3 \vee x_1^{-1})$ corresponds to a circuit that can be built with an input wire for each x_i and one overall output wire. In Figure 4.5 we show the circuit represented by the expression $(x_1 \vee x_2) \wedge (x_3 \vee x_1^{-1})$. Such a circuit is called a *combinatorial network* or *gate network*. A combinatorial network may be regarded as a directed graph whose vertices are the gates and the points where wires are split in two. We direct edges so that they are directed into gates where they are used as inputs and away from gates where they are used as outputs. A combinatorial network has no directed cycles (so-called feedback loops) and has only two incoming edges at an "and" gate or an "or" gate. Further, a vertex representing a split of a wire has only one incoming edge.

Figure 4.5

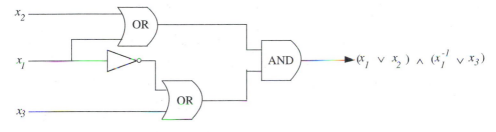

We can use the symbol 0 to stand for a device that accepts any number of input wires and produces low voltage regardless of the input. (A wire with a piece cut out or a circuit for $x_1^{-1} \wedge x_1$ is such a device.) We can use the symbol 1 to stand for a device that takes any number of inputs and produces high voltage regardless of the input. (The circuit for $x_1^{-1} \vee x_1$ is such a device.) The voltage table (4.6) corresponds directly with the tables (4.1, 4.2, and 4.3) we used to show that propositions form a lattice. Thus, using voltage tables in the same way that we used truth tables and considering two circuits equal if, based on the same inputs, they produce the same outputs, we get Theorem 4.3.

Theorem 4.3. The circuits built from inverters, "and" gates, and "or" gates with inputs $x_1, x_2, \ldots, x_n$ form a Boolean algebra.

We will give no proof for Theorem 4.3 since it uses the same ideas as Theorem 4.2.

Boolean Polynomials

The meaningful expressions we can build up from $x_1, x_2, \ldots, x_n$ using parentheses, $\vee$, $\wedge$, and inversion are called **Boolean polynomials** in the symbols $x_1, x_2, \ldots, x_n$. (Thus, except for the use of x^{-1} rather than $\neg x$ for the complement of x, Boolean polynomials are another name for symbolic statements made with the symbols x_1, x_2, and so on.) When we substitute the values 1 or 0 for the x_i's and use the rules

$$1 \vee 1 = 1 \wedge 1 = 1,$$
$$0 \vee 0 = 0 \wedge 0 = 0,$$
$$0 \vee 1 = 1 \vee 0 = 1,$$
$$1 \wedge 0 = 0 \wedge 1 = 0,$$

$0^{-1} = 1$, and $1^{-1} = 0$ of Boolean arithmetic, we are using our polynomials as **Boolean functions** of n variables. Thus, Boolean polynomials represent circuits and Boolean functions represent the relationship between the input and output of such circuits.

How can we determine whether two circuits accomplish exactly the same thing—i.e., are equivalent? Is there a less time-consuming way than writing down voltage tables? Determining such equivalences is one reason we develop algebraic manipulation rules for Boolean algebras. Our eventual goal will be to have a standard or **canonical form** for writing polynomials; two polynomials are then equivalent if and only if they have the same canonical form.

In the same way that the circuits we may build up from certain wires using Boolean operators form a Boolean algebra, we see that the set of Boolean polynomials in certain variables form an isomorphic Boolean algebra, an abstract version of the algebra we can build from circuits. Although what it means to put an inverter in front of a circuit is clear, what it means to form the complement $(p(x_1, x_2, \ldots, x_n))^{-1}$ of a Boolean polynomial is perhaps not so clear. It turns out that whenever we have an expression $(x \vee y)^{-1}$ or $(x \wedge y)^{-1}$, it can be reexpressed in terms of x^{-1} and y^{-1}.

DeMorgan's Laws

DeMorgan's laws tell us that in any Boolean algebra

$$(x \vee y)^{-1} = x^{-1} \wedge y^{-1}$$

and

$$(x \wedge y)^{-1} = x^{-1} \vee y^{-1}.$$

Theorem 4.4. DeMorgan's laws are valid in every Boolean algebra.

Proof. We must show that $x \vee y$ has $x^{-1} \wedge y^{-1}$ as its complement. Since $x \vee y$ can have only one complement, this proves that $(x \vee y)^{-1} = x^{-1} \wedge y^{-1}$. Thus, we form the join

$$
\begin{aligned}
(x \vee y) \vee (x^{-1} \wedge y^{-1}) &= ((x \vee y) \vee x^{-1}) \wedge ((x \vee y) \vee y^{-1}) \\
&= (x \vee x^{-1} \vee y) \wedge (x \vee y \vee y^{-1}) \\
&= 1 \wedge 1 = 1.
\end{aligned}
$$

Now we form the meet

$$
\begin{aligned}
(x \vee y) \wedge (x^{-1} \wedge y^{-1}) &= (x \wedge x^{-1} \wedge y^{-1}) \vee (y \wedge x^{-1} \wedge y^{-1}) \\
&= 0 \vee 0 \\
&= 0.
\end{aligned}
$$

DeMorgan's other law is proved similarly. ∎

By induction, we may extend DeMorgan's laws to read

$$
(x_1 \vee x_2 \vee \cdots \vee x_n)^{-1} = x_1^{-1} \wedge x_2^{-1} \wedge \cdots \wedge x_n^{-1} \tag{4.3}
$$

and

$$
(x_1 \wedge x_2 \wedge \cdots \wedge x_n)^{-1} = x_1^{-1} \vee x_2^{-1} \vee \cdots \vee x_n^{-1}. \tag{4.4}
$$

By applying DeMorgan's laws (4.3) and (4.4) repeatedly, we may take any Boolean polynomial and express it as a polynomial in the symbols $x_1, x_1^{-1}, x_2, x_2^{-1}, \ldots, x_n, x_n^{-1}$ such that the only operations used to build up the polynomial are join and meet. This is analogous to applying the binomial and multinomial theorems to ordinary polynomials built from multiplication, sum, and nonnegative integer exponentiation to get polynomials built up from symbols $x_1^{i_1}, x_2^{i_2}, \ldots, x_n^{i_n}$ by addition and multiplication. For ordinary polynomials, we have a canonical form: namely, every polynomial in variables $x_1, x_2, \ldots, x_n$ may be built up using a sum of numerical multiples of terms of the form $x_1^{i_1} \cdot x_2^{i_2} \cdots x_n^{i_n}$ where, of course, x_i^0 is to be interpreted as 1.

Disjunctive Normal Form

Our canonical form for Boolean polynomials, which goes by the name ***disjunctive normal form***, is given in Theorem 4.6. A preliminary result is Theorem 4.5, which just formalizes the fact that we distribute meet over join as we distribute multiplication over sum.

Theorem 4.5. Every Boolean polynomial may be written as a join of monomials of the form

$$x_1^{i_1} \wedge x_2^{i_2} \wedge \cdots \wedge x_n^{i_n} = \bigwedge_{j=1}^{n} x_j^{i_j},$$

where $i_j = +1$, -1, or 0, where $x_i^1 = x_i$, x_i^{-1} is the complement of x_i, and where $x_i^0 = 1$.

Proof. Rather than give a detailed formal proof, we explain by analogy how the proof works. By applying DeMorgan's laws, we may rewrite our polynomial as a polynomial in the terms $x_1, x_2, \ldots, x_n$ and $x_1^{-1}, x_2^{-1}, \ldots, x_n^{-1}$ built up from meet and join alone. Just as ordinary multiplication distributes over ordinary addition, meet distributes over join in a Boolean algebra. Thus, just as we may express an ordinary polynomial made from sum and product operations as a sum of products, we may write a Boolean polynomial made from join and meet operations as a join of meets. Since $x_i^0 = 1$, any meet of just some terms of the form x_i and x_j^{-1} may be written in the form $x_1^{i_1} \wedge x_2^{i_2} \wedge \cdots \wedge x_n^{i_n} = \bigwedge_{i=1}^{n} x_j^{i_j}$. ∎

Now in each monomial of the form

$$x_1^{i_1} \wedge x_2^{i_2} \wedge \cdots \wedge x_n^{i_n}$$

in which x_j^0 appears, we may replace x_j^0 by $x_j \vee x_j^{-1}$. We then distribute meet over join again as in

$$x_1 \wedge x_2 \wedge x_3^0 = x_1 \wedge x_2 \wedge (x_3 \vee x_3^{-1}) = (x_1 \wedge x_2 \wedge x_3) \vee (x_1 \wedge x_2 \wedge x_3^{-1}).$$

In this way, we may rid our expression of all terms with exponent 0 to get Theorem 4.6.

Theorem 4.6. (Disjunctive normal form) Every Boolean polynomial may be written as a join of monomials of the form

$$x_1^{i_1} \wedge x_2^{i_2} \wedge \cdots \wedge x_n^{i_n} = \bigwedge_{j=1}^{n} x_j^{i_j},$$

where each i_j is -1 or 1.

Proof. Replace each x_j^0 by $x_j \vee x_j^{-1}$ and distribute as shown previously. ∎

It may be surprising that each finite Boolean algebra is isomorphic to a Boolean algebra of sets. From results used to prove this fact, we will conclude that each of the joins of monomials is different from all the rest, i.e., that the disjunctive normal form is unique. Then to tell if two polynomials (or two circuits or two logical expressions) are equivalent, we put the two polynomials in disjunctive normal form and determine whether they are joins of the same monomials.

All Finite Boolean Algebras Are Subset Lattices

An **atom** in a lattice is an element that covers the bottom element 0. We shall show that in a finite Boolean algebra, each element can be represented uniquely as a join of atoms. From this, it will follow that a Boolean algebra whose atoms are a set A is isomorphic to the lattice of subsets of A. Our next two theorems are special cases of Garrett Birkhoff's unique representation theorems for distributive lattices in general and Marshall Stone's representation theorem for Boolean algebras in general (including infinite Boolean algebras).

> **Theorem 4.7.** Every nonzero element of a finite Boolean algebra B is a join of atoms.
>
> *Proof.* Let x be nonzero and let $a_1, a_2, \ldots, a_k$ be the atoms below or equal to x (there must be at least one or else x would be 0). Let $y = a_1 \vee a_2 \vee \cdots \vee a_k$. Now since $y \vee y^{-1} = 1$ and $x \geq y$, $x \vee y^{-1} = 1$. If there are any atoms below $x \wedge y^{-1}$, they are atoms below x, but no atom below x is below y^{-1} since $a_i \wedge y^{-1} = 0$ by the definition of y^{-1}. But if $x \wedge y^{-1}$ has no atoms below or equal to it, then $x \wedge y^{-1} = 0$. Now we apply the distributive law to
>
> $$x = x \wedge 1 = x \wedge (y \vee y^{-1})$$
>
> to get
> $$x = (x \wedge y) \vee (x \wedge y^{-1}) = (x \wedge y) \vee 0 = y.$$
>
> Thus, x must be the join of the atoms below it. ∎

Rather than using the distributive law in the proof, we could have observed that x and y are both complements to y^{-1}, and so by Theorem 4.1, they are equal.

We have just shown that each nonzero x is the join of the set of all atoms below it. If we show that x is not a join of a proper subset of these atoms, then we have shown that x is expressed uniquely as a join of atoms.

Theorem 4.8. Each nonzero element in a finite Boolean algebra can be represented in only one way as a join of atoms, namely as the join of all atoms below it.

Proof. Let $x = a_1 \vee a_2 \vee \cdots \vee a_k$ where the a_i's are all the atoms below x. Suppose $x = a_1 \vee \cdots \vee a_{k-1}$ as well. Then

$$
\begin{aligned}
a_k = a_k \wedge x &= a_k \wedge (a_1 \vee a_2 \vee \cdots \vee a_{k-1}) \\
&= (a_k \wedge a_1) \vee (a_k \wedge a_2) \vee \cdots \vee (a_k \wedge a_{k-1}) \\
&= 0 \vee 0 \vee \cdots \vee 0 \\
&= 0
\end{aligned}
$$

because the meet of two distinct atoms is 0. This contradiction means that x is not the join of any $k - 1$ of the atoms below it and thus not the join of any proper subset of the atoms below it. ∎

Theorem 4.9. The map $f(x) = \{a | a \text{ is an atom and } a \leq x\}$ is an isomorphism from a Boolean algebra with atoms A onto the lattice of subsets of A.

Proof. Theorems 4.7 and 4.8 show the map is one-to-one and onto. Both f and f^{-1} are order preserving, so f is an isomorphism between the lattices, understood as ordered sets; thus, f is an isomorphism of lattices and, therefore, is an isomorphism of Boolean algebras. ∎

Corollary 4.10. If a Boolean algebra has n atoms, it has 2^n elements.

Corollary 4.11. Every Boolean polynomial in n variables has a unique representation in disjunctive normal form.

Proof. In the lattice of Boolean polynomials, the atoms are the monomials $x_1^{i_1} \wedge x_2^{i_2} \wedge \cdots \wedge x_n^{i_n}$ where $i_k = \pm 1$. To see this, note that

$$
(x_1^{i_1} \wedge x_2^{i_2} \wedge \cdots \wedge x_n^{i_n}) \wedge (x_1^{j_1} \wedge x_2^{j_2} \wedge \cdots \wedge x_n^{j_n}) = 0,
$$

unless each i_k and j_k are equal, for if $i_k \neq j_k$, then $x_k \wedge x_k^{-1} = 0$ is in this meet. By the distributive law, the monomial $M = x_1^{i_1} \wedge x_2^{i_2} \wedge \cdots \wedge x_n^{i_n}$ will meet with a join of monomials to give 0 unless M is a monomial in that join. Then by Theorem 4.6, $M \wedge p(x_1, x_2, \ldots, x_n) = 0$ unless $M \leq p(x_1, x_2, \ldots, x_n)$. Thus there is no polynomial $q(x_1, x_2, \ldots, x_n)$ with $M > q(x_1, x_2, \ldots, x_n) > 0$. Therefore, each monomial is an atom. Since by Theorem 4.6 everything else is a join of monomials, the monomials are the only atoms. Then by Theorem 4.8, this representation of our Boolean polynomial is unique. ∎

Corollary 4.12. The number of nonequivalent Boolean polynomials in n variables is 2^{2^n}.

Proof. We have 2^n atoms, the monomials. Apply Corollary 4.10. ∎

Corollary 4.13. There are 2^{2^n} nonequivalent combinatorial networks (gate networks) with n inputs.

An important topic not covered in this section is the efficient representation of a Boolean polynomial, that is, finding a representation of the polynomial with as few connectives as possible. This is especially important in electrical network design. The books in the Suggested Reading by Birkhoff and Bartee, by Fisher, and by Prather all contain discussions of this problem, as do books on the design of digital circuits.

EXERCISES

1. Which of the lattices in Figure 4.6 are distributive?

Figure 4.6

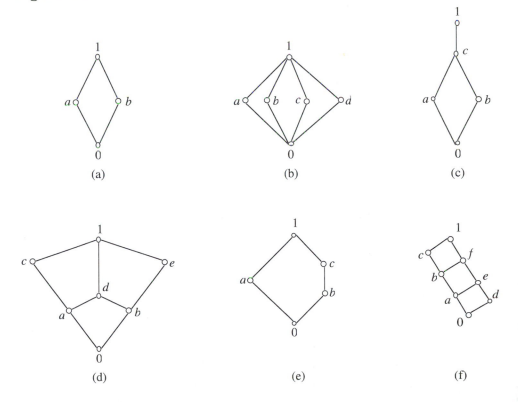

(a)   (b)   (c)

(d)   (e)   (f)

2. Which of the lattices in Figure 4.6 are complemented?

3. Which lattices in Figure 4.6 are Boolean algebras and why?

4. A lattice is "uniquely complemented" if, for each element a, there is a unique element c such that

$$a \vee c = 1$$

and

$$a \wedge c = 0.$$

Which lattices in Figure 4.6 are uniquely complemented?

5. Draw the Boolean algebra of all subsets of a four-element set.

6. By Corollary 4.13, there are 2^{2^2} combinatorial networks that can be built with two input wires. Some of these are trivial, for example, the network consisting of x_1 and x_2 going in and only x_1 coming out. The Boolean algebra has 16 elements. Draw the Boolean algebra and sketch two typical combinatorial networks corresponding to elements of height 1, height 2, and height 3, respectively.

7. Draw and label the Boolean algebra of all Boolean polynomials in two variables x and y.

8. Construct a truth table to show that the associative law for $\wedge$ holds for propositions.

9. Construct a truth table to show that the distributive law $p \wedge (q \vee r) = (p \wedge q) \vee (p \wedge r)$ holds for propositions.

10. Construct a truth table to show that the distributive law $p \vee (q \wedge r) = (p \vee q) \wedge (p \vee r)$ holds for propositions.

11. Explain why the lattice of nonnegative factors of 30 is a Boolean algebra while the lattice of nonnegative factors of 36 is not. (The ordering in both cases is by "is a factor of.")

12. Construct circuits that correspond to the following Boolean polynomials:
    (a) $(x_1 \vee x_2) \wedge x_3$.
    (b) $x_1^{-1} \vee x_2$.
    (c) $(x_1 \vee x_2)^{-1}$.
    (d) $(x_1 \wedge x_2^{-1}) \vee (x_1^{-1} \wedge x_2)$.
    (e) $(x_1 \vee x_2) \wedge x_3^{-1}$.
    (f) $(x_1 \wedge x_2 \wedge x_3) \vee (x_1 \wedge x_2^{-1} \wedge x_3^{-1}) \vee (x_1^{-1} \wedge x_2 \wedge x_3^{-1}) \vee (x_1^{-1} \wedge x_2^{-1} \wedge x_3)$.

13. It is conventional to use $x_i \cdot x_j$ to stand for $x_i \wedge x_j$, and to leave the dot out unless it is necessary to the understanding. It is also conventional to

use the symbol $+$ in place of $\vee$. Rewrite each of the Boolean polynomials in Exercise 12 using this convention.

14. The truth set of a statement about a set of arrangements is the set of arrangements that makes the statement true. In Figure 4.7, we show Venn diagrams of truth sets 1 and 2 corresponding to statements x_1 and x_2 and of truth sets 1, 2, and 3 corresponding to the statements x_1, x_2, and x_3. Copy the diagrams and shade in the truth sets of the statements corresponding to the polynomials in Exercise 12.

Figure 4.7

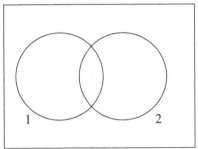 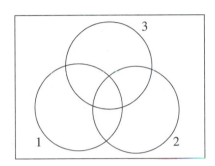

15. An element p of a lattice is called *prime* or *join-irreducible* if, whenever $p = x \vee y$, either $p = x$ or $p = y$.
    (a) Show that in a finite distributive lattice, every element is a join of join-irreducibles.
    (b) Is the word "distributive" necessary in part (a)?
    (c) An element p_i in a subset $\{p_1, p_2, \ldots, p_n\}$ of a lattice is redundant if $p_i \leq \vee_{j:j\neq i} p_j$ for some $j \neq i$. Show that each element of a finite distributive lattice may be represented uniquely as a join of a set of join-irreducible elements that has no redundant elements. (This is one of the general results of Birkhoff referred to in the text.)

16. Prove the DeMorgan law $(x \wedge y)^{-1} = x^{-1} \vee y^{-1}$.

17. Write out the DeMorgan laws for two variables in the notation of propositions and verify them by means of truth tables.

18. Addition in a Boolean algebra can be defined by

$$a \oplus b = (a \wedge b^{-1}) \vee (a^{-1} \wedge b).$$

    (a) Show that addition is commutative and associative and that $a \oplus 0 = a$ and $a \oplus a = 0$.
    (b) Show that meet distributes over addition.

(c) Does join distribute over addition?

19. Prove Theorem 4.1.

20. Prove that if an element of a distributive lattice has a complement, it has a unique complement.

21. (For those who have had linear algebra with finite fields.) Show that a Boolean algebra is a vector space over the integers (mod 2) with the addition defined in Exercise 18 and with $1 \cdot a = a$, $0 \cdot a = 0$.

22. The truth tables for the two propositions $p \to q$ and $p \leftrightarrow q$ (read as "if p then q" and "p if and only if q") are

p	q	$p \to q$
T	T	T
T	F	F
F	T	T
F	F	T

and

p	q	$p \leftrightarrow q$
T	T	T
T	F	F
F	T	F
F	F	T

Write down propositions involving p, q, $\neg$, $\wedge$, and $\vee$ that have each of these truth tables. Explain (in words) why the statements involving the three basic connectives "should be" equivalent to $p \to q$ and $p \leftrightarrow q$.

23. Write down schematic diagrams of electric circuits that are the analogs of the propositions "if x_1 then x_2" and "x_1 if and only if x_2" (see Exercise 22).

Section 5 Möbius Functions

A Review of Inclusion and Exclusion

One of the important enumeration principles we developed in Chapter 3 was the principle of inclusion and exclusion. A similar basic principle called Möbius inversion has been thoroughly developed in number theory. In solutions of various combinatorial problems there have been a number of similar computations involving ordered sets. These similarities led G.-C. Rota in the 1960s to develop an all-inclusive formulation of a general method known as Möbius inversion that unified all these results and has had far-reaching applications to partially ordered sets. We shall present the elements of this generalization and a few of its applications here.

In computing the number of functions from a set J onto a set K, we studied, for each subset I of K, the number $N_\geq(I)$ of functions from J to K that "skipped" I or some bigger subset of K. We also studied the number $N_=(I)$ of functions from J to K that skipped exactly the set I (in other words, functions that skip all of I but don't skip anything else). The number $N_=(\emptyset)$ is the number of functions that miss the empty set and nothing else; thus, it is the number of functions from J onto K. We found $N_=(\emptyset)$ by applying the principle of inclusion and exclusion. We could also find $N_=(\emptyset)$ by solving systems of linear equations. Note that a function that skips I or some bigger set must skip exactly a certain set S with $S \supseteq I$. This lets us write the symbolic equation:

$$N_\geq(I) = \sum_{S=I}^{K} N_=(S) \tag{5.1}$$

for each subset I of K. This gives a system of 2^k equations in 2^k unknowns; the unknowns are the numbers $N_=(S)$. We can see without too much effort that the number $N_\geq(I)$ is just $(k-i)^j$. The principle of Möbius inversion for sets will tell us that whenever we have a system of equations of the form (5.1), regardless of the actual values $N_\geq(I)$ and $N_=(S)$, the system of equations can be solved for $N_=(I)$ in terms of $N_\geq(S)$ by writing

$$N_=(I) = \sum_{S=I}^{K} (-1)^{|S|-|I|} N_\geq(S). \tag{5.2}$$

This is the same formula that we got for the principle of inclusion and exclusion. The difference is that there is no mention of properties here; instead the equations in (5.1) *define* how $N_=$ and $N_\geq$ are related.

Notice that Equation (5.1) has a counterpart for any finite partially ordered set (X, P) with top element t; namely, if N is a numerical function defined on the set X, we can write

$$N_\geq(x) = \sum_{y=x}^{t} N(y) = \sum_{y:y\geq x} N(y). \qquad (5.3)$$

Here $y \geq x$ means $y \geq x$ in P. (From here on we will use N rather than $N_=$ to remind us that it is Equation (5.3) that defines the relation between N and $N_\geq$ rather than some set of properties.) The second sum in Equation (5.3) is read as "the sum over all y such that y is greater than or equal to x" and the first sum is read as "the sum of $N(y)$ over all y from x to t." Since the notation of the second sum is more explicit and does not force us to assume that (X, P) has a top element t, we shall adopt it. Equation (5.3) then actually determines the values $N_\geq(x)$ of a numerical function $N_\geq$ defined on X. Just as we saw in the case of onto functions, $N_\geq(x)$ will often be a number we can compute while $N(x)$ may be difficult to compute directly. Thus, a result analogous to Equation (5.2) would be desirable. The principle of Möbius inversion is just this kind of result.

The Zeta Matrix

We show how to convert the equations shown in (5.3) into a matrix equation. By solving this matrix equation, we will get the desired analog to (5.2). In our discussion of graph theory, we presented the adjacency matrix of a graph. Since a partial ordering P of a set X is a directed graph on X, it has an adjacency matrix; in the context of partially ordered sets, it is traditional to call this the **Zeta matrix** of P. Specifically, given a listing $x_1, x_2, ..., x_n$ of X, we let

$$Z_{ij} = \begin{cases} 1 \text{ if } x_i \leq x_j \\ 0 \text{ otherwise.} \end{cases}$$

This is the matrix version of the function "Zeta" (The Greek letter zeta, denoted by ζ, is pronounced zay-ta) given by

$$\zeta(x, y) = \begin{cases} 1 \ \text{ if } x \leq y \\ 0 \text{ otherwise.} \end{cases}$$

Example 5.1. Compute the Zeta matrix for the lattice of partitions of a three-element set.

Figure 5.1

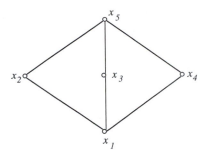

If we number the elements of the lattice of partition of a three-element set as shown in Figure 5.1, we get for that ordered set the Zeta matrix

$$
\begin{array}{c c c c c c}
 & 1 & 2 & 3 & 4 & 5 \\
1 & \begin{pmatrix} 1 & 1 & 1 & 1 & 1 \\ 0 & 1 & 0 & 0 & 1 \\ 0 & 0 & 1 & 0 & 1 \\ 0 & 0 & 0 & 1 & 1 \\ 0 & 0 & 0 & 0 & 1 \end{pmatrix}
\end{array}
$$

with row labels $1,2,3,4,5$. ∎

Given a column vector C of length n, the product ZC is another column vector D whose ith entry is given by

$$D_i = \sum_{j=1}^{n} Z_{ij} C_j. \tag{5.4}$$

Since Z_{ij} is 1 exactly when $x_i \leq x_j$, the sum in Equation (5.4) is the sum of the values C_j that correspond to x_j's greater than or equal to x_i. In symbols,

$$D_i = \sum_{x_j : x_j \geq x_i} C_j. \tag{5.5}$$

This leads us to a restatement of the equations in (5.3) as follows.

Theorem 5.1. Let C be the column vector whose entries are $N(x_1)$, $N(x_2)$, ..., $N(x_n)$. Let D be the column vector whose entries are $N_{\geq}(x_1)$, $N_{\geq}(x_2)$, ..., $N_{\geq}(x_n)$. Then

$$D = ZC.$$

Proof. For C_j we substitute $N(x_j)$ in Equation (5.5). For D_i we substitute $N_{\geq}(x_i)$. Then Equation (5.5) reads

$$N_{\geq}(x_i) = \sum_{x_j : x_j \geq x_i} N(x_j). \tag{5.6}$$

Equation (5.6) is the same as Equation (5.3) with x_i playing the role of x and x_j playing the role of y. Thus, the equation $D = ZC$ is just a restatement of the relation between N and $N_\geq$. ∎

The Möbius Matrix

Note that the matrix Z we computed for the lattice in Figure 5.1 is upper triangular and has 1's along the main diagonal. For any ordered set, there is a way to number the elements of the ordered set so the Zeta matrix will be upper triangular. From this, we can show that Z is invertible.

Theorem 5.2. For any ordered set (X, P), the Zeta matrix Z has an inverse M. Further, there is a way to list X as $x_1, x_2, ..., x_n$ so that Z and M are both upper triangular.

Proof. List X in the order of a linear extension of P. Use this listing to construct Z. Then if $i > j$, we know that $x_i \not\leq x_j$ so $Z_{ij} = 0$. Thus, Z is upper triangular. Further, $Z_{ii} = 1$ because $x_i \leq x_i$ for each i. Since Z is upper triangular and has ones on the main diagonal, it has an inverse M. The inverse of an upper triangular matrix is upper triangular, so M is upper triangular. ∎

The matrix M is called the **Möbius matrix** of the partially ordered set.

Example 5.2. What is the Möbius matrix for the lattice of partitions of a three-element set? By row reduction, we may compute that

$$M = \begin{pmatrix} 1 & -1 & -1 & -1 & 2 \\ 0 & 1 & 0 & 0 & -1 \\ 0 & 0 & 1 & 0 & -1 \\ 0 & 0 & 0 & 1 & -1 \\ 0 & 0 & 0 & 0 & 1 \end{pmatrix}.$$

(You may wish to check that $ZM = I$.) ∎

The fact that Z has an inverse allows us to solve our system of equations.

Theorem 5.3. Let C be the column vector whose entries are $N(x_1)$, $N(x_2)$, ..., $N(x_n)$. Let D be the column vector whose entries are $N_\geq(x_1), N_\geq(x_2), ..., N_\geq(x_n)$. Then

$$C = MD.$$

Proof. By Theorem 5.1, $D = ZC$. Thus

$$
\begin{aligned}
MD &= M(ZC) \\
&= (MZ)C \\
&= IC \\
&= C.
\end{aligned}
$$

∎

The Möbius Function

The **Möbius function** μ (the Greek letter mu, denoted by μ, is pronounced myou) of a partially ordered set (X, P) is a function of two variables given by

$$
\mu(x_i, x_j) = M_{ij}.
$$

Thus μ is defined from M in the same way that ζ is defined from Z. The analog of the inclusion–exclusion theorem is Theorem 5.4, the **principle of Möbius inversion**.

Theorem 5.4. Suppose (X, P) is a partially ordered set and let the functions N and $N_\geq$ be related by

$$
N_\geq(x) = \sum_{y:y \geq x} N(y).
$$

Then

$$
N(x) = \sum_{y:y \geq x} \mu(x, y) N_\geq(y).
$$

Proof. The fact that $N(x) = \sum_{y:y \in x} \mu(x, y) N_\geq(y)$ is a translation of Theorem 5.3 out of matrix notation. The remainder of the proof consists in showing that $\mu(x, y)$ is 0 unless $y \geq x$. Recall that Z_{ij} is 0 unless $x_i \leq x_j$; from this we can conclude that M_{ij} is 0 unless $x_i \leq x_j$. To see why this is so, note that

$$
MZ = I
$$

is equivalent to saying

$$
\sum_{k=1}^{n} M_{ik} Z_{kj} = \begin{cases} 1 \text{ if } i = j \\ 0 \text{ otherwise.} \end{cases} \tag{5.7}
$$

Assume there is a pair (i, k) such that $M_{ik} \neq 0$ but $x_i \not\leq x_k$. Since the inverse of an upper triangular matrix is upper triangular, M_{ik} must be

0 if $i > k$, so we may assume $i \leq k$. If $i = k$, then $x_i \leq x_k$ regardless of whether $M_{ik} = 0$. Thus, if $M_{ik} \neq 0$ but $x_i \nleq x_k$, we may assume $i < k$. Among all pairs (i, k) with these properties, pick a pair with $k - i$ as small as possible. In other words, pick i and k so that if $i \leq j < k$ and $M_{ij} \neq 0$, then $x_i \leq x_j$. Now by Equation (5.7) and the fact that M and Z are upper triangular

$$0 = \sum_{j=i}^{k} M_{ij} Z_{jk} = \sum_{j=i}^{k-1} M_{ij} Z_{jk} + M_{ik} Z_{kk}$$

$$= \sum_{j=i}^{k-1} M_{ij} Z_{jk} + M_{ik}$$

so that

$$\sum_{j=i}^{k-1} M_{ij} Z_{jk} = -M_{ik} \neq 0.$$

Thus for some j, both M_{ij} and Z_{jk} must be nonzero, so that, because we chose $k - i$ to be as small as possible, $x_i \leq x_j$ since $M_{ij} \neq 0$ and $j - i < k - i$; in addition, $x_j \leq x_k$ since $Z_{jk} \neq 0$. Then by transitivity $x_i < x_k$, a contradiction. Thus for all i and j, $M_{ij} \neq 0$ implies $x_i \leq x_j$. For this reason we can rewrite the conclusion of Theorem 5.3 as

$$N(x_i) = \sum_{j=1}^{n} M_{ij} N_{\geq}(x_j) = \sum_{x_j : x_i \leq x_j} M_{ij} N_{\geq}(x_j). \tag{5.8}$$

Now we define $\mu(x_i, x_j)$ to be M_{ij}; this allows us to rewrite Equation (5.8) as

$$N(x) = \sum_{y : x \leq y} \mu(x, y) N_{\geq}(y). \qquad \blacksquare$$

Theorem 5.5. Suppose (X, P) is a finite ordered set and μ is the Möbius function of (X, P). Whenever two functions N and $N_{\leq}$ are related by the equations

$$N_{\leq}(x) = \sum_{y : y \leq x} N(y)$$

for all x in X, then

$$N(x) = \sum_{y : y \leq x} \mu(y, x) N_{\leq}(y).$$

Proof. The proof is analogous to that of Theorem 5.4. ∎

Equations That Describe the Möbius Function

It would appear that the function μ depends on the order in which we list the elements of X since M depends on that order. The next theorem shows that this is not the case.

Theorem 5.6. μ is the unique function such that
(a) $\mu(x, x) = 1$ for all x
(b) $\sum_{z:x \leq z \leq y} \mu(x, z) = 0$ for all x and y with $x < y$ or
(b') $\sum_{z:x \leq z \leq y} \mu(z, y) = 0$ for all x and y with $x < y$
(c) $\mu(x, y) = 0$ if $x \not\leq y$.

Proof. We show first that μ satisfies these three conditions. We have already shown in our proof of Theorem 5.4 that $M_{ij} = 0$ unless $x_i \leq x_j$, so $\mu(x_i, x_j) = 0$ unless $x_i \leq x_j$, which proves statement (c). Statement (a) simply restates the fact that M has ones on the main diagonal. Alternatively, since $MZ = I$,

$$I_{ii} = \sum_{k=1}^{n} M_{ik} Z_{ki}$$

$$= \sum_{x_k : x_i \leq x_k \leq x_i} M_{ik} Z_{ki} = M_{ii} Z_{ii} = M_{ii} = \mu(x_i, x_i).$$

Since $I_{ii} = 1$, this proves statement (a).

To prove statement (b), note that, by the definition of the identity matrix and matrix multiplication, if $x_i < x_j$, then

$$0 = I_{ij} = \sum_{k=1}^{n} M_{ik} Z_{kj} = \sum_{x_k : x_i \leq x_k \leq x_j} M_{ik} Z_{kj}$$

$$= \sum_{x_k : x_i \leq x_k \leq x_j} \mu(x_i, x_k) \cdot 1 = \sum_{z : x_i \leq z \leq x_j} \mu(x_i, z).$$

Substituting x and y for x_i and x_j gives statement (b).

By using any listing of X in a linear extension of P, we may prove by induction on $j - i$ that the value of $\mu(x_i, x_j)$ is determined by Equations (a), (b), and (c) of Theorem 5.6. The idea behind this inductive proof is that the equation

$$\sum_{z : x \leq z \leq y} \mu(x, z) = 0$$

may be rewritten as

$$\mu(x, y) + \sum_{z:x \leq z < y} \mu(x, z) = 0,$$

so

$$\mu(x, y) = - \sum_{z:x \leq z < y} \mu(x, z).$$

The construction of the proof is left as an exercise. ∎

Example 5.3. Compute the Möbius functions for the ordered sets shown in Figure 5.2.

Figure 5.2

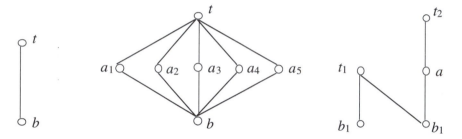

For the first ordered set we note $\mu(b, b) = \mu(t, t) = 1$ by Theorem 5.6, part (a). From part (b), note that $\sum_{z:b \leq z \leq t} \mu(b, z) = 0$, so $\mu(b, b) + \mu(b, t) = 0$ gives $\mu(b, t) = -1$. For the second ordered set, again $\mu(x, x) = 1$ for $x = b, t$, or a_i. Also, $\mu(b, b) + \mu(b, a_i) = 0$ by part (b) of Theorem 5.6, so $\mu(b, a_i) = -1$. Again applying part (b) of Theorem 5.6, we get

$$\mu(a_i, a_i) + \mu(a_i, t) = 0,$$

so $\mu(a_i, t) = -1$. Still another application of part (b) gives us

$$\mu(b, b) + \mu(b, a_1) + \mu(b, a_2) + \mu(b, a_3) + \mu(b, a_4) + \mu(b, a_5) + \mu(b, t) = 0$$

or

$$1 - 1 - 1 - 1 - 1 - 1 + \mu(b, t) = 0,$$

so

$$\mu(b, t) = 4.$$

Since $a_i \not\leq a_j$ unless $i = j$, $\mu(a_i, a_j) = 0$ by part (c) of Theorem 5.6 whenever $i \neq j$.

For the third ordered set, $\mu(x,x) = 1$ once again for each x, and $\mu(b_1, t_1) = \mu(b_2, t_1) = \mu(b_2, a) = \mu(a, t_2) = -1$ as before. Another application of Theorem 5.6, part (b) gives

$$\mu(b_2, b_2) + \mu(b_2, a) + \mu(b_2, t_2) = 0$$

$$1 - 1 + \mu(b_2, t_2) = 0,$$

so

$$\mu(b_2, t_2) = 0.$$

Note that this is our first example of an x and y with $x \le y$ but $\mu(x, y) = 0$. Thus as in Theorem 5.6, $\mu(x, y) \ne 0$ implies that $x \le y$, but $x \le y$ *need not imply* $\mu(x, y) \ne 0$. ∎

For small ordered sets, the preceding method is usually simpler than matrix inversion for computing μ.

The Number-Theoretic Möbius Function

In the next section of this chapter we shall give an elegant proof of the fact that in the lattice of positive integers, ordered by "is a factor of,"

$$\mu(m, n) = \begin{cases} 0 & \text{unless } n/m \text{ is a product of distinct primes} \\ (-1)^k & \text{if } n/m \text{ is a product of } k \text{ distinct primes.} \end{cases} \quad (5.9)$$

Notice that $\mu(m, n)$ depends only on the single integer n/m. The Möbius function studied in number theory is this one, regarded as a function of the one variable n/m. In number theory, factorization is a very important topic, so it is no surprise that the "is a factor of" ordering helps explain many ideas in number theory. For example, the greatest common divisor is a greatest lower bound or meet in this ordered set and the least common multiple is a join. Thus, lattice properties may be used to explain number-theoretic ideas.

There are many combinatorial problems that arise in number theory. For example, among the numbers between one and n, we might ask how many are factors of n, how many are relatively prime to n, or, more generally, how many have a specified greatest common divisor with n. Since the relation of having the same greatest common divisor with n is an equivalence relation, it divides the numbers between 1 and n into equivalence classes, one for each possible greatest common divisor. Since each divisor of n is a possible greatest common divisor, we may write

$$\{1, 2, \ldots, n\} = \bigcup_{k:k|n} \{j | j \le n \text{ and } \gcd(n, j) = k\}.$$

In more traditional notation, if the factors of n are k_1, k_2, ..., k_m, then we may write

$$\{1, 2, \ldots, n\} = \bigcup_{i=1}^{m} \{j | j \leq n \text{ and } \gcd(n, j) = k_i\}$$
$$= \{j | j \leq n \text{ and } \gcd(n, j) = k_1\} \cup \{j | j \leq n \text{ and } \gcd(n, j) = k_2\}$$
$$\cup \cdots \cup \{j | j \leq n \text{ and } \gcd(n, j) = k_m\}.$$

Thus, by the sum principle,

$$n = |\{1, 2, \ldots, n\}| = \sum_{i=1}^{m} |\{j | j \leq n \text{ and } \gcd(n, j) = k_i\}|$$
$$= \sum_{k:k|n} |\{j | j \leq n \text{ and } \gcd(n, j) = k\}|. \tag{5.10}$$

If we use $\varphi(k, n)$ to denote the number of elements j less than or equal to n whose greatest common divisor $\gcd(j, n)$ with n is k, then we may rewrite our last formula in Equation (5.10) in the form

$$n = \sum_{k:k|n} \varphi(k, n). \tag{5.11}$$

Although we have written just one symbolic equation, the equation is valid for all values of n, giving a system of equations we might be able to solve for $\varphi(k, n)$ by Möbius inversion. However, the theory of Möbius inversion applies to sums of functions of one variable, and we have defined φ as a function of two variables. Note, however, that j and n have greatest common divisor k if and only if j/k and n/k are relatively prime integers. Thus, we define the one variable function—called the **Euler phi function**—by letting $\varphi(i)$ be the number of integers between 1 and i that are relatively prime to i; with this definition we get that for any k dividing n,

$$\varphi(n/k) = \varphi(k, n).$$

Thus, Equation (5.11) becomes

$$n = \sum_{k:k|n} \varphi(n/k) = \sum_{r:r|n} \varphi(r). \tag{5.12}$$

(The second equality in (5.12) follows from the fact that n/k is a factor m of n exactly when k is.) By Theorem 5.5, we obtain

$$\varphi(n) = \sum_{k:k|n} k \cdot \mu(k,n). \tag{5.13}$$

Now recall that $\mu(k,n)$ is 0 unless n/k is a product of distinct primes $p_1, p_2, \ldots, p_r$, in which case $\mu(k,n) = (-1)^r$. Thus, we may write, using P for the set of all primes dividing n,

$$\varphi(n) = \sum_{\substack{\{p_1,p_2,\ldots,p_r\}: \\ p_i|n; p_i \neq p_j}} \frac{n}{p_1 p_2 \cdots p_r}(-1)^r = n \sum_{S=\emptyset}^{P} \frac{(-1)^{|S|}}{\prod_{p:p\in S} p}.$$

This formula appears complicated until we recognize that we may apply the formula

$$\prod_{i=1}^{n}(1 - x_i) = \sum_{S=\emptyset}^{\{x_1,x_2,\ldots,x_n\}} (-1)^{|S|} \prod_{i:x_i\in S} x_i$$

to write

$$\varphi(n) = n \prod_{\text{primes } p:p|n} \left(1 - \frac{1}{p}\right).$$

This is known as Euler's product formula for the phi function. This formula is one of the reasons for the introduction of the Möbius function in number theory.

The Number of Connected Graphs

Example 5.4. Show how to use the Möbius function of the partition lattice to compute the number of connected graphs on a vertex set V.

Each graph is a union of its connected components. Since the connected components are disjoint, they form a partition of the vertex set of g. We call this the connected component partition of the graph G. For each partition $\mathbf{P}$ of V, let $N(\mathbf{P})$ stand for the number of graphs G whose connected component partition is $\mathbf{P}$. The number of connected graphs is $N(\{V\})$. The following equation in which the $\mathbf{P}$ on the left-hand side runs over all partitions $\mathbf{P}$ of V

$$\sum_{\substack{\mathbf{P}: \ \mathbf{P} \text{ is a} \\ \text{partition of } V}} N(\mathbf{P}) = 2^{\binom{v}{2}}$$

holds because the right-hand side is the total number of graphs on the vertex set V with v elements while the left-hand side sums the number of graphs having a certain connected component partition over all possible connected component partitions. This gives us one equation in quite a number of unknowns; one of the unknowns, $N(\{V\})$, in which $\{V\}$ is the partition with just one part, is the number of connected graphs. If we could get the same number of equations as unknowns, then perhaps we could solve for all the unknowns, including the one we are interested in. The Möbius inversion theorems suggest that we look for equations involving summing over all partitions less than or equal to a given one or all partitions greater than or equal to a given one. If we add up $N(\mathbf{P})$ for all partitions $\mathbf{P}$ contained in the partition $\mathbf{Q}$, we should get the total number of graphs whose connected component partitions are contained in $\mathbf{Q}$, that is, the number of graphs each of whose connected components is a subset of some class of $\mathbf{Q}$. Thus, this sum is the number of graphs all of whose edges connect two points in one of the classes C_i of $\mathbf{Q}$. This, by the product principle, is the number of graphs all of whose edges are in C_1 multiplied by the number of graphs of all whose edges are in C_2 multiplied by, $\ldots$, etc. The number of graphs all of whose edges connect vertices in C_i is just the number of graphs whose vertex set is C_i. Thus, if C_i has size c_i, this number is

$$2^{\binom{c_i}{2}}.$$

Putting this all together gives

$$\sum_{\substack{\text{Partitions } \mathbf{P}: \\ \mathbf{P} \leq \mathbf{Q} = \{C_1, C_2, \ldots, C_k\}}} N(\mathbf{P}) = \prod_{i=1}^{k} 2^{\binom{c_i}{2}}.$$

By Theorem 5.5,

$$N(\mathbf{Q}) = \sum_{\substack{\text{Partitions } \mathbf{P}: \\ \mathbf{P} \leq \mathbf{Q} \\ \mathbf{P} = \{B_1, B_2, \ldots, B_j\}}} \mu(\mathbf{P}, \mathbf{Q}) \prod_{i=1}^{j} 2^{\binom{b_i}{2}}$$

where b_i stands for the size of the class B_i of $\mathbf{P}$. The number we want is $N(\{V\})$. Also, all partitions $\mathbf{P}$ of V are finer than $\{V\}$, so

$$N(V) = \sum_{\substack{\text{Partitions } \mathbf{P} \text{ of } V: \\ \mathbf{P} = \{B_1, B_2, \ldots, B_j\}}} \mu(\mathbf{P}, \{V\}) \prod_{i=1}^{j} 2^{\binom{b_i}{2}}.$$

This completes the example and suggests that we should try to compute $\mu(\mathbf{P}, \{V\})$ for the partition lattice. ∎

Many methods have been developed for computation of Möbius functions, and we have space only to "scratch the surface" here.

A General Method of Computing Möbius Functions of Lattices

Theorem 5.7. Let L be a lattice with an element b below all elements of L and an element t above all elements of L. Then for any element a of L with $a \neq t$

$$\mu(b, t) = - \sum_{\substack{x:x \wedge a = b \\ \text{and } x \neq b}} \mu(x, t).$$

Proof. We prove that

$$\sum_{x:x \wedge a = b} \mu(x, t) = 0.$$

The statement of the theorem follows because $x = b$ is one of the x's such that $x \wedge a = b$. For each $y \leq a$ in L, let $N(y) = \sum_{x:x \wedge a = y} \mu(x, t)$. Then, for each $w \leq a$ we let

$$N_{\geq}(w) = \sum_{y:y \geq w \text{ and } y \leq a} N(y).$$

We note that because every x above w meets with a to give some element between w and a,

$$N_{\geq}(w) = \sum_{y:y \geq w \text{ and } y \leq a} N(y) = \sum_{y:y \geq w \text{ and } y \leq a} \left(\sum_{x:x \wedge a = y} \mu(x, t) \right)$$

$$= \sum_{x:x \geq w} \mu(x, t) = 0$$

by Theorem 5.6 and the fact that $w \leq a \neq t$. Then, by applying Theorem 5.5 to the ordered set of all elements between 0 and a, we get

$$N(w) = \sum_{y:y \geq w \text{ and } y \leq a} \mu'(w, y) \cdot N_{\geq}y = \sum_{y:y \geq w \text{ and } y \leq a} \mu'(w, y) \cdot 0 = 0,$$

where μ' is the Möbius function for the ordered set of all elements between 0 and a. The substitution of b for w gives the desired equation. ∎

The Möbius Function of the Partition Lattice

Theorem 5.8. Let μ_n stand for the Möbius function of the lattice of partitions of an n-element set. Then

$$\mu_n(b, t) = (-1)^{n-1} \cdot (n-1)!.$$

Proof. Let $S = \{n | \mu_n(b, t) = (-1)^{n-1}(n-1)!\}$. Then 1 is in S because when $n = 1$ the lattice has just one element. Suppose $n - 1$ is in S. If the set being partitioned is $N = \{1, 2, ..., n\}$, let $\mathbf{P}$ be the partition with two classes given by $\mathbf{P} = \{\{1, 2, ..., n-1\}, \{n\}\}$. Theorem 5.7 tells us that

$$\mu_n(b, t) = - \sum_{\substack{\text{Partitions } \mathbf{Q} \neq b \\ \mathbf{Q} \wedge \mathbf{P} = b}} \mu_n(\mathbf{Q}, t). \tag{5.14}$$

Now if $\mathbf{Q} \wedge \mathbf{P} = b$, all intersections of classes of $\mathbf{P}$ and $\mathbf{Q}$ have size 1. However, for a set C to intersect $\{1, 2, ..., n-1\}$ in a set of size 1, C must be $\{i\}$ or $\{i, n\}$ for some $i \neq n$. Thus, any class C of $\mathbf{Q}$ of size more than 1 must be $\{i, n\}$, and since n can't be in two classes, $\mathbf{Q}$ can have exactly one nontrivial class $\{i, n\}$.

Theorem 5.6 tells us that $\mu_n(\mathbf{Q}, t)$ depends only on the ordered set consisting of partitions of N containing $\mathbf{Q}$. However, the ordered set of partitions of N containing $\mathbf{Q}$ is isomorphic to the ordered set of partitions of $N' = \{1, 2, ..., n-1\}$. (To see this, for each $\mathbf{R}$ containing $\mathbf{Q}$, let $\mathbf{R}'$ be the partition obtained from $\mathbf{R}$ by deleting n from the class of $\mathbf{R}$ containing i. This mapping is one-to-one, onto the partitions of N', and $\mathbf{R} \leq \mathbf{S}$ if and only if $\mathbf{R}' \leq \mathbf{S}'$, so the mapping is an isomorphism.) Theorem 5.6 may be used to show that isomorphic ordered sets have the same Möbius function values, so

$$\mu_n(\mathbf{Q}, t) = \mu_{n-1}(b, t). \tag{5.15}$$

Now there are $n - 1$ choices for the element i of $\mathbf{Q}$, so there are $n - 1$ choices for $\mathbf{Q}$. Thus, substituting Equation (5.15) into Equation (5.14) and using our inductive hypothesis, we get

$$\begin{aligned}
\mu_n(b, t) &= -(n-1)\mu_{n-1}(b, t) \\
&= -(n-1)(-1)^{n-2}(n-2)! \\
&= (-1)^{n-1}(n-1)!
\end{aligned}$$

since $n - 1$ is in S. Thus, n is in S and so the theorem is proved by the principle of mathematical induction. ∎

In the proof of the theorem, we used an isomorphism between the "interval" consisting of all elements between $\mathbf{Q}$ and the top t of the partition lattice of a set with the partition lattice of a smaller set. In fact, any interval from *any* partition $\mathbf{P}$ to the top t of the partition lattice is isomorphic to a partition lattice on a smaller set. In particular, if $\mathbf{P}$ has k classes, then any partition above $\mathbf{P}$ has unions of these classes as its classes. In particular, these unions partition the *classes* of $\mathbf{P}$. Two classes of $\mathbf{P}$ are partitioned together by $\mathbf{R}$ if their union lies entirely in a class of $\mathbf{R}$. This partitioning of classes of $\mathbf{P}$ gives an isomorphism between the partitions of $K = \{1, 2, ..., k\}$ and the partitions of N containing our partition $\mathbf{P}$ with k classes. We have proved the following theorem.

Theorem 5.9. If $\mathbf{P}$ has k classes, then

$$\mu(\mathbf{P}, t) = (-1)^{k-1}(k-1)!.$$

Proof. We have seen that the interval from $\mathbf{P}$ to t is isomorphic to the partitions of K, and thus $\mu(\mathbf{P}, t)$ must be $\mu_k(b, t)$. ∎

EXERCISES

1. By using condition 2 of Theorem 5.6, write down four equations involving the Möbius function $\mu(\emptyset, A)$ for subsets A of a two-element set. (You should have one equation for each subset.) Use these four equations to compute $\mu(\emptyset, A)$ for each subset A.

2. Use the method of Exercise 1 to compute $\mu(\emptyset, A)$ for the subsets A of a three-element set.

3. What is $\mu(x, y)$ for each x and y in $\{1, 2, 3, 4\}$ with the usual ordering?

4. Compute the Zeta matrix and its inverse for the ordered set of subsets of $\{1, 2, 3\}$ ordered by inclusion. (Note: These are 8×8 matrices.)

5. Complete the proof of Theorem 5.6.

6. Write out a proof of Theorem 5.6 in which condition (b) is replaced by the alternate condition (b').

7. Find $\mu(x, y)$ for all x and y in the ordered sets of Figure 5.3.

8. Write down a Zeta matrix for each ordered set in Exercise 7. Check your Möbius function from Exercise 7 by matrix multiplication.

9. Use an isomorphism argument similar to the one we used in discussion of the Möbius function of the partition lattice to explain why for subsets S and T of N with $S \leq T$, $\mu(S, T) = \mu(\emptyset, T - S)$.

Figure 5.3

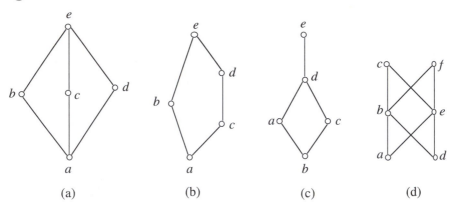

(a)   (b)   (c)   (d)

10. (Exercise 9 continued.) Prove that $\mu(S,T) = (-1)^{|T|-|S|}$ whenever $S \subseteq T$.

11. To do this exercise, review the *bond lattice* of a graph. Given a coloring of a graph with x colors, the *bond* of the coloring is the equivalence class partition of the relation "x is related to y if x and y are connected by a path all of whose vertices have the same color."
    (a) Show that the bond of a coloring is a bond in the sense of the bond lattice.
    (b) What is the bond of a proper coloring (i.e., a coloring in which adjacent vertices get different colors)?
    (c) Given a bond B whose partition has k classes, how many colorings have a bond which is B or contains B?
    (d) Find a formula for the number of proper colorings of the graph using x colors. (Your formula should make use of the Möbius function of the bond lattice and will involve summing over the bond lattice.)
    (e) Show the number you have computed is a polynomial $\sum a_i x^i$. What is the coefficient a_i? (This polynomial is called the *chromatic polynomial* of the graph.)

12. State and prove an analog of Theorem 5.7 that involves joins.

13. Determine the Möbius function of a linearly ordered set. What does the Möbius inversion theorem say in this case?

14. Determine the Möbius function of the positive factors of 12, ordered by "is a factor of." Do not simply apply the rule of Equation (5.9) given without proof in the text, but rather work out the Möbius function as the examples in the text were worked out.

*15. Show that, for the bond lattice of a connected graph, $(-1)^{h(B)}\mu(b, B) > 0$, where b is the bottom element of the lattice and $h(B)$ is the height of the bond B in the bond lattice. What does this say about the coefficients of the chromatic polynomial (Exercise 11)?

16. What can you say about $\mu(b, t)$ in a ordered set with a bottleneck, i.e. a pair of points $x \neq b$, $y \neq t$ such that y covers x and every other element of the ordered set is either above y or below x?

*17. Use Theorem 5.6 to verify the description we gave in the text for the Möbius function of the positive factors of an integer n, ordered by "is a factor of."

18. The element x of an ordered set is an *articulation point* if, for each y, either $y < x$ or $x < y$, i.e., x is comparable to all elements of the ordered set. What can you say about $\mu(b, t)$ in an ordered set with bottom b, top t, and with an articulation point different from b or t?

19. Two circular arrangements of the symbols $a_1, a_2, \ldots, a_n$ are equivalent if we obtain one from the other by rotating the first. It is a straightforward exercise in counting equivalence classes of equal size to find the number of circular arrangements if a_1 through a_n are distinct. However, in this exercise we want to consider the case in which the objects are not necessarily distinct. If one member of an equivalence class of sequences consists of $a_1 a_2 \ldots a_r$ repeated m times (so that $mr = n$), then we say that the class has period r. Let $N(r)$ be the number of circular arrangements of (smallest) period r, that is, of period r and no smaller period.

    (a) Show that if our circular arrangements are formed from n symbols, not necessarily distinct, chosen from a k-element set, then

$$\sum_{d:d|n} dN(d) = k^n.$$

    (b) Find a formula in terms of the number-theoretic Möbius function for $N(d)$.

*20. A three-by-n Latin rectangle is a three-by-n matrix with entries chosen from n distinct symbols so that each row is a permutation of the symbols and each column is a three-element permutation of the symbols. For the purposes of this exercise, define a three-by-n rectangle to be a three-by-n matrix with entries chosen from n distinct symbols so that each row is a permutation of the n symbols. Thus, a Latin rectangle is a rectangle, but a rectangle need not be a Latin rectangle.

    (a) If R is a three-by-n rectangle, let S_{ij} denote the set of columns in which row i and row j are equal. Show that

$$S_{12} \cap S_{13} = S_{12} \cap S_{23} = S_{13} \cap S_{23}.$$

(b) A list of three sets $(S_{12}, S_{13}, S_{23}) = S$ is called an *equality triple* if it satisfies the equalities of part (a). Show that the equality triples form an ordered set with the definition $S \leq T$ if $S_{ij} \subseteq T_{ij}$ for each i and j chosen from $\{1, 2, 3\}$. Find the Möbius function values $\mu((\emptyset, \emptyset, \emptyset), (S_{12}, S_{13}, S_{23}))$ of this ordered set.

(c) Use the Möbius function of part (b) to derive a formula for the number of three-by-n Latin rectangles.

21. An *orbit* of a function $f : X \to X$ is a set of the form $\{f^i(x) | \ i$ is a nonnegative integer$\}$. (Recall that the power function f^i means the composition of i functions all equal to f.) A *fixed point* of f is an x such that $f(x) = x$. Show that the number of fixed points of f^k is the sum, over all divisors d of k, of d times the number of orbits of f of size d. Find a formula for the number of orbits of f of a given size in terms of the number of fixed points of powers of f.

Section 6   Products of Orderings

The Idea of a Product

We began our study of ordered sets with the idea of developing a single ordering based on several other orderings. In particular, a professor with several rankings of students would rank person A above person B if and only if A ranks above B on each test. The same kind of construction can be used to put together several partially ordered sets into a new one. Given partially ordered sets $(X_1, P_1), (X_2, P_2), ..., (X_n, P_n)$, we define their *product*

$$\prod_{i=1}^{n}(X_i, P_i) = (\prod_{i=1}^{n}X_i, \prod_{i=1}^{n}P_i)$$

to be the set X of all lists $(y_1, y_2, ..., y_n)$ with y_i in X_i ordered by the rule

$$(y_1, y_2, ..., y_n) \leq (z_1, z_2, ..., z_n)$$

if and only if $y_i \leq z_i$ In P_i for all i. We call this ordering the *product* ordering. (To explain this terminology note that the set X of lists is the Cartesian product of the sets X_i.)

The Euclidean plane gives an example that shows how to visualize products of orderings. If we think of the ordinary x- and y-axes as ordered sets with the usual ordering for numbers, then the set of pairs (x, y) is an ordered set in which a point A is less than or equal to a point B if and only if A is below and to the left of B in the plane. When we have just two partially ordered sets, we normally use the notation $(X_1, P_1) \times (X_2, P_2)$ to stand for $\prod_{i=1}^{2}(X_i, P_i)$.

Example 6.1. Show that if $n = p_1^{k_1} p_2^{k_2} \cdots p_m^{k_m}$ and the p_i are distinct, then the lattice of factors of n is isomorphic to the product of $\{1, 2, \ldots, k_1\}$, $\{1, 2, \ldots, k_2\}$, ..., $\{1, 2, \ldots, k_m\}$, each of these sets ordered by the usual ordering on integers.

We define a function f from the factors of n to the product by defining for each number $r = p_1^{j_1} p_2^{j_2} \cdots p_m^{j_m}$ with $j_i \leq k_i$

$$f(r) = (j_1, j_2, \ldots, j_m).$$

Now if $r' = p_1^{j_1'} p_2^{j_2'} \cdots p_m^{j_m'}$ divides r, then the exponents j_i' for r' are less than or equal to the corresponding exponents j_i for r. Thus, the function is order preserving. It is a bijection because the exponents j_i determine r. The inverse is order preserving because, if $r = p_1^{j_1} p_2^{j_2} \cdots p_m^{j_m}$ and

$r' = p_1^{j_1'} p_2^{j_2'} \cdots p_m^{j_m'}$ are two integers with $j_i' \leq j_i$ for each i, then r' is a factor of r. Thus, f is an isomorphism. ∎

Example 6.2. If $\mathbf{P} = \{M, N\}$ is a partition of the set S into two classes M and N of size m and n, then the ordered set of all partitions contained in $\mathbf{P}$ (finer than $\mathbf{P}$) is isomorphic to $L_m \times L_n$, the product of the lattice of partitions of M and the lattice of partitions of N.

To see this, note that if $\mathbf{Q}$ is finer than $\mathbf{P}$, then every block of $\mathbf{Q}$ is a subset of M or N. Thus, if $f(\mathbf{Q})$ is the ordered pair consisting of a partition of M and a partition of N given by

$$f(\mathbf{Q}) = \left(\begin{matrix} \{C|C \text{ a class of} \\ \mathbf{Q} \text{ and } C \subseteq M\} \end{matrix}, \begin{matrix} \{D|D \text{ a class of} \\ \mathbf{Q} \text{ and } D \subseteq N\} \end{matrix} \right),$$

then f is a function from the ordered set of partitions finer than $\mathbf{P}$ to $L_M \times L_N$. We shall show that f is an isomorphism. First, we show that $\mathbf{Q} \leq \mathbf{R}$ if and only if $f(\mathbf{Q}) \leq f(\mathbf{R})$. If $\mathbf{Q}$ is finer than $\mathbf{R}$ (which is finer than $\mathbf{P}$), then every class of $\mathbf{Q}$ is a subset of a class of $\mathbf{R}$ (which in turn is a subset of M or N). Thus, $f(\mathbf{Q}) \leq f(\mathbf{R})$ in the product ordering. Further, if $f(\mathbf{Q}) \leq f(\mathbf{R})$ and if $\mathbf{Q}$ and $\mathbf{R}$ are both finer than $\mathbf{P}$, then every class of $\mathbf{Q}$ and every class of $\mathbf{R}$ is a subset of either M or N. But each class of $\mathbf{Q}$ in M must be a subset of some class of $\mathbf{R}$ in M if $f(\mathbf{Q})$ is to be less than or equal to $f(\mathbf{R})$ in the first coordinate. Similarly, since the only other classes of $\mathbf{Q}$ must be subsets of N, they must be contained in classes of $\mathbf{R}$ if $f(\mathbf{Q})$ is to be less than or equal to $f(\mathbf{R})$ in the second coordinate. We conclude that if $f(\mathbf{Q}) \leq f(\mathbf{R})$, then $\mathbf{Q} \leq \mathbf{R}$. In particular, if $f(\mathbf{Q}) = f(\mathbf{R})$, then $\mathbf{Q} \leq \mathbf{R}$ and $\mathbf{R} \leq \mathbf{Q}$, so $\mathbf{Q} = \mathbf{R}$. Therefore, the function f is one-to-one. Finally, f is onto, so we have a one-to-one onto function f such that $f(x) \leq f(y)$ if and only if $x \leq y$. Thus, f is an isomorphism. ∎

Recall that we have found $\mu(\mathbf{P}, t)$ for any partition $\mathbf{P}$ but still don't know $\mu(b, \mathbf{P})$ for a partition $\mathbf{P}$. If $\mathbf{P} = \{M, N\}$, then the interval of all partitions between b and $\mathbf{P}$ is isomorphic to $L_M \times L_N$. Thus, $\mu(b, \mathbf{P})$ is the same as the value of $\mu(b, t)$ in $L_M \times L_N$. One of the advantages to understanding products of ordered sets is that these products help us understand Möbius functions.

Products of Ordered Sets and Möbius Functions

Theorem 6.1. If $(X, P) = (Y, Q) \times (Z, R)$, and if μ_P, μ_Q, and μ_R stand for the corresponding Möbius functions, then

$$\mu_P((y_1, z_1), (y_2, z_2)) = \mu_Q(y_1, y_2)\mu_R(z_1, z_2).$$

Proof. Our strategy will be to show that the function μ defined by

$$\mu((y_1, z_1), (y_2, z_2)) = \mu_Q(y_1, y_2)\mu_R(z_1, z_2)$$

satisfies the conditions of Theorem 5.6 and hence is μ_P, the Möbius function of the product P of Q and R. Note that

$$\sum_{\substack{(t_1, t_2): \\ y_1 \leq t_1 \leq y_2 \\ z_1 \leq t_2 \leq z_2}} \mu_Q(y_1, t_1)\mu_R(z_1, t_2) = \sum_{\substack{t_1: \\ y_1 \leq t_1 \leq y_2}} \mu_Q(y_1, t_1) \sum_{\substack{t_2: \\ z_1 \leq t_2 \leq z_2}} \mu_R(z_1, t_2)$$

$$= 0 \cdot 0 = 0,$$

if $y_1 < y_2$ and $z_1 < z_2$. In fact, this sum is 0 if $y_1 = y_2$ and $z_1 < z_2$ or $y_1 < y_2$ and $z_1 = z_2$ because one of the factors of the product will be zero. Thus, the function μ_P defined by

$$\mu_P((y_1, z_1), (y_2, z_2)) = \mu_Q(y_1, y_2)\mu_Q(z_1, z_2)$$

satisfies condition 2 of Theorem 5.6 applied to $P = Q \times R$. But conditions 1 and 3 are immediate consequences of the formula used to define μ_P. Thus, by Theorem 5.6, μ_P must be the Möbius function of (X, P). ∎

Corollary 6.2. Let μ be the Möbius function of the partition lattice. Then for a partition $\mathbf{P}$ with k blocks of size $i_1, i_2, ..., i_k$,

$$\mu(b, \mathbf{P}) = \prod_{j=1}^{k} (-1)^{i_j - 1}(i_j - 1)!$$

Proof. By induction on the result of Theorem 6.1, the Möbius function of a product of k ordered sets will be the product of their individual Möbius functions. Thus, $\mu(b, \mathbf{P})$ is a product of the Möbius functions of k partition lattices. ∎

Example 6.3. What is the Möbius function $\mu(k', k)$ in the lattice of divisors of an integer n?

We have seen that if m is the number of distinct prime factors of n, then this lattice is the product of m linearly ordered sets. It is a direct result of Theorem 5.6 that in a linearly ordered set,

$$\mu(x, y) = \begin{cases} 1 & \text{if } x = y \\ -1 & \text{if } y \text{ covers } x \\ 0 & \text{otherwise.} \end{cases}$$

Suppose that $k = p_1^{j_1} p_2^{j_2} \cdots p_m^{j_m}$ and $k' = p_1^{j_1'} p_2^{j_2'} \cdots p_m^{j_m'}$. Then in each of the linear orderings,

$$\mu(j_i', j_i) = \begin{cases} 1 & \text{if } j_i' = j_i \\ -1 & \text{if } j_i = j_i' + 1 \\ 0 & \text{otherwise.} \end{cases}$$

Thus by Theorem 6.1, $\mu(k', k) = 0$ unless k/k' is a product of distinct primes and in this case if k/k' is a product of r distinct primes, then $\mu(k, k') = (-1)^r$. ∎

In the exercises you will see explicitly what you may have guessed already: that the lattice of subsets of a set is the product of two-element lattices. Up to isomorphism, there is only one two-element lattice, the set $\{0, 1\}$ with $0 < 1$, which is denoted by $\underline{2}$. In $\underline{2}$, $\mu(0, 1) = -1$. Thus, for any set K of size k, $\mu(\emptyset, K) = \prod_{i=1}^{k} \mu_i(0, 1) = (-1)^k$. This can be applied to the Möbius inversion theorem in Section 5 to obtain the inclusion–exclusion theorem of Chapter 2.

Products of Ordered Sets and Dimension

The notion of products is closely related to the dimension of a partial ordering. (Recall that the dimension of P is the minimum number of linear orderings whose intersection is P.) Suppose P is the intersection of the orderings $L_1, L_2, ..., L_n$ of X. Then there is a natural way to embed (X, P) in the product

$$\prod_{i=1}^{n} (X, L_i),$$

namely, send the element x to the n-tuple $(x, x, ..., x) = f(x)$. Then, $y \leq x \pmod{P}$ if and only if $(y, x) \in L_i$ for all i, so $y \leq x \pmod{P}$ if and only if $f(y) \leq f(s) \pmod{\prod_{i=1}^{n} Li}$. Recall that a *restriction* or *subposet* (X, P) of an ordered set (Y, Q) is a subset X of Y ordered by a relation P consisting of all pairs in Q that relate only points of X. Thus it consists of X and the restriction of Q to X. Therefore if an ordering is an intersection of n linear orderings, it is isomorphic to a restriction of the product of n linearly ordered sets by the map f we just described. (It is necessary only to check that the map f is one-to-one in order to show that it is an isomorphism from (X, P) to the image of f.) We summarize these remarks with a theorem.

Theorem 6.3. If (X, P) has dimension n, then it is isomorphic to a restriction of a product of n linearly ordered sets.

Proof. Given before the statement of the theorem. ∎

In this theorem, all the linear orderings are orderings of X. However, it turns out that if we can imbed (X, P) in a product of n linear orderings of n sets, then it happens that (X, P) has dimension no more than n. Thus, the dimension of (X, P) is the smallest n such that (X, P) can be imbedded in a product of n linear orderings. Pictorially, this means a finite poset has dimension n if and only if we can imbed it into n-dimensional Euclidean space with the "componentwise" partial ordering.

To simplify the proof of the converse of Theorem 6.3, we review the idea of the *transitive closure* $T(R)$ of a relation R. Recall that $T(R)$ is the intersection of all transitive relations containing R. We proved that (x, y) is in $T(R)$ if and only if y is reachable from x in the digraph of R.

Theorem 6.4. If a reflexive relation R has no directed cycles, then $T(R)$ is a reflexive partial ordering.

Proof. $T(R)$ is transitive, and $T(R)$ is reflexive because R is. If (x, y) and (y, x) were in $T(R)$ and $x \neq y$, then R would have a directed cycle because x would be reachable from y and y would be reachable from x. Thus $T(R)$ must be an antisymmetric relation as well. ∎

Theorem 6.5. If there is an isomorphism f from a poset (X, P) onto a restriction of

$$\prod_{i=1}^{n}(X_i, P_i),$$

with each P_i linear, then (X, P) has dimension n or less.

*Proof. All partial orderings referred to in this proof are assumed to be reflexive partial orderings. We must prove that P is an intersection of n linear orderings of X. We prove that there are n partial orderings Q_i of X, each containing P, such that if $x \not\leq y \pmod{P}$, then $y \leq x \pmod{Q_i}$ for some i. Then we choose arbitrary linear extensions L_i of the partial orderings Q_i. Their intersection contains exactly those ordered pairs in P, because any pair (v, w) of incomparable elements has $v \leq w \pmod{Q_j}$ and $w \leq v \pmod{Q_i}$ for some i and j. Thus, we will know P has dimension no more than n.

Now we define the orderings Q_i. Note that since the range of f is a subset of a product,

$$f(x) = (f_1(x), f_2(x), ..., f_n(x))$$

is an n-tuple of elements with $f_i(x)$ from X_i. Now if $x \leq y$, $f_i(x) \leq f_i(y)$ for all i, and if $f_i(x) \leq f_i(y)$ for all i, then $x \leq y$ since f is an isomorphism. Consequently, if $x \not\leq y$, then $f_i(y) < f_i(x)$ for some i. Now let

$$Q_i = T(P \cup \{(u, v) | f_i(u) < f_i(v)\}).$$

By our immediately preceding remarks, if $u \nleq v$, then $(v, u) \in Q_i$ for some i. However, the relation

$$P \cup \{(u, v) | f_i(u) < f_i(v)\}$$

has no directed cycles in its digraph. Assume to the contrary that

$$x_1 \ (x_1, x_2) \ \cdots \ (x_n, x_1) \ x_1$$

were such a directed cycle. In what follows, adopt the convention that (x_n, x_{n+1}) means (x_n, x_1). Then for each j, either $x_j < x_{j+1}$ in P or $f_i(x_j) < f_i(x_{j+1})$ in L_i.

However, since f is order-preserving, if $x_j \leq x_{j+1}$ in P then $f_i(x_j) \leq f_i(x_{j+1})$ in P_i. (It could be that $x_j < x_{j+1}$ and $f_i(x_j) = f_i(x_{j+1})$, though.) Thus, either $f_i(x_j) = f_i(x_{j+1})$ for all j, or the elements $f_i(x_j)$ form a cycle in the linear ordering L_i. But then since the linear ordering L_i has no cycles, all the ordered pairs (x_j, x_{j+1}) in the assumed cycle must be in P. However, the graph of P has no cycles. Therefore by Theorem 6.4, Q_i is an ordering. By the remarks preceding the construction of Q_i, the fact that Q_i is an ordering proves the theorem. ∎

Width and Dimension of Ordered Sets

This result lets us prove an interesting theoretical constraint on the dimension of an ordered set. Recall that the *width* of a poset is the largest size of any antichain of the poset.

Theorem 6.6. If (X, P) is a finite ordered set, then the dimension of P is less than or equal to its width.

Proof. Let P have width w. Assume first that X has an element $b \leq x \pmod{P}$ for all $x \in X$. By Dilworth's theorem (Theorem 2.3 in this chapter), X is a union of w chains C_i, each linearly ordered by P. Let $X_i = C_i \cup \{b\}$. Choose L_i to be the linear ordering of X_i that orders X_i in the same way as P does.

Now define a function f_i by

$$f_i(x) = \text{the highest element of } X_i \text{ below } x \pmod{P}.$$

Thus, if $x \leq y$, $f_i(x) \leq f_i(y)$. Define

$$f(x) = (f_1(x), f_2(x), ..., f_w(x)).$$

Then f is an order-preserving map of (X, P) onto a subset of $\prod_{i=1}^{w} X_i$. Suppose now that $f_k(x) = f_k(y)$ for all k. Let C_i be the chain containing x and C_j be the chain containing y. Then using $k = i$, we see that x is the highest element of X_i below y (mod P) and using $k = j$, y is the highest element of X_j below x (mod P). Thus, $x \leq y$ and $y \leq x$, giving $y = x$. Therefore, f is one-to-one. To show that f is an isomorphism from (X, P) to a restriction of the product, we must show that when $f_i(x) \leq f_i(y)$ for all i, then $x \leq y$.

However, if C_i is the chain containing x, then $f_i(x) = x$, and thus $x \leq f_i(y) \leq y$ (mod P), so $x \leq y$ (mod P). Thus, f is an isomorphism from (X, P) onto a restriction of $\prod_{i=1}^{w} (X_i, L_i)$. Now if (X, P) has no bottom element $b \leq x$ (mod P) for all $x \in X$, adding such an element will not change the dimension of (X, P). (The proof of this is elementary.) Thus, it suffices to show the theorem for ordered sets with bottom elements. Then by Theorem 6.5, we have shown that the dimension of (X, P) is less than or equal to w. ∎

Theorem 6.6 is just one illustration of how Dilworth's theorem may be used as a theoretical tool in the study of ordered sets.

EXERCISES

1. Show explicitly that the lattice of subsets of $\{1, 2, 3\}$ is the product of the lattices of subsets of $\{1\}, \{2\}$, and $\{3\}$.

2. Recall the function f_A given by

$$f_A(i) = \begin{cases} 0 & \text{if } i \notin A \\ 1 & \text{if } i \in A \end{cases}$$

that we used in Chapter 1 to compute the number of subsets of a set. Show that the function f given by

$$f(A) = (f_A(1), f_A(2), ..., f_A(n))$$

is an isomorphism from the lattice of subsets of a set $N = \{1, 2, ..., n\}$ to a product of the poset $\{0, 1\} = \underline{2}$ with itself n times.

3. Show that the poset of all positive factors of 12 ordered by "is a factor of" is a product of two smaller ordered sets by writing out the correspondence between ordered pairs and factors of 12 explicitly.

4. If the number $n = pqr$ where p, q, and r are primes, explain why the lattice of positive factors of n, ordered by "is a factor of," is a product of three two-element ordered sets. To what other well-known lattice is this lattice of factors isomorphic?

5. How did we use the assumption that $b \leq x$ for all x in Theorem 6.6?

6. Let G be a graph with two connected components. Prove that the bond lattice of G is the product of the bond lattices of its two connected components.

7. Generalize Exercise 6 appropriately.

8. Draw a diagram of the product of a chain of length 2 and a chain of length 3.

9. We say a multiset S is a multisubset of a multiset T if each x in S has smaller multiplicity in S than in T. Using multisubsets, give another description of a product of k chains of length $i_1, i_2, ...,$ and i_k. Describe another lattice we have studied that is isomorphic to this one.

10. Show that if the top element of L_1 is t_1 and the bottom element of L_2 is b_2, then the element $x = (t_1, b_2)$ of $L_1 \times L_2$ satisfies the distributive laws

$$x \wedge (y \vee z) = (x \wedge y) \vee (x \wedge z)$$

and

$$y \vee (x \wedge z) = (y \vee x) \wedge (y \vee z).$$

11. Discuss the relationship between products of positive integers and lattices of factors of positive integers.

12. If the poset P has height h and the poset Q has height k, what is the height of $P \times Q$?

13. What is the Möbius function of the lattice of multisubsets of a set S (see Exercise 9) ordered by $X \leq Y$ if each x in X has no higher multiplicity in Y than in X?

14. Find an example of a poset whose dimension is 2 and width is 3.

15. Find an ordering of width n whose width and dimension are both equal to half its number of elements.

*16. Find all six-element ordered sets whose dimension is equal to 3.

17. Show that the poset of 1- and $(n-1)$-element subsets of an n-element poset has its dimension equal to its width.

*18. Show that if M is a matroid that is the disjoint union of two matroids M_1 on X_1 and M_2 on X_2 (i.e., X_1 and X_2 are disjoint and its independent sets are exactly those sets that are disjoint unions of an independent set of M_1 and an independent set of M_2), then the lattice of flats of M is the product of the lattice of flats of M_1 and the lattice of flats of M_2. (See Exercise 25, Section 3.)

19. Discuss whether the ordering of the product ordered set $(X, P) \times (Y, Q)$ is, at least up to the order in which we write the symbols, equal to the Cartesian product of P and Q for both reflexive and irreflexive orderings P and Q, thought of as sets of ordered pairs.

Suggested Reading

Anderson, I. 1987. *Combinatorics of Finite Sets.* New York: Oxford University Press.

Bender, E. A., and J. R. Goldman. 1975. On the Applications of Möbius Inversion in Combinatorial Analysis. *American Mathematical Monthly* 789.

Birkhoff, Garrett. 1967. *Lattice Theory.* 3d ed. Providence, RI: American Mathematical Society.

Birkhoff, G., and T. Bartee. 1970. *Modern Applied Algebra.* New York: McGraw-Hill.

Crawley, Peter, and R. P. Dilworth. 1973. *Algebraic Theory of Lattices.* Englewood Cliffs, NJ: Prentice Hall.

Even, S. 1979. *Graph Theory Algorithms.* Rockville, MD: Computer Science Press.

Fisher, J. L. 1977. *Application Oriented Algebra.* New York: IEP, Crowell.

Ford, Lester and D. R. Fulkerson. 1962. *Flows in Networks.* Princeton, N.J: Princeton University Press.

Prather, R. 1976. *Discrete Mathematical Structures for Computer Science.* Boston: Houghton Mifflin.

Trotter, William T. 1992. *Combinatorics and Ordered Sets, Dimension Theory.* Baltimore: The Johns Hopkins University Press.

8
Enumeration under Group Action

Section 1   Permutation Groups

Permutations and Equivalence Relations

We have seen how the equivalence principle allows us to compute the number of equivalence classes of an equivalence relation *when the classes have the same size*. By analyzing the nature of the equivalence relations that arise in practice, we will be able to develop counting techniques that apply in many circumstances where equivalence classes do not have the same size. We begin the analysis with an example to which the equivalence principle applies.

Example 1.1. We wish to paint four different-colored vertical stripes, each covering a fourth of the area on the vertical surface of a wastebasket. How many different patterns of stripes may we distinguish?

We may think of numbering the four areas to be painted as areas 1, 2, 3, and 4. Then we may think of a paint job as a list of four colors; for example, we may use *RBGY* to stand for painting area 1 with red, area 2 with blue, area 3 with green, and area 4 with yellow. Some other lists will represent paint jobs we cannot distinguish from *RBGY*; for example, if we paint the wastbasket according to the list *BGYR* and then turn it a quarter turn clockwise, we cannot distinguish if from the *RBGY* paint job. This kind of analysis leads us to conclude that the lists may be organized into equivalence classes of four lists each, where two lists are equivalent if we can get from one to the other by turning the wastebasket. Thus, since there are 4! lists, the equivalence principle tells us there are 4!/4 distinguishable ways to paint the stripes. ∎

The reasoning in the previous example may be applied equally well to four people sitting at a round table for a game of cards or to four colored beads at the four corners of a square which is allowed to rotate in the plane. What is common to all these circumstances is that rotating a physical configuration in space changes the list of symbols we use to represent the object abstractly but does not change our perception of the object.

Of course, rotation is not the only sort of physical motion we might consider. If, for example, we consider four colored beads fastened at the vertices of a square and allow the square to be moved in three-dimensional space and returned to its original position, then a list of colors such as $RBYG$ may be converted to the reversed list $GYBR$ by flipping the square over in space and then returning it to its resting place. (The fact that we can imagine turning a square over in space was the reason Example 1.1 dealt with a wastebasket; we are unlikely to want to visualize turning it over! When we want particularly simple examples we will return to painting stripes on a wastebasket for just this reason.) Another problem of arrangement—such as connecting computers together in a communications network—could lead us to consider the results of rotations, rotations and flips, or some more general kind of rearrangements of lists of vertices as equivalent.

Thus, it seems that the equivalence relations that appear in practice include all those generated by "motions" in space that lead to "rearrangements" of lists. To understand how these ideas lead to equivalence relations, we will need to have a precise understanding of either "rearrangements" or "motions" or perhaps both. Although rotations and flips about diagonals or perpendicular bisectors are easy to visualize, it is possible that complex patterns of interconnections (say in a computer network) may lead to rearrangements that are harder to visualize or describe precisely with geometric language. On the other hand, we can visualize a list of n things as n objects in a row and we can visualize a rearrangement of the list in terms of picking up the objects and moving them. Thus, we shall try to develop a precise understanding of the idea of rearrangements of lists. Since we will want to be able to use our geometric intuition as much as possible, we begin by analyzing the rearrangements that correspond to geometric operations, specifically rotations of a square around its center in the plane. Figure 1.1(a) shows a square with its four vertices numbered and the four places where the vertices lie in the plane numbered in a corresponding way. Figure 1.1(b) shows the results of rotating the square through a quarter turn clockwise, a half turn, and three quarters of a turn. (Of course, part (a) shows the result of a full turn.) The quarter turn puts the vertex in place 4 into the place where vertex 1 was, the vertex in place 1 into place 2, etc., giving us 2341 as the list of which places the vertices have gone to.

We have called a list of distinct objects a *permutation* of these objects. We have seen that we may regard a permutation of n objects to be a one-to-one function from the set $1, 2, 3, \ldots, n$ to the set of objects. Using the Greek letter rho (ρ, pronounced *row*) as a shorthand for rotation, we may

Figure 1.1

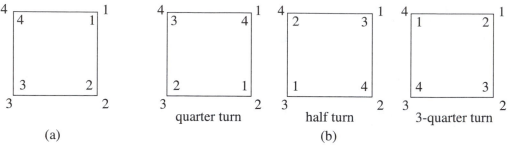

quarter turn   half turn   3-quarter turn

(a)   (b)

use functional notation to write

$$\rho(1) = 2, \quad \rho(2) = 3, \quad \rho(3) = 4, \quad \rho(4) = 1.$$

In this context another traditional notation for this permutation is

$$\rho = \begin{pmatrix} 1 & 2 & 3 & 4 \\ 2 & 3 & 4 & 1 \end{pmatrix}.$$

This is called the "two-line form" or "column notation form" of the permutation. (Notice that the choice of names depends on whether you choose to think of the representation as a matrix with two rows or as a matrix with n—in this case 4—columns; clearly, the name chosen is not important.) It turns out that standard functional notation is quite useful for studying the permutations corresponding to geometric motions. For example, the result of two quarter turns is a half turn. Since, for each j, $\rho(j)$ is the number of the vertex replaced by vertex j in one quarter turn, $\rho(\rho(j))$ is the number of the vertex replaced by vertex j after two quarter turns. But by definition of the composite function $\rho \circ \rho = \rho^2$, $\rho^2(j) = \rho(\rho(j))$, so ρ^2 is the permutation resulting from two quarter turns or a half turn. Similarly, ρ^3 is the permutation resulting from three quarter turns and ρ^4 is the permutation resulting from four quarter turns. Since a rotation through 360° returns each vertex to its original place, we see that ρ^4 must be the identity function given by

$$\rho^4(j) = j \quad \text{for all } j.$$

We use the Greek letter iota (denoted by ι, and pronounced eye-oh-ta) to stand for the identity function. That is,

$$\iota(j) = j \quad \text{for all } j,$$

so that $\rho^4 = \iota$.

The Group Properties

We can use properties of functional composition to verify the results of geometric reasoning. For example, geometry suggests that the result of seven quarter turns and the result of three quarter turns should be identical. Using the properties of function composition and the usual rule for adding exponents, we may write

$$\rho^7 = \rho^{4+3} = \rho^4 \circ \rho^3 = i \circ \rho^3 = \rho^3.$$

(Notice we are using the fact that composing the identity function with a second function gives the second function.) This suggests that arithmetic with powers of permutations can replace geometric insight in situations where geometry is irrelevant. It will turn out that even when geometry is relevant, arithmetic with permutations provides us with counting tools we cannot easily obtain from the geometry alone. Since we will make heavy use of the arithmetic of permutations, we shall give as a theorem enough of the basic properties of the arithmetic to allow us to derive any other properties we need. It is traditional to use the notation S_m to stand for the set of all permutations of $\{1, 2, \ldots, m\}$. (The significance of the letter S is that we have been using permutations to study symmetry in geometric figures, and they often come up in studies of symmetry.) It is also traditional to use the Greek letter sigma (σ, used because of the similarity to "S") to stand for a typical permutation.

> **Theorem 1.1.** The operation of function composition ($\circ$) on the set S, also denoted S_m, of permutations of $\{1, 2, \ldots, m\}$ has the following properties:
>
> (1) If $\sigma_1 \in S$ and $\sigma_2 \in S$, then $\sigma_1 \circ \sigma_2 \in S$. (The closure property)
> (2) If σ_1, σ_2, and $\sigma_3 \in S$, then $\sigma_1 \circ (\sigma_2 \circ \sigma_3) = (\sigma_1 \circ \sigma_2) \circ \sigma_3$. (The associative property)
> (3) There is an element $\iota \in S$ such that if $\sigma \in S$ then $\iota \circ \sigma = \sigma \circ \iota = \sigma$. (The identity property)
> (4) For each $\sigma \in S$ there is an element $\sigma^{-1} \in S$ such that $\sigma \circ \sigma^{-1} = \sigma^{-1} \circ \sigma = \iota$. (The inverse property)

Proof. Property 1 states that the composition of bijections is a bijection. Property 3 states that the identity function is a bijection *and* that composing the identity function with a second function will not change the second function. Property 4 restates the definition of an inverse function *and* states the fact that a bijection has an inverse function which is itself a bijection. These are familiar facts about function composition.

Property 2 states that the composition of permutations is an associative operation; in fact, the composition of functions in general is associative. To see why, let f, g, and h be three functions such that the compositions $(f \circ g) \circ h$ and $f \circ (g \circ h)$ are both defined. To say that these compositions are equal is to say they have the same domain and range and $(f \circ g) \circ h(x) = f \circ (g \circ h)(x)$ for each x in the domain. Both of these compositions have the same domain as h and the same range as f, and, by the definition of composition,

$$(f \circ g) \circ h(x) = f \circ g\big(h(x)\big) = f\big(g(h(x))\big)$$

and

$$f \circ (g \circ h)(x) = f\big(g \circ h(x)\big) = f\big(g(h(x))\big).$$

This shows that function composition is associative. ∎

The properties in Theorem 1.1 are the properties we use in grouping together and simplifying expressions when we solve equations and make other computations. We will refer to them as the *grouping properties* or the **group** properties for short. Because of the similarity of function composition with ordinary multiplication, people often write $\sigma\tau$ for $\sigma \circ \tau$. (The Greek letter tau, denoted τ, rhymes with either paw or pow, whichever you prefer.)

Example 1.2. Show how the grouping properties of the multiplication of nonzero real numbers are used in solving the equation $2x = 6$.

Clearly, we solve the equation by multiplying both sides by the fraction $\frac{1}{2}$ to get $x = 3$. In this mental computation we skip over the following steps!

$$\frac{1}{2}(2x) = \frac{1}{2} \cdot 6 \qquad \text{(existence of inverse, namely } \tfrac{1}{2}\text{, of 2)}$$

$$\left(\frac{1}{2} \cdot 2\right)x = \frac{1}{2} \cdot 6 \qquad \text{(associative law)}$$

$$1 \cdot x = \frac{1}{2} \cdot 6 \qquad \text{(inverse property of 2 and } \tfrac{1}{2}\text{)}$$

$$x = \frac{1}{2} \cdot 6 \qquad \text{(identity property)}$$

$$x = \frac{1}{2}(2 \cdot 3) \qquad \text{(definition of multiplication)}$$

$$= \left(\frac{1}{2} \cdot 2\right) \cdot 3 \qquad \text{(associative law)}$$

$$= 1 \cdot 3 \qquad \text{(inverse property of 2 and } \tfrac{1}{2}\text{)}$$

$$= 3 \qquad \text{(identity property).} \qquad \blacksquare$$

Example 1.3. Show which grouping properties are used in the demonstration given earlier that $\rho^7 = \rho^3$ for the permutation ρ which corresponds to a quarter turn of a square.

We write

$$\rho^7 = \rho^3 \circ \rho^4 \qquad \text{(addition law for exponents)}$$
$$= \rho^3 \circ \iota \qquad \text{(geometric fact)}$$
$$= \rho^3 \qquad \text{(identity property of } \iota \text{)}. \qquad \blacksquare$$

Notice that we used a rule, the addition law for exponents, which is not one of our grouping properties. In fact, we have been rather imprecise in our definition of exponents so that it would be difficult to derive the addition law for exponents from our grouping properties. A more precise definition will allow us to make a straightforward derivation of the law.

Powers of Permutations

We use the technique of mathematical induction to define σ^n for any permutation σ and any nonnegative integer n. First, we define σ^0 to be the identity function ι. Then we state that once σ^m has been defined, we shall define $\sigma^{m+1} = \sigma^m \circ \sigma$. By induction this defines σ^n for each nonnegative integer n.

Now although σ^n has been defined for all n, we had different descriptions of σ^1 and σ^2 earlier in the book. It is natural to ask if our new definition is consistent with our earlier ones. To see that it is, we write

$$\sigma^1 = \sigma^{0+1} = \iota \circ \sigma = \sigma$$

and

$$\sigma^2 = \sigma^{1+1} = \sigma^1 \circ \sigma = \sigma \circ \sigma.$$

This shows that σ^1 is σ and σ^2 is $\sigma \circ \sigma$ as we defined them earlier. With this definition we can derive the addition law for exponents.

Example 1.4. Derive the addition law for exponents.

We use induction on n to prove that $\sigma^{m+n} = \sigma^m \circ \sigma^n$ for all nonnegative n as follows. We consider the set S of integers n such that $\sigma^{m+n} = \sigma^m \circ \sigma^n$. To show that 0 is in S, we write

$$\sigma^{m+0} = \sigma^m = \sigma^m \circ \iota = \sigma^m \circ \sigma^0.$$

Notice that we are using the identity property of ι in this computation. Now we assume n is in S. To show that $n+1$ is in S we write

$$\sigma^{m+(n+1)} = \sigma^{(m+n)+1} = \sigma^{m+n} \circ \sigma$$
$$= (\sigma^m \circ \sigma^n) \circ \sigma \qquad \text{(inductive hypothesis)}$$
$$= \sigma^m \circ (\sigma^n \circ \sigma) \qquad \text{(associative law)}$$
$$= \sigma^m \circ \sigma^{n+1} \qquad \text{(inductive definition)}.$$

Thus, for all integers $n \geq 0$, $\sigma^{m+n} = \sigma^m \circ \sigma^n$. Notice that m plays no role in the induction, so the proof is valid for all m for which σ^m is defined. Had we wanted to, we could have inducted on m; however, this would have been less convenient for the application of the inductive definition of σ^n. ∎

There are two interesting features in the example. First, we used the associative law for addition as well as the associative law for function composition. Thus, to prove a fact about composition of permutations, we are using grouping properties of both addition and composition. Second, we did not use the inverse grouping property. It may not be surprising that this property comes into play in the proof of the addition law for exponents when one of the exponents is negative. In order to deal with negative exponents we define σ^{-1} as usual to be the inverse function to the permutation σ and we define σ^{-m} to be $(\sigma^{-1})^m$. Then function and grouping properties allow us to derive all the usual rules of exponents involving negative exponents. We describe the details of this and the multiplication rule for exponents in the exercises.

Permutation Groups

In Theorem 1.1 we described the four grouping properties the composition operation has on the set $S = S_m$. When the composition operation has these four properties on a set G of permutations, we call G a **permutation group**. It is traditional to call the permutation group S_m the **symmetric group** on m letters.

Example 1.5. Show that the set $R = \{\iota, \rho, \rho^2, \rho^3\}$ of permutations corresponding to quarter turns of a square is a permutation group.

By definition, to show that a set of permutations is a permutation group we need to show that the composition operation has the grouping properties (1)–(4) on that set. However, one of these properties is automatically satisfied. Property 2, the associative property, is a property of the composition of functions that has nothing to do with the set S, so it is satisfied for every set S of permutations. Property 1, the closure property, is verified by observing that following a rotation through m degrees by a rotation through n degrees gives a rotation through $m + n$ degrees. If m and n are both multiples of 90, then so is $m + n$, so a composition of powers of ρ must also be a power of ρ. We have seen that $\rho^4 = \iota$ and we have seen how to convert any higher power of ρ to a ρ^0, ρ^1, ρ^2, or ρ^3 by using the addition rule for exponents. Thus, R is closed under composition, so property 1 holds. The composition operation on R satisfies property 3 because ι, the identity permutation, is in R.

Now we show that the composition operation on R satisfies property 4. First, if ρ^i is a rotation through m degrees, then ρ^{-i} is a rotation through

m degrees in the opposite direction and $\rho^{-i} = \iota \circ \rho^{-i} = \rho^4 \circ \rho^{-i} = \rho^{4-i}$. But since ρ^{4-i} is in R whenever ρ^i is in R, and $\rho^i \circ \rho^{4-i} = \rho^{i+4-i} = \rho^4 = \iota$, the composition operation on R satisfies property 4. Therefore, R is a permutation group. ∎

Permutation groups will turn out to be fundamental in our study of counting, so it will be useful to have an efficient technique to use in order to show that a set of permutations is a permutation group. Surprisingly, we need only check that a finite set of permutations is closed under composition, that is, that it satisfies property 1 in Theorem 1.1, to show that it is a permutation group.

> **Theorem 1.2.** A finite set G of permutations is a permutation group if and only if $\sigma_1 \circ \sigma_2$ is in G whenever σ_1 and σ_2 are in G.
>
> *Proof.* Properties 1 and 2 of Theorem 1.1 are automatically satisfied. To show that property 3 is satisfied, we need only show that ι is in G. However, for any $\sigma \in G$, the sequence $\sigma, \sigma^2, \sigma^3, \ldots$, cannot consist entirely of distinct permutations because G is finite and each of these permutations is in G. However, if $\sigma^i = \sigma^j$ and $i < j$, then in S_m we may make the computation
>
> $$\iota = \sigma^{-i} \circ \sigma^i = \sigma^{-i} \circ \sigma^j = \sigma^{j-i}.$$

But σ^{j-i} is in G by the hypothesis of the theorem so that ι is in G.

Now to verify property 4, we must show that σ^{-1} is in G whenever σ is in G. But we have just shown that for each σ in G, $\sigma^m = \iota$ for some $m > 0$, so then $\sigma \circ \sigma^{m-1} = \iota = \sigma^{m-1} \circ \sigma$. By definition of the inverse property, this means that $\sigma^{m-1} = \sigma^{-1}$. Thus, since σ^{m-1} is in G by the hypothesis of the theorem, σ^{-1} is in G whenever σ is. Therefore, G is a permutation group. ∎

Associating a Permutation with a Geometric Motion

The geometric motions of a square provide us with our second example of a permutation group.

Example 1.6. Show that the set of all permutations of the four vertices of a square resulting from three-dimensional motions, including flips around a diagonal or edge-bisector as well as rotations, forms a permutation group.

Since the composition of two three-dimensional motions is a three-dimensional motion, we need only show that if σ_1 and σ_2 are permutations resulting from three-dimensional motions, then $\sigma_2 \circ \sigma_1$ is the permutation resulting from performing the first motion and then the second. Our discussion of rotations suggests that $\sigma(i) = j$ should mean that the vertex in place i in Figure 1.1 is moved to the vertex in place j. The places referred to are the numbered places in the plane where the vertices sit, not the places in the square that number the vertices themselves. Let us adopt this in general as the way to associate a permutation with a movement of a geometric object whose vertices determine a set of well-defined places in space. Then when we compose two motions, if the first motion sends j to k[†] and the second motion sends k to h, then the result sends j to h, as does the composition $\sigma_2 \circ \sigma_1$. (Since $\sigma_1(j) = k$ and $\sigma_2(k) = h$, we get $\sigma_2 \circ \sigma_1(j) = \sigma_2(k) = h$.) Thus, this set of permutations forms a group. This group is called the "dihedral group of the square" and is denoted by D_4. ∎

Though we have verified that we have a group and given it a name in the previous example, we know very little about the group. For example, how many permutations are in the group? In addition to the identity and the three nonidentity rotations, we could flip the square about either diagonal or about either of the bisectors of opposite sides. Think about how to show that these last four motions are different from the first four and from each other.

We see now that the dihedral group of the square has at least eight members. However, composing two of the members we have listed might give a member we have not listed. Also, there might be some other way of moving the square around in space. Thus, we might have 9, 12, 16, or some other number of permutations in our group. A theorem (which we will prove as Theorem 2.6 in the next section) from the theory of groups tells us that the number of elements of the dihedral group of the n-gon must be a factor of the number of permutations in the symmetric group on n letters, or 24 permutations in this case. This result will help us show that the dihedral group has exactly eight members.

Abstract Groups

While our use of group properties and their consequences will be limited largely to permutation groups, there are many other situations to which they apply, leading to many other families of mathematical structures known as groups. For example, the set of all integers (negative as well as nonnegative) satisfies properties 1–4 of Theorem 1.1 when we replace the "circle" representing function composition by addition. The set of positive rational numbers satisfies properties 1–4 of Theorem 1.1 when we replace the circle by multiplication. The set of congruence classes modulo m with

---

[†] This is a shorthand for "the first motion sends the vertex in place j to place k."

congruence class addition mod m also satisfies these properties. The set of invertible m-by-m matrices with the operation of matrix multiplication satisfies these rules as well. Group theory is devoted to the study of sets and operations satisfying these rules; the word **group** is defined to mean a set with an operation satisfying properties 1–4 defined on it.

A simple but important concept in the theory of groups is the idea of a subgroup; a group S is a **subgroup** of a group G if S is a subset of G and the rule we use to combine elements in S gives the same result as applying the rule of G to these elements. In this terminology, a permutation group is a subgroup of one of the symmetric groups S_m (consisting of all permutations of $\{1, 2, \ldots, m\}$). Thus, the rotation group R_4 and dihedral group D_4 are subgroups of the symmetric group S_4. Further, the rotation group R_4 is a subgroup of the dihedral group D_4. Similarly, the group of even integers under addition is a subgroup of the group of all integers under addition. Using the same computations as in Theorem 1.2, we may prove the following.

Theorem 1.3. A finite set S is a subgroup of a group G if and only if S is closed under the operation of G.

Proof. Essentially the same as the proof of Theorem 1.2, with permutations replaced by elements of the group G. ∎

A common way to describe a finite group is with an "operation table." Table 1.1 describes the rotation group.

Table 1.1

$\circ$	ι	ρ	ρ^2	ρ^3
ι	ι	ρ	ρ^2	ρ^3
ρ	ρ	ρ^2	ρ^3	ι
ρ^2	ρ^2	ρ^3	ι	ρ
ρ^3	ρ^3	ι	ρ	ρ^2

Table 1.2 describes a different four-element group. The table shows us that $\sigma \circ \sigma = \iota$ for every σ in the group; Table 1.1 shows that this is not the case for the group of rotations of a square.

The group in Table 1.2 is usually called the "four group." In this group we say the "product" of σ_1 with σ_2 is σ_3, as is the product of σ_2 with σ_1. We write $\sigma_1 * \sigma_2 = \sigma_3$. We observe that $\sigma_1 * \sigma_2 = \sigma_2 * \sigma_1$. In fact the operation of the four group is commutative (in standard mathematical terminology we

Table 1.2

$*$	ι	σ_1	σ_2	σ_3
ι	ι	σ_1	σ_2	σ_3
σ_1	σ_1	ι	σ_3	σ_2
σ_2	σ_2	σ_3	ι	σ_1
σ_3	σ_3	σ_2	σ_1	ι

say a group with a commutative operation, such as the four group, is *Abelian* (pronounced uh–*bee*–lian). Not every group operation is commutative. It is possible to find two permutations σ and τ in the symmetric group S_3 such that $\sigma\tau \neq \tau\sigma$; see Exercise 3.

EXERCISES

1. Consider an equilateral triangle with vertices labeled 1, 2, and 3. Using two-line form, write down the permutations of $\{1, 2, 3\}$ that correspond to rotations of the triangle.

2. Using two-line form, write down the members of S_3, the group of all permutations of $\{1, 2, 3\}$.

3. Pick two members σ and τ of S_3 in such a way that $\sigma\tau \neq \tau\sigma$. (In this way you will show that S_3 is not commutative.)

4. Write down in two-line form the permutations of $\{1, 2, 3, 4, 5\}$ that correspond to rotations of a regular pentagon with its vertices cyclically labeled from 1 to 5.

5. A tetrahedron is a four-sided, three-dimensional figure with four vertices and four triangular sides, i.e., a pyramid with a triangular base. A regular tetrahedron has four congruent equilateral triangles for its faces. Write down a permutation that is *not* a member of the group of all permutations of $\{1, 2, 3, 4\}$ which arise from three-dimensional motions of a regular tetrahedron with vertices labeled with 1, 2, 3, and 4.

6. How many permutations are there in the group of all permutations of $\{1, 2, 3, 4\}$ that correspond to three-dimensional motions of a tetrahedron whose corners are labeled with the labels 1, 2, 3, and 4?

7. Use the inductive definition of σ^m and the group properties to show that $\sigma^m \circ \sigma^{-1} = \sigma^{m-1}$ for any σ in a (permutation) group.

8. Use the definition that $\sigma^{-n} = (\sigma^{-1})^n$ and the inductive definition of exponents to prove that $\sigma^m \circ \sigma^{-n} = \sigma^{m-n}$. Consider the case $m \geq 0$ first, then the case $m = -k$, $k > 0$.

9. Use the inductive definition of exponents and the addition law to prove the product rule for exponents, namely

$$(\sigma^m)^n = \sigma^{mn}$$

for members σ of a (permutation) group.

10. Write down a multiplication table, using the symbols ι, ρ, ρ^2 as the group elements for the group you gave in Exercise 1.

11. Three of the permutations you gave in Exercise 2 are the permutations ι, ρ, ρ^2 from Exercise 10. The other permutations correspond to three-dimensional flips about bisectors of angles. Let φ_j denote the flip about the bisector of the angle at vertex j. (The Greek letter phi, denoted φ, is usually pronounced "fee," though sometimes "fie.") With this notation the symmetric group S_3 you wrote down in Exercise 2 consists of ι, ρ, ρ^2, φ_1, φ_2, and φ_3. Write a multiplication table for this group using the notation just given. Use your multiplication table to find two permutations that show that the group operation is not commutative.

12. Suppose that σ and τ are elements of a (permutation) group and $\sigma \circ \tau = \iota$, the identity element. Does this mean $\sigma = \tau^{-1}$ and $\tau = \sigma^{-1}$? Explain.

13. Show that for any elements σ and τ of a (permutation) group, $\tau^{-1} \circ \sigma^{-1}$ has the defining property of being an inverse for $\sigma \circ \tau$. Note that by Exercise 12 this means $\tau^{-1} \circ \sigma^{-1}$ is $(\sigma \circ \tau)^{-1}$.

14. In the group S_3, using the notation of Exercise 11, find permutations σ and τ such that $(\sigma \circ \tau)^{-1} \neq \sigma^{-1} \circ \tau^{-1}$.

15. The dihedral group of a pentagon consists of all permutations of the (cyclically numbered) vertices of a pentagon that correspond to three-dimensional motions (including both rotations and flips) of a pentagon. Show that the group of all permutations corresponding to rotations is a subgroup of this dihedral group.

16. Using the notation of Exercise 11 for S_3, find a three-element subgroup and a two-element subgroup of S_3. Explain why they are subgroups.

17. Is it possible to find a three-element subgroup of S_3 different from the one you gave in Exercise 15? Explain.

18. Can a group have two different identity elements? Explain. (Here the problem becomes rather trivial if you assume the group is a permutation group.)

19. Suppose a subgroup of the dihedral group of the square contains the flip about the 1-3 diagonal and the flip about the 2-4 diagonal. Is there

any other element of the dihedral group, other than the identity, which it must contain?

20. (a) Show that $\sigma^{-n} = (\sigma^n)^{-1}$.
    (b) Show that $\sigma^n = \iota$ if and only if $(\sigma^{-1})^n = i$.

21. Prove true or demonstrate false with a counterexample. In a (permutation) group, $(\sigma\tau)^n = \sigma^n\tau^n$.

22. (a) Show that if G is a (permutation) group then for any σ in G the set

$$\{\sigma^k \mid k \text{ is an integer}\}$$

is a subgroup of G (remember that an integer can be positive or negative).

    (b) Show that if G is finite then

$$\{\sigma^k \mid k \text{ is a positive integer}\}$$

is a subgroup of G.

23. (Due to R. A. Dean.) Let S be the set

$$S = \{\sqrt[3]{m} \mid m \text{ is an integer}\}.$$

For $a, b \in S$, define
$$a \circ b = \sqrt[3]{a^3 + b^3}.$$

Show that this operation satisfies the group properties so that with this operation S forms a group.

24. On the set of integers, define $m \circ k = m + k - 2$. Show that this "circle" operation satisfies the group properties.

25. Write out a proof of Theorem 1.3.

26. Show that Theorem 1.3 is false if G is infinite. Suppose S is a subset of a (possibly infinite) group G such that for each σ in S, σ^{-1} is in S, and for each σ and τ in S, $\sigma\tau^{-1}$ is in S. Is S a subgroup of G? (Prove or give a counterexample.)

Section 2 Groups Acting on Sets

Groups Acting on Sets

In our discussion of the movements of a square we have focused on one geometric aspect of the square, its vertices. However, we might, for geometric reasons, have been interested in some other aspect—perhaps the edges or the diagonals. Notice that a rotation sends a side to a side or a diagonal to a diagonal. Similarly, a three-dimensional movement of the square (which returns it to the original location) sends a side to a side, a diagonal to a diagonal, and so on. A given movement will not send a side to two different sides or a diagonal to two different diagonals, meaning that the movements define functions on the sides, diagonals, etc. Now two different sides cannot be moved to the same one, nor can two different diagonals be moved to the same one, so each movement of the square gives rise to a bijection of the sides, a bijection of the diagonals, etc. Further, as with the vertices, when we follow one motion by another, we get the composition of the bijections of the two motions as the new bijection.

We would say intuitively that a group of motions of the square *acts* on the sides of the square or the diagonals of the square, etc. We make this intuitive notion precise as follows. A group G (which you may think of as a permutation group) **acts** on a set X if, for each element σ of G, there is a bijection β_σ of X (using the Greek letter beta, pronounced *bay*-ta) with the property that

$$\beta_{\sigma \circ t} = \beta_\sigma \circ \beta_t.$$

Example 2.1. Among the rotations of a square, studied earlier, what is the effect on the side joining vertices 1 and 2 of the $90°$ rotation ρ and the $180°$ rotation ρ^2? How does this effect correspond to the defining property of the action of a group? What is the effect on the diagonal joining 1 and 3?

Because ρ takes 1 to 2 and 2 to 3, ρ takes the side s_{12} joining 1 and 2 to the side s_{23} joining 2 and 3. Since ρ takes 2 to 3 and 3 to 4, applying ρ twice takes 1 to 3 and 2 to 4, so ρ^2 takes the side joining 1 and 2 to the side joining 3 and 4. Since β_ρ takes the side s_{12} to the side s_{23} and the side s_{23} to the side s_{34}, $\beta_\rho \circ \beta_\rho$ takes s_{12} to s_{34}. Therefore, $\beta_\rho \circ \beta_\rho$ has the same effect as β_{ρ^2} on the side s_{12}, as it must if the group of rotations acts on the edges.

Because ρ takes 1 to 2 and 3 to 4, ρ takes the diagonal joining 1 and 3 to the diagonal joining 2 and 4. Because ρ^2 takes 1 to 3 and 3 to 1, ρ^2 takes the diagonal joining 1 and 3 to the diagonal joining 3 and 1, which is the same as the diagonal joining 1 and 3. ∎

Notice how β_{ρ^2} acts on the diagonals. We saw already that it takes the 1-3 diagonal to itself; similarly, β_{ρ^2} takes the 2-4 diagonal to itself. Thus, on the diagonals, β_{ρ^2} and β_ι are the same bijection; we say they have the *same action* on the diagonals. This shows that when a group acts on a set, it is possible for *different* group elements to act in the *same* way, i.e., to give the same bijection.

Figure 2.1

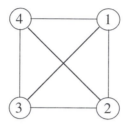

In Figure 2.1 we have drawn a square with its diagonals. We may also regard the drawing as a drawing of a complete graph on four vertices. The four edges and two diagonals are then the edges of the complete graph. In the preceding example we observed that because ρ takes 1 to 2 and 2 to 3, it must take the side joining vertices 1 and 2 to the side joining vertices 2 and 3. We reasoned similarly with the diagonal. Thus, we are using information about the action of ρ on single vertices to determine the action of ρ on two-element sets of vertices and then inferring information about the action of ρ on sides and diagonals. From the point of view of the complete graph on four vertices, the action of the rotation group on the two-element subsets themselves is equally interesting since these subsets *are* the edges of the graph. Choosing the edges used in Example 2.1 as typical edges, we may write

$$\beta_\iota(\{1,2\}) = \{1,2\} \qquad \beta_\rho(\{1,2\}) = \{2,3\}$$
$$\beta_{\rho^2}(\{1,2\}) = \{3,4\} \qquad \beta_{\rho^3}(\{1,2\}) = \{4,1\}$$

and

$$\beta_\iota(\{1,3\}) = \{1,3\} \qquad \beta_\rho(\{1,3\}) = \{2,4\}$$
$$\beta_{\rho^2}(\{1,3\}) = \{1,3\} \qquad \beta_{\rho^3}(\{1,3\}) = \{2,4\}.$$

The use of β_ρ for the bijection determined by ρ is a bit cumbersome. It is quite suggestive, in fact, to write instead

$$\rho(\{1,2\}) = \{\rho(1), \rho(2)\} = \{2,3\}.$$

In this notation we may write

$$\rho^0(\{1,2\}) = \{1,2\} \qquad \rho(\{1,2\}) = \{2,3\}$$
$$\rho^2(\{1,2\}) = \{3,4\} \qquad \rho^3(\{1,2\}) = \{4,1\}$$

and

$$\rho^0(\{1,3\}) = \{1,3\} \qquad \rho(\{1,3\}) = \{2,4\}$$
$$\rho^2(\{1,3\}) = \{1,3\} \qquad \rho^3(\{1,3\}) = \{2,4\}.$$

We will continue to omit the β in this way where doing so should not be confusing. Further, we shall write σx for $\sigma(x)$ when the parentheses would be cumbersome. Thus, for example, the preceding expressions become

$$\rho^0\{1,2\} = \{1,2\} \qquad \rho\{1,2\} = \{2,3\}$$
$$\rho^2\{1,2\} = \{3,4\} \qquad \rho^3\{1,2\} = \{4,1\}$$

and

$$\rho^0\{1,3\} = \{1,3\} \qquad \rho\{1,3\} = \{2,4\}$$
$$\rho^2\{1,3\} = \{1,3\} \qquad \rho^3\{1,3\} = \{2,4\}.$$

Notice that the action of the powers of ρ has divided the edge set of the graph into two classes,

$$\{\{1,2\}, \{2,3\}, \{3,4\}, \{4,1\}\} \quad \text{and} \quad \{\{1,3\}, \{2,4\}\}.$$

We would have gotten the same division if we had started with the edge $\{2,3\}$, $\{3,4\}$, or $\{4,1\}$ in place of $\{1,2\}$ and if we had started with the edge $\{2,4\}$ in place of $\{1,3\}$. These two classes are called "orbits" of the rotation group acting on the edge set. In general, if a group $G = \{\sigma_1, \sigma_2, \sigma_3, \ldots\}$ acts on a set X, the set

$$\{\sigma(x) \mid \sigma \in G\} = \{\sigma_1(x), \sigma_2(x), \sigma_3(x), \ldots\}$$

is the **orbit** of x under the action of G. Notice that in our example our orbits were disjoint sets whose union is the entire set X of edges. This suggests that in general orbits may be the equivalence classes of an equivalence relation on X.

Orbits as Equivalence Classes

Theorem 2.1. Suppose a group G acts on a set X. The relation that relates x to y if there is a σ in G such that $\sigma x = y$ is an equivalence relation. The equivalence classes are the orbits of G acting on X.

Proof. In this proof we use the notation β_σ for the bijection associated with σ. First, since $\iota \circ \iota = \iota$ we get that $\beta_\iota \circ \beta_\iota = \beta_\iota$, so that $\beta_\iota(\beta_\iota(x)) = \beta_\iota(x)$. Therefore, β_ι is the identity bijection. Because $\beta_\iota(z) = z$ for each z in X, the relation is reflexive. Since $\beta_\sigma^{-1} = \beta_{\sigma^{-1}}$, the relation is symmetric because $\beta_\sigma(x) = y$ if and only if $\beta_{\sigma^{-1}}(y) = x$.

The relation is transitive because composing the bijections associated with σ and τ gives the bijection associated with $\sigma \circ \tau$. By the definition of the orbit containing x, it must consist of all elements of X equivalent to x; thus, it is the equivalence class containing x. ∎

By now you may suspect that, as equivalence classes, orbits of a group acting on a set have something to do with equivalence class counting. This is the case; in almost all cases where we are able to make a general statement about the number of equivalence classes of an equivalence relation, these classes are the orbits of a group action. For example, the rotation group acts on the edges of the complete graph on four vertices giving us one equivalence class $\{\{1,3\},\{2,4\}\}$ of two edges and one equivalence class $\{\{1,2\},\{2,3\},\{3,4\},\{4,1\}\}$ of four edges. By studying how G acts on these two orbits we discover still more equivalence relations, this time on G, which help to explain why these orbits have different sizes.

Notice that the list

$$(\rho^0\{1,2\}, \rho\{1,2\}, \rho^2\{1,2\}, \rho^3\{1,2\}) = (\{1,2\},\{2,3\},\{3,4\},\{4,1\})$$

lists each member of the orbit once and that each member of the orbit is associated with exactly one member of R, the rotation group. On the other hand, the list

$$(\rho^0\{1,3\}, \rho^1\{1,3\}, \rho^2\{1,3\}, \rho^3\{1,3\}) = (\{1,3\},\{2,4\},\{1,3\},\{2,4\}) \quad (2.1)$$

lists each member of the orbit twice and associates each member of the orbit with two members of the group. The list shows us that it is natural to think of an orbit as a *multiset*. In particular, the multiplicity of $\{1,3\}$ in this multiset is two and the multiplicity of $\{2,4\}$ in this multiset is two. To help us keep track of when we are thinking of an orbit as a set and when we are thinking of it as a multiset, we will define the multiset

$$\{\sigma(x)|\sigma \in G\} = \{\sigma_1(x), \sigma_2(x), \sigma_3(x), \ldots\}$$

to be the **multiorbit** of x under the action of G. Because we use set braces to describe both sets and multisets, we will take care to make it clear when we are thinking of an orbit as a multiset.

The Subgroup Fixing a Point and Cosets

The earlier list marked (2.1) shows that ρ^0 and ρ^2 send $\{1,3\}$ to $\{1,3\}$ and that ρ^1 and ρ^3 send $\{1,3\}$ to $\{2,4\}$. Thus, ρ^0 and ρ^2 have an equivalent effect on $\{1,3\}$, giving it multiplicity 2 in the multiorbit, and ρ^1 and ρ^3 have an equivalent effect on $\{1,3\}$, giving it multiplicity 2 in the multiorbit. Keeping track of which elements of the group have the same action therefore divides the rotation group R into two disjoint classes. This suggests that defining two members of G to be related if they have an equivalent action on some member x of X may give an equivalence relation (this time an equivalence relation on G, not X). This is the case. Further, the equivalence classes have a straightforward description in terms of the permutations that *fix* x, that is, send x to itself. Notice that $\rho^2 \circ \rho^2 = \iota = \rho^0$, $\rho^0 \circ \rho^0 = \rho^0$, and $\rho^0 \circ \rho^2 = \rho^2 \circ \rho^0 = \rho^2$; the two permutations that fix $\{1,3\}$ form a subgroup of the rotation group. Now the equivalence class that sends $\{1,3\}$ to $\{2,4\}$, namely ρ^1 and ρ^3, does not form a subgroup. We may, however, write

$$\rho\{\rho^0, \rho^2\} = \{\rho \circ \rho^0, \rho \circ \rho^2\} = \{\rho^1, \rho^3\},$$

so that the equivalence class of permutations mapping $\{1,3\}$ to $\{2,4\}$ is obtained by "multiplying" the class fixing $\{1,3\}$ by the permutation ρ which sends $\{1,3\}$ to $\{2,4\}$. We see the same pattern whenever we have a group acting on a set.

Theorem 2.2. If G is a finite group acting on a set X, and x is a member of X, then the relation given by "σ is related to τ if $\sigma x = \tau x$" is an equivalence relation. The equivalence class containing the identity element of G consists of all permutations which fix x and is a subgroup H of G. The equivalence class containing a permutation σ is the set

$$\sigma H = \{\sigma \circ \tau \mid \tau \in H\}.$$

Proof. Since $\sigma x = \sigma x$, the relation is reflexive, and since $\sigma x = \tau x$ if and only if $\tau x = \sigma x$, the relation is symmetric. If $\sigma x = \tau x$ and $\tau x = \mu x$, then $\sigma x = \mu x$, so the relation is transitive as well.

Since the identity fixes x, the equivalence class containing the identity consists of all permutations that fix x. If the permutations σ and τ fix x, then so does the permutation $\sigma \circ \tau$, so by Theorem 1.3, this equivalence class is a subgroup of G.

To determine the equivalence class containing an arbitrary permutation σ, note that when $\sigma x = \tau x$, then $x = \sigma^{-1} \circ \tau x$, so $\sigma^{-1} \circ \tau$ is in H, that is, $\sigma^{-1} \circ \tau = \gamma \in H$. (The Greek letter gamma is denoted by γ.) Then $\tau = \sigma \circ \gamma$, so that $\tau \in \sigma H$ (recall how σH was defined in the statement of the theorem). Further, if $\tau \in \sigma H$, then there is a γ in H such that $\tau = \sigma \circ \gamma$ and therefore

$$\tau x = (\sigma \circ \gamma)x = (\beta_\sigma \circ \beta_\gamma)(x) = \beta_\sigma(\beta_\gamma(x)) = \sigma(\gamma x) = \sigma x.$$

Thus, $\tau x = \sigma x$ if and only if $\tau \in \sigma H$. ∎

In order to emphasize which element x is being fixed, we will often write G_x in place of H. We read G_x as the **subgroup fixing** x. The set $\sigma H = \sigma G_x$ which occurs in Theorem 2.2 is called the **left coset** of H determined by σ. The fact that each left coset of H is an equivalence class means that the left cosets of H are disjoint sets whose union is G. In our example of the rotation group, the left cosets of the subgroup $\{\rho^0, \rho^2\}$ were $\{\rho^0, \rho^2\}$ itself and $\{\rho^1, \rho^3\}$, both sets of the same size. The group properties ensure that all left cosets of a subgroup H have the same size as H.

Theorem 2.3. The size of a left coset σH of a finite subgroup H of a group G is the same as the size of H.

Proof. The function f from H to σH given by $f(\tau) = \sigma \circ \tau$ is a bijection, because if $\sigma \circ \tau_1 = \sigma \circ \tau_2$, then since $\sigma^{-1} \circ (\sigma \circ \tau_1) = (\sigma^{-1} \circ \sigma) \circ \tau_1 = \tau_1$, we may write

$$\tau_1 = \sigma^{-1} \circ (\sigma \circ \tau_1) = \sigma^{-1} \circ (\sigma \circ \tau_2) = (\sigma^{-1} \circ \sigma) \circ \tau_2 = \iota \circ \tau_2 = \tau_2. \quad ∎$$

Theorem 2.4. If the finite group G acts on a set X, then every element of the multiorbit $\{\sigma(x) | \sigma \in G\}$ has the same multiplicity, and this multiplicity is the size $|G_x|$ of the subgroup of G fixing X.

Proof. The multiplicity of x in the multiset $\{\sigma(x) | \sigma \in G\}$ is the number of elements σ such that $\sigma(x) = x$, which is exactly the size of G_x. For any other element y in the multiorbit of x, there is an element ρ in G such that $\rho(x) = y$. If $\sigma \in G_x$, then $\rho\sigma(x) = y$ as well, so for every ρ' in the left coset ρG_x, $\rho'(x) = y$.

However, if $\varphi(x) = y$, then the equation $\varphi(x) = \rho(x)$ implies that $\rho^{-1}\varphi(x) = x$, so that $\rho^{-1}\varphi \in G_x$, and consequently $\varphi \in \rho G_x$. Thus the set of elements that send x to y is a left coset of G_x. By Theorem 2.3 this coset has the same size as G_x, and this proves that the multiplicity of y is $|G_x|$. ∎

Corollary 2.5. If the finite group G acts on a set X and G_x is the subgroup fixing the element x of X, then the size of the orbit of G containing x is $|G|/|G_x|$.

Proof. Since each element of the orbit containing x has multiplicity $|G_x|$ in the multiorbit $\{\sigma(x) | \sigma \in G\}$ containing x, the size of the multiorbit is $|G_x|$ times the size of the orbit. However, the size of the multiorbit $\{\sigma(x) | x \in G\}$ is $|G|$, so the size of the orbit must be $|G|/|G_x|$. ∎

Notice that although our definition of equivalence was for a group G acting on a set X and our equivalence classes were cosets of the subgroup

H, usually written as G_x, fixing our element x, the definition of a left coset has nothing to do with fixing an element, and the theorem that the left cosets of H all have the same size as H has nothing to do with fixing an element. In fact, we can show (as you are asked to in Exercise 32) that *for any subgroup H of any group G the relation of determining the same left coset of H (that is, the relation given by "σ is related to τ if and only if $\sigma H = \tau H$") is an equivalence relation on G*. This gives us useful information about both the size of H and the number of left cosets of H.

The Size of a Subgroup

> **Theorem 2.6.** If H is a subgroup of a group G, then the size of G is a multiple of the size of H and the number of left cosets of H is $|G|/|H|$.
>
> *Proof.* Our preceding remarks imply that the left cosets of H form a partition of G. From this the proof is a direct application of the equivalence principle. ∎

This theorem, usually attributed to Lagrange (who essentially proved it in the case that G is a symmetric group), is the one that allows us to show that the dihedral group D_4 of the square has order 8, that is, has eight elements. In group theory, **order** is sometimes a synonym for size.) Since the dihedral group contains the rotation group R_4 as a subgroup, the size of D_4 must be a multiple of 4. Since the group D_4 is a subgroup of the symmetric group S_4 of all $4! = 24$ permutations, the size of D_4 must be a factor of 24. The only numbers bigger than 4 that are factors of 24 and have 4 as a factor are 8, 12, and 24. Since some permutations of $\{1, 2, 3, 4\}$ do not correspond to a rigid motion in space, the only two possible sizes for D_4 are 8 and 12. (For example, exchanging vertices 1 and 2 but leaving vertices 3 and 4 fixed cannot be achieved by a rigid motion in space.) A group of order 12 cannot have a subgroup of order 8, so if we can show that the dihedral group must either be of order 8 or have a subgroup of order 8, we can conclude that it must be of order 8.

> **Example 2.2.** Let φ_{13} (the Greek letter phi, denoted by φ, is usually pronounced *fee*) stand for the flip of a square about the diagonal joining vertices 1 and 3. Show that the set of elements of the form $\rho^i \circ \varphi_{13}^j$ is closed under composition. Explain why this means that D_4 has order 8.
>
> We must show that the set of elements is closed under the operation of composition. Note that φ_{13} has the form $\rho^0 \varphi_{13}^1$ and ρ has the form $\rho^1 \varphi_{13}^0$. Thus, one of the many possible compositions to consider is $\varphi_{13} \circ \rho$ (since φ_{13} has the form $\rho^0 \circ \varphi_{13}^1$ and ρ has the form $\rho^1 \varphi_{13}^0$). We can show that $\varphi_{13} \circ \rho = \rho^3 \circ \varphi_{13}$. Remarkably, we can then apply this equation to see that

the entire set is closed under composition. First we show why the equation we gave is true. Note that

$$(\varphi_{13} \circ \rho)(1) = \varphi_{13} \circ (\rho(1)) = \varphi_{13}(2) = 2$$

and

$$\rho^3 \circ \varphi_{13}(1) = \rho^3 \circ (\varphi_{13}(1)) = \rho^3(3) = 2.$$

The three other domain values are treated similarly. Thus we have

$$\varphi_{13} \circ \rho(x) = \rho^3 \circ \varphi_{13}(x)$$

for each x in our domain, so we conclude that

$$\varphi_{13} \circ \rho = \rho^3 \circ \varphi_{13}.$$

From the equation $\varphi_{13} \circ \rho = \rho^3 \circ \varphi_{13}$ we may derive (with some work) the equations

$$\varphi_{13}^j \circ \rho = \rho^{3^j} \circ \varphi_{13}^j$$

and

$$\varphi_{13}^j \circ \rho^i = \rho^{i3^j} \circ \varphi_{13}^j.$$

Then we may write

$$(\rho^{i_1} \circ \varphi_{13}^{j_1}) \circ (\rho^{i_2} \circ \varphi_{13}^{j_2}) = \rho^{i_1} \circ \rho^{i_2 3^{j_1}} \circ \varphi_{13}^{j_1} \varphi_{13}^{j_2} = \rho^{i_1 + i_2 3^{j_1}} \circ \varphi_{13}^{j_1 + j_2}.$$

Since there are four distinct elements of the form ρ^i, namely $\rho^0 = \iota$, ρ, ρ^2, and ρ^3, and two distinct elements of the form φ_{13}^j, namely $\varphi_{13}^0 = \iota$ and φ_{13} (because $\varphi_{13}^2 = \iota$), there are exactly eight elements of the form $\rho^i \circ \varphi_{13}^j$. Since these elements are closed under composition, they form an eight-element subgroup of D_4 by Theorem 1.3. Since 8 is not a factor of 12, D_4 cannot have order 12, so by our previous computations, the group D_4 has order 8. ∎

The Subgroup Generated by a Set

In our example we showed how to express every element of D_4 in terms of ρ and φ_{13}. We refer to ρ and φ_{13} as "generators" of D_4 and say that D_4 is the subgroup of S_4 generated by ρ and φ_{13}. Notice that any subgroup of S_4 that contains ρ and φ_{13} contains all permutations of the form $\rho^i \circ \varphi_{13}^j$ as well. Then any such subgroup of S_4 is at least as big as the subgroup D_4. This seems to give us a precise way to define the subgroup generated by a subset of a group. If S is a subset of a (finite) group G, then the subgroup generated by S is the smallest subgroup of G containing all the elements of S. However, it is conceivable that there are several subgroups of the same smallest size containing S. In this case our definition would not make sense.

There is a nice way to show that our definition does make sense; in fact, it leads to an alternate way to say the same thing. We can show (as you are asked to in Exercise 33) that the intersection of subgroups of G is again a subgroup of G. In the finite case, $H_1 \cap H_2$ will be no larger than H_1 or H_2. Therefore, the intersection of all subgroups containing S will be the smallest subgroup containing S. Thus, it is appropriate to define the subgroup of G **generated** by a set S to be the intersection of all subgroups of G containing S. We now have a definition that captures our idea of "smallest" without the potential complication we pointed out earlier. Notice that this definition makes perfect sense when G is infinite.

Our examples show that the rotation group R_4 is the subgroup of S_4 generated by ρ and that the dihedral group is the subgroup of S_4 generated by ρ and φ_{13}. A subgroup of a group G generated by a single element is called **cyclic**. To see why, think of an element γ (the Greek letter gamma) of a finite group G. We can show that there is a number m such that the sequence

$$(\gamma^0, \gamma, \gamma^2, \ldots, \gamma^{m-1}, \gamma^m, \gamma^{m+1}, \ldots)$$

equals the sequence

$$(\gamma^0, \gamma, \gamma^2, \ldots, \gamma^{m-1}, \gamma^0, \gamma, \gamma^2, \ldots, \gamma^{m-1}, \gamma^0, \gamma, \ldots).$$

This second sequence keeps repeating in "cycles" of powers of γ. The smallest positive m such that $\gamma^m = \gamma^0$ is called the **order** of γ. Because the set $\{\gamma^0, \gamma, \ldots, \gamma^{m-1}\}$ is closed under multiplication, it is a subgroup of G by Theorem 1.3. Since $\{\gamma^0, \gamma, \ldots, \gamma^{m-1}\}$ is the subgroup generated by γ, the order of γ is the size of the subgroup generated by γ. Notice the similarity between the subgroup generated by *any* element of *any* finite group and the rotation group. A subgroup generated by two elements of a group need not be so similar to the dihedral group. In the exercises we will show two permutations σ and τ in S_4 such that the subgroup they generate is all of S_4, not just the set of elements of the form $\sigma^i \circ \tau^j$.

The Cycles of a Permutation

A given group G can act on many different sets X, and each of these actions can give us some information about G. In particular, a permutation group consisting of permutations of $\{1, 2, \ldots, m\}$ acts on the set $\{1, 2, \ldots, m\}$; each permutation σ sends i to $\sigma(i)$. By studying how elements of S_m and the subgroups they generate act on $\{1, 2, \ldots, m\}$, we can learn more about permutations. We begin by analyzing the action of the rotation group R_4 on $\{1, 2, 3, 4\}$. The permutation ρ sends 1 to 2, 2 to 3, 3 to 4, and 4 to 1. We may visualize this as

$$1 \to 2 \to 3 \to 4 \to 1 \to 2 \to 3 \to 4 \to 1 \to 2 \to 3 \cdots$$

where the pattern cycles around again and again. A permutation which we can visualize in this way is called a "cycle." In the standard shorthand for cycles, we would write the notation (1 2 3 4) to denote this cycle. Thus we may write $\rho = $ (1 2 3 4). On the other hand, ρ^2 sends 1 to 3 and 3 back to 1 and ρ^2 sends 2 to 4 and 4 back to 2. We say this permutation has the two cycles (1 3) and (2 4).

We can visualize a permutation σ of $\{1, 2, \ldots, n\}$ by drawing a digraph with vertex set $\{1, 2, \ldots, n\}$ and an arrow from i to j if $i \neq j$ and $\sigma(i) = j$. Figure 2.2(a) shows the digraph for the cycle $\rho = $ (1 2 3 4) and Figure 2.2(b) shows the digraph for $\rho^2 = $ (1 3)(2 4). Figure 2.2(c) shows the permutation, usually denoted by (1 2 3), which sends 1 to 2, 2 to 3, 3 to 1, and does nothing to 4. The first and third permutations are called cyclic because their digraphs have exactly one directed cycle. The second permutation is not called a cycle because its digraph has two nontrivial cycles. More precisely, the permutation γ is *cyclic with* the **cycle** (or *k-cycle*) $(i_1\ i_2\ \ldots\ i_k)$ (a list of distinct integers) if $\gamma(i_r) = i_{r+1}$ for $r = 1, 2, \ldots, k-1$ and $\gamma(i_k) = i_1$ but $\gamma(j) = j$ for all other j. When there is no likelihood of confusion, a cyclic permutation is called a cycle.

Figure 2.2

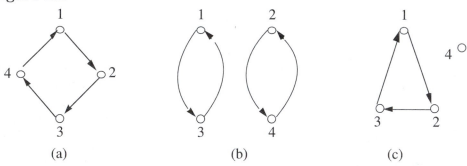

(a)   (b)   (c)

Example 2.3. If γ is cyclic with cycle (1 3), what is $\gamma(i)$ for $i = 1, 2, 3, 4$?

By definition, $\gamma(1) = 3$, $\gamma(2) = 2$, $\gamma(3) = 1$, and $\gamma(4) = 4$. ∎

The permutation ρ is cyclic with the cycle

$$(1\ 2\ 3\ 4) = \left(1\ \rho(1)\ \rho^2(1)\ \rho^3(1)\right).$$

This shows us that the cycle is a special way to write down the orbit containing 1 for the rotation group. However, the sets $\{1,3\}$ and $\{2,4\}$ that come from the cycles $(1\ \ 3)$ and $(2\ \ 4)$ of ρ^2 are not orbits of the rotation group. They are, however, orbits of the subgroup $\{1, \rho^2\}$.

We have said what it means for a permutation to be cyclic; whether σ is cyclic or not, we say a list of distinct integers $(i_1\ i_2 \ldots i_k)$ **is a** *cycle* of σ if $\sigma(i_r) = i_{r+1}$ for $r = 1, 2, \ldots, k-1$ and $\sigma(i_k) = i_1$. Notice that the cycle $(1\ 3)$ tells us what ρ^2 does to 1 and 3 and the cycle $(2\ 4)$ tells us what ρ^2 does to 2 and 4. Even though ρ^2 is not cyclic, knowledge of its cycles allows us to describe it completely.

When σ is cyclic with the cycle $(i_1\ i_2 \ldots i_k)$ it is traditional to write

$$\sigma = (i_1\ i_2 \ldots i_k)$$

and call σ a cycle. With this notation, the identity permutation may be written as the cycle (j) for any j in its domain. In this notation the fundamental rotation ρ may be written as

$$\rho = (1\ 2\ 3\ 4) = (2\ 3\ 4\ 1) = (3\ 4\ 1\ 2) = (4\ 1\ 2\ 3).$$

By thinking of using the cycles $(1\ 3)$ and $(2\ 4)$ to stand for the corresponding cyclic permutations of $\{1, 2, 3, 4\}$ we may write

$$\rho^2 = (1\ 3) \circ (2\ 4) = (2\ 4) \circ (1\ 3) = (4\ 2) \circ (1\ 3) = \cdots .$$

It is traditional to ignore the circle for composition and write

$$\rho^2 = (1\ 3)(2\ 4)$$

in this context.

We see now that the cycles of ρ^2 are listings of the orbits of the group $\{\iota, \rho^2\}$. Of course, the cycle $(1\ 2\ 3\ 4)$ of ρ is just a list of the single orbit of the subgroup generated by ρ, namely, the entire rotation group. This relationship between cycles and orbits shows how we may always factor noncyclic permutations into compositions of cyclic ones, as we did with ρ^2.

Theorem 2.7. If σ is a permutation in a group G of permutations of $N = \{1, 2, \ldots, n\}$, then there is a bijection between the cycles of σ and the orbits of the subgroup generated by σ acting on N with the property that each cycle is a list of the members of the orbit to which it corresponds.

Proof. Suppose that $(i_1\ i_2 \ldots, i_k)$ is a cycle of σ. Then $\sigma^{k-1}(i_1) = i_k$, $\sigma^k(i_1) = i_1$, $\sigma^{k+1}(i_1) = \sigma(i_1) = i_2$, and so on. Therefore, we can prove by induction that $\sigma^j(i_1) \in \{i_1, i_2, \ldots, i_k\}$ for all integers j. Since $\sigma^{j-1}(i_1) = i_j$, this means that, by definition,

$$\{i_1, i_2, \ldots, i_k\}$$

is the orbit containing i_1 of the subgroup generated by σ (consisting of all powers of σ).

Now suppose that

$$R = \{\sigma^0(i), \sigma(i), \sigma^2(i), \ldots, \sigma^n(i)\}$$

is the orbit containing i of the subgroup generated by σ. Since there are $n+1$ elements listed and N has only n elements, there must be some repeats among those listed. Choose the smallest j between 0 and n such that $\sigma^j(i) = \sigma^k(i)$ for some $k < n$. Since $\sigma^0(i) = \sigma^{-j} \circ \sigma^j(i) = \sigma^{-j} \circ \sigma^k(i) = \sigma^{k-j}(i)$ as well, we may assume j is 0. Again we may prove by induction that for each integer r, $\sigma^r(i) = \sigma^s(i)$ for some s with $0 \le s < k$. Thus,

$$\left(i \ \sigma(i) \ \sigma^2(i) \ldots \sigma^{k-1}(i) \right)$$

is a cycle of σ.

We have shown that each cycle corresponds to an orbit and each orbit corresponds to a cycle; it is clear that these correspondences are inverses to each other so we have demonstrated a bijection. ∎

Corollary 2.8. The cycles (including the one-element cycles) of a permutation of N are lists of disjoint sets whose union is N.

Proof. This is a consequence of the fact that orbits are equivalence classes. ∎

Theorem 2.9. A permutation is equal to the composition, in any order, of the cyclic permutations corresponding to its cycles.

Proof. We say i is a member of the cycle $\gamma = (i_1 \ i_2 \ \ldots \ i_k)$ if $i = i_j$ for some j. Suppose $\gamma_1, \gamma_2, \ldots \gamma_n$ are the cycles of σ. Then for any number n in γ_i, n is not in γ_j for $j \ne i$ by Corollary 2.8. But then $\gamma_j(n) = n$ for $j \ne i$ and $\gamma_i(n) = \sigma(n)$. Since $\sigma(n)$ is in the cycle γ_i, we also have $\gamma_j(\sigma(n)) = \sigma(n)$ for $j \ne i$. Thus we have

$$\gamma_1 \gamma_2 \cdots \gamma_m(n) = \gamma_i(n) = \sigma(n)$$

for each n in the domain of σ. Therefore $\sigma = \gamma_1 \gamma_2 \cdots \gamma_m$. ∎

We may summarize the previous corollary and theorem in less precise but suggestive language as follows:

Every permutation may be written as the product of disjoint cycles.

EXERCISES

1. The complete graph on six vertices may be drawn as a regular hexagon together with the chords joining nonadjacent vertices. Using S_{ij} to stand for the side *or* chord joining vertex i and j, write down the orbits of the action of the group of all rotations of the hexagon in the plane on the edges of the complete graph on six vertices.

2. The dihedral group of a hexagon is the set of permutations corresponding to three-dimensional motions that return the hexagon to its starting place. Explain why the size of the dihedral group must be a factor of 720 and why the size must have 6 as a factor.

3. A natural guess for the size of the dihedral group of the hexagon (see Exercise 2) is 12. Describe geometrically 12 motions of the hexagon that correspond to different members of the dihedral group.

4. (a) A tetrahedron is described in Exercise 5 of Section 1 of this chapter. Find an element of order 3 in the group of permutations of $\{1, 2, 3, 4\}$ corresponding to movements of the tetrahedron in space. Give a geometric description of this element of the group.

   (b) From the geometric description in part (a) you should be able to describe eight such elements of the group. Describe them geometrically.

   (c) The eight elements described in part (b) together with the identity element form a nine-element set. Is this set a subgroup (perhaps even the whole group) of the group of permutations arising from motions of the tetrahedron?

   (d) Find an element of order 2 in the group of permutations arising from motions of the tetrahedron.

   (e) Find the order of the group of permutations arising from motions of the tetrahedron.

5. Describe the subgroup fixing vertex number 1 in the dihedral group of the square.

6. Describe the coset of members of the dihedral group on four vertices that map vertex number 1 to vertex number 4. (One way to describe it is to give its members in two-row form.)

7. The rotation group of the square is a subgroup of the symmetric group on four vertices. Find the left coset of the rotation group that contains the permutation $\left(\begin{smallmatrix} 1 & 2 & 3 & 4 \\ 3 & 2 & 1 & 4 \end{smallmatrix}\right)$.

8. Explain how the equivalence principle may be used to prove Theorem 2.6 of this section.

9. Let ρ stand for a rotation through $30°$ in the rotation group of the 12-gon.

   (a) Describe a subgroup of order 3 by listing its members as powers of ρ.

   (b) In the rotation group of the 12-gon, describe the left cosets of the subgroup of part (a).

10. Derive the equations $\varphi_{13}^j \circ \rho = \rho^{3^j} \circ \varphi_{13}^j$ and $\varphi_{13}^j \circ \rho^i = \rho^{3^j i} \circ \varphi_{13}^j$ of Example 2.2 of this section.

11. Deal with the three domain values not dealt with in Example 2.2 of this section in showing that $\varphi_{13}\rho = \rho^3\varphi_{13}$.

12. Which of the following permutations are cycles?

   (a) $\sigma = \begin{pmatrix} 1\ 2\ 3\ 4 \\ 4\ 3\ 1\ 2 \end{pmatrix}$

   (b) $\sigma(1) = 5, \quad \sigma(2) = 4, \quad \sigma(3) = 1, \quad \sigma(4) = 3, \quad \sigma(5) = 2$

   (c) $\sigma = \begin{pmatrix} 1\ 2\ 3\ 4\ 5\ 6 \\ 2\ 5\ 1\ 6\ 3\ 4 \end{pmatrix}$

   (d) $\sigma(1) = 4, \quad \sigma(2) = 3, \quad \sigma(3) = 6, \quad \sigma(4) = 5, \quad \sigma(5) = 1, \quad \sigma(6) = 2$

13. Draw the digraphs of the permutations of the preceding exercise and use them to explain your answers.

14. Write the following permutations as the composition of their cycles. (You need not give the cycles in two-line form or in functional notation.)

   (a) $\begin{pmatrix} 1\ 2\ 3\ 4\ 5\ 6\ 7\ 8 \\ 3\ 4\ 2\ 1\ 6\ 5\ 8\ 7 \end{pmatrix}$

   (b) $\varphi(1) = 2, \quad \varphi(2) = 1, \quad \varphi(3) = 4, \quad \varphi(4) = 3, \quad \varphi(5) = 7, \quad \varphi(6) = 6,$
   $\varphi(7) = 8, \quad \varphi(8) = 5$

   (c) $\begin{pmatrix} 1\ 2\ 3\ 4\ 5\ 6\ 7\ 8\ 9 \\ 9\ 1\ 5\ 8\ 3\ 7\ 4\ 6\ 2 \end{pmatrix}$

   (d) $\sigma(1) = 7, \quad \sigma(2) = 9, \quad \sigma(3) = 8, \quad \sigma(4) = 10, \quad \sigma(5) = 1, \quad \sigma(6) = 2,$
   $\sigma(7) = 5, \quad \sigma(8) = 3, \quad \sigma(9) = 6, \quad \sigma(10) = 4$

15. Draw the digraphs of the permutations of Exercise 14 and use them to explain your answers.

16. We say a number i is *moved* by the permutation σ if $\sigma(i) \neq i$. We say the permutations σ and τ are *disjoint* if every element moved by σ is fixed by τ and every element moved by τ is fixed by σ. Show that σ and τ are disjoint permutations if and only if

$$\{i \mid i \text{ is moved by } \sigma\} \cap \{j \mid j \text{ is moved by } \tau\} = \emptyset.$$

Explain why every permutation may be written as a composition of disjoint cyclic permutations. (Notice how this exercise gives more meaning to the informal statement of Theorem 2.9.)

17. Show that the dihedral group of the hexagon has a subgroup of order 12. (In fact it has 12 elements, but this is not quite so easy to see.)

18. Suppose we number the vertices of a regular hexagon in clockwise fashion. Show that a flip around the perpendicular bisector of edge $\{1\ 2\}$ (which sends 1 to 2, 3 to 6, and 4 to 5) may be written as the composition of a rotation with the flip around the diagonal joining vertices 1 and 4.

19. A group G is said to act *faithfully* on a set X if whenever σ and τ are members of G with $\sigma \neq \tau$, then $\beta_\sigma \neq \beta_\tau$ as well.
    (a) Show that the set F of all elements of G fixing *every* element of X is a subgroup of G (regardless of whether G acts faithfully).
    (b) Show that a group acts faithfully if and only if the subgroup F of part (a) consists of the identity element alone.
    (c) Show that a dihedral group of a polygon acts faithfully on the edges.
    (d) Show that if G acts faithfully on X, then

    $$|G| = |\{\beta_\sigma | \sigma \in G\}|.$$

    (e) Show that by determining where two adjacent edges are taken by a geometric (distance-preserving) motion, we determine where every edge of a regular polygon is taken.
    (f) Explain why the dihedral group of an n-gon has $2n$ elements.

20. In this exercise we describe an element of order 2 and an element of order 3 in S_4 such that the subgroup they generate is all of S_4 rather than a six-element subgroup.
    (a) Show that the composition of the cycle $(1\ 2\ 3)$ and the cycle $(3\ 4)$ is the rotation $(1\ 2\ 3\ 4)$.
    (b) Show that the subgroup of S_4 generated by $(1\ 2\ 3)$ and $(3\ 4)$ contains the cycles $(1\ 2)$, $(2\ 3)$, and $(1\ 4)$ as well.
    (c) Show that the dihedral group D_4 is a subgroup of the group generated by $(1\ 2\ 3)$ and $(3\ 4)$.
    (d) Show an element of the group generated by $(1\ 2\ 3)$ and $(3\ 4)$ that is not in the dihedral group D_4.
    (e) Why may you conclude that the group generated by $(1\ 2\ 3)$ and $(3\ 4)$ is all of S_4?

21. Explain why the only way to write a permutation as a product of disjoint cycles is the way described in Theorem 2.9. (This shows, in formal language, that, up to order, a permutation may be written *uniquely* as a product of disjoint cycles.)

22. Give an example that shows that when σ and τ are cycles that are not disjoint, then $\sigma \circ \tau$ and $\tau \circ \sigma$ need not be equal.

23. Is it possible to find an example of nonidentity cycles σ and τ that are not disjoint or equal but for which $\sigma \circ \tau = \tau \circ \sigma$ nonetheless? If so, find such an example, and if not, explain why not.

24. When cycles used to factor a permutation are not disjoint, then, in contrast to the previous exercise, there are many ways to factor a permutation as a composition of cyclic permutations. Show that each permutation may be written as a composition of *transpositions*, cyclic permutations with one cycle of size two.

25. A permutation σ may be represented by a matrix $M(\sigma)$ with $M(\sigma)_{ij} = 1$ if $\sigma(j) = i$ and $M(\sigma)_{ij} = 0$ otherwise. Show that $M(\sigma \circ \tau) = M(\sigma)M(\tau)$.

26. The *sign* of a permutation σ may be *defined* as the determinant of the matrix $M(\sigma)$ of the previous exercise. Explain why the sign of σ is always 1 or -1.

27. The sign of a permutation is defined in the previous exercise. Show that the set of permutations with sign $+1$ is a subgroup of S_m and the permutations with sign -1 form a coset of that subgroup. What is the order of this subgroup?

28. The group of three-dimensional motions of a cube, called the *symmetry group* of the cube, may be thought of as a permutation group on the 8 vertices. This permutation group acts on both the 12 edges and the 6 faces. Use either of these actions to explain why the symmetry group of the cube has 24 members.

29. An octahedron is a regular solid with 6 vertices, 8 sides, each of which is an equilateral triangle, and 12 edges. (Think of the result of fastening two square pyramids together along the squares.) The group of permutations resulting from three-dimensional motions of an octahedron is called the *symmetry group* of the octahedron. Show that the symmetry group of an octahedron has 24 members.

30. Show that the number of permutations of $\{1, 2, \ldots, n\}$ which are products of k (but no fewer than k) cycles is the absolute value of the Stirling number $s(n, k)$ of Chapter 2.

31. Does Theorem 2.2 remain true if the group G is infinite? Why or why not?

32. Show that the relation of determining the same left coset of the subgroup H of G (i.e., σ is related to τ if $\sigma H = \tau H$) is an equivalence relation

and that all its equivalence classes have the same size. What are the equivalence classes?

33. Show that the intersection of a set of subgroups of a group G is again a subgroup of G.

Section 3 Pólya's Enumeration Theorem

The Cauchy–Frobenius–Burnside Theorem

Our goal in the study of group theory is to develop tools to compute the number of orbits of a group acting on a set. Our first such tool is a result of Frobenius which appeared in 1900 but was implicit in work of Cauchy 30 years earlier and given again by Burnside in 1910 in his pioneering book on group theory. We call the result the **Cauchy–Frobenius–Burnside theorem** or the **CFB theorem** for short. (It used to be called *Burnside's lemma*.)

Recall that when we first introduced the equivalence principle we said that if we have an equivalence relation on a set X of size n and all the equivalence classes have size m, then there are n/m equivalence classes. Thus, for example, if we are arranging four distinct beads at the corners of a square free to move in three-dimensional space, then two arrangements are equivalent if we get one from the other by picking the square up, moving it around, and returning it back to its original place (though not necessarily returning the corners to their own original places). Our set X is the set of all lists of the four beads. A list is equivalent to eight others, so we have $4!/8 = 3$ arrangements of the beads. However, if we have two red and two green beads, some equivalence classes have two elements and others have four elements. Thus we cannot use the equivalence principle to compute the number of equivalence classes. However, the equivalence classes are orbits under the action of the dihedral group on four elements, the four places where the corners of the square sit.

If, instead of orbits, we think of multiorbits, then all our multiorbits have the same size, namely 8. When we take their "union" we should get a multiset with a multiple of 8 elements. Of course, this union will not be the set of all lists, because in this union elements will occur with multiplicities. We used quotation marks around the word union because we have not yet defined what we mean by the union of two multisets. The *elements* of a **union of two multisets** are the elements of the two multisets, and the *multiplicity* of an element in the union is the maximum of its multiplicities in the two multisets. Since two different multiorbits have no elements in common, the size of the union of the multiorbits is the sum of the multiplicities of all our lists. We saw in the last section that the multiplicity of an element in a multiorbit is the number of group elements that fix it (which happens to be the size of the set $|G_x|$). Thus the size of the union in our preceding example is the number of ordered pairs (x, σ) such that x is a list and σ is a group element that fixes x. It turns out that in many practical cases this number can be computed with relative ease; the

CFB theorem takes advantage of this to give us a formula that often makes computing the number of orbits straightforward.

Theorem 3.1. (Cauchy–Frobenius–Burnside) If the finite group G acts on a finite set X, then the number of orbits of G acting on X is given by

$$\frac{1}{|G|} \sum_{\sigma:\sigma\in G} |F_\sigma| = \frac{1}{|G|} \sum_{\sigma:\sigma\in G} \# \text{ of elements fixed by } \sigma.$$

Proof. As in the preceding example, consider the union of the multiorbits. The size of the union is the sum of the multiplicities $|G_x|$ of the element x of X, which is the number of ordered pairs (x, σ) where σ fixes x. This is, however, simply the sum over all σ of the number of elements fixed by σ.

However, if there are n multiorbits, the size of the union is $n|G|$. Setting these last two computations of the size of the multiorbit union equal and dividing by $|G|$ gives the desired formula. ∎

Example 3.1. Before doing a new example, let us recompute the number of orbits of the edges of the complete graph on four vertices under the action of the rotation group $\{\iota, \rho, \rho^2, \rho^3\}$ on the vertices. The number of edges fixed by the identity ι is six; that is, all six edges are fixed by the identity. No edge is fixed by ρ or ρ^3 and the two diagonal edges are fixed by ρ^2. Thus,

$$\# \text{ orbits} = \frac{1}{4}(6 + 0 + 2 + 0) = 2.$$

You may recall that just before Theorem 2.1 of this chapter we explicitly constructed these orbits and found there was one orbit of size two and one orbit of size four. ∎

Example 3.2. The outside of a wastebasket is to be painted with four vertical stripes using exactly two colors of paint. How many distinct paint jobs are possible? (We will allow adjacent strips to be the same color as long as all four strips aren't the same color.)

A paint job may be described by a list of the colors of the four stripes. Suppose our colors are red and blue. However, changing the list a bit and rotating the wastebasket could give us an apparently identical paint job. Thus, the paint jobs are the orbits under the action of the rotation group on lists of colors. The lists have four entries, each one of two colors, say R and B. We can have one, two, or three R's in the list and still have two colors. The total number of such lists is

$$\binom{4}{1} + \binom{4}{2} + \binom{4}{3} = 4 + 6 + 4 = 14.$$

Under the action of the rotation group, all 14 lists are fixed by the identity, no list is fixed by ρ or ρ^3 (because each color would have to equal the adjacent color), and only lists with every other color equal are fixed by ρ^2. Thus, two lists are fixed by ρ^2, so the number of paint jobs is

$$\frac{1}{4}(14 + 0 + 2 + 0) = 4.$$ ∎

Enumerators of Colorings

We see from the preceding example how "coloring" problems may be treated by using the Cauchy–Frobenius–Burnside theorem. Notice that the final equation in the example appears to take no account of whether a coloring uses one, two, or three R's. In typical coloring problems we may be interested in knowing the number of colorings with each possible combination of colors. That is, we may wish to know the number of colorings using one red and three blue stripes, the number using two of each, and so on. We could deal with this situation by solving a separate problem for each possible combination; however, by using an appropriate generating function we can solve just one problem to obtain the answers. Recall that in order to write down symbolic series describing pennies, nickels, dimes, and quarters we used the letters P, N, D, and Q as variables and in order to write down symbolic series describing fruits we used letters standing for the fruits as variables. We need to introduce similar symbols and symbolic series for colorings. For example, a good symbolic description of a coloring that uses three reds and one blue would be $R^3 B$.

In the symbolic series we used for nickels, for example, we used

$$N^0 + N^1 + N^2 + \cdots = \sum_{i=0}^{\infty} N^i$$

to list, or *enumerate*, all the possible choices of nickels. Similarly, we will want to enumerate all possible colorings of a set of objects, and though we will have powers of variables involved, they will not necessarily fall into a pattern we would think of as a series. Thus we will have a set of colored objects, say solid-color blocks, wastebaskets with stripes painted on their sides, or cubes with their sides painted with several different colors, for example. We will have a symbolic description $D(y)$ (D is for description) of each colored object y (like R for a red object, B for a blue object, perhaps $R^3 B^2$ for an object with three red stripes and two blue ones, etc.), and we will want to enumerate all these possible colored objects.

It will be easier to understand symbolic descriptions if we begin without having any group action involved at all. Later on, we shall see that the theorems we develop without thinking about a group apply perfectly well to orbits of groups as well.

Example 3.3. If we are painting four stripes on a rectangle in red and blue, we could use R as the description of a red stripe and B as the description of a blue stripe. For one particular stripe we will say our symbolic enumerator is $R + B$ (which says symbolically that this particular stripe is either red or blue). We will use $R^i B^j$ as the symbolic description of a painting with i red stripes and j blue stripes. Then we will say that our symbolic enumerator for the descriptions of all four stripes would be

$$R^4 + \binom{4}{3} R^3 B + \binom{4}{2} R^2 B^2 + \binom{4}{3} R B^3 + B^4.$$

The coefficient of $R^i B^{4-i}$ is the number of ways to choose the i stripes that we will paint red. By the binomial theorem this enumerator is equal to $(R + B)^4$. Notice this is the result of multiplying four terms, each equal to $R + B$, the enumerator for painting just one stripe. ∎

More generally, given symbolic descriptions of all the objects in a set $S = \{x, y, z, \ldots\}$, we define the *symbolic enumerator* of S to be the sum

$$D(x) + D(y) + D(z) + \cdots = \sum_{s:s \in S} D(s),$$

so long as it makes sense to write down that sum.

The pattern we saw in the previous example, that we multiply together enumerators for sets to get the enumerator for sequences of elements of these sets, was no accident.

Theorem 3.2. (Product principle for enumerators) Suppose we have symbolic descriptions $D_1(x)$ for the elements x of a set X and $D_2(y)$ for the elements y of a set Y, and we define the symbolic description of the ordered pair (x, y) to be $D_1(x)D_2(y)$. Let $E(X)$ stand for the enumerator of X and $E(Y)$ stand for the enumerator of Y. Then the enumerator for the Cartesian product $X \times Y$ is

$$E(X \times Y) = E(X)E(Y).$$

Proof. By repeated application of the distributive law,

$$\sum_{x:x \in X} D_1(x) \sum_{y:y \in Y} D_2(y) = \sum_{(x,y):(x,y) \in X \times Y} D_1(x)D_2(y).$$

Mathematical induction allows us to extend Theorem 3.2 to enumerators of n-tuples.

Corollary 3.3. If D_i is the symbolic description of elements of the set X_i, and we define the symbolic description of $(x_1, x_2, \ldots, x_n) \in X_1 \times X_2 \times \cdots \times X_n$ to be $D_1(x_1)D_2(x_2) \cdots D_n(x_n)$, then

$$E(X_1 \times X_2 \times \cdots X_n) = E(X_1)E(X_2) \cdots E(X_n).$$

Proof. See Exercise 23. ∎

Example 3.4. Use Corollary 3.3 to solve Example 3.3.

We defined the symbolic description of a list of stripes to be the product of the symbolic descriptions of the stripes. We noted already that the symbolic enumerator for one stripe is $R + B$. Thus by Corollary 3.3, the symbolic enumerator for all four stripes is $(R + B)^4$. ∎

Generating Functions for Orbits

We introduced enumerators of colorings in order to describe colorings. However, as we have seen in many cases, what we are interested in is not simply colorings but equivalence classes of colorings under the action of a group. For example, R^3B might describe the colors of four beads at the vertices of a square. If we pick up the square, then no matter how we move the square before we return it to the place where we picked it up, its symbolic description will still be R^3B. Thus every coloring in an orbit will have the same symbolic description, so we can use this symbolic description of our colorings in an orbit as a symbolic description of the orbit.

Example 3.5. Let us return to the example of painting a circular wastebasket with four vertical stripes each red or blue. Let us allow any combination of colors, so the wastebasket could be all blue or all red. The paint job with all red stripes is in an orbit all by itself and has symbolic description R^4 and the all-blue paint job is in an orbit by itself with symbolic description B^4. The paint jobs (which you can visualize by imagining wrapping our previous striped rectangle around the wastebasket!) $RBBB$, $BRBB$, $BBRB$, and $BBBR$ are another orbit, and they all have symbolic description RB^3, so we adopt this symbolic description of the orbit. Similarly, we have four paint jobs in an orbit with symbolic description R^3B. The paint jobs with two red and two blue stripes are more interesting. The paint jobs $RBRB$, and $BRBR$ form an orbit for which the natural symbolic description is R^2B^2, while the paint jobs $RRBB$, $RBBR$, $BBRR$, and $BRRB$ form another orbit whose natural symbolic description is R^2B^2. All the possible paint jobs are accounted for in one of the orbits we described, so a good symbolic description of all the possible orbits is

$$R^4 + R^3B + 2R^2B^2 + RB^3 + B^4.$$ ∎

So now in analogy with our example, we suppose that we have a group G acting on a set X, that we have a symbolic description of each element in X, and that all elements of an orbit Q have the same symbolic description $D(Q)$. (We will use Q to stand for an orbit because O might be confused with 0.) Then we define the **orbit enumerator** $Orb(G, X)$ to be the sum of the symbolic descriptions of the orbits; in symbols

$$Orb(G, X) = \sum_{Q:Q \text{ is an orbit}} D(Q).$$

Example 3.6. Suppose we attach identical red beads and identical blue beads at the corners of a square free to move in space. There is just one orbit with all red beads. A natural symbolic description for it is $D(R) = R^4$. Placing three red beads and one blue one gives just one orbit with symbolic description $D(R) = R^3B$. Placing two red beads side by side gives one orbit; placing them opposite each other gives another orbit. Both these orbits have symbolic description $D(R) = R^2B^2$. Similarly, we have one orbit with three blue beads and one red bead and we have one orbit with four blue beads. Thus, our orbit enumerator is

$$Orb(G, X) = R^4 + R^3B + 2R^2B^2 + R^1B^3 + B^4.$$ ∎

It is interesting that the orbit enumerators in Examples 3.5 and 3.6 are the same even though the groups are different. Perhaps a good way to make sense of this is to observe that turning the wastebasket upside down doesn't make the possible paint jobs look any different (though it might have other consequences).

The technique used in proving the Cauchy–Frobenius–Burnside theorem lets us derive a useful formula for the orbit enumerator of a group acting on a set. Just as the *number* of orbits may be expressed in terms of the *number* of fixed points, the *enumerator* of orbits may be expressed in terms of an "enumerator" of fixed points. In particular, we define $Fix(\sigma)$, the **fixed point enumerator** of σ, to be the polynomial

$$Fix(\sigma) = \sum_{x:x \text{ is fixed by } \sigma} D(x).$$

Example 3.7. Again imagine four beads, either red or blue, fastened at the corners of a square free to move in space. The group of motions in space that put the square back down where it was before is the dihedral group D_4. The identity element fixes all ways to place the beads. We have the choice $R + B$ at each corner, so by Corollary 3.3, the fixed-point enumerator for the identity is given by

$$Fix(\iota) = (R + B)^4.$$

If the vertices of the square are numbered clockwise 1,2,3,4, then the permutation $\varphi = (1)(3)(24)$, which fixes vertices 1 and 3, flips the square around the diagonal joining vertices 1 and 3. For a coloring to be fixed by this permutation, vertices 2 and 4 must be the same color. Thus the enumerator of the possible colors of those two vertices is $R^2 + B^2$, which says that either both are red or both are blue. Vertex 1 may be either color and vertex 3 may be either color. Thus by Corollary 3.3, the fixed-point enumerator for φ is

$$Fix(\varphi) = (R^2 + B^2)(R + B)(R + B) = (R^2 + B^2)(R + B)^2. \qquad \blacksquare$$

In the terminology of orbit enumerators and fixed-point enumerators, the Cauchy–Frobenius–Burnside theorem becomes

Theorem 3.4. (Orbit–fixed point lemma) Suppose the finite group G acts on the finite set X. Suppose further we have symbolic descriptions of the elements of X such that all elements of an orbit have the same symbolic description. Then the orbit enumerator may be expressed in terms of the fixed-point enumerator by

$$Orb(G, X) = \frac{1}{|G|} \sum_{\sigma:\sigma \in G} Fix(\sigma) = \frac{1}{|G|} Fix(G, X)$$

(in which the second equality defines $Fix(G, X)$).

Proof. Simply a translation of the proof of the Cauchy–Frobenius–Burnside theorem. $\blacksquare$

Using the Orbit–Fixed Point Lemma

To see how we use the orbit–fixed point lemma, as we shall call the previous theorem, we repeat Example 3.5 using this new technique. We introduce a useful new way of thinking about paint jobs, namely as functions from the set of places to be painted to the set of colors to be used.

Example 3.8. For each possible choice of i reds and $4 - i$ blues, how many ways are there of painting four solid-color stripes evenly spaced around the outside of a wastebasket using either red or blue for each stripe?

The set X is now the set of all possible paint jobs, that is, functions f from $\{1, 2, 3, 4\}$ to $\{R, B\}$, and the group is the rotation group. There are 16 such functions. As before, the natural symbolic description of a function f is $f(1)f(2)f(3)f(4)$; if the corrresponding paint job has i reds and $4 - i$ blues, our symbolic description of the function is $R^i B^{4-i}$. Since applying a group member to a function simply rotates the pattern corresponding to that function, if $R^i B^{4-i}$ is our symbolic description for f, then $R^i B^{4-i}$ is our symbolic description for $\sigma(f)$ as well. Thus, all functions in an orbit have the same symbolic description. By the orbit–fixed point lemma, our orbit enumerator is $\frac{1}{|G|}$ times the sum of the fixed-point enumerators.

The fixed-point enumerator for the identity will enumerate all functions (because the identity fixes all functions), that is, all lists of the four possible colors. Thus, as we have already seen, the fixed-point enumerator for the identity permutation ι is

$$Fix(\iota) = \sum_{i=0}^{4} \binom{4}{i} R^i B^{4-i} = (R + B)^4.$$

For a function to be fixed by the rotation ρ, each color must equal the adjacent one, so the "all-reds" coloring and the "all-blues" coloring will be the only ones fixed by ρ. Recall that the symbolic description for the all-reds coloring is R^4 and the symbolic description for the all-blues coloring is B^4. This gives us

$$Fix(\rho) = R^4 + B^4,$$

and $Fix(\rho^3)$ similarly has the same value. Since ρ^2 fixes those colorings in which each color is the same as the color two places away, it fixes the colorings $RBRB$ and $BRBR$ as well as the solid-color colorings. Thus we get

$$Fix(\rho^2) = R^4 + 2R^2 B^2 + B^4 = (R^2 + B^2)^2.$$

Thus, by substitution into the orbit–fixed point lemma we see that the coefficient of $R^i B^{4-i}$ in

$$\frac{1}{4}\left[(R + B)^4 + R^4 + B^4 + (R^2 + B^2)^2 + R^4 + B^4\right]$$

is the number of paint jobs with i red stripes and $4-i$ blue stripes. Expanding the polynomial gives

$$\frac{1}{4}(R^4 + 4R^3 B + 6R^2 B^2 + 4RB^3 + B^4 + R^4 + B^4 + R^4 + 2R^2 B^2 + B^4 + R^4 + B^4)$$

$$= \frac{1}{4}(4R^4 + 4R^3 B + 8R^2 B^2 + 4RB^3 + 4B^4)$$

$$= R^4 + R^3 B + 2R^2 B^2 + RB^3 + B^4.$$

Thus, as in Example 3.5, there are two ways to paint two red and two blue stripes on the wastebasket but only one way to use any other combination of colors. ∎

Orbits of Functions

It probably seems that we obtained the conclusion of the preceding example with less work in Example 3.5, but the methods we introduced illustrate some quite general principles that save us considerable work when we have bigger groups, bigger objects, and more colors. The problem we solved is a prototype of many problems we encounter in studying equivalence classes determined by a group acting on a set. We were studying functions that assigned colors to particular places on a wastebasket. In more general applications we might be assigning one of several different types of computers to the positions of a network or assigning one of several different kinds of chemical radicals to carbon bonds in a molecule. We would want to know how many different networks are possible or how many different kinds of molecules are possible. In each situation we have a set of "places" that we think of as numbered one through n and a permutation group on the set $N = \{1, 2, \ldots, n\}$ which we think of as moving the places around in a permissible fashion. In each situation we may represent our physical objects (colored wastebaskets, computer networks, molecules, etc.) as functions, often called *colorings*, from N to a set C of constituent parts (colors of paint, computers, atoms), often called *colors*. Two functions are equivalent under the action of the permutation group if they represent the same physical object. We visualize our permutation group as acting on the functions from N to C and ask for the number of orbits of this action or perhaps for the number of orbits having certain properties.

We shall develop general tools for solving this kind of problem. In order to do so, we will first give a real definition of the idea of a group acting on a set of functions. Next we will develop a general way to define symbolic descriptions and orbit enumerators for the action of G on functions. Then we will compute the fixed-point enumerator for individual permutations of G acting on functions. Finally, we will apply the orbit–fixed point lemma.

To have a solid foundation for the theory, we begin by showing that whenever we have a group G acting on the domain N, then there is a natural action of G on the set, denoted by C^N, of functions from N to C. For example, if we think of the rotation group acting on the corners (numbered 1,2,3,4 clockwise) of a square, and if we think of red, yellow, and blue beads fastened at the corners, we may describe an assignment of beads to corners by using a function f from $\{1, 2, 3, 4\}$ to $\{R, Y, B\}$ in which $f(i)$ is the color of the bead assigned to corner number i. For example, $f(1) = R$,

$f(2) = Y$, $f(3) = B$, $f(4) = B$, which we can describe as $RYBB$ for short, is a typical such function.

Then when we rotate the square through a quarter turn clockwise, the color of the bead at the corner in place i is the color that *used* to be at corner in place $\rho^{-1}(i)$. For example, rotating $RYBB$ gives $BRYB$, so the red bead now at position 2 is the bead that used to be at position 1. Thus, after the rotation the placement of beads may be described by the function g with

$$g(i) = f(\rho^{-1}(i)).$$

In the notation of function composition,

$$g = f \circ \rho^{-1}.$$

Recall that in order to say a group acts on a set X (in this case the set C^N of functions from N to C) we must show that for each member γ of G, there is a bijection β_γ of X such that

$$\beta_\gamma \circ \beta_\alpha = \beta_{\gamma \circ \alpha}.$$

Theorem 3.5. If G is a permutation group on the set N, then defining $\beta_\gamma(f) = f \circ \gamma^{-1}$ determines an action of G on the set C^N of functions from N to a set C.

Proof. We just saw that we should define $\beta_\gamma(f)$ to be $f \circ \gamma^{-1}$. To show that β_γ is a bijection we must show that if $\beta_\gamma(f) = \beta_\gamma(g)$, then $f = g$. Assuming $\beta_\gamma(f) = \beta_\gamma(g)$ gives us

$$f \circ \gamma^{-1} = g \circ \gamma^{-1},$$

then composition of functions gives us

$$(f \circ \gamma^{-1}) \circ \gamma = (g \circ \gamma^{-1}) \circ \gamma$$
$$f \circ (\gamma^{-1} \circ \gamma) = g \circ (\gamma^{-1} \circ \gamma)$$
$$f \circ \iota = g \circ \iota$$
$$f = g.$$

Thus, because a one-to-one function from a finite set to itself is onto, β_γ is a bijection. To show that $\beta_\gamma \circ \beta_\alpha = \beta_{\gamma \circ \alpha}$, we observe that

$$\begin{aligned}(\beta_\gamma \circ \beta_\alpha)(f) &= \beta_\gamma(\beta_\alpha(f)) \\ &= \beta_\gamma(f \circ \alpha^{-1}) \\ &= (f \circ \alpha^{-1}) \circ \gamma^{-1} \\ &= f \circ (\alpha^{-1} \circ \gamma^{-1}) \\ &= f \circ (\gamma \circ \alpha)^{-1} \\ &= \beta_{\gamma \circ \alpha}(f).\end{aligned}$$

∎

Notice how lucky it was in our proof that the natural way to define the action of γ on f was to use $f \circ \gamma^{-1}$. This made things work out perfectly because the inverse of a composition of two permutations is the composition of the inverses *in reverse order*. Had we not had the inverse, then γ and α would have ended up in reverse order from the way they should be for a group action.

The Orbit Enumerator for Functions

Recall how we used $R^i B^{(4-i)}$ in Example 3.8 as the symbolic description of a function using i reds and $4 - i$ blues in the orbit enumerator to get an enumerator for the orbits according to the number of reds and blues they use. The important point that made this possible was the fact that all the functions in an orbit had the same symbolic description. The reason why all the functions had the same symbolic description was that the symbolic description we used for a function was the product of the symbolic descriptions of its values. That is, if the function assigned R to three stripes and B to one stripe, we got its description by multiplying three Rs and a B. Our next lemma tells us that this technique works in general, so that we can always use the symbolic description of any function in an orbit as the symbolic description of that orbit.

Lemma 3.6. Suppose the finite group G acts on the finite set N. Suppose we have a symbolic description $D(y_k) = c_k$ for every element y_k of the set C. Suppose we use $\prod_{i=1}^{n} D(f(i))$ as the symbolic description for a function $f : N \to C$. Then

(1) Every function in an orbit of the group acting on the set of functions C^N has the same symbolic description and

(2) Furthermore, the coefficient of $c_1^{i_1} c_2^{i_2} \cdots c_m^{i_m}$ in the the orbit enumerator for G acting on the functions from N to C (thought of as colorings) is the number of orbits whose functions use c_1 as a color i_1 times, c_2 as a color i_2 times, ..., and c_m as a color i_m times.

Proof. To see that two functions in the same orbit have the same symbolic description, imagine listing all the elements of N, then replacing each element i by $D(f(i))$, the symbolic description of $f(i)$. The resulting product will be

$$\prod_{i=1}^{n} D(f(i)).$$

Now imagine listing $\sigma^{-1}(i)$ for each i. Since σ is a permutation, this is just a list of N, perhaps in a different order. Now replace each element

$j = \sigma^{-1}(i)$ by $D(f(j))$, the symbolic description of $f(j)$. This product will be equal to the original product and will also be

$$\prod_{i=1}^{n} D(f \circ \sigma^{-1}(i)).$$

Thus all elements of the orbit containing f have the same symbolic description. This proves (1). In particular, since the symbolic representation of y_k is c_k, if i_k is the number of i such that $f(i) = y_k$, then this product is $c_1^{i_1} c_2^{i_2} \cdots c_m^{i_m}$. This proves (2). ∎

Note that item (2) in Lemma 3.6 is really included for emphasis; we could conclude the same thing just from the definition of the orbit enumerator.

How Cycle Structure Interacts with Colorings

We are now going to examine fixed points of permutations acting on functions in the way we have described. To begin, we review some salient points from Example 3.8. Notice that in Example 3.8 the permutations ρ and ρ^3 both had the same fixed-point enumerator, namely, the polynomial $R^4 + B^4$, which may be interpreted as the statement that a function fixed by ρ or ρ^3 is either "all-red" or "all-blue." The element ρ^2 has fixed-point enumerator $(R^2 + B^2)^2$, which may be interpreted by Theorem 2.2 as saying that positions 1 and 3 may each be painted the same color—both red (R^2) or both blue (B^2)—and then positions 2 and 4 may each be painted the same color—both red (R^2) or both blue (B^2). Thus our fixed-point enumerator for ρ^2 is $(R^2 + B^2)(R^2 + B^2)$. The identity element ι allows us to choose either R or B for each of the four positions to get a fixed function giving by Corollary 3.3 the fixed-point enumerator $(R + B)^4$ for the identity. Notice that ρ and ρ^3 seem to be the same kind of permutation while ι and ρ^2 seem different. Is there something about the permutations themselves that explains these differences?

We recall that ρ is the cycle $(1\ 2\ 3\ 4)$ and ρ^3 is the cycle $(1\ 4\ 3\ 2)$, while ρ^2 is the composition of two cycles, the two-cycles $(1\ 3) \circ (2\ 4)$, and the identity $\rho^0 = \iota$ is the composition of four cycles, all one-cycles. This suggests that the cycle decomposition of a permutation might tell us how to write down the fixed-point enumerator of a permutation—and it does. To say that a function f is fixed by a permutation σ is to say $f \circ \sigma^{-1}(i) = f(i)$ for $i = 1, 2, 3, 4$. Thus, if $\sigma(i) = j$, the "colors" $f(i)$ and $f(j)$ given to i and j must be the same. Similarly, if $\sigma^2(i)$ or $\sigma^k(i)$ is j, then the colors given to i and j must be the same. Thus, for every j in a cycle

$$\left(i\ \ \sigma(i)\ \ \sigma^2(i)\ \ \ldots\ \ \sigma^k(i) \right)$$

the color $f(j)$ assigned to j by the function f must be the same as the color $f(i)$ assigned to i by the function f. This observation gives us an important lemma.

Lemma 3.7. A function f is fixed by a permutation σ if and only if f is "constant" on the cycles of σ in the following sense: For each i, $f(i) = f(j)$ for every j in the same cycle as i.

Proof. We've already shown that if f is fixed by σ then f is constant on the cycles of σ in this sense. Thus, we assume f is constant on the cycles of σ. Then for any i, $\sigma^{-1}(i)$ and i are in the same cycle of σ, so $f(\sigma^{-1}(i)) = f(i)$. Thus f is fixed by σ. ∎

Our lemma tells us that to specify a coloring fixed by σ we choose a "color" (member of C) for each cycle of σ. If the permutation σ is a k-cycle, then that color is applied to all k members of σ. This lets us express the fixed-point enumerator for the action of a permutation of N on functions from N to C (that is, colorings of N by C) in terms of the cycle structure of the permutation.

Theorem 3.8. Suppose a permutation group G on N acts on the set C^N of functions from N to C (in the usual way). Suppose we use $D(y_k) = c_k$ for the symbolic description of the element y_k of C. If the permutation σ of N has j_h cycles of size h for $h = 1, 2, \ldots, n$, then the fixed-point enumerator of σ, $Fix(\sigma)$, for the action of sigma on functions from N to C is

$$\prod_{k=1}^{n} (c_1^k + c_2^k + \cdots + c_m^k)^{j_k}.$$

Proof. We assume σ has j_1 cycles of size 1, j_2 cycles of size 2, $\ldots$, j_n cycles of size n. We think of the functions as colorings. Since a cycle of size 1 may have one color assigned to its one member in one way, the symbolic enumerator for colorings of that cycle of size 1 is simply $c_1 + c_2 + \cdots + c_m$. Now the product principle for enumerators tells us that the symbolic enumerator for colorings of the j_1 cycles of size one is

$$(c_1 + c_2 + \cdots + c_m)^{j_1}.$$

Now the symbolic enumerator for colorings of one k-cycle so that all members receive the same color is

$$c_1^k + c_2^k + \cdots + c_m^k,$$

because the symbolic description of using color i on all elements of that cycle is c_i^k. By the product principle for enumerators, the symbolic enumerator for colorings, constant on cycles, of the elements of j_k disjoint k-cycles is the product

$$(c_1^k + c_2^k + \cdots + c_m^k)^{j_k}.$$

One more application of the product principle for symbolic enumerators gives us that the symbolic enumerator for colorings constant on cycles of σ is

$$\prod_{k=1}^{n} (c_1^k + c_2^k + \cdots + c_m^k)^{j_k}.$$

Since a coloring is constant on the cycles of σ if and only if σ fixes that coloring, this proves the theorem. ∎

The Pólya–Redfield Theorem

Now we can apply the orbit–fixed point lemma to give one version of the fundamental theorem of Pólya (developed by Pólya in a series of papers that culminated in 1937) and Redfield (who independently described the essence of the result in 1927) which describes the orbits of a group of permutations of N acting on the set C^N of functions from N to C. This theorem has been used by Pólya to solve a wide variety of problems in chemistry and graph theory and has been a fundamental tool used in solving a wide variety of counting problems. See the books by Pólya and Read and by Harary and Palmer in the Suggested Reading for more details on uses of the Pólya–Redfield theorem. In the statement of this theorem, we recognize that symbolic enumerators are just special kinds of generating functions.

Theorem 3.9. (Pólya–Redfield) Suppose a permutation group G on N acts on the set C^N of functions (thought of as colorings) from N to C (in the usual way). Use the symbolic descriptions of functions of Lemma 3.6. Then the generating function in which the coefficient of $c_1^{i_1} c_2^{i_2} \cdots c_m^{i_m}$ is the number of orbits of the permutation group G acting on C^N for which color h is used i_h times in each element of the orbit is

$$\frac{1}{|G|} \sum_{\sigma:\sigma \in G} \prod_{k=1}^{n} (c_1^k + c_2^k + \cdots + c_m^k)^{j_k(\sigma)},$$

where $j_k(\sigma)$ is the number of k-cycles of σ.

Proof. Lemma 3.6 tells us that the hypotheses of the orbit–fixed point lemma are satisfied. Apply the orbit–fixed point lemma with the fixed point enumerators of Theorem 3.8. ∎

The usual form of the Pólya–Redfield theorem uses a polynomial called the cycle-index polynomial of G to express the result more compactly. The cycle-index polynomial is a polynomial in n variables Z_1 through Z_n. This polynomial is denoted by $Z(G; Z_1, Z_2, \ldots, Z_n)$ or just $Z(G)$ when the variables are clear and is defined by

$$Z(G) = \frac{1}{|G|} \sum_{\sigma:\sigma \in G} Z_1^{j_1(\sigma)} Z_2^{j_2(\sigma)} \cdots Z_n^{j_n(\sigma)},$$

where $j_k(\sigma)$ is the number of k-cycles of σ. Then the Pólya–Redfield theorem may be expressed as follows.

Theorem 3.10. If the permutation group G on a set N acts (in the usual way) on the functions from N to C, then the generating function in which the coefficient of $c_1^{i_1} c_2^{i_2} \cdots c_m^{i_m}$ is the number of functions in which color h is used i_h times may be obtained from the cycle-index polynomial by substituting

$$c_1^k + c_2^k + \cdots + c_m^k$$

for each variable Z_k.

Example 3.9. If we place yellow beads, green beads, and blue beads at the corners of a square that is free to move in space, in how many ways may we make the placement of two yellow, one green, and one blue bead?

Since the square is free to move in space, the group of symmetries that acts on the square is the dihedral group D_4. To apply the Pólya–Redfield theorem in either form we need to know the possible cycle structure of the various elements of D_4. The four rotations may be classified as the identity, with four cycles of size 1, ρ and ρ^3 with one cycle of size 4, and ρ^2 with two cycles of size 2. The other elements are flips. Two of these flips are about diagonals; these fix the endpoints of the diagonals so they have one cycle of size 2 and two cycles of size 1. The other two flips are about side bisectors and so they have two cycles of size 2. All these remarks about cycle sizes may be neatly summarized in the cycle index

$$Z(G; Z_1, Z_2, Z_3, Z_4) = \frac{1}{8}(Z_1^4 + 2Z_4^1 + Z_2^2 + 2Z_1^2 Z_2^1 + 2Z_2^2)$$

$$= \frac{1}{8}(Z_1^4 + 2Z_4 + 3Z_2^2 + 2Z_1^2 Z_2).$$

Now we are in a perfect position to use our last theorem. We substitute $Y + G + B$ for Z_1 (since we can use one bead of one color in a cycle of size 1); we substitute $Y^2 + G^2 + B^2$ for Z_2 (since we use two beads, necessarily of the same color, to get a cycle of size 2 all of the same color); and we substitute $Y^4 + G^4 + B^4$ for Z_4 (since we use four beads, necessarily of the same color, to give the same color for all the members of a cycle of size 4). This gives us the generating function

$$f(Y, G, B) = \frac{1}{8}\left[(Y + G + B)^4 + 2(Y^4 + G^4 + B^4) + 3(Y^2 + G^2 + B^2)^2\right.$$
$$\left. + 2(Y + G + B)^2(Y^2 + G^2 + B^2)\right]. \tag{3.1}$$

To answer the question of the example, we need the coefficient of $Y^2 GB$; we will compute the $Y^2 GB$ part from each of the four terms, getting

$$\frac{1}{8}\left[\binom{4}{2\ 1\ 1}Y^2 GB + 0 + 0 + 2Y^2 \cdot 2GB\right] = \frac{1}{8}\left[12Y^2 GB + 4Y^2 GB\right] = 2Y^2 GB.$$

Thus, there are two colorings with two yellow, one green, and one blue bead. ∎

We answered the last question by choosing the $Y^2 GB$ terms. Suppose now instead we want to know the total number of colorings regardless of how many times each color is used. Since the coefficient of $Y^i G^j B^k$ is the number of functions with i yellows, j greens, and k blues, the sum of *all* the coefficients of all the $Y^i G^j B^k$ terms is the total number of colorings with three colors. If we substitute 1 for Y, 1 for G, and 1 for B in Equation (3.1), the result will be the sum of all the coefficients of all the terms, so $f(1, 1, 1)$ is the total number of colorings.

Example 3.10. In how many ways may we place beads chosen from identical yellow, green, and blue ones at the corners of a square free to move in space? We compute $f(1, 1, 1) =$

$$\frac{1}{8}\left[3^4 + 2 \cdot 3 + 3 \cdot 3^2 + 2(3)^2(3)\right] = \frac{1}{8}\left[81 + 6 + 27 + 54\right] = \frac{168}{8} = 21. \quad ∎$$

Notice that $f(1, 1, 1)$ is the same as $Z(G; 3, 3, 3, 3)$. In fact, it always happens that $Z(G; c, c, \ldots, c)$ is the total number of distinct colorings with c colors, that is, the number of orbits of functions from X to a set with c elements.

Theorem 3.11. If G is a group of permutations of the set X, then the number of orbits of functions from X to a set C may be obtained by substituting $|C|$ for each variable Z_j in the cycle index.

EXERCISES

1. Use the CFB theorem to count the number of orbits of edges of the complete graph on four vertices under the action of the dihedral group of the square.

2. Use the CFB theorem to compute the number of orbits of the edges of the complete graph on six vertices under the action of the rotation group of the hexagon.

3. In how many ways may we place beads of two different colors, using both colors, at the vertices of a hexagon free to rotate in the plane?

4. In how many ways may we place six beads of two different colors, using both colors, at the vertices of a hexagon free to move in space?

5. The symmetry group of a tetrahedron is the group of 12 permutations of four vertices described in Exercise 5 of Section 1 and Exercise 4 of Section 2. The tetrahedron has four vertices, six edges, and four faces.

   (a) Use the CFB theorem to show that there is, in essence, only one way to place two red beads and two blue beads at the vertices of a tetrahedron free to move in space.

   (b) Use the CFB theorem to determine the number of ways to color the edges of a tetrahedron free to move in space so that two edges are red and four edges are blue. Explain the result geometrically.

6. In how many ways may we paint the sides of a square using six different colors of paint (so that we have a different color on each side)?

7. In how many ways may we paint the sides of a square using three different colors of paint (so that one color is used exactly twice and no other is repeated)?

8. In how many ways may we paint the sides of a square using two different colors of paint if both colors are in the paint job?

9. Write down the cycle index for the rotation group of a hexagon.

10. Write down the cycle index for the dihedral group of a hexagon.

11. Write down the cycle index for the group of symmetries of a tetrahedron, thought of as a permutation group on the vertices.

12. Write down the cycle index for the group of symmetries of a cube, thought of as a permutation group on the vertices.

13. Imagine numbering the six sides of a cube (rather than the vertices of the cube) and think of the group of symmetries of the cube as a permutation group on the (numbered) sides rather than as a permutation group on the vertices. Write down the cycle index for this permutation group.

14. In the graph in Figure 3.1, each vertex is connected by an edge to the two vertices *not* directly opposite it and not adjacent to it. A symmetry of this graph is a permutation σ of the vertex set so that if i and j are connected by an edge then $\sigma(i)$ and $\sigma(j)$ are connected by an edge. Find the cycle index of the group of symmetries of this graph.

Figure 3.1

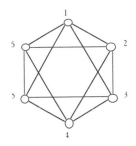

15. In how many ways may we place beads chosen from otherwise identical beads of three different colors at the corners of a hexagon free to rotate in the plane?

16. Write down the generating function according to the number of beads of each color for the number of ways to place the beads in the previous exercise.

17. Write down the generating function according to the number of ways to place beads chosen from otherwise identical beads of two colors at the vertices of a hexagon free to move in space. How many ways are there to place two red and four blue beads? (For those with some experience with chemistry, this is the same question as asking for the number of isomers of benzyl alcohol with two hydroxyl radicals.)

18. What is the generating function for the number of ways to place beads of three different colors at the vertices of a tetrahedron? In how many ways may we place the beads with two red and two blue? With two red, one blue, and one green? Explain your numerical answers geometrically. (In chemistry the four beads may be viewed as radicals being attached to a single carbon atom.)

19. Write down the generating function for the number of ways to place beads chosen from identical beads of three different colors at the vertices of a cube free to move in space. How many ways are there to place four red, two blue, and two green beads?

20. Write down the generating function for the number of ways to paint the six sides of a cube with colors chosen among red, blue, and green. How many ways use two of each color?

21. Write down the generating function for the number of ways to place computers chosen from two different models at the nodes of a hexagonal computer network with connections between computers as shown in Figure 3.1 of problem 14

22. Two graphs on the set $N = \{1, 2, \ldots, n\}$ with edge sets E and E' are said to be *isomorphic* if there is a permutation σ such that $\{i, j\}$ is in E if and only if $\{\sigma(i), \sigma(j)\}$ is in E'. Two graphs are said to be different if they are not isomorphic. By labeling the two-element subsets of N with the two labels "edge," "nonedge" (we label $\{i, j\}$ with "edge" if it is an edge and with "nonedge" if it is not), we may think of the construction of a graph as a coloring problem. By thinking of S_4 as a permutation group on the six different two-element subsets of $\{1, 2, 3, 4\}$ *rather than* as a permutation group on $\{1, 2, 3, 4\}$ itself, find the number of different graphs on four vertices.

23. Use mathematical induction on n to prove Corollary 3.3.

Suggested Reading

Baglivo, Jenny A., and Jack E. Graver. 1983. *Incidence and Symmetry in Design and Architecture.* New York: Cambridge University Press.

Dean, Richard A. 1966. *Elements of Abstract Algebra.* New York: Wiley.

De Bruijn, N. G. 1964. Polya's Theory of Counting. In *Applied Combinatorial Mathematics,* ed. Edwin F. Beckenach. New York: Wiley.

Harary, Frank, and Edgar Palmer. 1973. *Graphical Enumeration.* Orlando, FL: Academic Press.

Pólya, George, and R. C. Read. 1987. *Combinatorial Enumeration of Groups, Graphs, and Chemical Compounds.* New York: Springer.

Rotman, Joseph J. 1984. *An Introduction to the Theory of Groups,* 3d ed. Newton, MA: Allyn and Bacon.

Tucker, Alan. 1984. *Applied Combinatorics,* 2d ed. New York: Wiley.

Answers to Exercises

CHAPTER 1

Section 1.1, pages 8–13

1. $7 \cdot 6 = 42$ **3.** (a) $10^5 = 100,000$ (b) $10 \cdot 9 \cdot 8 \cdot 7 \cdot 6 = 30,240$
(c) $(10 \cdot 9/2)^5 = 45^5$ (d) $(10 \cdot 9/2 + 10)^5 = 55^5$

5. $3 \cdot 3 + 3 \cdot 4 + 3 \cdot 4 \cdot 3 = 57, \ 2 \cdot 57 = 114$ **7.** $2^8 = 256$

9. 9^{10} (assuming you don't start with 0) and $9 \cdot (10^9 - 9^9)$ respectively.

11. $\left(\frac{8 \cdot 7}{2} + 8\right) \cdot 4 \cdot 2 \cdot 3 = 864$ **13.** $4! = 24$ **15.** (a) $(3!)^2 \cdot 2 = 72$
(b) $2 \cdot 3! \cdot 3! = 72$

16. $Y - X$ and X are two disjoint sets whose union is Y, so $|Y| = |Y - X| + |X|$ by the sum principle. Thus $|Y - X| = |Y| - |X|$.

17. $A \cup B = \left(A - (A \cap B)\right) \cup \left(B - (A \cap B)\right) \cup (A \cap B)$. The three sets on the right are disjoint, so by the sum principle and Exercise 16 $|A \cup B| = |A| - |A \cap B| + |B| - |A \cap B| + |A \cap B| = |A| + |B| - |A \cap B|$.

19. 546   **21.** (a) and (b) are applications of calculus.
(c) $\sum_{i=1}^{n} \log i - \frac{1}{2} \log n \leq \int_1^n \log x \, dx$ so $\sum_{i=1}^{n} \log 1 \leq \int_1^n \log x \, dx + \frac{1}{2} \log n$. Part (b) gives the second inequality.
(d) This follows by subtracting $\sum_{i=1}^{n} \log i$ from each part of the inequality in (c).
(e) Another application of calculus.
(f) Since exponentiation is a continuous function $e^d = n^{n+\frac{1}{2}} e^{-n+1}/n!$ so $e^{d-1} = \lim_{n \to \infty} n^{n+\frac{1}{2}} e^{-n}/n!$
(g) Since $0 \leq d_n \leq \frac{1}{2} \log \frac{3}{2}$, $-1 \leq d - 1 \leq \frac{1}{2} \log \frac{3}{2} - 1$. Then $e^{-1} \leq e^{d-1} \leq e^{\frac{1}{2} \log \frac{3}{2} - 1} = e^{-1} \sqrt{\frac{3}{2}}$, and since $c = 1/e^{d-1}$, we have $e \geq c \geq e\sqrt{2/3}$. It is interesting to note that to one decimal place this is $2.7 > 2.5 > 2.2$.

Section 1.2, pages 23–26

1. $\{(0,1), (1,1)\}$; $\{(0,2), (1,2)\}$; $\{(0,3), (1,3)\}$; $\{(0,1), (1,2)\}$; $\{(0,1), (1,3)\}$; $\{(0,2), (1,1)\}$; $\{(0,2), (1,3)\}$; $\{(0,3), (1,1)\}$; $\{(0,3), (1,2)\}$.

3. $\emptyset$; $\{(0,a)\}$; $\{(0,b)\}$; $\{(1,a)\}$; $\{(1,b)\}$; $\{(0,a), (1,a)\}$; $\{(0,b), (1,b)\}$;
$\{(0,a), (1,b)\}$; $\{(0,b), (1,a)\}$; $\{(0,a), (0,b)\}$; $\{(1,a), (1,b)\}$;
$\{(0,a), (0,b), (1,a)\}$; $\{(0,a), (0,b), (1,b)\}$; $\{(0,a), (1,a), (1,b)\}$;
$\{(0,b), (1,a), (1,b)\}$; $\{(0,a), (0,b), (1,a), (1,b)\}$.

Figure A.1

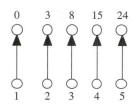 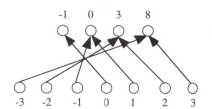

5. $\{(1,0),(2,3),(3,8),(4,15),(5,24)\}$.

7. See Figure A.1 The function of Exercise 5 is one-to-one.

8. (a) A function.
   (b) Not a function because 1 is the first element of two ordered pairs.
   (c) A function.
   (d) Not a function since 4 is the first element of two ordered pairs, as is 1.
   (e) A function.

9. $\{(1,12),\ (2,12),\ (3,12),\ (4,12),\ (6,12),\ (1,11),\ (1,10),\ (2,10),\ (5,10),$
   $(1,9),\ (3,9),\ (1,8),\ (2,8),\ (4,8),\ (1,7),\ (1,6),\ (2,6),\ (3,6),\ (1,5),$
   $(1,4),\ (2,4),\ (1,3),\ (1,2)\}$.

11. $(10)^4 = 10,000$; $(10)_4 = 5040$.

13. $10^3 = 1000$; $10^3 - 10 = 990$; $(10)_3 = 10 \cdot 9 \cdot 8 = 720$.

15. f^{-1} is a function if and only if whenever (y, x_1) and (y, x_2) are in f^{-1}, then $x_1 = x_2$, which is true if and only if f is one-to-one. The domain of f^{-1} is the range of f if and only if f is an onto function, so f^{-1} is a function whose domain is the range of f if and only if f is one to one and onto, i.e., a bijection.

17. (a) The function is one-to-one but not onto.
    (b) The function is onto but not one-to-one.

19. Suppose f is a one-to-one and onto function. Then (x_1, y) and (x_2, y) cannot be in f unless $x_1 = x_2$, so the relation f^{-1} is a function. The function f^{-1} is one-to-one because no two pairs of f have the same first entry, meaning that no two pairs of f^{-1} have the same second entry. The function f^{-1} is onto because every member of the domain of f appears as a first element of a pair in f. The domain of f^{-1} is the range of f because every member of the range of f appears as the second element of an ordered pair of f.

21. Every element of the range of g is an image of some element of the domain of g and the domain of g is equal to the image of f. Thus every element of the range of $g \circ f$ is the image of some element in the domain of $g \circ f$.

23. (a) Any number ending in five is divisible by five, so the power cannot end in 5. Since it cannot end in an even digit either, there are only four different odd digits in which the power could end.
    (b) Yes, because it has the form $10n + r$, where r is 1, 3, 7, or 9 and the fourth power of each of these numbers ends in a 1.
    (c) No, consider the prime 3 as an example.

25. There are more than n numbers whose decimal representation has only ones, so two of them must give the same remainder when divided by n. Then their

difference is a multiple of n whose decimal representation has only zeros and ones.

27. Consider one person, say A. In the group, A either has three acquaintances or three nonacquaintances. In the first case, either two of these three people are acquainted and so with A they form a threesome of acquainted people, or the three people are mutually nonacquainted, in which case they serve as the desired threesome. In the second case, if two of the people are nonacquainted, then with A they form a set of three mutual strangers; otherwise the three are acquainted and form the desired threesome.

29. Consider the set S of ordered pairs (P, Q) of people with P acquainted with Q. Since the acquaintance relation is symmetric, this set has an even number of ordered pairs. But S is the union over all people P of those ordered pairs P, Q such that P is acquainted with Q. Thus if we had an odd number of people each acquainted with an odd number of people, our set S would, by the sum principle, have an odd number of elements. Thus someone must have an even number of acquaintances (and thus an even number of nonacquaintances).

31. Imagine people numbered 1 through 8 seated around a table so that each person knows the two people on the left and right and the two people who are the second to the left or the second to the right. We ask first if there is a set of 4 mutually acquainted people. By symmetry we may assume 1 is in this set, so the set must be chosen from 1 and acquaintances of 1, i.e., from $\{1, 2, 3, 7, 8\}$. For the set to have four elements, it must include either 2 or 8, and by symmetry we may assume it is 2. But 2 does not know 8, so the set must be $\{1, 2, 3, 7\}$. But 3 does not know 7, so there is no such set.

   If there were three mutual nonacquaintances, we could again assume 1 is one of them, so the others would have to be chosen from 4, 5, and 6. But 4, 5, and 6 are mutually acquainted, so there is no such set.

Section 1.3, pages 35–39

1. $C(8; 6) = 28$

3. $C(4; 2) = 6$; $\{A, B\}$, $\{A, C\}$, $\{A, D\}$, $\{B, C\}$, $\{B, D\}$, $\{C, D\}$

5. $C(5; 3)C(5; 3) = 10^2 = 100$

7. Note that row 4 has the same property; so does row 2^i for each integer i. In particular, row 16 is an example.

9. $(20)_4 \binom{16}{3} = \binom{20}{3}(17_4)$; $(20)_4 \binom{20}{3}$; $(20)_4 \binom{16}{3} \cdot 3 = (20)_5 \binom{15}{2} = \binom{20}{2}(18)_5$; $(20)_5 \binom{20}{2} = (20)_4 \binom{20}{3} \cdot 3$

11. $C(4; 2)C(5; 2)C(8; 5)C(3; 2)C(6; 3) = 201{,}600$

13. $C(20; 3) \cdot C(3; 1) \cdot C(17; 3) = 2{,}325{,}600$ with exactly one member in common; total number of possible choices is $5{,}038{,}800 = C(20; 3)C(17; 4) + C(20; 3)C(3; 1)C(17; 3)$.

15. There are $C(n; k)$ ways to choose the k objects with label 0. The remainder of the objects get label 1.

17. **(a)** First, $\frac{(n-1)!}{(n-k-1)!} + \frac{k(n-1)!}{(n-k)!}$ simplifies to $\frac{n!}{(n-k)!}$
   For a more interesting proof, a k-element permutation is either a permutation

of the first $n-1$ members of $N = \{1, 2, \ldots, n\}$ or else it maps exactly one of the k numbers $1, 2, \ldots, k$ to n and maps the remaining $k-1$ numbers in a one-to-one fashion into the first $n-1$ members of N.

(b) 1
      2   2
      3   6   6
      4 12 24   24
      5 20 60 120 120

19. **(a)** k-tuples   **(b)** None of these (n-tuples)
(c) k-element permutations
(d) None of these (n-element permutations)   **(e)** None of these
(f) None of these   **(g)** k-element subsets
(h) None of these (n-element subsets)

21. The left-hand side of the formula counts the number of ways of choosing an ordered pair of one element x of $N = \{1, 2, \ldots, n\}$ and a $(k-1)$-element subset not containing x; the right-hand side counts the number of ways to choose k elements and divide them into an ordered pair consisting of one element x and a $(k-1)$-element subset not containing x. Clearly the sets of ordered pairs described are equal.

23. Both sides of the formula count the number of ways to choose an ordered pair consisting of a k-element subset K of $N = \{1, 2, \ldots, n\}$ and an m-element subset of N containing K.

25. Note that $\binom{m}{j} = \binom{m}{m-j}$. (This trick is the reason for the star.) Now the left- and right-hand sides count the total number of ways to choose $m + k$ elements out of the union of an m-element set and a disjoint set of n elements.

27. To choose $m + 1$ elements out of $n + 1$, choose the largest first and then choose the remaining elements. The sum on the left adds the number of ways of making the remaining choices for each possible largest element.

29. We had the wrong number of digits. Stirling's approximation gives 3,303,436 while the actual value is 3,268,760.

31. The prettiest description is geometric. In Table 1.2, locate position (m, n). Draw a parallelogram from this position by drawing lines parallel to the two all-one sides of the Pascal triangle until these lines intersect the sides of the triangle. We need to compute the values of $C(j; k)$ lying inside or on the boundary of this parallelogram. One of these lines may be described by the equality $k = n$. The other line may by described by the equality $j - k = m - n$. Thus we must compute $C(j; k)$ for $0 \le k \le n$ and $0 \le j \le m - n + k$.

33. Applying recursion blindly (without an optimizing compiler) gives a program that takes far longer than the iterative program.

35. For each nonnegative integer n, row $2^n - 1$ consists entirely of odd numbers and so row 2^n has even numbers in every position except the ends. To see how a proof of this goes, it is instructive to write out Pascal's triangle substituting a blank for each even number and a one for each odd number. What you can see is that in rows 16 to 32, we have two triangles of ones and blanks which are both identical to the triangle of ones and blanks we have in rows 1 to 16. On closer observation you will see that rows 8 to 15 consist of two triangles that are identical to the triangle in rows 1 to 7, and the same thing occurs within rows 0–7. Then we assume inductively that row $2^i - 1$ is a row of 2^n odd numbers so that row 2^i consists of a 1 followed by $2^i - 1$ even numbers followed by a 1. We then argue that in the next $2^i - 1$ rows below the 1 followed by even numbers we repeat the even–odd pattern of the first $2^i - 1$ rows of Pascal's triangle and below and to the right of the 1, we repeat the same pattern again.

Section 1.4, pages 50–53

1. $x^4 + 4x^3y + 6x^2y^2 + 4xy^3 + y^4$ (or in reverse order).
3. $16x^4 - 96x^3 + 216x^2 - 216x + 81$ (or in reverse order).
5. $(2x)^6 - 6(2x)^5y + 15(2x)^4y^2 - 20(2x)^3y^3 + 15(2x)^2y^4 - 6(2x)y^5 + y^6 = 64x^6 - 192x^5y + 240x^4y^2 - 160x^3y^3 + 60x^2y^4 - 12xy^5 + y^6$.
7. 169
9. 65.944160601201, 65.9440, a bit more than .00016 from the correct value (answer may vary for last part).
11. **(a)** Yes, note that $1 + \binom{1000}{1}(.001) = 2$.
    (b) Yes; again the first two terms sum to 2.
13. $x^3 + y^3 + z^3 + 3x^2y + 3xy^2 + 3x^2z + 3xz^2 + 3y^2z + 3yz^2 + 6xyz$.
15. 15,120   17. 1680
19. 1680; $C(9; 3, 3, 3) + 6C(9; 2, 3, 4) + 3C(9; 2, 2, 5) = 11508$
21. Since if the sum started at 0 we would have $(1+3)^{10}$, the result is $4^{10} - 3^{10} = 989,527$.
23. **(a)** Since $x^0 + C(n; 1)x^1 + C(n; 2)x^2 + C(n; 3)x^3 + \cdots = (x+1)^n$, taking the derivative of both sides gives $C(n, 1)x^0 + 2C(n, 2)x + 3C(n, 3)x^2 \cdots = n(x+1)^{n-1}$. Now let $x = 1$ to get the desired equality.
    (b) Repeat the process of part **(a)** with $(-1 + x)^n$.
25. Multiply $(x+y)^n$ by $(x+y)^m$ and compute the coefficient of either $x^{m+k}y^{n-k}$ or $x^{n-k}y^{m+k}$.
27. This formula is just a complicated way to say that $2^n \cdot 2^n = 2^{2n}$; see Example 4.3. Alternatively, use the fact that $\binom{n}{i}^2 = \binom{n}{i}\binom{n}{n-i}$ and the binomial theorem to expand $(1+x)^n(1+x)^n = (1+x)^{2n}$; then substitute $n = 1$.
29. Since $(x-1)^n = \sum_{i=0}^n \binom{n}{i}(-1)^i x^{n-i}$, substitute 1 for x and move all negative terms to the zero side of the equation.
31. Use the argument used to prove the Pascal relation; namely when we label the integers zero through n with m labels, n is labeled with one of the m labels. The number of labelings that label n with label number i is the same as the number of labelings of the first $n - 1$ integers in which the number of integers receiving label i is one less.
33. The sequence must start with a 1. Removing the one gives a sequence of ones and negative ones so that the number of ones is at least as large as the number of negative ones at each stage. That is, we get the Catalan number $\frac{1}{n}\binom{2n}{n}$.
34. Since at each stage we must have as many left parentheses as right, this is another description of the Catalan numbers.

35. Suppose we keep track of the votes at each stage as an ordered pair (x, y) where y is the number of votes for the winning candidate and x is the number of votes for the losing candidate. Then the number of counting sequences in which the winner is always ahead is the number of lattice paths from $(0,0)$ to (m, n) which never touch the line $y = x$ after $(0,0)$. Thus this is the number of paths from $(0, 1)$ to (m, n) which never touch the line $y = x$, and we computed this number to be $\binom{m+n}{n}\frac{n-m+1}{n+1}$. Since the total number of lattice paths is $\binom{m+n}{m}$, the fraction is $\frac{n-m+1}{m+1}$.

Section 1.5, pages 64–68   (Note: Only outlines of proofs are given.)

1. When $n = 1$, we have $1 = \frac{1(2)}{2} = 1$. We let S be the set of n making the formula true and from $n - 1 \in S$ we write $\sum_{i=0}^{n-1} i = \frac{(n-1)(n)}{2}$. Now to show $n \in S$, note that $\sum_{i=0}^{n} i = \sum_{i=0}^{n-1} i + n = \frac{(n-1)(n)}{2} + n = \frac{n(n+1)}{2}$.

3. Probably the most natural guess for the right-hand side is $\binom{n+1}{3}$. When $n = 1$ the formula $\sum_{i=1}^{n} \binom{i}{2} = \binom{n+1}{3}$ says $0 = 0$, which is true. For larger values of n, add $\binom{n+1}{2}$ to both sides and apply the Pascal relation to the right-hand side to get $\binom{n+2}{3}$.

5. The natural guess is that the right-hand side should be $\binom{n+1}{k+1}$; the technique described for Exercise 3 gives this to you.

7. $n^3 = 6\binom{n}{3} + 6\binom{n}{2} + \binom{n}{1}$. $\sum_{n=1}^{m} n^3 = 6\binom{m+1}{4} + 6\binom{m+1}{3} + \binom{m+1}{2} = \frac{m^2(m+1)^2}{4}$

9. Let S be the set of integers $n \geq 1$ such that 6 is a factor of $n^3 + 5n$. Note that 1 is in S. Assume n is in S, note that $(n+1)^3 + 5(n+1) = n^3 + 5n + 3(n^2 + n) + 6$, and since the binomial in parentheses is divisible by 2 (see Exercise 1 for a reason), we have a sum of three summands, each divisible by 6.

11. By assuming that $n^3 > 2n^2$, we have $(n + 1)^3 = n^3 + 3n^2 + 3n + 1 > 2n^2 + 3n^2 + 3n + 1 = 5n^2 + 3n + 1 > 2n^2 + 4n + 2$ (since $n > 1$ implies $n^3 > 2n^2$ and $n^2 > n$). Thus $(n + 1)^3 > 2(n + 1)^2$.

13. This uses the second version of the induction principle. The number of prime factors of 2 is $1 = \log_2 2$. Assume the number of prime factors of i is less than $\log_2 i$ for each i between 2 and $n - 1$. Either n is prime or $n = mk$, in which case the number of prime factors of n is no more than the sum of the number of prime factors of m and the number of prime factors of k, which is no more than $\log_2 m + \log_2 k = \log_2 mk$.

15. Inductively we assume there are $(n)_{k-1}$ lists of $k - 1$ distinct elements; now there are $n - (k - 1) = n - k + 1$ choices for the kth position. By the product principle, this gives us $(n)_{k-1}(n - k + 1) = (n)_k$ lists.

17. Use $A_1, A_2, \ldots, A_n$ for the sets. Inductively assume $\cup_{i=1}^{k} A_i$ has mk elements. Then by the sum principle $\cup_{i=1}^{k+1} A_i$ has $mk + m = m(k + 1)$ elements. Proceed as in Example 5.5 (except here you need less notation).

19. This is like number 15 or 16, except we have $m_1 \cdots m_{k-1}$ lists of length $k - 1$ and m_k choices for the kth place in the list.

21. If $n = 0$ we need make no comparisons and $0 \le 0 \cdot 2^0$. Suppose now we know 2^n numbers may be sorted with no more than $n \cdot 2^n$ comparisons. Given a list of 2^{n+1} numbers, split it into two lists of 2^n numbers, its first and second half. Each can be sorted with $n \cdot 2^n$ comparisons, giving $2n \cdot 2^n$ comparisons so far. Now select the smaller of the two first numbers of their lists, put it first in a third list, and strike it from its list. Repeat this selection process on the first two lists (as just modified) until one or the other runs out. This takes no more than 2^{n+1} comparisons, and we need make no more comparisons to finish building the third list in sorted order. Thus we have made no more than $2n2^n + 2^{n+1} = n2^{n+1} + 2^{n+1} = (n+1)2^{n+1}$ comparisons.

23. We assume $\sum_{j=0}^{n-1} \binom{j}{k} = \binom{n}{k+1}$. We must thus show $\binom{n}{k+1} + \binom{n}{k} = \binom{n+1}{k+1}$. But this is exactly the relation of Pascal's triangle, so we may complete our inductive proof. (This is just a variant on Exercises 1–5.)

25. The inductive step is

$$A \cap \cup_{i=1}^n B_i = A \cap \left((\cup_{i=1}^{n-1} B_i) \cup B_n \right) = (A \cap \cup_{i=1}^{n-1} B_i) \cup (A \cap B_n)$$
$$= (\cup_{i=1}^{n-1} A \cap B_i) \cup (A \cap B_n)$$
$$= \cup_{i=1}^n (A \cap B_i).$$

27. For the inductive step assume that $\sum_{i=0}^{n-1} a_i = a_0 \frac{1-r^n}{1-r}$. Add $a_n = a_0 r^n$ to both sides to get $\sum_{i=0}^n a_i = a_0 \frac{1-r^n}{1-r} + \frac{1-r}{1-r} a_0 r^n = a_0 \frac{1-r^{n+1}}{1-r}$.

29. 8 is $3 + 5$ is our base step. Further 9 is $3 \cdot 3$ and 10 is $2 \cdot 5$. Now if $n > 10$, then $n - 3 > 7$ but $n - 3 < n$ so that by our inductive hypothesis, $n - 3 = a \cdot 3 \mid b \cdot 5$ so that $n = (a+1) \cdot 3 + b \cdot 5$. Notice that we showed that 9 and 10 were in the set making the statement true by a different technique than the one used for the larger n values. With $n = 9$ or $n = 10$ we cannot apply the inductive hypothesis to $n - 3$ because $n - 3$ is not greater than 7. You might say we needed three base steps here.

33. **(a)** This is not an inductive proof. It is, however, based on the proof of Theorem 5.1. We let T be a set of size $R(m-1, n) + R(m, n-1)$. We let f be a symmetric function defined on pairs of elements of T. Choose an element x of T. For each other element y of T, $f(x, y)$ is an A or S. Thus by the sum principle there are either $R(m-1, n)$ elements y such that $F(x, y) = A$ or $R(m, n-1)$ elements y such that $f(x, y) = S$. This gives us two cases. In the first case we have $R(m-1, n)$ elements y such that $f(x, y) = A$. Thus we have either a set T_1 of size $m-1$ of elements of T on whose pairs f has the value A or a set T_2 of size n on whose pairs f has the value S. If we have the set T_1 then on the pairs of $T_1' = T_1 \cup \{x\}$ f still has the value A. Therefore in case 1 we have either a set T_1' of size m on whose pairs f has the constant value A or a set T_2 of size n on whose pairs f has the constant value S. In case 2 we similarly either have a set T_1 of size m on whose pairs f has the constant value A or a set T_2' of size n on whose pairs f has the constant value S.

Therefore for any set of size $R(m-1, n) + R(m, n-1)$ and any symmetric function f defined on this set, we have either a subset of size m on whose pairs

f has the constant value A or a subset of size n on whose pairs f has the constant value S. This proves that $R(m,n) \leq R(m-1,n) + R(m,n-1)$.
(b) This tells us that $R(5,3) \leq R(4,3) + R(5,2) = 9 + 5 = 14$.
(c) By symmetry, if there were a set of three mutual acquaintances, we could asume person 1 is in it. However, then the set must be a subset of $\{1, 2, 13, 6, 9\}$ and no three people in this set are mutually acquainted. If there were a set of five people no two of whom are acquainted, we could again assume it contains 1. Then it cannot contain 13, 2, 6, or 9. If it contains neither 12 nor 3, then it is easy to check there are not four more elements for it to contain. Thus by symmetry we can asume it contains 1 and 3. But that means the other three members must be chosen from $\{5, 7, 10, 12\}$ but 5 and 10 can't be in the set together, nor can 7 and 12, so there is no such set.
(d) $R(5,3) = 14$

CHAPTER 2

Section 2.1, pages 79–82

1. $\{(1,1), (1,2), (2,2), (2,1), (2,3), (3,3), (3,2), (3,4), (4,4), (4,3), (4,5), (5,5), (5,4)\}$. Not an equivalence relation (not transitive).
3. No; No; Yes; Yes.
5. $\{(1,1), (2,2), (3,3), (4,4), \ldots, (12,12), (1,2), (1,3), \ldots, (1,12), (2,4), (2,6), (2,8), (2,10), (2,12), (3,6), (3,9), (3,12), (4,8), 4,12), (6,12)\}$; No.
7. $(x,x) \in R$ for each x since P is a partition. If x and y are in the same class, then y and x are in the same class. If x is in the same class as y, and y is in the same class as z, then x, y and z are all in the same class because the classes are disjoint. Thus $(x,z) \in R$. Therefore R is an equivalence relation and its equivalence classes are the sets C_i. We have proved that if we have a partition, then it is the equivalence class partition of an equivalence relation R. Now we need to show there is only one equivalence relation with P as its equivalence class partition. But if P is the partition of some equivalence relation R', then $(x,y) \in R'$ means x and y are in the same class C_i of R' by part 4 of Theorem 1.1 and x and y are in the same class C_i of R' means $(x,y) \in R'$ by condition 3 of Theorem 1.1. Therefore $R = R'$. This completes the proof.
9. **(a)** 3   **(b)** $3! = 6$ 11. $6!/(6 \cdot 2) = 60$ 13. $(k-1)!/2$
15. $(k-1)!k!$, $(k-1)!k!/2$ 17. 15   19. n^2, n
20. $\{(0,0), (1,1), (-1,1), (1,-1), \ldots, \text{etc.}\}$, which is the set of all ordered pairs $(r,+r)$ and $(r,-r)$. The equivalence classes are sets of the form $\{r,-r\}$. that is, each number is equivalent to itself and its negative.
21. n, k.
23. Given a partition of type $(j_1, j_2, \ldots, j_n)$, we can construct a labeling from it in $j_1! j_2! \ldots j_n!$ ways. This proves Theorem 1.6.
25. $\frac{15!}{(4!)^2 \cdot 2!(2!)^3 \cdot 3!}$; $\frac{15}{2!3!4!5!}$
27. a is related to a since $0 = n \cdot 0$. If $a - b$ is kn, then $b - a$ is $-kn$. If $a - b = kn$ and $b - c = jn$, then $a - c = (j+k)n$. In the case $n = 2$ we get the even and odd numbers.

29. **(a)** Use I to stand for this intersection. If (x, y) and (y, z) are in I, then they are in each of the relations being intersected, and since these relations are transitive, (x, z) is in each of them, meaning that (x, z) is in I.
(b) If T is one of the transitive relations containing R, then it is one of the relations being intersected, so $I \subseteq T$.
(c) When everything is related to at least one thing, perhaps itself.

Section 2.2, pages 96–98

1. $\binom{10+4-1}{10} = 286$ **3.** $\binom{n-nj+k-1}{k-nj}$

5. m is zero for each letter not given. $m(e) = 2$, $m(i) = 2$, $m(l) = 2$, $m(m) = 1$, $m(o) = 2$, $m(r) = 2$, $m(t) = 2$, $m(u) = 1$, $m(v) = 1$

6. The multiset with multiplicity m' is a multisubset of the multiset with multiplicity m if $m'(x) \leq m(x)$ for each x in the first multiset.

7. $\binom{6+3-1}{3} = \binom{8}{3} = 56$ **9.** $\binom{4+6-1}{6} = \binom{9}{6} = 84$ **11. (a)** k-tuples; n^k
(b) n-tuples; k^n **(c)** k-element permutations; $(n)_k$
(d) n-element permutations; $(k)_n$ **(e)** k-element multisets; $\binom{n+k-1}{k}$
(f) n-element multisets; $\binom{k+n-1}{n}$ **(g)** k-element subsets; $\binom{n}{k}$
(h) n-element subsets; $\binom{k}{n}$

13. $\frac{(n+r+s+t+2)!}{n!r!s!t!2!} = C(n+r+s+t+2; n, r, s, t, 2)$

15. $\binom{6+k-1}{k}\binom{2+n-1}{n} = \binom{k+5}{k}(n+1)$.

17. For row 10, if the objects and recipients are identical, then there either is or is not a way to give one or zero objects to each recipient; if each recipient must receive one, then we must have the same number of objects and recipients; finally if there are more objects than recipients, then one recipient would receive more than one, so there are no distributions satisfying the given conditions in this case. For column 2 of rows 5 and 8, there is a distribution of identical objects in which each recipient receives one if and only if the number of objects equals the number of recipients, and all such distributions are equivalent because the objects are identical.

19. Consider two ordered distributions of the k-element set $K = \{1, 2, \ldots, k\}$ to $N = \{1, 2, \ldots, n\}$ as equivalent if we obtain one from the other by rearranging the objects without changing the number any recipient receives. Given an equivalence class of ordered distributions let $f(x_i)$ be the number of objects recipient i receives. Each function arises from such a distribution; simply give the first $f(i)$ members of K to the first member of N and so on. For two equivalence classes to be different, some recipient must receive a different number of objects in the two distributions. Thus we have a bijection.

21. Consider two ordered distributions to be equivalent if we get one from the other by rearranging the recipients. We have $(k)_n(k-1)_{k-n}$ ordered distributions of k objects to n recipients so that each recipient gets at least one. Dividing by $n!$ gives us the number of equivalence classes of distributions; and

$$(k)_n(k-1)_{k-n}/n! = \binom{k}{n}\frac{(k-1)!}{((n-1)!} = \binom{k}{n}(k-1)_{k-n}.$$

24. First make a list of the k objects in one of $k!$ ways, then choose $n-1$ places between members of the list as breakpoints where you will break the list. This breaks the list into n nonempty segments; give segment i to recipient i. The number of ways to carry out the process is the product of the number of ways to make the list and the number of ways to choose the breakpoints.

25. (Partial answers) (a) You get transitivity by composing two bijections of K.
(b) Distributions of four identical objects to three recipients.
(c) The equivalence class of a constant function will have just one member. However, the equivalence class containing $aabb$ will also contain $abab$ and $bbaa$ (and other functions).
(d) Since the classes are not the same size, the equivalence principle does not apply.

Section 2.3, pages 106–109

1. The first four rows are in Table 3.1. Rows 5 through 8 are $0, 1, 15, 25, 10, 1$;
$0, 1, 31, 90, 65, 15, 1$; $0, 1, 63, 301, 350, 140, 21, 1$;
$0, 1, 127, 966, 1701, 1050, 266, 28, 1$

3. $s(0,0) = 1$; $s(1,0) = 0$; $s(1,1) = 1$; $0, -1, 1$; $0, 2, -3, 1$; $0, -6, 11, -6, 1$;
$0, 24, -50, 35, -10, 1$; $0, -120, 274, -225, 85, -15, 1$;
$0, 720, -1764, 1624, -735, 175, -21, 1$; $0, -5040, 13068, -13132, 6769,$
$-1960, 322, -28, 1$

5. $x^4 = (x)_1 + 7(x)_2 + 6(x)_3 + (x)_4$

7. $1/234$; $2/134$; $3/124$; $4/123$; $12/34$; $13/24$; $14/23$; Yes.

9. $S(8,5) \cdot 5! = 126,000$; $S(8,5) = 1050$ 11. $S(8,4) \cdot 4! = 40824$

13. $S(m, m-1)$ is the number of ways of partitioning an m-element set into $m-1$ parts, one of size 2 and the rest of size 1. Thus it is the number of ways to select that two-element set, which is $\binom{m}{2}$.

15. To partition an m-element set into two parts, we select the part containing a certain element, say element 1. To fill out this part, we have a choice of any subset of the remaining $m-1$ objects except for the choice of all of them. Thus we have $2^{m-1} - 1$ choices.

17. Suppose we are breaking a permutation of $\{1, 2, \ldots, m\}$. Either m is in a class by itself and we have $L(m-1, n-1)$ such broken permutations, or m is not, in which case it could occur at the end of any of the n permutations we get by removing it, or before any of the $m-1$ remaining elements in whatever permutation they belong to. Thus $L(m, n) = L(m-1, n-1) + (m+n-1)L(m-1, n)$.

19. By applying Theorems 3.4 and 3.6, we write $x^m = \sum_{n=0}^{m} S(m,n)(x)_n = \sum_{n=0}^{m} S(m,n) \sum_{j=0}^{n} s(n,j)x^j = \sum_{j=0}^{m} \left(\sum_{n=j}^{m} S(m,n)s(n,j) \right) x^j = \sum_{j=0}^{m} (\sum_{n=0}^{m} S(m,n)s(n,j))x^j$. Equating coefficients of powers of x shows that the coefficient of x^j must be zero unless $j = m$, giving us $\delta(m,j)$ for the sum in parentheses. However, the sum stops at the upper limit j if $j < m$. The inner sum is the formula for the (m, j) entry of a product of two matrices, so

if we take the matrices of the values of $S(n,k)$ and $s(n,k)$ for $0 \le k \le n \le m$ for some fixed m, then they are inverses of each other.

21. **(a)** We may assume that $f^i(a) = f^j(a)$ for some $j < i$. If $j \neq 0$ (note we assume $f^0(a) = a$), then act with f^{-1} to get $f^{i-1}(a) = f^{j-1}(a)$ and contradict the fact that $f^i(a)$ is the first repeated element.

(b) Each a is in the cycle $\big(a, f(a), \ldots, f^{i-1}(a)\big)$ described in the statement of the exercise. If a is also in the cycle $\big(b, f(b), \ldots, f^{j-1}(b)\big)$, then each $f^k(a)$ is in the cycle b determines, so the cycle determined by a is a subset of the cycle determined by b. Thus a is $f^k(b)$ for some k. But $f^{-1}(a) = f^{i-1}(a)$, so $b = f^{-k}(a) = f^{k(i-1)}(a)$, so b is in the cycle a determines. Thus the cycle determined by b is a subset of the cycle determined by a. Therefore these cycles are equal.

(c) There are $c(m-1, n-1)$ permutations in which m is in a cycle by itself; the number of permutations in which m shares a cycle with some other elements is $(m-1)c(m-1, n-1)$, because placing m immediately *before* any of the other $m-1$ elements in that element's cycle gives a different permutation, and if m occurs in a cycle the cycle may be written so that n occurs before some element of the cycle. Adding gives the desired recursion relation.

(d) They are equal.

23. Rewrite the Stirling recurrence as $S(m,n) = S(m+1, n+1) - (n+1)S(m, n+1)$. Now substitute $S(m, n+1) = S(m+1, n+2) - (n+2)S(m, n+2)$ into the previous formula and iterate. This shows how an inductive proof of the formula would go.

25. Write $\sum_{n=0}^{m} L'(m,n)((x))^n = (x)_m = (x)_{m-1}(x - m + 1) = \sum_{i=0}^{m} L'(m - 1, i)((x))^i(x - m + 1) = \sum_{i=0}^{m} L'(m - 1, i)((x))^i[(x + i) + (1 - m - i)] = \sum_{i=0}^{m} L'(m-1, i)((x))^{i+1} - \sum_{i=0}^{m} L'(m-1, i)(m+i-1)((x))^i$ and the coefficient of $((x))^n$ on the right-hand side is $L'(m - 1, n - 1) - L'(m - 1, n)(m + n - 1)$.

Section 2.4, pages 115–118

1. $P(7,3) = 4$, $P(7,3) + P(7,2) + P(7,1) = 8$

3. Integer being partitioned is **(a)** 34; see Figure A.2.

Figure A.2   **Figure A.3**   **Figure A.4**

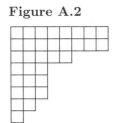   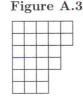

(b) 24; see Figure A.3.   **(c)** 15; see Figure A.4.

5. In type vector notation **(a)** $(0, 3, 2, 0, 0, 1, 1, 0, 1)$. See Figure A.5.

Figure A.5   **Figure A.6**   **Figure A.7**

(b) $(0, 1, 0, 1, 0, 3)$. See Figure A.6.

(c) $(0, 2, 2, 0, 1)$. See Figure A.7.

7. The conjugate of a partition of m into n parts is a partition with largest part n and vice versa. This gives a one-to-one correspondence between the two sets of partitions.

9. $P(20, 10) = \sum_{i=1}^{10} P(10, i) = 42$

11. If n is even, the smallest part can be any number from 1 to $n/2$. If n is odd, the smallest part may be any number up to $(n-1)/2$. In both cases the smallest part determines the partition and the number of smallest parts is the number of partitions; in each case this gives the desired number.

13. The number of partitions of m with smallest part 1 is $P(m-1)$ by Theorem 4.4; therefore, $P(m) - P(m-1)$ is the number of partitions whose smallest part is not 1.

15. The number of ways of listing the parts of a partition of m into n parts is no more than $n!$; thus $n!P(m, n)$ is at least as big as the number of lists of n positive integers that add to m. This is the number of lists of n nonnegative integers that add to $m - n$ (subtract 1 from each entry in the list). By Theorem 2.5 (or Exercise 14a of Section 2 of Chapter 2) the number of such lists is $C(m-1; n-1)$.

17. Subtract one from each part of a partition of $2m$ into m parts and you will get a partition of m. Add one to each part of a partition of m with n parts and create $m - n$ more parts all of size 1 and you get a partition of $2m$ into m parts. The two mappings we have described are inverses of each other, and so we have a bijection between the two sets of partitions; therefore there are equal numbers of partitions of each type.

19. Given the decreasing list $n_1, n_2, \ldots, n_j$, write n_i as $m_i 2^{a_i}$ where m_i is odd. (Note that different n_i's may have the same m_i.) The sum $2^{a_1} m_1 + 2^{a_2} m_2 + \cdots + 2^{a_j} m_j$ may be regarded as a partition whose parts are m_1 used 2^{a_1} times, m_2 used 2^{a_2} times, and so on. (Thus, the number of times a certain m_i is used is the sum of all the 2^{a_k} with $m_k = m_i$.) This is a partition of n into odd parts. To reverse the process, start with the type vector $(k_1, k_2, \ldots, k_n)$, with $k_r = 0$ if r is even. Write k_r in its binary decomposition as a sum of powers of 2. Then r becomes one m_i and the 2^i's appearing with it are the powers of 2 appearing in the binary decomposition of k_r. Construct a new partition whose parts are r times the powers of 2 appearing in the binary decomposition of k_r. The partition that arises in this way has unequal parts, for each multiple of an odd number by a power of 2 appears at most once. This last correspondence is bijective because the first correspondence reverses it. (Note, however, that the first correspondence is defined but not bijective on the set of all partitions because it sends the partition with decreasing list (1,1) and the partition with decreasing list (2) to the same partition, the one with two parts of size 1.

21. $\binom{m-1}{n-1}$ is the number of ordered lists of n positive numbers that add to m (think of distributing apples to children so each gets one). This includes all lists of sets of n distinct numbers adding to m, each set occurring in $n!$ orders. Thus dividing by $n!$ gives a number at least as big as the number of partitions of m into n distinct parts.

23. Each row of a Ferrers diagram of a partition of a number into distinct parts will be a different length. Thus each column of the Ferrers diagram of the conjugate will be a different length. Therefore, in the conjugate, as we move down the diagram from row to row, the rows either stay the same length or become one square shorter. Thus the number of partitions of m into distinct parts equals the number of partitions of m whose parts form a multiset of consecutive integers.

25. To expand on the hint, we consider the ordered pairs of the form (π, i) where π is a partition of n and i is an integer between 1 and $a(\pi)$. These ordered pairs form a set A. The number of elements of A is the number of ones among all partitions of n. We consider also a set B of ordered pairs (π, j) where π is a partition of n and j is an integer between 1 and $b(\pi)$. The size of A is the number $A(n)$, and the size of B is $B(n)$, the sum of $b(\pi)$ over all partitions of n. The hint says that the set A and the set B are each in a one-to-one correspondence with the set of all partitions of numbers less than or equal to n. To establish the bijection between this set and A, remove all but i parts of size one from π to get the partition corresponding to the ordered pair (π, i). Given a partition π' of an integer k less than or equal to n, let the partition π be the partition whose parts are those of π' together with $n - k$ additional ones and let i be the number of parts of π' of size 1. The correspondence with B involves subtracting one from each of parts of the j largest sizes that appear in π.

CHAPTER 3

Section 3.1, pages 134–137

1. $3^5 - \binom{3}{1}2^5 + \binom{3}{2}1^5 = 150$

3. Number of functions from a 12-element set onto a 6-element set $= 6^{12} - \binom{6}{1}5^{12} + \binom{6}{2}4^{12} - \binom{6}{3}3^{12} + \binom{6}{4}2^{12} - \binom{6}{5}1^{12} = 953,029,440.$

5. $8 - 4 - 5 + 3 = 2$; $\mathbf{N}_=(\{PB\}) + \mathbf{N}_=\{PB, PS\}) = 2 + 5 - 3 = 4$

7. $\mathbf{N}_=(\{N\}) + \mathbf{N}_=(\{P\}) + \mathbf{N}_=(\{K\}) = 6$ **9.** 8

11. The number of plots needed for the experiment would be negative (-4).

13. Let property i be "This element of $A_1 \cup A_2 \cup \cdots \cup A_n$ is a member of A_i." Then use Corollary 1.4. Alternatively, by inclusion–exclusion, the alternating sum of the number of elements with each set of properties is zero. The number of elements with properties $\{i_1, i_2, \ldots, i_k\}$ is the size of $A_{i_1} \cap A_{i_2} \cap \cdots \cap A_{i_k}$. Substitute and solve the resulting equation for $|A_1 \cup \cdots \cup A_n|$.

15. This is done in the same way as Example 1.7 with the pairs of beads playing the roles of the married couples. However, we now arrange beads whose diameter is two units (representing a pair of identically shaped beads) or one unit on a necklace so the term $(2n - i - 1)!$ is replaced by $(2n - i - 1)!/2$. Thus there are $\sum_{i=0}^{n} (-1)^i C(n; i) 2^{i-1} (2n - i - 1)!$ necklaces.

17. (a) Let property j be "person j receives the correct hat." Then for an i-element set I, $N_{\geq}(I) = (n - i)!$. (To people outside I, pass out the hats at random.) Then $N_{\geq}(\emptyset) = \sum_{i=0}^{n} (-1)^i \binom{n}{i} (n - i)!$ where the $\binom{n}{i}$ arises from collecting equal terms $N_{\leq}(I)$.

(b) Fraction is $\sum_{i=0}^{n} (-1)^i \frac{1}{i!}$.

(c) The limit of this sum is $e^{-1} = \frac{1}{e}$, or a bit more than a third.

19. Let property j be "child j gets more than four pieces." Note $N_{\geq}(I) = \binom{7-4i}{2}$. Thus $N_{=}(\emptyset) = 21 - 3 \cdot 3 + 0 - 0 = 12$.

21. Let property j be "shelf j gets m or more books." To compute $N_{\geq}(I)$, argue as follows: first choose mi books, line them up, put the first m on the first shelf in I, and so on. Now arrange the remaining $n - mi$ books however you want; there are $\frac{(n-mi+k-1)!}{(k-1)!}$ ways to do this. Thus $N_{\geq}(I) = \binom{n}{mi}(mi)!\frac{(n - mi + k - 1)!}{(k - 1)!}$.

Therefore $N_{\geq}(\emptyset) = \sum_{i=0}^{n} (-1)^i \binom{n}{i}\binom{n}{mi}(mi)!\frac{(n - mi + k - 1)!}{(k - 1)!}$.

23. Given a set I of people who are to return to their seats, the number of ways for them to return to their own chairs and for everyone else to take some seat is $N_{\geq}(I) = (m - i)_{n-i} = (m - i)!/(m - n)!$ Now substitute into the inclusion–exclusion formula.

25. $N_{\geq}(\emptyset) = (n - 1)!(n)!$ and if $I \neq \emptyset$, placing the first couple (which can be done in $2 \cdot 2n/2n = 2$ ways) leaves $i - 1$ "tied together" couples, $n - i$ men and $n - i$ women to be seated. First choose the places for the "tied together" couples in $\binom{2n-i-1}{i-1}$ ways and assign them to these places in $(i - 1)!$ ways. Now there are $n - i$ places left to be filled with men in $(n - i)!$ ways and $n - i$ places left to be filled with women in $(n - i)!$ ways. Thus we have $2 \cdot \binom{2n-i-1}{i-1}(i-1)!(n-i)!^2 = \frac{(2n-i-1)!}{(2n-2i-1)!}(n-i)!(n-i-1)!$ arrangements in which the couples in I sit together; note if $i = 0$ this agrees with our observation for $N_{\geq}(\emptyset)$. Thus $N_{=}(0) = \sum_{i=0}^{n} (-1)^i \binom{n}{i}\frac{(2n - i - 1)!}{(2n - 2i - 1)!}(n - i)!(n - i - 1)! =$

$$\sum_{i=0}^{n}(-1)^i n!\binom{2n-i-1}{i}(n-i-1)!.$$

27. We let property j be "person j has the same person to the right in both seatings." If person a is to the right of person b and person b is to the right of person c, then a, b, and c are in a row. If we have a set I of people who should sit with the same person to the right both times, we call a row of people that sit together in both arrangements a *run* of I if we cannot increase it by adding another member of I to it. If I has i elements and r runs, then I tells us how $i+r$ people must sit; namely the people in I and the people just to the right of a run in the first arrangement. The other $u = n-i-r$ people are unpaired and may sit wherever they choose. To compute $N_{\geq}(I)$, we seat the r runs (along with the people to their immediate right) and the u unpaired people in a circle in $(r+u-1)!$ ways. Since $i+r+u = n$, $r+u = n-i$, so $(r+u-1)! = (n-i-1)!$, which is $N_{\geq}(I)$. Thus the number of ways of seating people so that no one has the same right-hand neighbor is $\displaystyle\sum_{i=0}^{n}(-1)^i\binom{n}{i}(n-i-1)!.$

Section 3.2, pages 149–153

1. $B + B^2 + B^3$ 3. $(x+x^2+x^3)^5$

5. Substitute x^{20} for B and A, x^{30} for P and T, and x^{25} for N. $(x^{20}+x^{40}+x^{60})^2(x^{30}+x^{60}+x^{90})^2(x^{25}+x^{50}+x^{75})$. No fruit selection costs a dollar. Four fruit selections cost \$2.50.

7. $(P^0+P+P^2+P^3)(C^0+C+C^2+C^3)(N^0+N+N^2+N^3)(T^0+T+T^2+T^3)$. The polynomial is $(1+x+x^2+x^3)^4$. The coefficient of x^7 is 40.

9. $\left(\sum_{i=0}^{4}A^i\right)\left(\sum_{i=0}^{4}B^i\right)\left(\sum_{i=0}^{4}C^i\right)\left(\sum_{i=0}^{4}D^i\right)\left(\sum_{i=0}^{4}E^i\right)\left(\sum_{i=0}^{4}F^i\right)\left(\sum_{i=0}^{4}G^i\right)\left(\sum_{i=0}^{4}H^i\right)$. Substitute x for A, B, C, D, x^2 for E, F, G, and x^4 for F. This gives the generating function $(1+x+x^2+x^3+x^4)^4(1+x^2+x^4+x^6+x^8)^3(1+x^4+x^8+x^{12}+x^{16})$.

11. $\sum_{i=0}^{\infty}A^{2i}\sum_{i=0}^{\infty}T^i = \sum_{i=0}^{\infty}x^{2i}\sum_{i=0}^{\infty}x^i = \frac{1}{1-x^2}\frac{1}{1-x}$

13. The product is 1.

15. Because the typical term $F_1^{i_1}F_2^{i_2}\cdots F_n^{i_n}$ may be regarded as saying "Take i_1 pieces of fruit 1, take i_2 pieces of fruit 2, ..., take i_n pieces of fruit n." $(1+x+x^2+x^3)^n$

17. $(x+x^2+x^3+x^4+x^5+x^6)^n$

19. Let $f(x) = \sum_{i=0}^{\infty}a_i x^i$, $g(x) = \sum_{i=0}^{\infty}b_i x^i$, $h(x) = \sum_{i=0}^{\infty}c_i x^i$. Then $f(x)g(x) + h(x)g(x) = \sum_{i=0}^{\infty}(\sum_{j=0}^{i}a_j b_{i-j})x^i + \sum_{i=0}^{\infty}(\sum_{j=0}^{i}c_j b_{i-j})x^i = \sum_{i=0}^{\infty}(\sum_{j=0}^{i}(a_j b_{i-j}+c_j b_{i-j}))x^i = \sum_{i=0}^{\infty}(\sum_{j=0}^{i}(a_j+c_j)b_{i-j})x^i = (f(x)+g(x))h(x)$

21. $C(9+7-1;7) = \binom{15}{7} = 6435$ 23. $C(n+6;7)$

25. $C(6+5-1;5) = C(10;5) = 252$ 27. $\binom{10}{8} = 45$

29. $(1+x)^4$. Each coin has zero or one head when it is tossed. The product gives the number of 4-tuples with a total of i heads. (The result can be factored

because Theorem 2.1 applies; each factor corresponds to each coin being heads or tails.)

31. $(x + x^2 + x^3 + x^4 + x^5 + x^6)^n$

33. $\frac{x^2}{1-x^2} \cdot \frac{x}{1-x^2} = \frac{x^3}{(1-x^2)^2}; x^3 \sum_{k=0}^{4}(k+1)x^{2k}$

35. $\frac{1}{(1-x)^5}; \frac{x^5}{(1-x)^5}; \binom{19}{15} = 3876; \binom{14}{10} = 1001$ **37.** $\frac{1}{1-x^2} \cdot \frac{1}{1-x}$

39. The generating function is $(x + x^2 + \cdots)^n = \left(\frac{x}{1-x}\right)^n = x^n \sum_{k=0}^{\infty} C(n+k-1, k)x^k = \sum_{k=0}^{\infty} C(n+k-1; k)x^{k+n}$. The coefficient of x^k is $C(k-1, k-n)$.

40. $a_i = f^{(i)}(0)/i!$ where $f^{(i)}$ stands for the ith derivative of f.

41. $\left(\frac{1}{1-x}\right)^{(r)} = (-1)^r r! \frac{1}{(1-x)^{r+1}}$. Thus $(-1)^r x^r \left(\frac{1}{1-x}\right)^{(r)} = \sum_{n=0}^{\infty}(n)_r x^n$.

43. The sum is $(1)_n$, which is 0 if $n > 1$.

45. Expand both sides of $(1+x)^m(1+x)^n = (1+x)^{m+n}$ and equate the coefficient of x^k.

47. Let $v_1, v_2, \ldots, v_n$ be numerical functions defined on sets S_i. Let a_{ji} be the number of objects z_j in S_j with $v_j(z_j) = i$. Then $\prod_{j=1}^{n}\sum_{i=0}^{\infty} a_{ji}x^i$ is the generating function for the sequence c_k, where c_k is the number of n-tuples $(z_1, z_2, \ldots, z_n) \in \times_{j=1}^{n} S_j$ with $\sum_{j=1}^{n} v_j(z_j) = k$. The proof is an inductive proof in which the inductive step is analogous to the proof of Theorem 2.1.

Section 3.3, pages 165–168

1. $\frac{1}{(1-x^{10})(1-x^{20})(1-x^{50})}$ **2.** $\frac{x^{40}}{(1-x^5)(1-x^{10})(1-x^{25})}$

3. The last row is given by $d_i = c_i + d_{i-50}$; thus $d_{50} = 11$ and $d_{100} = 40$.

4. We now have $a_i = 1$ for all $i \geq 0$, $b_i = a_i + b_{i-5}$, $c_i = b_i + c_{i-10}$, $d_i = c_i + d_{i-25}$ with $a_i = b_i = c_i = d_i = 0$ for $i < 0$. See Table A.1, first four rows.

5. We now have the recurrences of Exercise 4 as well as $e_i = d_i + e_{i-50}$ and $e_i = 0$ for $i < 0$. See last row of Table A2.1.

Table A.1

1	1	1	1	1	1	1	1	1	1	1	1	1	1	1	1	1	1	1	1	1
1	2	3	4	5	6	7	8	9	10	11	12	13	14	15	16	17	18	19	20	21
1	2	4	6	9	12	16	20	25	30	36	42	49	56	64	72	81	90	100	110	121
1	2	4	6	9	13	18	24	31	39	49	60	73	87	103	121	141	163	187	213	242
1	2	4	6	9	13	18	24	31	39	50	62	77	93	112	134	159	187	218	252	292

7. $\prod_{i=1}^{\infty} \frac{1}{1-x^{2i}}$ **8.** $\prod_{i=1}^{\infty} \frac{1}{1-x^{2i}}$

9. The number of partitions of n into even parts is the number of partitions of n into parts each used an even number of times.

11. The result is $\prod_{i=1}^{\infty} \frac{1}{1-x^{2i-1}}$. This means that the number of partitions of n into parts each used zero or one times, i.e., distinct parts, equals the number of partitions of n into odd parts.

13. $\prod_{i=1}^{\infty}(1 + x^i + x^{2i} + \cdots + x^{i^2})$

15. $g(t,x) = \prod_{i=1}^{\infty}\sum_{j=0}^{\infty} t^j x^{ij}$. A typical monomial in this product is a monomial of the form $t^{j_1} x^{1 \cdot j_1} t^{j_2} x^{2 \cdot j_2} \cdots t^{j_k} x^{k \cdot j_k}$. This represents taking j_1 ones, j_2 twos, up through j_k parts equal to k. (Note that some of the j_i's could be zero for $i < k$; further all j_i's are zero for $i > k$.) Thus the total number of parts we take is the sum of the j_i's; this is the exponent of t, and the size of the number we are partitioning is the sum of the $k j_k$'s; this sum is the exponent of x.

17. On the basis of Exercise 16, the coefficient of x^{2n} in $\frac{x^n}{(1-x)(1-x^2)\cdots(1-x^n)}$ is the number of partitions of $2n$ with n parts. However, this is also the coefficient of x^n in $\frac{1}{(1-x)(1-x^2)\cdots(1-x^n)}$. However, this coefficient is also the number of partitions of n into any number of parts by Theorem 3.2.

19. (a) The exponent on t counts the number of rows of the Ferrers diagram of a partition corresponding to a $t^j x^k$ term. The exponent on x is the total number of squares. Nothing corresponds to the number of columns.

(b) k will be the maximum possible part size, so it will be the maximum possible number of columns. However, we might choose the term 1 from the factor $\frac{1}{1-x^k}$ of the product; this corresponds to taking no parts of size k. If we replace $\frac{1}{1-x^k}$ by $\frac{x^k}{1-x^k}$, this will give us the generating function for the number of partitions of an integer with parts of size no more than k and at least one part of size k. This corresponds to multiplying the original generating function by x^k; once we do this, k is the number of columns of the Ferrers diagram.

21. Since $p(x)\sum_{i=0}^{\infty} a_i x^i = \sum_{i=0}^{\infty} b_i x^i$, multiplying out the term on the left and equating coefficients gives $b_n = \sum_{i=0}^{k} c_i a_{n-i}$. Solving for a_n gives the desired equation.

23. The coefficient of x^{100} in $\left(\sum_{i=0}^{10} x^{5i}\right)\left(\sum_{j=0}^{5} x^{10j}\right)\left(\sum_{k=0}^{4} x^{25k}\right)$ is 14.

25. The generating function is $(1 + x + x^2 + \cdots + x^m)^n$. However, $(1 + x + x^2 + \cdots + x^m) = (1 - x^{m+1})/(1 - x)$.

27. $\prod_{n=1}^{k}\sum_{i=j}^{m} x^{ni} = x^{jk(k+1)/2}\prod_{n=1}^{k}\frac{1-x^{n(m-j+1)}}{1-x^n}$

28. $N_{\geq}^{+}(i) = \binom{n}{i}(n-i)!$; $\sum_{i=0}^{\infty} N_{=}^{+}(i)x^i = \sum_{i=0}^{\infty} N_{\geq}^{+}(i)(x-1)^i = f(x)$. $f(0) = \sum_{i=0}^{n}(-1)^i\binom{n}{i}(n-i)!$

29. $N_{\geq}^{+}(i) = \binom{n}{i}\binom{n+k-mi-1}{n-1}$. $f(0) = \sum_{i=0}^{\infty}(-1)^i\binom{n}{i}\binom{n+k-mi-1}{n-1}$.

31. In Exercise 28, replace the $(n-i)!$ by n^{n-i}.

33. Apply Theorem 3.4 to the level sum $\binom{n}{i}(n-i)^m$ you get by analyzing Theorem 1.2.

Section 3.4, pages 180–183

1. (a) $\frac{1}{1-\frac{1}{3}x} = \frac{3}{3-x}$; $\frac{3}{3+x}$ (b) $a_i = \frac{1}{3}a_{i-1}$; $a_i = -\frac{1}{3}a_{i-1}$; in both cases $a_0 = 1$.

3. The condition on the a_i's makes $a_1 - 2a_0$ cancel out, so we get $\sum_{n=0}^{\infty} a_n x^n = \frac{1}{(1-x)^2} = \sum_{k=0}^{\infty}(k+1)x^k$, and $a_n = n+1$.

5. $\sum_{n=0}^{\infty} a_n x^n = \frac{a_0 + a_1 x - 4a_0 x}{(1-2x)^2} = a_0 + a_1 x - 4a_0 x \sum_{n=0}^{\infty} (n+1)2^n x^n$. Thus we have $a_n = (2-n)2^{n-1}$; $a_n = 2^n$; $a_n = (2+n)2^{n-1}$.

7. $a_{n+2} = a_{n+1} + 3a_n$; $a_n = (5 + \frac{5}{\sqrt{13}})(\frac{1+\sqrt{13}}{2})^n + (5 - \frac{5}{\sqrt{13}})(\frac{1-\sqrt{13}}{2})^n$.

9. $a_n = (c+1)a_{n-1} - ca_{n-2}$; If $c \neq 1$ then $a_n = \frac{1}{c-1}c^n + \frac{c-2}{c-1}$. In this case $c = 2$, so that $a_n = 2^n$.

11. $\sum_{n=0}^{\infty} a_n x^n = \frac{a_0}{1-bx} + \frac{dx}{(1-bx)(1-x)}$; $a_n = (a_0 - \frac{d}{1-b})b^n + \frac{d}{1-b}$.

13. $\sum_{n=0}^{\infty} a_{n+2} x^{n+2} + b \sum_{n=0}^{\infty} a_{n+1} x^{n+2} + c \sum_{n=0}^{\infty} a_n x^{n+2} = 0$; $\sum_{i=0}^{\infty} a_i x^i - a_0 - a_1 x + bx(\sum_{j=0}^{\infty} a_j x^j - a_0) + cx^2 \sum_{n=0}^{\infty} a_n x^{n+2} = 0$. Now replace all three dummy variables by the same thing and solve for the power series.

15. $\sum_{n=0}^{\infty} a_{n+2} x^{n+2} - 2r \sum_{n=0}^{\infty} a_{n+1} x^{n+2} + r^2 \sum_{n=0}^{\infty} a_n x^{n+2} = 0$; $\sum_{i=0}^{\infty} a_i x^i - a_0 - a_1 x - 2rx(\sum_{j=0}^{\infty} a_j x^j - a_0) + r^2 x^2 \sum_{n=0}^{\infty} a_n x^{n+2} = 0$. This simplifies to $(1 - 2rx + r^2 x^2) \sum_{n=0}^{\infty} a_n x^n = a_0 + (a_1 - 2ra_0)x$. Solving for the generating function and determining the coefficient of x^n gives $a_n = a_0 r^n + \frac{a_1 - a_0 r}{r} n r^n$.

17. The idea is to move the top $n-1$ rings to the post not chosen if $n > 1$ and then move the bottom ring to the desired post. After moving the bottom ring, put the $n-1$ top rings on the desired post. If $n = 1$, just move the ring to the desired post. The remainder of the programming consists of deciding how the output is to look.

19. The generating function for the solution is $\sum_{n=0}^{\infty} a_n x^n = \frac{2-4x}{(1-2x)^2} + \frac{x^2}{(1-2x)^3}$. This gives $a_n = 2^{n+1} + (n^2 - n)2^{n-3}$.

21. $\frac{a_0 + (a_1 - a_0)x}{(1-2x)(1+x)} + \frac{x^2}{(1-2x)^2(1+x)} = \frac{2+2x}{(1-2x)(1+x)} + \frac{x^2}{(1-2x)^2(1+x)} = \frac{2}{1-2x} + \frac{x^2}{(1-2x)^2(1+x)}$

22. Let $b_k = a_{2^k}$. Then $b_{k+1} = 2b_k + 2^{k+1} - 1$; $b_0 = 0$. For $k > 0$ we can prove by induction that $b_k = k \cdot 2^k - 1$, so $a_n = n \log_2(n) - 1$, for $n \geq 2$.

23. We will write a recurrence for a_{n+1} by considering an $(n+3)$-gon. Label the vertices cyclically as 0 through $n+2$, and call the edge from 0 to $n+2$ the base. Any triangulation must have a triangle containing the base, and the third vertex of the triangle can be any number from 1 to $n+1$. If it is vertex 1 then the other triangles of the triangulation actually triangulate the polygon $n+2, 1, 2, \ldots, n$ (there are a_n such triangulations) and if it is vertex $n+1$, then the other triangles of the triangulation triangulate the polygon $0, 1, 2, \ldots, n+1$ (there are a_n such triangulations). If the third vertex of the triangle is any other vertex i of the polygon, then the triangulation splits into the triangle $n+2, 0, i$ and the triangulations of both the polygon $0, 1, \ldots, i$ and the polygon $n+2, i, i+1, \ldots, n+1$ (there are $a_{i-1}a_{n-i}$ such triangulations). Thus $a_{n+1} = 2a_n + \sum_{i=2}^{n} a_i a_{n-i-1}$, or $a_{n+1} = 2a_n + \sum_{i=1}^{n-1} a_i a_{n-i}$. Taking $a_0 = 1$, we may write $a_{n+1} = \sum_{i=0}^{n} a_i a_{n-i}$. (In particular this gives $a_1 = a_0 a_0 = 1$ so taking $a_0 = 1$ does not lead to an inconsistency between our definition of a_n for $n > 0$ and our recurrence.) Multiplying both sides of our recurrence by x^{n+1}, adding, and letting $y = \sum_{i=0}^{\infty} a_i x^i$, we get $y - 1 = xy^2$, because $\sum_{n=0}^{\infty} \sum_{i=0}^{\infty} a_i a_{n-i} x^{n+1} = x \sum_{i=0}^{\infty} a_i x^i \sum_{j=0}^{\infty} a_j x^j$. By the quadratic formula, $y = \frac{1 \pm \sqrt{1-4x}}{2x}$. To get a power series in x we must take the negative square root

and we get $y = \frac{-1}{2x}\sum_{i=1}^{\infty}\binom{1/2}{i}(-4x)^i = 2\sum_{j=0}^{\infty}\binom{1/2}{j+1}(-4x)^j$. The coefficient of x^n is $2\binom{1/2}{n+1}(-4)^n = \frac{(2n)!}{n!(n+1)!} = \frac{1}{n+1}\binom{2n}{n}$, which is the nth Catalan number.

25. Let $a_n = \sum_{k=0}^{n}\binom{n-k}{k}$. Substitute the Pascal relation into the formula this gives for a_{n+2}. Show that the result is the formula for a_{n+1} plus most of the formula for a_n plus another sum which adds to the remainder of the formula for a_n. Since $a_0 = a_1 = 1$, this is a formula for the nth Fibonacci number.

Section 3.5, pages 196–198

1. We have n books and one divider that we can arrange in $(n+1)!$ ways. If each shelf must have a book, we have $(n-1)$ places for the divider once the books are arranged, so the generating function is $\sum_{n=0}^{\infty}(n-1)n!x^n$.

3. Values of x between -1 and 1.   5. $\left(\sum_{n=0}^{10} x^n\right)^2$

7. Generating function is $x^n/(1-x)^n = x^n\sum_{i=0}^{\infty}C(n+i-1;i)x^i = \sum_{k=n}^{\infty}C(k-1,k-n)x^k$. Thus there are $k!C(k-1,k-n)$ arrangements of k books.

9. $\left(\sum_{n=1}^{10} x^n\right)^k$ or $\left(\frac{1-x^{11}}{1-x}\right)^k$

11. Using $f^{(k)}$ for the kth derivative of f, we get $a_k = f^{(k)}(0)$.

13. $\sum_{n=0}^{\infty}\frac{1}{n!}S(n,k)x^m = \frac{1}{k!}(e^x-1)^k$.

15. Now the generating function is $(e^x-x-1)^3 = e^{3x}-3e^{2x}(x+1)+3e^x(x+1)^2-(x+1)^3$. Now the x^{10} term is $\frac{(3x)^{10}}{10!} - 3\left\{\frac{(2x)^{10}}{10!} + x\frac{(2x)^9}{9!}\right\} + 3\left\{x^2\frac{x^8}{8!} + 2x\frac{x^9}{9!} + \frac{x^{10}}{10!}\right\}$.

Thus the number of distributions is $10!\left(\frac{3^{10}}{10!} - 3\frac{2^{10}}{10!} - 3\frac{2^9}{9!} + 3\frac{1^8}{8!} + 6\frac{1^9}{9!} + 3\frac{1^{10}}{10!}\right)$ which is $40,950$.

17. $(k)_n(k-1)_{k-n}$ 18. $(e^x-x-1)^k$

19. $\sum_{i=0}^{\infty}\frac{x^{2i}}{2i!}$ is the generating function for the number of ways of distributing an even number of objects to one person. The kth power of the hyperbolic cosine is the number of ways to distribute an even number of distinct objects to k people.

21. In general, the coefficient of x^n on the right-hand side will be a sum of terms of the form $\frac{b_{i_1}}{i_1!}\frac{b_{i_2}}{i_2!}\cdots\frac{b_{i_k}}{i_k!}$, where the sum ranges over all k-tuples of nonnegative integers whose sum is n. Thus, the coefficient of x^n is $\sum\binom{n}{i_1,i_2,\dots i_k}b_{i_1}b_{i_2}\cdots b_{i_k}$, where the sum ranges over the same k-tuples as before. This is the number of ways to first divide an n-element set up into parts of size $i_1, i_2, \dots, i_k$ and then select an element x of S_{i_h} with value J_h for each i_h. This is the number of lists of the desired type.

23. (a) $a_n = a_0(n!)^2 + \sum_{i=1}^{n-1}\left((n)_i\right)^2$ (b) $a_n = (a_0+n)n!$.

25. $a_n = (n!)^r a_0$.

27. Let $y = \sum_{n=0}^{\infty}B_n\frac{x^n}{n!}$. Then $\frac{dy}{dx} = \sum_{n=0}^{\infty}B_{n+1}\frac{x^n}{n!}$. Using the recurrence, we get $\sum_{n=0}^{\infty}B_{n+1}\frac{x^n}{n!} = \sum_{n=0}^{\infty}\sum_{i=0}^{n}\binom{n}{i}B_{n-i}\frac{x^n}{n!} = \sum_{n=0}^{\infty}\sum_{i=0}^{n}B_{n-i}\frac{x^{n-i}}{(n-i)!}\cdot\frac{x^i}{i!} = ye^x$. Therefore $\frac{dy}{dx} = ye^x$, so $\log y = e^x + c$ and $y = e^{e^x+c}$. To determine c, note that

the coefficient of x^0 is found by substituting $x = 0$ into $\sum_{n=0}^{\infty} B_n \frac{x^n}{n!}$, and that coefficient must be $B_0 = 1$. Therefore $e^{e^0+c} = 1$, which implies that $c = -1$. Therefore $\sum_{n=0}^{\infty} B_n \frac{x^n}{n!} = e^{e^x-1}$.

29. Subtract nD_{n-1} from both sides of the recursion for D_n developed in this section of the text. Note that the recurrence lets us substitute $-(D_{n-2} - (n-2)D_{n-3})$ for $D_{n-1} - (n-1)D_{n-2}$. Repeating this substitution gives $(-1)^{n-2}(D_2 - 2D_1) = (-1)^{n-2}$, which must equal the left-hand side of the original recurrence.

31. In a list, either the element to the right of n is not one more than the element to the left (including the possibility that there is either no element to the right or no element to the left); in this case deleting n gives a length $n-1$ list of the type we want, and each such list arises from $n-1$ deletions (we could put n in any place except right after $n-1$). If the element to the right of n is one more than the element to the left, then deleting n and the element to the left of it and subtracting 1 from everything to the right of n gives a length $n-2$ list of the type we want. Each such list arises from $n-2$ such deletions, because when we reinsert n we must do it with at least one element to its right. Therefore $R_n = (n-1)R_{n-1} + (n-2)R_{n-2}$. We can prove that $R_n = D_n + D_{n-1}$. If $R(x)$ is the generating function, then $R'(x) = e^{-x}/(1-x^2)$ and $R(0) = 0$ so $R(x) = \int_0^x \frac{e^{-t}}{(1-t)^2} dt$.

CHAPTER 4

Section 4.1, pages 208–210

1. See Figure A.8   3. 6

Figure A.8

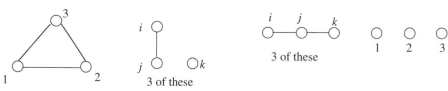

5. There is a way to give each vertex even degree by removing two edges (say $\{a,b\}$ and $\{c,d\}$) so the answer is yes. It does matter (remove both $\{c,d\}$'s). If you don't have to start and end at the same place, removing $\{b,d\}$ and $\{c,d\}$ will give an affirmative answer.

7. $7, 6, 5, 3, 1, 2, 3, 4, 2, 6, 4, 5$.

9. The sum of the degrees is twice the number of edges, for each edge contributes a one to the sum twice.

11. This is clearly true for graphs with one or two vertices. Suppose inductively it is true for graphs on $n-1$ vertices and let G have n vertices. Suppose the vertices x and y are connected by a walk $x = x_0, \{x_0, x_1\}, \ldots, \{x_{m-1}x_m\}, x_m = y$ of shortest length. If x_0 and x_{m-1} are disconnected when we delete y from G,

then y appears earlier in the walk, so it is not of shortest length. But taking y from G leaves a graph with $n-1$ vertices in which x and x_{m-1} may be connected by a walk with at most $n-2$ edges, so x and y may be connected by a walk with at most $n-1$ edges. This is true for multigraphs as well.

13. Let m be the number of edges and n be the number of vertices. Clearly, $m \geq n-1$ if G is connected and m is 0 or 1, so assume the result inductively for graphs with fewer than m edges. Remove an edge. If G is connected, then $m-1 \geq n-1$ by assumption. Otherwise the resulting graph has two connected components with n_1 vertices and m_1 edges and with n_2 vertices and m_2 edges. Note $m_1 < m$ and $m_2 < m$. Thus by induction, $m_1 \geq n_1 - 1$, $m_2 \geq n_2 - 1$ and $m = m_1 + m_2 + 1 \geq n_1 - 1 + n_2 - 1 + 1 = n_1 + n_2 - 1 = n - 1$.

15. If the vertices are not connected, then G has two components with only one vertex of odd degree and by Exercise 9 this is impossible. (Alternate proof: Start a path at one vertex of odd degree. Keep going until you must stop. When this happens, you will be at the other vertex of odd degree.)

17. $C(n-1;2)$, because this is the maximum number of edges we could get by drawing edges between all vertices connected by a path in a disconnected graph.

19. **(a)** Yes (any sequence of vertices will do). In fact, it has a closed Hamiltonian walk. **(b)** We prove there is a closed Hamiltonian walk. Among all graphs on n vertices satisfying the conditions, assume there are some without closed Hamiltonian walks and choose one with the maximum possible number of edges. This graph is not complete. So there are two nonadjacent vertices x and y. Connecting them with an edge gives a graph with more edges, so it has a closed Hamiltonian walk. Thus x and y are connected in G by a Hamiltonian path $x = v_0 v_1 \ldots v_{n-1} v_n = y$. If, for some i, $\{x, v_i\}$ and $\{y, v_{i-1}\}$ are edges, then the walk $x v_i v_{i+1} \cdots v_{n-1} v_n v_{i-1} \cdots v_1 x$ is a Hamiltonian cycle. Thus for each v_i adjacent to x, v_{i-1} is not adjacent to y. Thus if x is adjacent to k vertices, v_y is not adjacent to k vertices among the $n-1$ not equal to it. Thus the sum of the degrees of x and y is at most $n-1$, a contradiction. **(c)** Consider the multigraph on $\{1, 2, 3, 4\}$ in which the edges $\{1, 2\}$, $\{1, 3\}$, and $\{1, 4\}$ each have multiplicity 2. This is a counterexample.

21. Draw two pentagons labeled 1,2,3,4,5 cyclically. Connect the vertices labeled 5 with an edge. Now inside each pentagon draw an edge between vertices 1 and 3 and vertices 2 and 4. This is a counterexample since once a walk leaves a pentagon it cannot return without using a vertex labeled 5 a second time.

23. $2^{C(n;2)}$

25. To have a maximum number of edges they should all be in one component so the maximum is $\binom{n-k+1}{2}$. To have the minimum a component with k vertices should have $k-1$ edges, so the minimum is $n-k$.

Section 4.2, pages 222–225

1. 2. The graphs are a four-vertex path and a vertex of degree three surrounded by three vertices of degree one.

3. You have either a path of length 3 and an isolated vertex or, if you don't have a path of length 3, you have two pairs of vertices, each pair connected by an edge. With three connectivity classes, there is only one possibility: an edge and two isolated vertices.

5. If a class has n vertices, it has $n-1$ edges. Thus $n_1-1+n_2-1+n_3-1+\cdots = 4$ and $n_1 + n_2 + \cdots = 7$. There must be three negative ones in the first sum for this to work out, so there are three classes.

Figure A.9

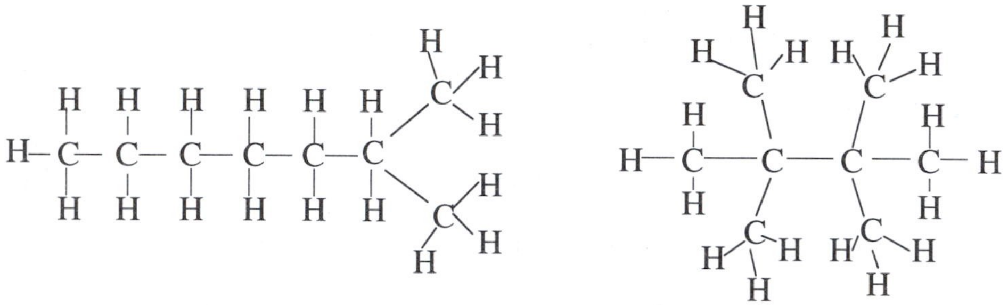

7. See Figure A.9. The smallest value is four C's.

9. We must show it is connected. Each connected component is a tree, for it has no cycles. If component i has n_i vertices, then it has $n_i - 1$ edges. Thus $n_1 - 1 + n_2 - 1 + n_3 - 1 + \cdots = n - 1$, for the sum is the total number of edges. Thus there can be only one -1 in the sum, so there is only one connected component.

11. The tree is centered at vertex 6 and has edges from 6 to 1, 2, 3, 4, and 5. One with the longest path is the path 6 1 2 3 4 5.

13. The edge e connects two vertices in G already connected by a path in T_2. Let e' be any edge of that path. Then replacing e' by e in the edge set of T_2 gives a connected graph with n vertices and $n - 1$ edges; this is the desired spanning tree T_3.

15. $v - e$. Use the fact that the number of vertices minus the number of edges is the number of connected components, a fact that you can derive from applying Theorem 2.2 to each component.

17. Use a proof by induction on the number of times step 2 is used. For the inductive step you need to note that you are connecting a new vertex to one in an already connected graph, and since the new edge has exactly one endpoint in common with edges already there, adding that edge cannot give a cycle.

19. When you apply Spantree to a tree, you get the original tree. (Why?) The last vertex added by Spantree has degree 1. (Why?) When you remove a vertex of degree one from a tree you get a tree (why?), and either this tree has one vertex or you may apply the inductive hypothesis to it. In the second case you have two vertices of degree one by the inductive hypothesis; adding the new vertex increases the degree of at most one vertex but gives you a new degree one vertex. The other case is straightforward.

21. Assume G is a tree. If a closed walk has no edge used twice, then the portion from the first repeated vertex back to its previous appearance in the walk is a

cycle. This contradicts the assumption that G is a tree, so (a) implies (b). Now assume (b) and let $x_1 e_1 x_2 e_2 \cdots e_n x_0$ be a shortest closed walk in which some edge e_i is not used twice. If the edge $e_j = \{x_{j-1}, x_j\}$ is used twice, shorten the path by deleting all edges and vertices beginning with the first appearance of e_j and ending right before the last. This shortens the path, therefore no edge is used twice, contradicting (b). Thus (b) implies (c). Now (c) implies G has no cycles, so (c) implies (a). Thus all three statements are equivalent.

23. Show that 1 and 2 imply 1 and 3, which imply 2 and 3, which imply 1 and 2 and you will show that all three unordered pairs of statements are equivalent. 1 and 2 imply 1 and 3 because removing one edge of the unique cycle gives a tree. To see that 1 and 3 imply 2 and 3 note that a spanning tree would have all but one of the edges. Now adding that edge back in can create exactly one cycle (why?). To show that 2 and 3 imply 1 and 2, argue that if there is more than one connected component, then either two connected components have more vertices than edges (and hence a cycle) or one component has at least two more vertices than edges (and hence two or more cycles.)

Section 4.3, pages 238–241

1. It starts out at a point spreading out as a path centered at that point. If n is odd, it has $(n-1)/2$ edges on each side of the starting point. If n is even, the starting side of the path has $n/2$ edges; the other side has $(n-2)/2$ edges.

3. Step 1 is carried out once. Step 4 is carried out $n-1$ times. Step 5 will be encountered after each use of step 4 and step 6 will be encountered once. (Thus essentially all the work of the program is in step 4, and the total amount of time taken will depend on the way the algorithm chooses the smallest j.)

5. $V_0 = \{x\}$, $E_0 = \emptyset$; $V_1 = \{x, a\}$, $E_1 = \{\{x, a\}\}$; $V_2 = \{x, a, d\}$, $E_2 = \{\{x, a\}, \{a, d\}\}$; $V_3 = \{x, a, d, b\}$, $E_3 = \{\{x, a\}, \{a, d\}, \{a, b\}\}$; $V_4 = \{x, a, d, b, e\}$, $E_4 = \{\{x, a\}, \{a, d\}, \{a, b\}, \{d, e\}\}$; $V_5 = \{x, a, d, b, e, c\}$, $E_5 = \{\{x, a\}, \{a, d\}, \{a, b\}, \{d, e\}, \{a, c\}\}$; the last edge added could have been $\{d, c\}$.

7. In step 4, choose the smallest j such that x_j is adjacent to a vertex in V_i other than its parent, or if no such vertex exists proceed with step 4 normally. (You must repeat this process from each x; from some x's you will find closed walks, but the shortest one will be a cycle.) Other answers are possible.

9. To find which vertex to add next, examine each vertex y not in the tree and choose the one for which $D(L(y)) + w(x, y)$ is a minimum. The number of steps required for this is the number of vertices not in V. If we had had to consider all vertices in the tree adjacent to each such y, the number of steps would have been on the order of n^2 for this one stage. To compute the new values of $L(v)$ when we add a vertex y to our tree, we check each vertex to see if this new y should be its $L(v)$, requiring one step each. Thus there are n stages, each requiring on the order of n steps, so the total number of steps is on the order of n^2 for the entire process. (It would have been on the order of n^3 if we hadn't introduced the idea of $L(v)$.)

11. This program can use the same breadth-first search routine as the previous one. Here, however, it is useful to record breadth-first numbers as you build a tree starting at a vertex. After building the tree centered at x, work back from each vertex y along tree edges to smaller breadth-first numbers until you reach x. The number of steps you take is the distance from y to x.

12. Many answers are possible depending on which direction the person starts out from x. One numbering is shown in Figure A.10.

Figure A.10

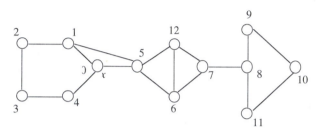

13. $L(0) = 0$, $L(1) = 0$, $L(2) = 0$, $L(3) = 0$, $L(4) = 0$, $L(5) = 0$, $L(6) = 5$, $L(7) = 5$, $L(8) = 8$, $L(9) = 8$, $L(10) = 8$, $L(11) = 8$, $L(12) = 5$. (These values of L are for the depth-first numbering given in Figure A.10. Other depth-first numberings will give different least vertices.) The descendants of 8 (9, 10, and 11) are not adjacent to any vertex with depth-first number less than 8.

14. $L(x_0) = 0$, $L(x_1) = 0$ $L(x_3) = 0$, $L(x_4) = 0$, $L(x_5) = 0$, $L(x_6) = 5$, $L(x_7) = 5$, $L(x_8) = 8$, $L(x_9) = 8$, $L(x_{10}) = 8$, $L(x_{11}) = 8$, $L(x_{12}) = 5$, $L(x_{13}) = 0$, $L(x_{14}) = 0$.

15. You visit all squares in and to the right of the column containing x and all squares in the bottom row.

17. Answer depends on which hand you use on which wall.

19. If there is an edge from v_j to v_{j+k} in the graph, then when we add v_{j+k} it will be added from v_j or a vertex with higher depth-first number $h < j + k$. Since all descendants of v_j will be examined in step 2 before any nondescendants are, this gives the inductive step of an inductive proof.

21. The decision graph has 500 vertices and follows the pattern of the figure directly, so it will not be reproduced here.

23. Suppose every walk from x to y includes $\{x, y\}$. Then removing $\{x, y\}$ disconnects G. Conversely, suppose $\{x, y\}$ is a bridge. If there were a walk from x to y not including $\{x, y\}$, then G would be connected after we remove $\{x, y\}$. Thus $\{x, y\}$ would not be a bridge, a contradiction.

25. **(a)** There can be no more than two adjacent central vertices. If there were three or more central vertices on a path, since some two must be connected by a path of length at least 2, one can be shown not to be central. There will be at least one central vertex on a maximal path.
(b) If the diameter is twice the radius, a maximal path has a single central vertex v_c. Any other maximal path must intersect the former maximal path at v_c, otherwise one of the paths can be shown not to be maximal. If the

diameter is not twice the radius, then there are at least two points with the same (minimal) eccentricity and hence no unique central vertex.

Section 4.4, pages 249–251

1. Each graph is isomorphic to itself so the relation is reflexive. If G is isomorphic to H by f, then H is isomorphic to G by f^{-1}. Thus the relation is symmetric. If G is isomorphic to H by f and H is isomorphic to K by g, then G is isomorphic to K by the composition $g \circ f$. Thus the relation is transitive, so it is an equivalence relation.

3. G_1 and G_2 are isomorphic, but G_3 is not isomorphic to them. (Note that adjacent degree four vertices in the first two graphs have a common degree three neighbor, but this is not the case with G_3. This does not show that G_1 and G_2 are isomorphic, but it shows that G_3 is not isomorphic with either G_1 or G_2.)

5. If $x_1 x_2 x_3 x_4$ is a cycle, then $f(x_1)f(x_2)f(x_3)f(x_4)$ is a cycle. No, four is not special; in fact the number of cycles of size k in G_1 and in G_2 must be the same for all k.

7. No, because $K_{3,3}$ is not planar.

9. If each face has at least four edges, the $4f \le 2e$ from the (edge,face) counting. Therefore, $8 + 4e - 4v \le 2e$, so $e \le 2v - 4$.

11. There are 10 nonisomorphic forests on five vertices. Three are trees. Two have two connected components. Two have three connected components. One has four connected components and one has five.

13. $n(n-1)/2$, $n!$, 2

15. Since $e \le 3v - 6$, the sum of the degrees, which is $2e$, is no more than $6v - 12$. If all vertices had degree 6 or more, this would be impossible.

17. If the graph is planar, each face has at least five edges, so that $5f \le 2e$, which leads to $3e \le 5v - 10$ for a planar graph. This equation is not satisfied by the Peterson graph.

19. No. Delete the top vertex, then convert the degree two vertices and their two edges each into one edge and you get a copy of $K_{3,3}$. Alternatively, note that the fact that there are no cycles of size five or less and the fact that each vertex has degree three implies $v \ge 20$, which is false.

21. To show the isomorphism, label the top vertex in Figure 4.7 with $\{1,2\}$ and from left to right label its neighbors with $\{3,4\}$, $\{3,5\}$, and $\{4,5\}$. Label the remaining neighbors of $\{3,5\}$ with $\{2,4\}$ and $\{1,4\}$ from left to right. Now there is one and only one way of extending this labeling to an isomorphism.

23. If we had a planar drawing of $K_{1,2,3}$, we could erase two edges and get a planar drawing of $K_{3,3}$, so the answer is no.

25. $I(0,2)$ is planar, $I(0,3)$ is not.

27. The complete graph on five vertices is regular of degree four and a five cycle is regular of degree two. A regular graph of degree three would have to have $15/2$ edges, so there is none.

29. For each vertex v in a graph, let S_v consist of all edges incident with v. Then $S_x \cap S_y$ is nonempty if and only if x and y are adjacent.

Section 4.5, pages 259–262

1. 2^9, 2^{100}. If loops are not allowed, then answers would be 2^6, 2^{90}. In all cases the reason is that the digraph is determined by a subset of the ordered pairs of distinct vertices.

3. $2^{C(n;2)}$ **5.** $n!$

7. The sum of the indegrees is the number of edges. So is the sum of the outdegrees.

9. (We can't simply take an Eulerian walk in the underlying graph because we might go the wrong way.) However, we mimic the corresponding proof for undirected graphs as follows. Start at a vertex x. Follow a strong walk until you reach a vertex with no outgoing edge you haven't used. If you started at a vertex with equal in- and outdegree, you will end there. Remove this strong walk and in each connected component of the remaining graph, each vertex has equal indegree and outdegree so that, inductively, you may assume these components have closed Eulerian walks and connect all the walks together to get a closed walk.

11. Suppose the tournament is transitive. Let x be a vertex with minimum possible indegree. If (y, x) is an edge, then (y, z) is also an edge whenever (x, z) is an edge. Thus the outdegree of y is one more than the outdegree of x. But the indegree plus the outdegree of a vertex is one less than the number of vertices. Thus the indegree of y is less than the indegree of x, so there is no such y. That is, x has indegree 0. Now remove x and the resulting tournament is transitive so, by induction, it has one and only one ranking. The one and only way to add x to the ranking is above all other vertices. Now assume the tournament has a ranking. Then the top ranked vertex x has arrows to all vertices. Remove it and you may assume the resulting tournament is transitive because it has a ranking. If (x, y) and (y, z) are in the original tournament, then (x, z) is in the original tournament, and since the transitive law also holds when x is not involved, the original tournament must be transitive. (We needn't check pairs involving (y, x) because there are none.)

13. Draw a digraph on $\{1, 2, \ldots, n\}$ with edge (i, j) if $p_j \le b_i$. Then a Hamiltonian path will ensure that the printing time of each book is no more than the binding time of the preceding book.

15. Since $D \subseteq T(D)$, (V, D) has no cycles if $(V, T(D))$ has no cycles. Conversely, suppose $(V, T(D))$ has a cycle $v_1(v_1, v_2)v_2, \ldots, v_k = v_1$. Then each v_{i+1} is strictly reachable from v_i, so there is a strong path from v_i to v_{i+1}. Putting all these paths together gives a strong closed walk which contains (or perhaps is itself) a directed cycle.

17. See Figure A.11. The orientation began with a depth-first search centered at vertex e.

19. A "complete directed graph," the digraph with all ordered pairs of vertices for edges.

21. 7, 46 (a loop is counted as a directed cycle).

Figure A.11

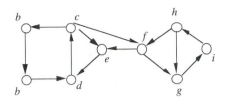

23. Start a strong walk at vertex x. When you get to a vertex y you haven't visited before, you can continue your walk unless the outdegree of y is 0. Since you cannot return to a vertex previously used (because this gives a directed cycle), you will reach a vertex of outdegree zero since the process must stop when (or before) you run out of vertices.

25. One, because we are allowed to repeat only the initial vertex as we search, and that only once. We will get this situation if the digraph we are searching has a directed cycle.

Section 4.6, pages 272–276

1. $2, 2$ 3. $2, 3$

5. 3 if the cycle has an even number of vertices; 4 if the cycle has an odd number of vertices.

7. The outside region of the graph is adjacent in the dual to each triangular region inside the wheel, and adjacent triangles give the adjacencies in the dual. Thus the dual is a wheel. Since the number of triangular regions in the original graph is the number of vertices on the cycle, the dual has the same number of vertices as the original. Therefore, they are isomorphic.

9. We know we have a vertex x of degree 5 or less. Delete x and inductively you may color the remainder of the graph with six colors. Only five of these colors are used on vertices previously adjacent to x, so we may reinsert x and color it with the sixth color.

11. Let v be the number of vertices, let i be the independence number. Let c be the chromatic number of G. Then coloring G with c colors gives c independent sets (the sets with the c colors) whose union is the vertex set of G. Each of these sets has size no more than i, so their union has no more than $c \cdot i$ elements. Thus $v \le c \cdot i$.

13. (a) $x(x-1)^{n-1}$ (b) $(x)_n$ (c) $x(x-1)^{n-1}$

15. (It is interesting to note that this follows from Dilworth's theorem in Chapter 7.) A more direct approach is the following. If one interval intersects all the rest, we remove it and apply induction. Otherwise, choose a real number x that lies to the right of one interval and to the left of another. Let L be the set of all intervals whose left-hand endpoint is to the left of x and let R be the set of all intervals whose right-hand endpoints lie to the right of x. The intersection of these two sets is the set of intervals containing x. No interval in R which is not in the intersection is adjacent to an interval in L which is not in the intersection. Thus a clique cannot contain both vertices of L not

in R and vertices of R not in L. Thus any clique of G lies entirely in R or entirely in L. Both these sets are smaller sets of intervals, so the theorem may be applied inductively to them. Now in L all the colors applied to intervals that include x are distinct, as they are in R, so we may permute the colors in L so that the intervals including x in both L and R have the same colors in each coloring. Thus the two colorings together form a proper coloring of the interval graph, because if an interval in L is adjacent to an interval in R, then they both contain the number x. Since both colorings have no more colors than the largest clique in their graph, the number of colors needed for G, which is the maximum of the number of colors used for R and the number of colors used for L, is the size of a largest clique in G.

17. Remove a vertex of degree one and apply induction.

19. $K_{3,3}$ is a counterexample.   **21.** 3 if $n \geq 4$.

23. Assume all vertices have degree 5 or more. By summing degrees $2e \geq 5v$ so $e \geq \frac{5}{2}v$. Thus we get $\frac{5}{2}v < 30$ since $e < 30$ which gives us $v < 12$. Now show that in a planar graph with every vertex having degree 5 or more, we have at least 12 vertices. We can turn all faces into triangles by adding edges but not vertices. Starting with a vertex of degree 5 or more, you may show by construction that at least 12 vertices are needed. Now prove the four-color theorem as the five-color theorem was proved in the text.

25. A wheel on an odd number of vertices is four-color critical. This gives infinitely many examples. Remove vertices so long as you can do so without reducing the chromatic number. The resuling graph is color critical.

27. Assign the colors as shown in the first display in Example 6.2.

29. $\emptyset$, $1R$, $1R2G$, $1R2G3R$, $1R2G3R4G$, $1R2G3R4G5B$, $1R2G3R4B$, $1R2G3R4B5G$

31. $\chi_{G-e}(x)$ is the number of colorings of G in which all vertices except perhaps the endpoints of e are properly colored. $\chi_{G/e}(x)$ is the number of colorings of G in which all vertices except the endpoints of e are properly colored and the endpoints of e are the same color. This proves the formula. We may assume by induction on the number of edges that χ_{G-e} and $\chi_G/e(x)$ are both polynomial functions of x, and the formula tells us that $\chi_G(x)$ is as well.

33. Delete a vertex v of minimum degree from G to get G'. We may assume inductively that the chromatic number of G' is the maximum of $1 + \delta(H)$ for any subgraph of G'. Now v is adjacent to at most $\delta(G)$ vertices, and so even if v is adjacent to $\delta(G)$ vertices all of different colors, all of G can be colored in the larger of $1 + \delta(G)$ and the maximum of $1 + \delta(H)$ (for all subgraphs H of G') colors, and this is no more than the maximum of $1 + \delta(H)$ over all subgraphs of G. The chromatic number of G' will equal the chromatic number of G unless $1 + \delta(G)$ is larger than the chromatic number of G'. Since $1 + \delta(G) \geq 1 + \delta(H)$ for any subgraph of H containing v, in either case the chromatic number of G will equal the maximum over all subgraphs H of G of $1 + \delta(H)$.

Section 4.7, pages 285–290

1. See Figure A.12   3. $\begin{pmatrix} 0 & 2 & 0 & 0 & 1 \\ 2 & 1 & 1 & 0 & 0 \\ 0 & 1 & 1 & 2 & 0 \\ 0 & 0 & 2 & 0 & 2 \\ 1 & 0 & 0 & 2 & 0 \end{pmatrix}$ 5. 34

Figure A.12

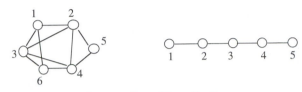

7. Two walks of length 3 and no walks of length 4.

9. In a directed graph, the trace is three times the number of directed triangles and D is transitive if and only if it has no triangles.

11. (a) $\begin{pmatrix} 0 & 0 & 0 & 1 & 1 & 0 \\ 1 & 0 & 1 & 0 & 0 & 0 \\ 1 & 0 & 0 & 1 & 1 & 0 \\ 0 & 1 & 1 & 0 & 0 & 0 \\ 0 & 1 & 0 & 0 & 0 & 1 \\ 0 & 0 & 1 & 0 & 0 & 0 \end{pmatrix}$ (b) 4   12. (a) $12 \cdot 2 \cdot 2 = 48$

    (b) $14 \cdot 6 = 84$

13. In Theorem 7.4 and 7.5 we need to compute the nth power of one matrix M. We can compute $M^2, M^4, \ldots, M^{2^k}$ in k multiplications. If $n = 2^k$, then $k = \log_2(n)$. Now $n \le 1 + 2 + 4 + \cdots + 2^k$, so m^n is a product of no more than $k+1$ of the matrices $M, M^2, \ldots, M^{2^k}$. But we can multiply $k+1$ matrices with k multiplications, so in general we need no more than $2\log_2(n)$ multiplications to compute M^n. For Theorem 7.3 we need to compute all the powers (without using a shortcut as in Theorem 7.4), and so we need one multiplication for each power larger than the first.

15. Because we can start and end an Eulerian walk at that vertex, so the "tree of last passage" must be directed into that vertex.

17. Not every orientation into a vertex i of every tree in an undirected graph gives the tree of last passage of an Eulerian walk.

19. Regard Warshall's algorithm as proceeding in stages where stage k is all the steps of the algorithm that occur when $j = k$. Note that at the end of stage 1, if x and y are connected by a path whose only intermediate vertex is 1, then $A_{xy} = 1$. Now assume inductively that if, at the end of stage k, x and y are connected by a path whose intermediate vertices are in the set $\{1, 2, \ldots k\}$, then $A_{xy} = 1$. Suppose that x and y are connected by a path whose only intermediate vertices are in the set $\{1, 2, \ldots k+1\}$. Then at stage $k+1$, A_{xy} is set equal to 1 when $i = x$. This proves by induction that for all m if x and y are connected by a path whose intermediate vertices are in $\{1, 2, \ldots m\}$, then after stage m, $A_{xy} = 1$. Taking $m = n$ proves Warshall's algorithm works.

21. At stage k we find the shortest path from i to j passing through some (or none) of the vertices $1, \ldots, k$.

23. Take the cycle on $\{1, 2, 3, 4\}$ with 1 adjacent to 2 adjacent to 3 adjacent to 4 adjacent to 1, and all edges having weight 1. Then the minimum total weight of a path from vertex 1 to vertex 3 is -2, while Floyd's algorithm gives -56.

25. Since the distance from 0 to j is j, there are no walks of length i from 0 to j if $i < j$. However, A_{0j}^j is positive. The vectors $(A_{01}^0, A_{02}^0, \ldots, A_{0d}^0)$, $(A_{01}^1, A_{02}^1, \ldots, A_{0d}^1)$, $\ldots$, $(A_{01}^d, A_{02}^d, \ldots, A_{0d}^d)$ are therefore linearly independent so that $\sum_{i=0}^d c_i A^i = 0$ only if all $c_i = 0$. Therefore the dimension of the space generated by the powers of A is at least $d+1$ and A has at least $d+1$ distinct eigenvalues.

CHAPTER 5

Section 5.1, pages 308–313

1. The condition states that $|R(S)| \geq |S|$ for each set S of boys; therefore, there is a complete matching from the boys to the girls.

3. Let the top vertices be a, b, c, d, e from left to right and let the bottom vertices be A, B, C, D, E from left to right. $\{C, e\}, \{E, a\}, \{A, d\}, \{D, b\}, \{B, c\}$, is a complete matching. (Other answers are possible.)

5. (Other answers are possible.) Use the notation suggested in Exercise 3.
(b) $\{b, c, A, E\}$ is a vertex cover. The vertical lines are a maximum partial matching.
(d) $\{a, d, D, E\}$ is a vertex cover. $\{a, A\}, \{b, D\}, \{d, C\}, \{e, E\}$ is a maximum partial matching.
(e) $\{a, b, c, C\}$ is a vertex cover. $\{a, A\}, \{b, B\}, \{c, E\}, \{C, e\}$ is a maximum partial matching.
(f) $\{a, B, c, D\}$ is a vertex cover. $\{a, A\}, \{b, B\}, \{c, C\}, \{e, D\}$ is a maximum partial matching.

7. (Other answers are possible except in part d.)   (a) $1, 4, 6, 8, 3$
(b) $1, 2, 3, 4, 5$ (c) No   (d) No   (e) $1, 6, 3, 9, 7, 2, 5, 8$.

8. Let S_i be the set of integers between 1 and n not in column i. There must be exactly $k(m-n)$ pairs (i, j) with $j \in S_i$ for k selected sets S_i. Each j appears in only $n - m$ sets S_i, so in order to have $k(n-m)$ pairs (i, j), we must have at least k distinct j's among our k selected sets S_i. Thus the union of any k sets S_i has at least k elements. An SDR for the sets S_i gives one integer that can be added to each column to extend the Latin rectangle by one more row.

9. The point is to use SDRs to show there is a complete matching from V_1 to V_2 if $|X| \leq |R(X)|$ for each subset X of V_1. For each member i of V_1, let S_i be the set of elements of V_2 adjacent to i. Then if $|X| \leq |R(X)|$, for all X, the union of any k sets S_i has at least k elements, so there is an SDR $x_1, x_2, \ldots, x_n$ for the sets $S_1, S_2, \ldots, S_n$. Then the set of edges $\{i, x_i\}$ is a complete matching.

11. (Answers may vary.) (a) $\{1, 2\}, \{\{1, 2\}\}$.
(b) $\{1, 3, 5, 6\}, \{\{1, 4\}, \{2, 5\}, \{3, 7\}\}$.   (c) $\{1, 3, 4\}, \{\{1, 2\}, \{3, 4\}\}$.

13. The tree has edge set $\{(3,6),\ (6,2),\ (2,7),\ (2,8),\ (2,9),\ (9,4),\ (4,10),$ $(4,11),\ (11,5)\}$. It contains several alternating paths of the type desired; one is $3,6,2,7$; another is $3,6,2,9,4,10$. Enlarging the matching by using the first path gives the matching $\{(2,7),\ ((3,6),\ (4,9),\ (5,11)\}$.

15. One such matching is $\{j,k\},\ \{i,h\},\ \{f,g\},\ \{b,c\},\ \{a,d\}$.

17. The proof is by induction on the number of nonzero entries. There must be at least m nonzero entries, all in different rows or columns, or else some row or column sum is zero. If these are the only nonzero entries, then they are all equal to k and the matrix is k times a permutation matrix. Otherwise these positions determine a permutation matrix P and these numbers have a smallest value s. The matrix $M - sP$ has (at least one) fewer nonzero entries than M and has equal row and column sums. By an appropriate inductive hypothesis, the resulting matrix is a combination of permutation matrices. Therefore M also is a combination of permutation matrices. Since every one of the permutation matrices has exactly one nonzero entry with value 1 in the first row, the sum of the coefficients must be the sum of the first row, which is 1.

19. Draw a bipartite graph with the classes of P and Q as vertices and a line between two classes if they have a vertex in common. From the condition that the union of k classes of P is not contained in the union of $k-1$ classes of Q, we get a complete matching from the classes of P into the classes of Q. This gives the system of common representatives.

21. (a) Suppose there is no alternating path for M from $x \in V_1$ to any other unmatched vertex. Suppose M' is created from M by augmenting along the alternating path $v_0 v_1 \ldots v_m$ between unmatched vertices $v_0 \in V_1$ and $v_n \in V_2$ for M. Finally, suppose $x = x_0 x_1 x_2 \ldots x_k$ is an alternating path for M' from x to another unmatched vertex (for M'). Suppose $x_i x_{i+1}$ is the first M' matching edge on the x-path that is not an edge of M. Since $x_i x_{i+1}$ is an M'-matching edge but not an M-matching edge, $x_i x_{i+1} = x_j x_{j+1}$, and x_i and v_j are both in V_1 and hence equal. Then $x = x_0 x_1 x_2 \ldots x_i v_{j+1} \ldots v_n$ is an alternating walk (some vertices and/or edges may be repeated) from x to an unmatched vertex. Since G is bipartite the walk may be shortened to a path. This contradiction means there is no such alternating path $x = x_0 x_1 \ldots x_k$ for M'.

(b) By part (a), once we have failed to find an alternating path from an unmatched vertex, we need never try again to find an alternating path from x. Thus the maximum number of searches we will need to make is v, the number of vertices. A search takes no more than e steps, so the number of steps in searches is no more than $e|V_1|$, where V_1 and V_2 are the parts of G. We may assume $|V_1| \leq |V_2|$, so the number of steps in searches is no more than $ev/2$. Beginning from the empty matching, the first alternating path we find would have 2 vertices, the next no more than 4, the next no more than 6, and so on, so the number of steps in creating alternating paths from search trees is at most $2 + 4 + 6 + \cdots + 2|V_1| = 2\frac{|V_1|(|V_1|-1)}{2} = |V_1|(|V_1|-1) < \frac{v^2}{4}$ since $V_1 \leq \frac{v}{2}$ by our preceding assumption. Thus the number of steps is no more than $ve/2 + v^2/4$.

22. Consider the triangle with vertices $1, 2, 3$ and an extra vertex and edge $\{0, 1\}$. Use the matching $\{1, 3\}$. Start the breadth-first search at vertex 2. This gives the edges $\{1,2\}$ and $\{2,3\}$ and no more. Thus we do not get the alternating path 0,1,3,2 that we can get by starting at vertex 0 instead. Add the edge $\{0,3\}$ to the graph and then no matter where we start the alternating breadth-first search, we do not find the alternating path 0,1,3,3.

23. We get a blossom $\{B_9, l, m, n, p\}$ with B_9 as base. This shrinks to a mega-vertex B_{14} attached by an edge to vertex j. We could then label w, x, and y if they are present. Now the graph has a seven-element blossom. When we examine the untested edge $\{k, B_{14}\}$, we again return to case 4 of the algorithm and discover the seven-vertex blossom and shrink it to B_{15}. Finally, we can label z if it is present and complete the process essentially as before. Suppose, for example, that z is present and we work back from it. In constructing the path back to a, when we replace B_{15} by its cycle, we discover that the path to a is $z, b, B_{14}, , k, j, i, h, a$. Now when we replace B_{14} by its cycle, we observe that the edge (B_{14}, k) on our path must be replaced by the long path around the blossom with base B_9; we find this out by noting that to get to k from the blossom on a nonmatching edge, we must have entered l along a matching edge.

25. The only possible labeling takes us along the path $hijklmnpgfcb$ at which point we find that the endpoints of $\{a,b\}$ have the same parity and collapse our blossom (or we find the path to the unmatched vertex x or z along the way and short-circuit the process).

27. When the path enters the blossom, it does so along a matching edge, the one entering the base, or else x is the base of the blossom. When the path leaves the blossom, it must do so on a nonmatching edge, for all matching edges touching blossom vertices other than the base are blossom edges. Thus if we replace the portion of the path from the blossom base to the last vertex the path has in the blossom by the megavertex, this new path is an alternating path from x to the unmatched vertex. Conversely, if there is a path from x to the unmatched vertex y using the megavertex, and y' is the vertex immediately after the megavertex on this path, then y is adjacent by a nonmatching edge to some vertex z of the blossom. We replace the megavertex in the path by the part of the cycle around the blossom that leaves z on a matching edge and goes to the base. Of course, if there is an alternating path from x to y that does not use the megavertex, then there is still the same path in the larger graph.

29. The preceding two exercises tell us that there is an alternating path from x to y before a blossom is shrunk if and only if there is one after the blossom is shrunk, and from the alternating path in the smaller graph we may recover an alternating path (perhaps not the original one) in the bigger graph. Thus we will be able to find an alternating path to use to extend the matching if and only if we find one after all our blossom shrinking is complete.

Section 5.2, pages 324–327

1. Yes   3. Yes   5. No   7. $\{1,2,3,4,5,6,8\}$
9. $\{a,h\}, \{a,f\}, \{d,e\}, \{d,b\}, \{d,c\}, \{e,g\}, \{f,g\}$.   11. a,b,e,k,i,f,l

13. c, d, l, i, h, g, j

15. $\{k, l\}$, $\{g, k\}$, $\{a, f\}$, $\{e, j\}$, $\{e, f\}$, $\{j, k\}$, $\{e, i\}$, $\{g, c\}$, $\{c, d\}$, $\{b, f\}$ $\{g, h\}$.

17. Clearly, the subset property is satisfied. Suppose X and Y are independent sets and $|X| < |Y|$. Clearly X can be expanded if $|X| = 1$. If $|X| = 2$, then no three points of Y lie on a line, so one is not on the line X generates and therefore can be added to X to give an independent set. If $|X| = 3$, then the points of Y don't all lie in the plane X generates, so X can be expanded to a larger independent set. (Note that $|X| = 4$ is impossible.)

19. Part (a) is a special case of part (b), which we do here. The subset property is clear: if I has fewer than $n/2$ elements, then so does any subset of I. For the expansion property, suppose I, J are in $\mathbf{I}$. If $|I| < |J|$, then J has at least one more element x than I, so $|I \cup \{x\}| \leq |J|$, and so $I \cup \{x\} \in \mathbf{I}$.

21. Note that the dimension of the subspace sum of a set K of one-dimensional subspaces is no more than $|K|$. Suppose $I \subseteq J$, and J is independent. Use $\overline{K}$ to stand for the subspace sum of the members of K. Note that $\overline{J} = \overline{I} + \overline{J - I}$. Thus $\dim \overline{J} \leq |I| + |J - I|$, so $|I|$ must equal $\dim(\overline{I})$. Thus I is independent, so we have the subset property. Suppose now I and J are independent and $|I| < |J|$. If, for every $x \in J$, the dimension of the subspace sum of $I \cup \{x\}$ is only $|I|$, then $J \subseteq I$, a contradiction. Thus we have the expansion property.

23. To check e requires only one table lookup. At most $n - 1$ vertices are connected to each end of e when we add it, so no more than $2n$ steps are used.

25. Let $I \subseteq B$ and $J \subseteq B'$ be independent sets with $|I| < |J|$. Choose $y_1 \in B$ with $y_1 \notin I$. Then there is an element $z_1 \in B'$ such that $B_1 = B - \{y_1\} \cup \{z_1\}$ is in $\mathbf{B}$, and $I \subseteq B_1$. Repeat the process inductively, and eventually, since $|I| < |J|$, you will get a z_i in J. Then $I \cup \{z_i\}$ is independent.

27. The edge sets of cycles. (Removing one edge from a cycle gives a tree. Any set containing a cycle is dependent and no other sets are, so cycles are the minimal dependent sets.)

29. Suppose I and J are independent sets with $|I| < |J|$ and induct on the size of I. If $|I| = 0$, then any member of J may be added to I to give an independent set. Suppose I has size i and suppose every element x_k of J forms a circuit C_k with elements of I. If there is an element of I in one or none of these circuits, we may delete it, delete the member of J that determines the circuit, and apply induction. Otherwise choose an element $x \in C_1 \cap I$ and for each other circuit C_m containing x form the circuit C'_m guaranteed by property $(2')$ not containing x but containing x_m. Now delete x from I and x_1 from J and apply induction.

31. You have to choose a method of testing for independence. Probably reperesenting the family as a bipartite graph and using alternating search trees to extend the matching is best. Use the greedy method for the modification.

32. We need to show that if $I, J \in \mathbf{I}$ and $|I| < |J|$, then there is an $x \in J - I$ such that $I \cup \{x\} \in \mathbf{I}$. For this purpose we choose a positive number $s < \frac{1}{|X||J|}$ (s is for small). Weight each element of I with weight $1 + s$, each element of $J - I$ with weight 1, and each other element of X with weight 0. Then J is an independent set of weight less than the weight of I, but unless there is an element x of J such that $I \cup \{x\} \in \mathbf{I}$, the greedy algorithm selects I as the minimum weight set of maximum size, contradicting our assumption that it selects a subset which is truly of maximum size. Therefore such an x exists. Therefore $\mathbf{I}$ defines a matroid on X.

Section 5.3, pages 340–344

1. No   **3.** 3   **5.** $s(s, a)a(a, g)g(g, t)t$. Other answers are possible.

7. 2; $f(s, c) = f(c, d) = f(d, t) = f(s, a) = f(a, g) = f(g, t) = 1$, $f(e) = 0$ for all other e. Other answers are possible.

9. $f(b, g) = f(g, t) = 2$; $f(s, b) = 3$, $f(b, a) = 1$; $f(s, a) = 3$; $f(a, d) = 4$; $f(d, t) = 4$; $f(e) = 0$ all other edges.

11. The cut s/all else and t/all else must each have the same capacity as the cut given by $\{s, g, d\}/\{b, a, c, t\}$. But the capacity of this last cut is at least the sum of the capacities of the first two, so all three capacities can't be both nozero and equal. However, the first two capacities must be nonzero to have a nonzero flow.

13. One million.

15. The flow we get by using the labeling algorithm and the flow-augmenting paths of integer worth the algorithm gives will make the flow along each edge an integer. Give each edge in Figure 3.3 capacity 2. Send flow 1.5 along edges (s, c) and (c, d), flow .5 along edges (s, b) and (b, d), and flow 2 along every other edge but (d, a).

17. In Figure 3.2, delete edge (a, c), give edges (s, b) and (d, t) capacity one and give all other edges capacity $N - 1$. The flow of one from s to b to d to t is maximal.

19. **(a)** a and b are in layer 1, c, d, and g are in layer 2, and t is in layer 3. Delete edges (b, a), (c, s), (c, g), (d, b), and (g, d).

(b) The graph consists of two paths from s to t with nothing but s and t in common.

(c) $f(s, b) = 3$, $f(s, a) = 3$, $f(b, g) = 2$, $f(g, t) = 2$, $f(b, a) = 1$, $f(a, d) = 4$, $f(d, t) = 4$.

21. **(a)** Each time you choose a flow-augmenting path by this method, you may remove, but not add, edges which are available to let flow go from one layer to the next. Thus if you had to back up from a vertex in one search, you would have to back up from it in any search. Thus removing the edges touching vertices you had to back up from simply removes some dead ends and no potential paths. Since you continue a search even after removing edges, if there is a path from the source to the sink at any stage, depth-first search will find that path or another one. Augmenting by that path and repeating the process until you find no augmenting path produces a blocking flow.

(b) The flow in at least one edge reaches capacity (or zero if it is a reverse edge) each time we augment, and we can never use it in the opposite direction; therefore the maximum number of augmenting paths we can find is the number of edges.

(c) The number of flow-augmenting paths is no more than m and the number of vertices per path is no more than n.

(d) The layer a vertex is in can only increase, and from this you can show that the number of layers must increase at each stage. However, the number of layers is no more than the number of vertices.

23. In step 2 of the labeling algorithm, change the directed edges (u, v) and (v, u) to $\{u, v\}$. Once you have flow in an edge, you have effectively directed it. You may still need to redirect that flow to another vertex and then the edge is used as a reverse edge to reduce its flow.

Section 5.4, pages 356–358

1. 2

3. The two edges coming into b or the horizontal edge leaving a and the edge parallel to it are two examples.

5. Any two vertex-disjoint directed paths will suffice.

7. The two rectangular paths from a to b and the path consisting of two diagonals will do, as will any other set of three edge-disjoint paths.

9. $1, 2, 5, 6, 3$ or any other SDR of total cost 17.

11. Use Menger's theorem. Suppose a is n-edge connected to b , and b is n-edge connected to c. If we remove fewer than n edges, then there is a path from a to b and a path from b to c, so there is a path from a to c. The relation of being n-edge connected is therefore transitive. It is clearly reflexive and symmetric, so it is an equivalence relation.

13. If a minimum-sized $u - v$ vertex cutset has n edges, then u is n vertex connected to v. To prove this, replace each vertex x by vertices x_1 and x_2 and the directed edge (x_1, x_2), replace each edge $\{x, y\}$ by directed edges (x_2, y_1) and (y_2, x_1). Give each edge capacity 1 and find a maximum flow. Now a flow that uses one vertex in the pair (x_1, x_2) must use both because (x_1, x_2) is the only edge leaving x_1 and the only edge entering x_2. Further, any use of (y_2, x_1) and (x_2, y_1) in one path may be eliminated. If (x_2, y_1) were used in one path and (y_2, x_1) were used in another, then both must use (x_1, x_2), and this is impossible. This gives the n vertex-disjoint paths.

15. This can be done by translating the problem to a network flow problem.

17. The edges of the network to which we apply the potential algorithm are each edges along which we may augment the flow, and the path we find is therefore flow-augmenting. Further, any flow-augmenting path of the original network corresponds to a path in the new network. Therefore, by Theorem 4.5, we must get a flow-augmenting path of minimum cost.

19. This is the same kind of network we used to compute matchings from flows, but we have prices on edges rather than vertices. Thus a minimum-price maximum flow corresponds to a maximum-sized matching of minimum price. The least expensive matching is (a, c), (b, d).

20. $f(s, a) = f(s, b) = f(b, c) = 2$; $f(a, c) = f(a, d) = f(d, t) = 1$; $f(c, t) = 3$; cost $= 28$.

21. $h, \{h, f\}, f, \{f, d\}$, $d, \{d, c\}, c$.

23. You must know that the graph has no directed cycles whose length is positive. This graph has no directed cycles whatever.

25. Note that now your costs are on edges, and so a minimum-cost flow algorithm will find a minimum-cost matching. In the ordering of jobs from left to right, one assignment is $a\ b\ f\ k\ c\ i\ e$.

27. In the digraph with vertices a, b, c, d, e and edges (a, b), (b, c), (c, d), (d, b), (a, c), (c, e), (e, d) with prices $10, 2, 3, 1, 4, 1, 1$ in order, vertex b gets price 10, and then 8 when all edges are traversed once in one breadth-first order, but the eventual price of b is 7.

29. Use the potential algorithm in place of the labeling algorithm in a network flow algorithm for matchings.

CHAPTER 6

Section 6.1, pages 377–378

1.

1	2	3	4		1	2	3	4		1	2	3	4		1	2	3	4
2	1	4	3		2	1	4	3		2	4	1	3		2	3	4	1
3	4	1	2		3	4	2	1		3	1	4	2		3	4	1	2
4	3	2	1		4	3	1	2		4	3	2	1		4	1	2	3

3.

0,0	1,1	2,2
1,2	2,0	0,1
2,1	0,2	1,0

5. $2 \cdot 3 \equiv 0 \pmod 6$ but neither 2 nor 3 is congruent to 0.

7. The class $\underline{a}(\underline{b} + \underline{c})$ is the class of the number $a(b + c)$ and the class of $\underline{a}\ \underline{b} + \underline{a}\ \underline{c}$ is the class of the number $ab + bc$ and the two numbers are the same.

8. **(a)** If $\underline{a}\ \underline{i} = \underline{a}\ \underline{j}$, then $ai - aj \equiv 0 \bmod n$ so that $a(i - j) = kn$. However, $i - j$ is strictly between $-n$ and n and a is strictly between 0 and n, so $a(i - j)$ cannot have n as a prime factor.
(b) Since there are $n - 1$ distinct classes listed in part (a) and none of them are the class $\underline{0}$ because n is prime, each nonzero class mod n, including $\underline{1}$, must be listed in part (a). Thus $\underline{a} \cdot \underline{b} = \underline{1}$ for some $\underline{a} \cdot \underline{b}$ in the list.

9. Let $\underline{a} = \underline{3}$. Then since $\underline{5} \cdot \underline{3} = \underline{0}$, if $\underline{3} \cdot \underline{b}$ were $\underline{1}$, then $\underline{5} \cdot \underline{3} \cdot \underline{b}$ would be both $\underline{0}$ and $\underline{5}$.

11. Apply Theorem 1.7 to four mutually orthogonal squares of order five (see Corollary 1.5) and four mutually orthogonal squares of order seven. Then use Theorem 1.7.

13. We may assume one of the squares is listed in Exercise 1. We may also assume the first row of the second square is 1 2 3 4. Then experimenting shows us that the squares

1	2	3	4		1	2	3	4
2	1	4	3		4	3	2	1
3	4	1	2	and	2	1	4	3
4	3	2	1		3	4	1	2

are orthogonal.

15. If $n = 12k + 1$, then $n - 1 = 12k + 0 = 3(4k + 3)$. But by Theorem 1.4 with $n = 3$ and $s = 2$, there is a pair of mutuallly orthogonal $4k + 3$ by $4k + 3$ Latin squares. Thus with $m = 4k + 3$, we have $n = 3m + 1$, so that Theorem 1.11 applies.

17. Note that we got our two orthogonal squares in Exercise 13 by moving rows 2 and 3 down one unit and bringing row 3 up to the second place. If we also shift everything 2 units down and around back up, we get the third square

1 2 3 4
3 4 1 2
4 3 2 1 which you may check is orthogonal to the two given.
2 1 4 3

19. When congruence is modulo a prime number.

21. Assume we have k mutually orthogonal squares. Changing the names of the symbols in one of a pair of orthogonal Latin squares does not change their orthogonality. Thus we may assume all k squares have 1 to k in order for their first row. But then there are only $k - 1$ different choices for row 2, column 1. Therefore, some pair i, i must appear in one pair of squares in row 2, column 1. But it also appeared in row 1, column i, so these squares are not orthogonal after all.

23. See the solution to Exercise 8 of Section 1 of Chapter 5.

25. Any power of two larger than the first is a multiple of a power of four and a power of eight (including, perhaps, the zeroth power). There is therefore a pair of orthogonal Latin squares of order 2^k for $k > 1$ by Theorem 1.7. There are pairs of orthogonal Latin squares of other prime power orders by Corollary 1.5. Thus one more application of Theorem 1.7 gives the desired result.

27. This is a straightforward application of backtracking.

28. **(a)** For matrices with two rows and six columns, let property i be "the entries in column i are equal." We want the number of matrices with none of these properties. Given a set I of properties, $N_\geq(I) = (6 - i)!$. Thus
$$N_=(\emptyset) = \sum_{i=0}^{6}(-1)^i \binom{6}{i}(6 - i)! = 6! \sum_{i=0}^{6} \frac{(-1)^i}{i!}.$$

29. Straightforward backtracking.

Section 6.2, pages 388–390

1. $k = 3$, $\lambda = 2$.   **3.** Since $r \cdot (k - 1) = \lambda(v - 1)$, if $r = \lambda$, then $v = k$.

5. There are $C(k; 2) \cdot b$ pairs consisting of two elements in a block together; there are also $C(v; 2) \cdot \lambda$ pairs consisting of two elements in a block together. Thus $C(v; 2) \cdot \lambda = C(k; 2) \cdot b$. Therefore $\frac{v(v-1)}{2}\lambda = \frac{k(k-1)}{2}b$.

7. Multiply the matrix shown times $(r - \lambda)I + \lambda J$.

9. **(a)** A row of N has r ones, thus if it is multiplied by an "all-ones column," we get k. Therefore $NJ = rJ$.   **(b)** A column of N has k entries. $JN = kJ$.

11. **(a)** No.   **(b)** No.

13. Take complements of the sets in the $(7, 7, 3, 3, 1)$ design.

15. The set of varieties, using the complement of Example 2.2 as our $(7, 7, 4, 4, 2)$ design, is $\{0, 1, 4, 5\}$ and the blocks are $\{4, 5\}$, $\{1, 5\}$, $\{1, 4\}$, $\{0, 4\}$, $\{0, 5\}$, $\{0, 1\}$.

16. Yes (show that the complements form a $(7, 7, 3, 3, 1)$ design or try to construct the design and find that, up to labeling, you get only one).

17. There is only one, all two-element subsets of a four-element set.

19. $\{0, 1, 2, 3, 4\}$, $\{0, 1, 5, 6, 7\}$, $\{0, 2, 5, 8, 9, \}$, $\{0, 3, 6, 8, 10\}$, $\{0, 4, 7, 9, 10\}$, $\{1, 2, 6, 9, 10\}$, $\{1, 3, 7, 8, 9\}$, $\{1, 4, 5, 8, 10\}$, $\{2, 3, 5, 7, 10\}$, $\{2, 4, 6, 7, 8\}$, $\{3, 4, 5, 6, 9\}$. (List these sets vertically to see a pattern.) Also in the terminology of Section 3, construct a cyclic design mod 11 with base $\{0, 2, 3, 4, 8\}$.

21. $\{0, 1, 2\}$, $\{0, 1, 3\}$, $\{0, 2, 4\}$, $\{0, 3, 5\}$, $\{0, 4, 5\}$, $\{1, 2, 5\}$, $\{1, 3, 4\}$, $\{1, 4, 5\}$, $\{2, 3, 4\}$ $\{2, 3, 5\}$.

23. False. Consider $\{1, 2\}, \{1, 3\}, \{1, 4\}, \{2, 3, 4\}$.

25. True because then $\lambda(v - 1) = 2r$ is even.

27. The design is balanced because the inner product of row i of N and row j of N is the number of blocks containing both i and j. The design is regular for a similar reason; this is, however superfluous (see Exercise 22).

29. (a) Since $6 - 2 = 4$ is a square, we must try to construct a $16, 16, 6, 6, 2$ design. If we can prove the construction is impossible, there is no design; if we succeed in the construction, there is a design.

(b) As our varieties we take the 16 sequences of 4 zeros and ones. One block of the design is

$$\{(1, 0, 0, 0), (0, 1, 0, 0), (0, 0, 1, 0), (0, 0, 0, 1), (1, 1, 0, 0), (0, 0, 1, 1)\}.$$

Take each of the 15 nonzero sequences of 4 zeros and ones and add them, one at a time, to all 6 vectors in this block. This gives us the 16 blocks of our design.

Section 6.3, pages 406–408

1.

0	1	3		2	6	7		5	10	12		4	8	11
1	2	4		3	7	8		6	11	0		5	9	12
2	3	5		4	8	9		7	12	1		6	10	0
3	4	6		5	9	10		8	0	2		7	11	1
4	5	7		6	10	11		9	1	3		8	12	2
5	6	8		7	11	12		10	2	4		9	0	3
6	7	9		8	12	0		11	3	5		10	1	4
7	8	10		9	0	1		12	4	6		11	2	5
8	9	11		10	1	2		0	5	7		12	3	6
9	10	12		11	2	3		1	6	8		0	4	7
10	11	0		12	3	4		2	7	9		1	5	8
11	12	1		0	4	5		2	8	10		2	6	9
12	0	2		1	5	6		4	9	11		3	7	10

3. There is one up to isomorphism, all three-element subsets of the four-element set.

5. If a design is resolvable, then v/k is an integer, so there is none.

7. We want a resolvable triple system with $v = 9$ and $\lambda = 1$. Then $\lambda(v - 1) = r(k - 1)$ gives $r = 4$ and $bk = vr$ gives $b = 12$.

1	2	3	1	4	7	1	5	9	1	6	8
4	5	6	2	5	8	2	6	7	2	4	9
7	8	9	3	6	9	3	4	8	3	5	7

We could also have applied Theorem 3.7 to two one-set triple systems on three-element sets.

9. $\{1, 2, 4\}$; $\{2, 3, 5\}$; $\{3, 4, 6\}$; $\{4, 5, 0\}$; $\{5, 6, 1\}$; $\{6, 0, 2\}$; $\{0, 1, 3\}$. By Exercise 8, this must be isomorphic to the design of Section 2 even though it has different blocks.

11. A Steiner triple system on nine vertices is a $(12, 9, 4, 3, 1)$ design. Suppose the vertices are the numbers 1 through 9. Vertex 1 is in four blocks, each time with two other elements. Vertex 1 must appear with eight other elements, so up to relabeling we may assume four blocks of the design are $1, 2, 3$; $1, 4, 5$; $1, 6, 7$; $1, 8, 9$. Now 2 must appear in three more blocks and there are exactly six elements it has not yet appeared with. Up to relabeling, we may assume one of these blocks is $2, 4, 6$; then the other two must be $2, 5, 9$ and $2, 7, 8$ (up to the possibility of switching 8 and 9). Now 3 must still appear in three more blocks and with the same elements that 2 appeared with. Also 4 must still appear with $3, 7, 8$, and 9. The combinations $4, 7, 8$ and $4, 8, 9$ are impossible because the $7, 8$ and $8, 9$ combinations have occurred. Thus $3, 4, 8$ and $4, 7, 9$ are the two remaining blocks including 4. Now 3 must appear with $5, 6, 7$, and 9, and 5 must appear with $3, 6, 7$, and 8. Also, each of 5 and 6 must appear twice and $7, 8$, and 9 must appear once. Thus the only way this can happen is to have blocks $5, 6, 8$; $3, 5, 7$ and $3, 6, 9$. Thus, up to labeling, there is only one way to have a $(12, 9, 4, 3, 1)$ design.

13. Since every member of V appears in r blocks of the original design, every element of V is *not* in the remaining $b - r$ blocks, so the complement is a regular design in which each element is in $b - r$ blocks.

15. True. Take a k-element subset of a p-element set and add one to each of its elements p times and you will get the same set back. Since p is a prime it is impossible to get the same set back earlier, because if you got the same set back after j additions, j would have to be a factor of p. Thus, since the sets which are not blocks of the design must be closed under the successive additions of one, they must also form a cyclic design.

17. Each pair of varieties may be increased to a set of size k in $\binom{v-2}{k-2}$ ways. Since λ of these sets are in the design, $\binom{v-2}{k-2} - \lambda$ are not in the design and thus are blocks of D^*.

19. Because ∞ appears twice as a difference, λ would have to be 2. However, 3 appears as a difference in each of the first three sets, so they cannot form a base.

21. There is a resolvable triple system on nine varieties, see Example 3.4, and there is a $(13,4,1)$ design as in the previous exercise. Now apply the construction of Theorem 3.16.

23. Does not work with $k = 4$. Consider the two complete designs $\{1, 2, 3, 4\}$ and $\{a, b, c, d\}$ on four elements. Then among the type 3 blocks, we would get $1a, 2b, 3c, 4d$ and $2a, 1b, 3c, 4d$. Thus the pair $3c, 4d$ will appear twice.

25. v must be a multiple of n, and the multiple is the number m of base blocks. There will be k elements in a base block and $k(k-1)$ differences per base block. Thus $k(k-1)m = \lambda(n-1)$ because $n-1$ is the number of nonzero differences and each must occur λ times. Thus for n, you try factors of v and for m you try factors of v which are multiples (probably small because λ is probably small) of $(n-1)/k(k-1)$.

27. This proof consists of detailed verification that each difference appears twice. Note that 1 appears in the second type of block as $i - 0$ with $i = 1$ and as $2t + 3 - i - i$ with $i = t + 1$. Other odd values between 1 and $t + 1$ appear similarly. Even numbers between 0 and $t + 1$ appear once as a difference $i - 0$ in a type 2 block and as a $2i - 0$ in a type 3 block. The even numbers between $t+1$ and $2t$ also appear as $2i-0$ values and as $2t+3-i-0$ values when i is odd between 1 and $t + 1$. Continue this sort of analysis to show that all differences up to $3t + 2$ appear positively, then work on their negatives. You will need to note that the difference $0 - (3t+3+i)$ is also $6t+5 - (3t+3+i) = 3t+2-i$.
    (b) $\{ax, ax^4, bx^0\}$, $\{ax^2, ax^3, bx^0\}$, $\{bx, bx^4, cx^0\}$, $\{bx^2, bx^3, cx^0\}$, $\{ax^0, bx^0, cx^0\}$, $\{cx, cx^4, ax^0\}$, $\{cx^2, cx^3, ax^0\}$. This design is not resolvable.

29. Since two vertices lie in a unique block, two different blocks could not have two or more vertices in common for this would make them equal. However, the four blocks in a parallel class have all 16 vertices, so there must be 16 size-one intersections between the blocks of one parallel class and the blocks of a second. But there are exactly 16 pairs of blocks with one member of the pair in one parallel class and one member in a (fixed) second class. Thus each of these pairs must have an intersection of size one. Suppose we have two classes, class one and class two. Number the blocks of class two as blocks one, two, three, and four. For one block of class one, label its unique intersection with block one (of class 2) with the label one, its unique intersection with block two with the label two, and so on. Pick a second block of class one. Label its unique intersection with block one with five, its unique intersection with block two with label six, and so on. Repeat this similarly for the other blocks of class one, and this gives the eight blocks listed in the problem. Write these eight blocks as the first two rows of an array of blocks in the order given. Now put three more blocks, $\{1, 6, 10, 14\}$, $\{1, 7, 11, 15\}$, and $\{1, 8, 12, 16\}$, in the first column of the array as in Example 3.5 and proceed as in Example 3.5, filling in blocks under $\{2, 6, 10, 14\}$ with blocks beginning with 2, etc.

31. In fact the design T' is balanced with $\lambda = 1$. To see this, we need to examine how many times pairs of the form $\{0, x_i\}$, $\{x_i, x_j\}$, and $\{x_i, y_j\}$ appear together in blocks of T'. Note that $\{0, x_i, y_i\}$ is a triple in the copy of T on every H' such that $i \in H$. However, $\{0, x_i, y_i\}$ appears once in the set union of all these blocks, so 0 appears with each of x_i and y_i exactly once. Since $\{0, a_i, b_i\}$ is the only triple of T in which a_i and b_i appear together, x_i and y_i appear together in exactly one triple of T'. If $i \neq j$, then a pair $\{i, j\}$ will appear in exactly one block H of D. When it does, the pair $\{x_i, x_j\}$ will appear in exactly one triple of the triple system on H'. The only kind of triple of D' that can contain $\{x_i, x_j\}$ will be a triple $\{x_i, x_j, z\}$ from some H'. But for $\{x_i, x_j, z\}$ to be a triple of H', $\{i, j, k\}$ must be a triple of H and therefore there is exactly one H' containing exactly one triple of the form $\{x_i, x_j, z\}$, so x_i and x_j appear together in exactly one triple of D'. The argument for a pair $\{x_i, y_j\}$ is the same. Therefore D' is balanced with $\lambda = 1$.

Section 6.4, pages 422–423

1. 30, 5, 6   **3.** $(13, 13, 4, 4, 1)$ **5.** $(21, 21, 5, 5, 1)$

7. $3n$ if $n > 2$, 4 if $n = 2$; $4n$ if $n > 3$, 9 if $n = 3$, and 4 if $n = 2$.

9. Since there must be three points on a line in a projective plane, the order of the plane would be 2, so the Steiner system would have to be the $(7, 7, 3, 3, 1)$ design. If there were an affine plane its order would be three and so it would have to be a $(12, 9, 4, 3, 1)$ design, the affine plane over the integers (mod 3).

11. Because of Theorem 4.14, the design of Exercise 22 of Section 3 is the desired design.

13. No.

15. Choose a line that does not include the given point P and there will be lines from P to each point on that line, giving n lines of this type by Theorem 4.1. By Axiom 3 there is exactly one more line through P, the one parallel to the line we chose.

17. Since there are n^2 points, each contained in $n + 1$ lines, there are $n^2(n + 1)$ point–line pairs with the point on the line. Since each line has n points, there must be $n(n + 1)$ lines in order to have this number of point–line pairs.

19. The method is the same as if B_1 is type 1 and B_2 is type 3, except rather than $i = r$, we have $j = r$.

21. The design you get consists of all two-element subsets of the four-element set $(1, 1)$, $(1, 2)$, $(2, 1)$, $(2, 2)$.

23. It leads to the construction of an affine plane on nine points with three points per line.

25. Our orthogonal Latin squares are

$$
\begin{array}{ccc} a_1 & a_2 & a_3 \\ a_2 & a_3 & a_1 \\ a_3 & a_1 & a_2 \end{array}
\qquad
\begin{array}{ccc} a_1 & a_2 & a_3 \\ a_3 & a_1 & a_2 \\ a_2 & a_3 & a_1 \end{array}.
$$

Our vertices are the ordered pairs 11, 12, 13, 21, 22, 23, 31, 32, and 33. The type 1 blocks are $\{11, 12, 13\}$ $\{21, 22, 23\}$, and $\{31, 32, 33\}$. The type 2 blocks are $\{11, 21, 31\}$, $\{12, 22, 32\}$, and $\{13, 23, 33\}$. The type 3 blocks are the sets $\{11, 32, 23\}$, $\{12, 21, 33\}$, $\{13, 22, 31\}$, $\{11, 22, 33\}$, $\{12, 23, 31\}$, $\{13, 21, 32\}$. These are the blocks and they are already resolved into four parallel families of three blocks each. Now we introduce the symbols f_1, f_2, f_3, f_4, the block $\{f_1, f_2, f_3, f_4\}$, and rebuild our original 12 blocks as

$$\begin{array}{lll} \{11,12,13,f_1\} & \{21,22,23,f_1\} & \{31,32,33,f_1\} \\ \{11,21,31,f_2\} & \{12,22,32,f_2\} & \{13,23,33,f_2\} \\ \{11,32,23,f_3\} & \{12,21,33,f_3\} & \{13,22,31,f_3\} \\ \{11,22,33,f_4\} & \{12,23,31,f_4\} & \{13,21,32,f_4\} \end{array} \cdot$$

This gives us our 13 lines of four points each and thus our projective plane.

Section 6.5, pages 438–440

1. Using w_1, w_2, w_3, and w_4 for the words, $d(w_1,w_2) = 5$, $d(w_1,w_3) = 4$, $d(w_1,w_4) = 3$, $d(w_2,w_3) = 3$, $d(w_2,w_4) = 4$, $d(w_3,w_4) = 5$.

3. 1 1 1 1 1 1 1   1 1 1 0 1 0 0   1 1 0 1 0 1 0   1 1 0 0 0 0 1
    1 0 1 1 0 0 1   1 0 1 0 0 1 0   1 0 0 1 1 0 0   1 0 0 0 1 1 1
    0 0 1 1 1 1 0   0 0 1 0 1 0 1   0 0 0 1 0 1 1   0 0 0 0 0 0 0
    0 0 1 1 1 1 0   0 0 1 0 1 0 1   0 0 0 1 0 1 1   0 0 0 0 0 0 0

5.
$$\begin{pmatrix} 1 & 1 & 1 & 1 & 1 & 1 & 1 & 1 \\ 1 & 1 & 1 & 0 & 1 & 0 & 0 & 0 \\ 1 & 1 & 0 & 1 & 0 & 1 & 0 & 0 \\ 1 & 1 & 1 & 1 & 0 & 0 & 1 & 0 \end{pmatrix}$$
This matrix does not have an identity matrix as its last four columns.

$$\begin{pmatrix} 1 & 1 & 1 & 0 & 1 & 0 & 0 & 0 \\ 1 & 1 & 0 & 1 & 0 & 1 & 0 & 0 \\ 1 & 1 & 1 & 1 & 0 & 0 & 1 & 0 \\ 0 & 1 & 1 & 1 & 0 & 0 & 0 & 1 \end{pmatrix}$$

7. $1 + 2 \cdot 11 + 4 \cdot \binom{11}{2} = 243 = 3^5$. A code consisting of 3^5 11-tuples of integers mod 3.

9. $\{2,3,4\}$, $\{1,2,7\}$, $\{1,3,6\}$, $\{1,4,5\}$, $\{2,5,6\}$, $\{3,5,7\}$, $\{4,6,7\}$;
   $(0,1,1,1,0,0,0)$, $(1,1,0,0,0,0,1)$,$(1,0,1,0,0,1,0)$,$(1,0,0,1,1,0,0)$,
   $(0,1,0,0,1,1,0)$,$(0,0,1,0,1,0,1)$,$(0,0,0,1,0,1,1)$.

11. $\{1,2,5,6,7\}$, $\{1,2,3\}$, $\{3,5,6,7\}$, $\emptyset$, $\{1,2,3,4,5,6,7\}$

15. Each i-element set is in $\binom{v-i}{t-i}$ different t-element sets, each of which is in λ blocks. An i-element subset of a block may be extended to a t-element subset of the block in $\binom{k-i}{t-i}$ ways. Thus the number of blocks in which a given i-element set lies is $\lambda\binom{v-i}{t-i}/\binom{k-i}{t-i}$.

17. If x and z are different in position i, then either x and y or y and z must be different in position i. Thus the total number of places where y differs from x plus the total number of places where y differs from z must be at least the number of places where x differs from z. This proves the inequality.

19. Two rows cannot have more than $t-1$ ones in common or else two blocks of the design would contain the same t-set. Therefore the two rows differ in at least $2k - 2(t-1) = 2(k-t+1)$ places.

21. (a) The alphabet would have size six, the length would be seven, the number of codewords would be 6^5, and the minimum distance would be three.
    (b) There are 6^3 choices for the first three entries of a codeword; for each such choice there must be 6^2 ways to complete the choice to a codeword. The set of choices of the last four places with a given first three will be a code of length

four, size 36, and minimum distance three (because there were no differences in the first three places).

(c) By Theorem 5.8, if there were such a code, there would be a pair of mutually orthogonal six by six Latin squares.

23. (a) Because the four rows of either matrix in Exercise 5 are all codewords as described in Exercise 4, and because any codeword is orthogonal to every row of a parity check matrix, these four codewords are orthogonal to all codewords. Thus all sums of two, three, or four of these codewords are orthogonal to all codewords (if $wv^t = 0$ and $uv^t = 0$ then $(w + u)v^t = 0$). The code is closed under addition and all these sums are distinct, giving us that the 15 nonzero vectors of the code are orthogonal to every member of the code. Therefore the code is self-orthogonal. However, any vector orthogonal to all codewords is orthogonal to the four rows of these matrices and so is in the code. Thus the code is self-dual.

(b) Neither. A code consists of all vectors orthogonal to the rows of its parity check matrix. Further, if a row vector is orthogonal to every word in the code, we can add it to the parity check matrix without changing the code. Thus a row vector orthogonal to every word in the code corresponds to a new parity check equation that is a combination of the original parity check equations defining the code. In other words, a row vector orthogonal to every code word is a sum of the rows of the parity check matrix. For an extended Hamming code, there are 2^{m+1} such sums, but $2^{2^m - m - 1}$ members of the code, and for $m > 3$, $2^{2^m - m - 1} > m + 1$.

CHAPTER 7

Section 7.1, pages 454–457

1. A is not older than A. If A is older than B, then B is not older than A. If A is older than B and B is older than C, then A is older than C. A linear ordering.

2. Substitute "weighs less than" for "is older than" in Exercise 1. Making the change to "less than or equal to" gives a reflexive relation which is not antisymmetric, so it is neither a strict partial ordering nor a reflexive partial ordering. Notice that the associated reflexive partial ordering is "weighs less than or is the same person as."

Figure A.13

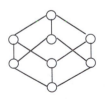

Figure A.14

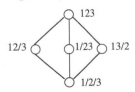

3. See Figure A.13.   **5.** See Figure A.14.

7. Every element of A is an element of A, so $A \subseteq A$. If $A \subseteq B$ and $B \neq A$, then some element of B is not in A. Therefore $B \not\subseteq A$. If each element of A is in B and each element of B is in C, then each element of A is in C. Thus the $\subseteq$ relation is reflexive, antisymmetric, and transitive, so it is a partial ordering.

9. Suppose y covers a minimal element x and let z be another minimal element. Since $x \not< z$, then $z < y$. By the same argument, if $y > y_1$, and y_1 covers z, then $y > y_1 > x$, so y must *cover* all minimal elements. To define our function f, we are going to construct a sequence of sets X_i. We start with $X_1 = X$. We let M_1 be the minimal elements of X_1. We define $f(x) = 1$ for all elements of M_1. Each minimal element of $X_1 - M_1$ covers each minimal element of X_1, so define $f(y) = 2$ for all minimal elements y of $X_1 - M_1 = X_2$. Given X_i, let M_i be its minimal elements and define $f(y) = i$ for each $y \in M_i$. When you run out of elements of X, f has the desired properties. This means the diagram of (X, P) is in layers with each element of a layer covering all the elements of the layer below and only those elements.

11. We prove there is at least one maximal element. Let $x_0 \in X$. If x_0 is maximal, we are done. Otherwise, let $x_1 \in X$ with $x_1 > x_0$. In general, given a sequence $x_0 < \ldots < x_i$, if x_i is not maximal, let x_{i+1} be a member of X with $x_{i+1} > x_i$. By the transitive law, $x_{i+1} > x_j$ for $j = 0, 1, \ldots, i$ so it is different from all other x_j's. Since X is finite, this process must stop, i.e., some x_i must be maximal. Proving there is at least one minimal element is similar.

Figure A.15

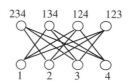

13. See Figure A.15.

15. Since $x < y$ if (x, y) is a cover, $(x, y) \not< (x, y)$ for this would require $y < x$. If $(x_1, y_1) < (x_2, y_2) < (x_3, y_3)$ then $y_1 < x_2 < y_2$ and $y_2 < x_3$, so $y_1 < x_3$. Since condition 2 follows from 1 and 3, this proves we have a partial ordering.

Figure A.16

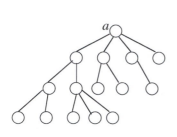
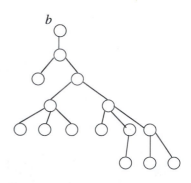

17. See Figure A.16.

19. Restrict the ordering of the subsets of $\{a, b\}$ to the two sets $\emptyset$ and $\{a, b\}$.

21. **(a)** The element t is in $\mathrm{Pred}(Z)$ if and only if $t < z$ for each $z \in Z$, which is true if and only if $t \in \cap_{z:z\in Z}\mathrm{Pred}(z)$. The proof for $\mathrm{Suc}(Z)$ is similar.
   (b) Take $Z = \{x\}$ for some minimal element x.   **(c)** Take $Z = \emptyset$.
   (d) If $x < y$ and $z \in \mathrm{Pred}(x)$, then $z < x < y$ implies that $z \in \mathrm{Pred}(y)$.
   (e) $\mathrm{Pred}(Y \cup Z) = \mathrm{Pred}(Y) \cap \mathrm{Pred}(Z)$, for t precedes all elements of $Y \cup Z$ if and only if t precedes all elements of Y and all elements of Z, which happens if and only if $t \in \mathrm{Pred}(Y)$ and $t \in \mathrm{Pred}(Z)$, which is true if and only if $t \in \mathrm{Pred}(Y) \cap \mathrm{Pred}(Z)$.
   (f) No, because a t may precede each element in $Y \cap Z$ without preceding each element in Y or each element in Z.

23. If $\mathrm{Suc}(x) \neq \mathrm{Suc}(y)$, then $\mathrm{Pred}(\mathrm{Suc}(x)) \neq \mathrm{Pred}(\mathrm{Suc}(y))$ since $z \in \mathrm{Pred}(\mathrm{Suc}(z))$. This implies that f is one to one. Now, $h(\mathrm{Pred}(\mathrm{Suc}(x))) < h(\mathrm{Pred}(y))$ if and only if $\mathrm{Pred}(\mathrm{Suc}(x)) \subseteq \mathrm{Pred}(y)$ since the predecessor sets are linearly ordered. However, this is not the case if and only if there is a predecessor z of y which is not a predecessor of any successor w of x. Thus, since $x < w$ and $z < y$, we may conclude that $x < y$ if and only if $\mathrm{Pred}(\mathrm{Suc}(x)) \subseteq \mathrm{Pred}(y)$.

25. The length of the interval assigned to w must be the same as the length of the intervals assigned to x and y, so that the interval assigned to w can overlap both those of x and y or both those of y and z, but not all three. Thus, either the interval for w is entirely below that of z or entirely above that for x.

Section 7.2, pages 471–476

1. $\emptyset$, $\{1\}$, $\{2\}$, $\{3\}$, $\{1, 2\}$, $\{1, 3\}$, $\{2, 3\}$, $\{1, 2, 3\}$.   **3.** No.

Figure A.17

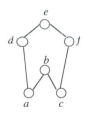

5. See Figure A.17. Width is 3. Height is 2.   **7.** 3

9. $\emptyset$, $\{1\}$, $\{1, 2\}$, $\{1, 2, 3\}$, $\{1, 2, 3, 4\}$; $\{2\}$, $\{2, 3\}$, $\{2, 3, 4\}$; $\{1, 4\}$; $\{3\}$, $\{1, 3\}$, $\{1, 3, 4\}$; $\{3, 4\}$; $\{4\}$, $\{2, 4\}$, $\{1, 2, 4\}$

10. If two chains have x in common, delete x from one of them. Repeat the process until no two chains have any elements in common.

11. Height $= 3$; width $= 6$.

13. The smallest number of antichains in a partition of a set X ordered by P into antichains is the maximum number of elements in any chain of P. Use the antichain of elements of each height from 0 to $h(P)$. You need at least this many antichains because each element of the longest chain of P must be in a different antichain.

15. There is no edge from the minimum element of either chain directed downward to any element of either chain, so there are no decreasing walks.

17. Simply put b at the bottom of each linear extension in a set of linear extensions whose intersection is P.

19. $\{\emptyset, \{1\}\}$ is a symmetric chain decomposition for the subsets of a one-element set. Assuming we have a symmetric chain decomposition for the subsets of $\{1, 2, \ldots, n-1\}$, for each chain $C = \{S_1, S_2, \ldots, S_k\}$, we construct two new chains $C' = \{S_2, \ldots, S_k\}$ (when C is not a one-element chain) and $C'' = \{S_1, S_1 \cup \{n\}, S_2 \cup \{n\}, \ldots, S_k \cup \{n\}\}$, which are both symmetric saturated chains. These chains include all subsets of $\{1, 2, \ldots, n\}$. While this is not part of the answer, note that when $n = 2m - 1$ the number of chains we get is

$$2\binom{2m-1}{m} = \binom{2m-1}{m-1} + \binom{2m-1}{m} = \binom{2m}{m},$$

which is the width of the set of all subsets of a zero-element set. When $n = 2m$, a slightly more complicated argument shows that we get the right number of chains.

20. A maximum-sized chain can include at most one member of an antichain A. Thus, for any antichain A, $\sum_{a \in A} |\{c | c$ is a maximum-length chain containing $a\}| \leq$ total number of maximum length chains of (X, P). In the poset of all subsets of an n-element set, the right-hand side is $n!$. (Each list of N tells an order to put elements of N together in subsets leading from $\emptyset$ to N.) If K is a set of size k, there are $k!$ chains of maximum length from $\emptyset$ to K and $(n-k)!$ chains of maximum length from K to N. Thus, by the multiplication principle, the inequality becomes $\sum_{K \in A} k!(n-k)! \leq n!$. Now let B be the biggest value of the binomial coefficients $\binom{n}{i}$. Then $\frac{n!}{k!(n-k)!} \leq B$, so $k!(n-k)! \geq \frac{n!}{B}$. Thus

$$\sum_{K \in A} \frac{n!}{B} \leq n!, \quad \text{or } |A|\frac{n!}{B} \leq n! \text{ so that } |A| \leq B.$$

21. **(a)** If the ordered pairs (a, b) and (b, c) are in relation p_i, then $a < b < c$ (mod P), so that (a, c) is not a cover. Thus no cover can be in the transitive closure of the P_i's unless it is in one of the P_i's.
(b) If $(w, x) < (x, y) < (y, z)$ (mod R), then $x < y$, so that $(w, x) < (y, z)$ (mod R). Since $y \leq x$ (mod P) is impossible when $x < y$ (mod P), the relation R must be irreflexive.
(c) Every cover must be in a chain for the chains to generate R. Assume we use maximal (or saturated) chains (so that they need not be disjoint); then each corresponds to a chain of covers relative to R, and the minimum number of chains we need to include all covers is the size of a maximum antichain in the ordered set of covers.

23. If it is a path and the root is at one end, then the tree is a chain and has dimension 1. Otherwise, remove the root and you get some smaller number

k of rooted trees, each one an intersection of two linear orderings. Thus tree i is the intersection of L_i and L'_i. Thinking of L_i and L'_i as lists, let $L = L_1 L_2 \cdots L_k \text{Root}$ and let $L' = L'_k L'_{k-1} \cdots L'_1 \text{Root}$ and the intersection of L and L' will be the tree.

Figure A.18

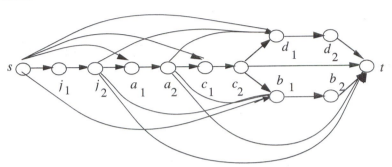

25. **(a)** See Figure A.18.   **(b)** Zero
   (c) If j of the flows f_i send a unit of flow into vertex k, then the same j flows f_i send a unit of flow out of vertex k. Therefore the sum of the flows satisfies the balance condition at each vertex.
   (d) The flow $f(s, d_1) = 1$, $f(d_1, d_2) = 1$, $f(d_2, t) = 1$; $f(s, j_1) = 1$, $f(j_1, j_2) = 1$, $f(j_2, b_1) = 1$, $f(b_1, b_2) = 1$, $f(b_2, t) = 1$; $f(s, a_1) = 1$, $f(a_1, a_2) = 1$, $f(a_2, c_1) = 1$, $f(c_1, c_2) = 1$, $f(c_2, t) = 1$; all other edges have flow zero. This flow has value 3.
   (e) $f(s, c_1) = 1$, $f(c_1, c_2) = 1$, $f(c_2, d_1) = 1$, $f(d_1, d_2) = 1$, $f(d_2, t) = 1$; $f(s, j_1) = 1$, $f(j_1, j_2) = 1$, $f(j_2, a_1) = 1$, $f(a_1, a_2) = 1$, $f(a_2, b_1) = 1$, $f(b_1, b_2) = 1$, $f(b_2, t) = 1$; all of the other edges have flow zero. This flow has value 2.
   (f) A chain decomposition problem with a minimum number of chains corresponds to a minimum flow in which each edge (x_1, x_2) has necessity one.
27. The minimum flow subject to certain necessities is equal to the maximum necessity of any cut. (The necessity of a cut is the sum of the necessities of the edges from the source-set to the sink-set in an acyclic network; an ordered set gives an acyclic network.) The cut of maximum necessity is a maximum antichain; the flow of minimum value gives the chain decomposition; flow-augmenting paths of worth one give the chains of the decomposition.

Figure A.19

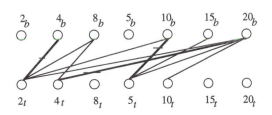

29. (a) See Figure A.19

(b) See Figure A.20. $\{2_b, 2_t, 4_b, 4_t, 20_b, 20_t\}$, $\{8_b, t_t\}$, $\{5_b, 5_t, 10_b, 10_t\}$, $\{15_b, 15_t\}$ are the connected components. They correspond to the chains $\{2, 4, 20\}$, $\{8\}$, $\{5, 10\}$, $\{15\}$.

Figure A.20

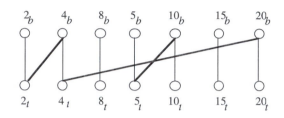

(c) An alternating path (in part (a)) is $10_t 20_b 4_t, 8_b$.

(d) The new matching is $\{2_t, 4_b\}$, $\{4_t, 8_b\}$, $\{10_t, 20_b\}$, $\{5_t, 10_b\}$. It gives the chain decomposition $\{2, 4, 8\}$, $\{5, 10, 20\}$, $\{15\}$.

(e) If we find a maximum matching in the Fulkerson graph for the $<$ relation, the connected components in the Fulkerson graph for the matching or equal relation give the chains of a chain decomposition with a minimum number of chains.

31. Konig's maximum matching–minimum vertex cover theorem. The number of chains in a matching or equal to graph is n minus the number of matching edges; this is a minimum when the matching has maximum size. The complement of an antichain is a vertex cover of the edges (the two vertices covering an edge of the ordering cannot both be in an antichain), so that a minimum-sized vertex cover is the complement of a maximum-sized antichain.

Section 7.3, pages 484–487

1. If x is the larger of x and y, then $x \geq x$ and $x \geq y$ and for any z with $z \geq x$ and $z \geq y$, it is the case that $z \geq x$. Thus x is the least upper bound of x and y.

Figure A.21

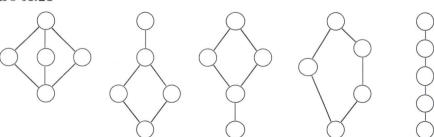

3. Yes, the greatest common divisor.   4. See Figure A.21. There are 5.

5. If $x = p_1^{i_1} \cdot p_2^{i_2} \cdots p_n^{i_n}$ and $y = p_1^{j_1} \cdot p_2^{j_2} \cdots p_n^{j_n}$, let k_m be the larger of i_m and j_m. $x \vee y = p_1^{k_1} \cdot p_2^{k_2} \cdots p_n^{k_n}$. (Note that we are assuming implicitly that the set $\{p_1, p_2, \ldots, p_n\}$ is the set of all primes used in factoring either x or y. Thus if p_m does not appear in the factorization of x, we have $i_m = 0$.) The product is clearly the smallest number that has both x and y as factors.

7. Since the intersection of two equivalence relations is an equivalence relation (it is necessary to check that the intersection of two equivalence relations is reflexive, symmetric, and transitive), it is their greatest lower bound. (Since it is the largest relation contained in both and is an equivalence relation, it must also be the largest equivalence relation contained in both.)

9. The lattice of partitions of Y. We let $f(R)$ be the equivalence class partition of R for each equivalence relation R. The relationship studied in Chapter 2 between equivalence class partitions and equivalence relations shows that this function is a bijection; the fact that the equivalence classes of the intersection of two equivalence relations are the intersections of the equivalence classes shows $f(R) \leq f(R')$ in the ordering of partitions if and only if $R \subseteq R'$ in the subset ordering.

11. See Figure A.22.

Figure A.22

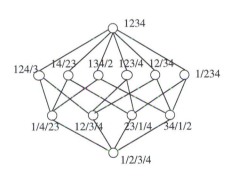

13. The partition lattice on a four-element set.

15. It is isomorphic to the lattice of subsets of a three-element set.

17. (a) We must show that if two partitions of V are bonds, then so is their join (in the partition lattice). However, two vertices in the same class of a bond are joined by a walk in that class, and if two classes overlap, then all vertices in their union are connected to each other by walks that lie in their union. Thus, as we repeatedly take unions of overlapping classes, all vertices in the union are connected to each other by walks in the union. Thus the join of two bonds in the partition lattice is a bond, so it is the least upper bound of these two bonds. (Why is it the *least*?)

(b) F and R are as in the example. Then R is a bond finer than both P and Q, so it is a lower bound to P and Q in the bond lattice. Now suppose R' is a bond finer than both P and Q. Then any two vertices in a class of R' are connected by a walk using edges in F. Thus they lie in a class of the connected component partition of (V, F). Thus $R' \leq R$. Thus R is the greatest lower bound of P and Q.

19. The proof is just like the first paragraph of the proof of Theorem 3.4.

21. We do part (a); part (b) is essentially the same. Since $X \cap Y \subseteq Y$, $X \cap Y \subseteq Y \cup Z$. Also $X \cap Y \subseteq X$. Therefore, the union $(X \cap Y) \cup (X \cap Z)$ must be a subset of $X \cap (Y \cup Z)$. In other words, every element of $(X \cap Y) \cup (X \cap Z)$ is an element of $X \cap (Y \cup Z)$. If we also show each element of $X \cap (Y \cup Z)$ is an element of $(X \cap Y) \cup (X \cap Z)$, then we have shown the two sets have exactly the same elements, so they are equal. To this end, suppose that $w \in X \cap (Y \cup Z)$. Then $w \in X$ and $w \in Y \cup Z$, so either $w \in X$ and $w \in Y$ or else $w \in X$ and $w \in Z$. Thus either $w \in X \cap Y$ or $w \in X \cap Z$, so that $w \in (X \cap Y) \cup (X \cap Z)$ Therefore each element of $X \cap (Y \cup Z)$ is in $(X \cap Y) \cup (X \cap Z)$. Therefore the two sets are equal.

23. The meet of all the elements in the lattice $(x_1 \wedge x_2 \wedge \cdots \wedge x_n$ is defined in analogy with $x_1 + x_2 + \cdots x_n)$ is a lower bound to all elements in the lattice and has no elements below it, so it is a bottom. Similarly, the join of all the elements in the lattice is a top.

25. We show that the intersection of two flats is a flat, so the flats form a meet semilattice. Suppose our matroid is defined on a set X and G and H are flats. Let I be a maximum-sized independent set in $G \cap H$. By the expansion property, we know there are independent sets I_1 and I_2 such that $I \cup I_1$ is a maximum-sized independent subset of G and $I \cup I_2$ is a maximum-sized independent subset of H. We want to show that $F(I) = F(I \cup I_1) \cap F(I \cup I_2)$. Note that if x depends on I, then x depends on $I \cup I_1$ and $I \cup I_2$. Thus $F(I) \subseteq F(I \cup I_1) \cap F(I \cup I_2)$. Now suppose x depends on $I \cup I_1$ and $I \cup I_2$. To show that x depends on I, suppose the contrary, that $I \cup \{x\}$ is independent. Since G and H are flats, $G = F(I \cup I_1)$ and $H = F(I \cup I_2)$. (This actually requires proof, because conceivably $G = F(J)$ for some other independent set J.) Thus $x \in G \cap H$, so I is not a maximum-sized independent set in $G \cap H$. Thus x depends on I, so $x \in F(I)$. Thus $F(I) = G \cap H$. However, the intersection of two flats G and H will contain any flat contained in both G and H, so $G \cap H$ is the greatest lower bound to G and H in the ordered set of flats. Since X is a flat, the ordered set has a top element, so by Theorem 3.3, the ordered set is a lattice.

27. (a), (c), and (f) are distributive; (a), (b), (c), and (f) are modular.

29. We must show that for $t_1, t_2 \in [x \wedge y, x]$, if $t_1 \vee y = t_2 \vee y$, then $t_1 = t_2$. To do so, note that if $t_1 \vee y = t_2 \vee y$, then $t_1 \vee t_2 \vee y = t_1 \vee y$. Thus we would have a $t < t'$, both in $[x \wedge y, x]$, such that $t \vee y = t' \vee y$. Then, by the modular law,

$$t' = t' \wedge (t \vee y) = t \vee (t' \wedge y) \leq t \vee (x \wedge y) = t,$$

so that $t = t'$. Therefore f is one-to-one. We must also show that f is onto; for this purpose note that if $y \leq w \leq x \vee y$, then, by the modular law, $(x \wedge w) \vee y = w \wedge (x \vee y) = w$. We may conclude that the two intervals have the same size.

31. We can show that the inverse for the mapping $f(t) = t \vee y$ of Exercise 21 is $g(w) = w \wedge z$. We must show that the mappings $f(t) = t \vee y$ and $g(w) = w \wedge x$

are order preserving; however, if $t_1 \geq t_2$, then $t_1 \vee y \geq t_2 \vee y$, and if $w_1 \leq w_2$ then $w_1 \wedge x \leq w_2 \wedge x$.

32. 13/24 is crossing. 12/34 is not and 14/23 is not. We show that if P and Q are noncrossing partitions, then so is $p \wedge Q = R$, where the meet is taken in the partition lattice. Let p and q be in the block B of R, and r and s be in the block C of R with $p < q < r < s$. Then p and q are in the same block of P as r and s. Since P is noncrossing, p, q, r, and s are in the same block of P. Similarly, all four are in the same block of Q. Then all four are in the same block of $R = p \wedge Q$, so $B = C$. Therefore R is noncrossing. This shows that the noncrossing partitions form a meet semilattice; since the partition with one block is noncrossing the semilattice is a lattice. The join operation is not the same; 13/2/4 and 24/1/3 join to a noncrossing partition in the partition lattice.

Section 7.4, pages 501–504

1. a, c, f are distributive.

3. Only a is a Boolean algebra because it is the only complemented distributive lattice.

5. See Figure A.23.

Figure A.23

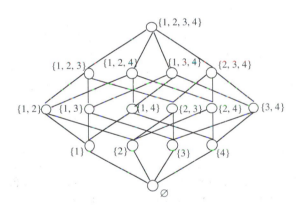

7. This has the same picture as Exercises 5 and 6.

9. See Table A.2.

11. The lattice of factors of 30 is isomorphic to the lattice of subsets of $\{2, 3, 5\}$. Thus it is a Boolean algebra. However, 2 has no complement in the lattice of factors of 36, because there is no element whose join with 2 is 36.

13. (a) $(x_1 + x_2) \cdot x_3$, which is also $x_1 x_3 + x_2 x_3$.   (b) $x_1^{-1} + x_2$
   (c) $(x_1 + x_2)^{-1}$ (which is also $x_1^{-1} \cdot x_2^{-1}$)   (d) $x_1 \cdot x_2^{-1} + x_1^{-1} \cdot x_2$
   (e) $(x_1 + x_2) \cdot x_3^{-1}$
   (f) $x_1 x_2 x_3 + x_1 x_2^{-1} x_3^{-1} + x_1^{-1} x_2 x_3^{-1} + x_1^{-1} x_2^{-1} x_3$

15. (a)   If some elements are not joins of join-irreducibles, let x be a minimal such element. Then x is not join-irreducible so $x = x_1 \vee x_2$ with $x_1 x_2 < x$.

Table A.2

p	q	r	+ p	∧	(q	+ ∨	r)	(p	+ ∧	q)	∨	(p	+ ∧	r)
T	T	T	T	T	T	T	T	T	T	T	T	T	T	T
T	T	F	T	T	T	T	F	T	T	T	T	T	F	F
T	F	T	T	T	F	T	T	T	F	F	T	T	T	T
T	F	F	T	F	F	F	F	T	F	F	F	T	F	F
F	T	T	F	F	T	T	T	F	F	T	F	F	F	T
F	T	F	F	F	T	T	F	F	F	T	F	F	F	F
F	F	T	F	F	F	T	T	F	F	F	F	F	F	T
F	F	F	F	F	F	F	F	F	F	F	F	F	F	F
			1	3	1	2	1	1	2	1	3	1	2	1

By minimality of x, x_1 and x_2 are joins of join-irreducibles, so x_1 is a join of join-irreducibles, a contradiction.

(b) No.

(c) Since, by part (a), every element is a join of join-irreducibles, x is the join of a set of join-irreducible elements. In fact, x is the join of the maximal join-irreducibles $p_1, p_2, \ldots p_n$ below x. If x is a join of a proper subset of these join-irreducibles, then some p_i, say p_1, is redundant. If so, then $n > 3$ because $n = 1$ would imply $x = p_1$ and $n = 2$ would imply $p_1 \leq p_2$, and the maximality of p_1 and p_2 would imply $p_1 = p_2$ and $n = 1$. Thus $n \geq 3$ and $p_1 \leq p_2 \vee \cdots \vee p_n$ so that $p_1 \leq (p_1 \wedge p_2) \vee \cdots \vee (p_1 \wedge p_n)$. Since p_1 is join-irreducible, we have $p_1 = p_1 \wedge p_i$ for some i, so $p_1 \leq p_i$. By the maximality of p_1 and p_i, $p_1 = p_i$, a contradiction. Now suppose $x = q_1 \vee q_2 \vee \cdots \vee q_m$, all the q_i are join-irreducible, and the maximal join-irreducible p_1 is not one of the q_i's. Then $p_1 = p_1 \wedge x = (p_1 \wedge q_1) \vee \cdots \vee (p_1 \wedge q_m)$, and by the join-irreducibility of p_1, we have $p_1 = p_1 \wedge q_i$ for some i. Thus $p_i \leq q_i$ and by the maximality of p_1, $p_1 = q_i$, a contradiction. Thus the only irredundant set of join-irreducibles whose join is x is $\{p_1, \ldots, p_n\}$. This proves that each element has a unique irredundant representation as a join of join-irreducible elements.

17. $\sim (p \vee q) = \sim p \wedge \sim q$ and $\sim (p \wedge q) = \sim p \vee \sim q$. See Table A.3. The other truth table is similar.

Table A.3

p	q	~	(p	∨	q)	~	p	∧	~	q
T	T	F	T	T	T	F	T	F	F	T
T	F	F	T	T	F	F	T	F	T	F
F	T	F	F	T	T	T	F	F	F	T
F	F	T	F	F	F	T	F	T	T	F
		3	1	2	1	2	1	3	2	1

19. Suppose x_1 and x_2 are both complements to the element x of a Boolean algebra. Then $x_1 = x_1 \wedge (x \vee x_2) = (x_1 \wedge x) \vee (x_1 \wedge x_2) = x_1 \wedge x_2$. By the same kind of computation $x_2 = x_1 \wedge x_2$. Thus $x_1 = x_2$. Therefore each element of a Boolean algebra has a unique complement.

21. Addition is commutative and associative, has a zero element, and each element of the Boolean algebra has an additive inverse. For each of the four ways to choose r and s, we may verify that $(r + s)a = ra + sa$ and $r(sa) = (rs)a$; further, $1 \cdot a = a$ is given. Therefore, every Boolean algebra is a vector space.

23. See Figure A.24.

Figure A.24

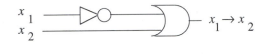

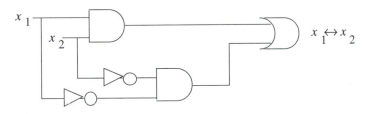

Section 7.5, pages 519–522

1. $\mu(\emptyset, \emptyset) = 1$; $\mu(\emptyset, \{a\}) + \mu(\emptyset, \emptyset) = 0$; $\mu(\emptyset, \{b\}) + \mu(\emptyset, \emptyset) = 0$; $\mu(\emptyset, \emptyset) + \mu(\emptyset, \{a\}) + \mu(\emptyset, \{b\}) + \mu(\emptyset, \{a, b\}) = 0$; $\mu(\emptyset, \{a\}) = \mu(\emptyset, \{b\}) = -1$; $\mu(\emptyset, \{a, b\}) = 1$.

3. $\mu(i, i) = 1$; $\mu(1, 2) = \mu(2, 3) = \mu(3, 4) = -1$; $\mu(1, 3) = \mu(2, 4) = \mu(1, 4) = 0$.

5. To complete the proof, note that when $i - j = 0$, Equation (a) determines $\mu(x_i, x_j)$. Now suppose inductively that the equations determine the value of $\mu(x_i, x_j)$ whenever $i - j < k$ and suppose $i = j + k$. Then $\displaystyle\sum_{z : x_i \leq z \leq x_j} \mu(x_i, z) = 0$

so that $\mu(x_i, x_j) = \displaystyle\sum_{x_h : x_i \leq x_h < x_j} \mu(x_i, x_h) = \sum_{h=i}^{i+k-1} \mu(x_i, x_h)$. But the right-hand side is determined by Equations 1–3 by the inductive hypothesis, so $\mu(x_i, x_j)$ is determined as well.

7. **(a)** $\mu(x, x) = 1$; $\mu(a, b) = \mu(a, c) = \mu(a, d) = \mu(b, e) = \mu(c, e) = \mu(d, e) = -1$; $\mu(a, e) = 2$.

(b) $\mu(x, x) = 1$, $\mu(a, b) = \mu(a, c) = \mu(b, e) = \mu(c, d) = \mu(d, e) = -1$; $\mu(a, d) = \mu(c, e) = 0$; $\mu(a, e) = 1$.

(c) $\mu(x, x) = 1$; $\mu(b, a) = \mu(b, c) = \mu(a, d) = \mu(c, d) = \mu(d, e) - 1$; $\mu(b, d) = 1$; $\mu(b, e) = \mu(a, e)\,\mu(c, e) = 0$.

(d) $\mu(x, x) = 1$ $\mu(a, b) = \mu(a, e) = \mu(b, c) = \mu(b, f) = \mu(d, b) = \mu(d, e) = \mu(e, c) = \mu(e, f) = -1$; $\mu(a, f) = \mu(a, c) = \mu(d, c) = \mu(d, f) = 1$.

9. The ordered set of subsets of $T - S$ is isomorphic to the ordered set of subsets of T containing S by the function $f(X) = S \cup X$. But isomorphic ordered sets have the same value for $\mu(b, t)$. Thus $\mu(S, T) = \mu(\emptyset, T - S)$.

11. **(a)** Two vertices in the same class are connected by a path of vertices in the class (because the path is entirely the same color as the other vertices in the the class).

(b) The "singleton partition" into parts all of size 1, the bottom element of the partition lattice.

(c) x^k (The coloring need only assign the same color to each member of a class, so we color k classes with x colors.)

(d) $\displaystyle\sum_{\pi:\pi \geq b} \mu(b, \pi) x^{k(\pi)}$ where $k(\pi)$ is the number of classes of π and b is the bottom element of the bond lattice of the graph. Begin by proving that $\mu(x, y) = \mu(1, y/x)$. Then prove that if n has k distinct prime factors, then $\mu(a, n) = (-1)^k$ using induction. (Outline: the lattice has $\binom{k}{i}$ elements x with i distinct factors, and by induction, $\mu(1, x) = (-1)^i$. Thus for an element y with k distinct prime factors, $\sum_{j=0}^{k} -1(-1)^j (\binom{k}{j}) + \mu(1, y) = 0$; by the binomial theorem $\mu(1, y) = (-1)^k$.) Now prove by induction on n that if p^2 is a factor of n for one prime p, then $\mu(1, n) = 0$. To do this let m be the product of the distinct prime factors of n. Then the sum of $\mu(1, x)$ over all values of x that are factors of m is 0, and $\mu(1, x)$ is 0 by induction for any factor x of n that is not a factor of m. Then by part (b) of Theorem 5.6, $\mu(1, n) = 0$.

(e) The sum of $\mu(b, \pi)$ over all bonds with i classes is the coefficient a_i of x^i, so letting a_i be this sum, we may write the number of proper colorings as $\displaystyle\sum_{i=0}^{n-1} a_i x^i$ where n is the number of vertices of the graph. This is a polynomial in x.

13. $\mu(i, i) = 1$; $\mu(i, i+1) = -1$, $\mu(i, j) = 0$ if $j < i$ or $j > i+1$. If $f(k) = \sum_{j=0}^{i} g(j)$, then $g(k) = f(k) - f(k - 1)$.

15. If $h(B) = 1$, then $\mu(b, B) = -1$, so $(-1)^{h(B)} \mu(b, B) = 1 > 0$. Proceed by induction on the height of B, assuming the theorem is true when $h(B) < k$. Let B be a bond of height k. Let G' be the graph on V whose edges connect two points lying in the same class of B. Then $\mu(b, B) = \mu(b, t)$ in the bond lattice L' of G'. Let e be an edge of G'. Then the partition π with e as one class and the other classes as single vertices is an atom in L'. If $\sigma \vee \pi = B$, then if $\sigma \neq B$, e must overlap exactly two classes of σ that have a class of B as their union, and all other classes of σ must be classes of B. Thus σ has one fewer class than B. Since the height of a bond is its height in the partition lattice (this requires some proof), all such bonds σ have height $h(B) - 1$. Then applying Exercise 12 we get $\mu(b, B) = \sum_{\substack{\sigma:\sigma \neq B \\ \sigma \vee \pi = B}} \mu(b, \sigma)$. But $(-1)^{h(\sigma)} \mu(b, \sigma) > 0$ by our inductive hypothesis, so $(-1)^{h(B)} \mu(b, B) = \sum_{\substack{\sigma:\sigma \neq B \\ \sigma \vee \pi = B}} (-1)^{h(\sigma)} \mu(b, \sigma) > 0$. This exercise can also be done using an argument somewhat like the proof of Theorem 5.8 and applying Theorem 5.7 rather than Exercise 12.

16. $\mu(b, t) = 0$

17. $\mu(x, x) = 1$; $\mu(x, y) = 0$ unless y/x is an integer which is a product of distinct primes; $\mu(x, y) = (-1)^k$ if y/x is a product of k distinct primes.

19. **(a)** Each function from the n-element set N to the k-element set K gives a list of n elements of K. A circular arrangement of period d may be listed in d different ways, corresponding to where we start listing members as we go through a period. Thus the total number of lists is both k^n and the sum given.

(b) $N(m) = \dfrac{1}{m} \displaystyle\sum_{d:d\mid m} \mu(d, m) k^d$

21. If $f^k(x) = x$, then the multiset $\{f(x), f^2(x), \ldots, f^k(x)\}$ is either an orbit or a multiset listing each member of an orbit of size d the same number h of times, so that $dh = k$. Thus $f^k(x) = x$ if and only if x is in an orbit of size d and d is a divisor of k. Let $N(d)$ be the number of orbits of f of size d. There are $dN(d)$ elements of orbits of size d, so summing over all numbers d that divide k gives the number of fixed points of f^k. Now proceed as in Exercise 19.

Section 7.6, pages 529–531

1. The product of the three lattices consists of all triples of the form (S_1, S_2, S_3) where S_i is either empty or $\{i\}$. Define $f(S_1, S_2, S_3) = S_1 \cup S_2 \cup S_3$. Then f maps the product onto the lattice of subsets of $\{1, 2, 3\}$, and $f(S_1, S_2, S_3) \geq f(T_1, T_2, T_3)$ if and only if each T_i is a subset of S_i. In particular $f(S_1, S_2, S_3) = f(T_1, T_2, T_3)$ if and only if each S_i equals T_i.

3. Let $X_1 = \{1, 2, 4\}$ ordered by "is a factor of" and $X_2 = \{1, 3\}$ ordered by "is a factor of." Then $f(a, b) = a \cdot b$ is an isomorphism from the product of (X_1, P_1) and (X_2, P_2) onto the positive factors of 12.

5. If we had not been able to add b to each chain C_i, then in a chain C_i there might have been *no* element less than or equal to some x, so $f_i(x)$ would not have been defined.

6. Each bond of G is finer than the connected component partition of G. Thus each bond is the join of two bonds, each a bond of the graph on one of G's connected components. Let V_1 and V_2 be the vertex sets of the connected components. Each bond B has a restriction B_1 to V_1 and B_2 to V_2, and each class of B is a class of B_1 or B_2. It is straightforward to check that $F(B) = (B_1, B_2)$ is the desired isomorphism.

7. The bond lattice of a graph is isomorphic to the product of the bond lattices of its connected components.

9. The product of chains of length $i_1, i_2, \ldots, i_k$ is isomorphic to the lattice of multisubsets of the multiset chosen from $\{1, 2, \ldots, k\}$ in which element j has multiplicity i_j. The ordered set is isomorphic to the factors of $n = p_i^{i_1} \cdot p_2^{i_2} \cdot \ldots \cdot p_k^{i_k}$ ordered by "is a factor of" if $p_1, p_2, \ldots, p_k$ are distinct primes.

11. If m and n have no factors in common, then the lattice of factors of mn is the product of the lattice of factors of m and the lattice of factors of n. In particular, if $p_1, p_2, \ldots, p_k$ are distinct primes, then the lattice of factors of $p_i^{i_1} \cdot p_2^{i_2} \cdot \ldots \cdot p_k^{i_k}$ is the product of the lattices of factors of $p_j^{i_j}$ for $j = 1, 2, \ldots, k$.

13. $\mu(X, Y) = 0$ unless each element of Y has multiplicity equal to zero more or one more than its multiplicity in X. If $\mu(X, Y) \neq 0$, then $\mu(X, Y) = (-1)^k$, where k is the number of elements of Y with multiplicity one more than their multiplicity in X.

15. The 1 element and $(n-1)$-element subsets of an n-element set.

17. In linear extension i, put the set $\{i\}$ above the set $N - \{i\}$ and order the other elements in any way that gives a linear extension. Such a set of n linear extensions will have the given partial ordering as its intersection. Now note that each set $\{i\}$ must be above $N - \{i\}$ in some linear extension, and when $\{i\}$ is above $N - \{i\}$, we cannot also have $\{j\}$ above $N - \{j\}$ (why?). Thus the partial ordering has dimension n.

18. We want to show that there is an order-preserving one-to-one correspondence between flats F of M and ordered pairs (F_1, F_2) of flats F_1 of M_1 and F_2 of M_2. If $F = F(I)$, we know that $I = I_1 \cup I_2$ with $I_1 \in M_1$ and $I_2 \in M_2$. Thus it seems F should correspond to $(F(I_1), F(I_2))$. For this to make sense, we must know that if $F(I) = F(I')$ and $I' = I_1' \cup I_2'$, then $F(I_1) = F(I_1')$ and $F(I_2) = F(I_2')$. But if M_1 is a matroid on X_1, then $I \cap X = I_1$, $I' \cap X = I_1'$. Now $F(I_1)$ must be $F(I) \cap X$ and $F(I_1')$ must also be $F(I) \cap X$. (That requires proof.) Thus $F(I_1) = F(I_1')$ and similarly, $F(I_2) = F(I_2')$. Because $F_1 = F \cap X_1$, it is clear that $F \subseteq G$ if and only if $(F_1, F_2) \leq (G_1, G_2)$. Because $X = X_1 \cup X_2$ and $X_1 \cap X_2 = \emptyset$, it is clear that if $(F_1, F_2) = (G_1, G_2)$ then $F_1 = G_1$ and $F_2 = G_2$. Since each pair of independent sets I_1, I_2 leads to an independent set $I_1 \cup I_2$ of X, the map from F to (F_1, F_2) is onto. Thus it is an isomorphism.

19. When P and Q are reflexive, we say that $(x_1, y_1) \leq (x_2, y_2)$ if and only if $x_1 \leq x_2 \pmod{P}$ and $y_1 \leq y_2 \pmod{Q}$, that is, if and only if $(x_1, x_2) \in P$ and $(y_1, y_2) \in Q$. Thus the pair of pairs $((x_1, y_1), (x_2, y_2))$ is in the product ordering if and only if the pair of pairs $((x_1, x_2), (y_1, y_2))$ is in $P \times Q$. Thus, except for the order in which the symbols appear, the reflexive product ordering is $P \times Q$. However, when P and Q are irreflexive, we would say that $(x_1, y_1) < (x_2, y_2)$ if $x_1 < y_1$ and $y_1 = y_2$. However, pairs in the product of irreflexive relations would correspond only to saying that $x_1 < y_1$ and $y_1 < y_2$. Thus in the irreflexive case, the product $P \times Q$ is not equal to the product ordering, even if we reorder the symbols.

CHAPTER 8

Section 8.1, pages 542–544

1. $\begin{pmatrix} 1 & 2 & 3 \\ 3 & 1 & 2 \end{pmatrix}$ $\begin{pmatrix} 1 & 2 & 3 \\ 2 & 3 & 1 \end{pmatrix}$ $\begin{pmatrix} 1 & 2 & 3 \\ 1 & 2 & 3 \end{pmatrix}$

3. $\begin{pmatrix} 1 & 2 & 3 \\ 3 & 1 & 2 \end{pmatrix}$, $\begin{pmatrix} 1 & 2 & 3 \\ 2 & 1 & 3 \end{pmatrix}$ (Any rotation and any flip will work.)

5. $\begin{pmatrix} 1 & 2 & 3 & 4 \\ 2 & 1 & 3 & 4 \end{pmatrix}$ is not in the group, nor is any other permutation which interchanges two vertices and fixes two vertices.

7. If $m = 0$, then $\sigma^m \sigma^{-1} = \sigma^0 \sigma^{-1} = \iota \sigma^{-1} = \sigma^{-1} = \sigma^{0-1}$. Assuming inductively that $\sigma^{m-1} \sigma^{-1} = \sigma^{m-2}$, multiply on the left by σ to get $\sigma^m \sigma^{-1}$, and on the right-hand side you get $\sigma \sigma^{m-2} = \sigma^{m-2+1} = \sigma^{m-1}$.

9. We give the inductive step of the proof for $n \geq 0$. $(\sigma^m)^n = (\sigma^m)^{(n-1)+1} = (\sigma^m)^{n-1} \cdot (\sigma^m)^1 = \sigma^{m(n-1)} \cdot \sigma^m = \sigma^{mn-m+m} = \sigma^{mn}$. For $n < 0$ we simply apply the fact that $(\sigma^m)^{-n} = ((\sigma^m)^{-1})^n = ((\sigma^{-1})^m)^n$.

10.

	ι	ρ	ρ^2
ι	ι	ρ	ρ^2
ρ	ρ	ρ^2	ι
ρ^2	ρ^2	ι	ρ

11.

	ι	ρ	ρ^2	φ_1	φ_2	φ_3
ι	ι	ρ	ρ^2	φ_1	φ_2	φ_3
ρ	ρ	ρ^2	ι	φ_3	φ_1	φ_2
ρ^2	ρ^2	ι	ρ	φ_2	φ_3	φ_1
φ_1	φ_1	φ_2	φ_3	ι	ρ^2	ρ
φ_2	φ_2	φ_3	φ_1	ρ	ι	ρ^2
φ_3	φ_3	φ_1	φ_2	ρ^2	ρ	ι

Note that $\varphi_1 \rho = \varphi_2$ but $\rho \varphi_1 = \varphi_3$.

13. $(\sigma\tau)\tau^{-1}\sigma^{-1} = \sigma(\tau\tau^{-1})\sigma^{-1} = \iota$ and similarly, $(\tau^{-1}\sigma^{-1})(\sigma\tau) = \iota$. These are the defining equations that tell us $\sigma\tau$ and $\tau^{-1}\sigma^{-1}$ are inverses of each other, and by Exercise 12, no elements but inverses of each other may satisfy these equations.

15. The rotations $\{\iota\ \rho, \rho^2, \rho^4, \rho^5\}$ form a five-element subgroup because they are closed under composition.

16. The rotations $\{\iota\ \rho, \rho^2\}$ form a three-element subgroup because they are closed under composition. Any flip and the identity form a subgroup because they are closed under composition.

17. No. If it has one rotation, it has them all; if it has two flips, it has all three flips plus the identity.

19. Yes, it must also have the 180 degree rotation.

21. False; take $\sigma = \varphi_1$, $\tau = \rho$, $n = 2$ in the group S_3 whose table is given in the solution to Exercise 11. Then $\sigma^2 \tau^2 = \iota \rho^2 = \rho^2$, but $\sigma \tau \sigma \tau = \varphi_1 \rho \varphi_1 \rho = \varphi_2 \varphi_2 = \iota$.

23. The element 0 is an identity for $\circ$, and $a \circ (-a) = \sqrt[3]{a^3 - a^3} = 0$. The set is closed under the operation because $a \circ b$ is always a cube root of an integer. For associativity,

$$(a \circ b) \circ c = \sqrt[3]{\left(\sqrt[3]{(a^3 + b^3)}\right)^3 + c^3} = \sqrt[3]{a^3 + b^3 + c^3} = a \circ (b \circ c).$$

25. Since S is closed under the operation and the operation is associative for all triples of elements in G, it is associative for all triples of elements in S. If $\sigma \in S$, then the sequence $\sigma^0, \sigma, \sigma^2 \ldots \sigma^n, \ldots$ has finitely many values. Therefore, $\sigma^i = \sigma^j$ for some $i > j$. Then $\sigma^{i-j} = \sigma^i \sigma^{-j} = \sigma^i (\sigma^i)^{-1} = \iota$, so ι is in S. This means that S has an identity element. Also $\sigma^{-1} = \sigma^0 \sigma^{-1} = \sigma^{i-j-1}$, and therefore each σ in S has an inverse in S. Thus S is a group, and since its operation is the same as that of G, S is a subgroup of G.

26. Consider the set of integers with the operation of addition. It is a group, but neither the positive integers nor the negative integers form a subgroup (no inverses). However, if G is a group and S is a subset of G such that whenever σ and τ are in S, then so is $\sigma\tau^{-1}$, then also $\sigma(t^{-1})^{-1} = \sigma\tau$ is in S for each σ and τ in S. Also, there is at least one σ in S since it is nonempty, and $\sigma\sigma^{-1}$ is in S so ι is in S. Finally, for each σ in S, $\iota\sigma^{-1}$ is in S, so σ^{-1} is in S. Thus S has an associative operation which is the same as the operation in G, and this operation satisfies the identity and inverse properties, so S is a subgroup of G.

Section 8.2, pages 557–561

1. $\{S_{12}, S_{23}, S_{34}, S_{45}, S_{56}, S_{61}\}$ $\{S_{13}, S_{24}, S_{35}, S_{46}, S_{51}, S_{62}\}$ $\{S_{14}, S_{25}, S_{36}\}$

3. There are rotations through 0, 60, 120, 180, 240, 300, and 360 degrees, flips about the bisectors of edges $\{1,2\}$, $\{2,3\}$, and $\{3,4\}$, and flips around the diagonals from 1 to 4, from 2 to 5, and from 3 to 6.

5. The identity and the flip about the diagonal from 1 to 3.

7. $\begin{pmatrix} 1 & 2 & 3 & 4 \\ 3 & 2 & 1 & 4 \end{pmatrix}$, $\begin{pmatrix} 1 & 2 & 3 & 4 \\ 2 & 1 & 4 & 3 \end{pmatrix}$, $\begin{pmatrix} 1 & 2 & 3 & 4 \\ 1 & 4 & 3 & 2 \end{pmatrix}$, $\begin{pmatrix} 1 & 2 & 3 & 4 \\ 4 & 3 & 2 & 1 \end{pmatrix}$.

9. (a) ρ^0, ρ^4, ρ^8 **(b)** $\{\rho^0, \rho^4, \rho^8\}$, $\{\rho, \rho^5, \rho^9\}$, $\{\rho^2, \rho^6, \rho^{10}\}$, $\{\rho^3, \rho^7, \rho^{11}\}$

11. $(\varphi_{13} \circ \rho)(2) = \varphi_{13}(\rho(2)) = \varphi_{13}(3) = 3;$ $\rho^3 \circ \varphi_{13}(2) = \rho^3(4) = 3.$
$(\varphi_{13} \circ \rho)(3) = \varphi_{13}(\rho(3)) = \varphi_{13}(4) = 2;$ $\rho^3 \circ \varphi_{13}(3) = \rho^3(3) = 2.$
$(\varphi_{13} \circ \rho)(4) = \varphi_{13}(\rho(4)) = \varphi_{13}(1) = 1;$ $\rho^3 \circ \varphi_{13}(4) = \rho^3(2) = 1.$

13. The digraphs of (a) and (b) are cycles; those of c and d are disjoint unions of two cycles.

15. In each case, the directed cycles of the graph we give are the cycles we have given for the permutation. **(a)** See Figure A.25

Figure A.25

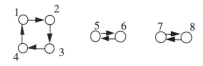

Figure A.26

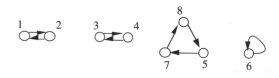

Figure A.27

Figure A.28

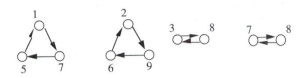

(b) See Figure A.26   **(c)** See Figure A.27   **(d)** See Figure A.28

17. If ρ is a rotation and φ is a flip (any flip), then $\varphi\rho = \rho^5\varphi$. From this you can show that the elements of the form $\varphi^i\rho^j$ are closed under multiplication, so they form a subgroup. Finally, elements of the group of the form $\varphi^i\rho^j$ form a 12-element set.

19. **(a)** If σ_1 and σ_2 fix all x in X, then $\sigma_1\sigma_2(x) = \sigma_1(x) = x$ and so the elements fixing all x in X form a subgroup.

(b) First, ι is in the subgroup. If σ is in the subgroup, then $\beta_\sigma(x) = x = \beta_\iota(x)$ for all x and therefore $\beta_\sigma = \beta_\iota$. Thus the group acts faithfully if and only if no $\sigma \neq i$ is in the subgroup.

(c) If a member of the group fixes the edges $\{x, y\}$ and $\{y, z\}$, then it must send y to y. Thus if σ fixes all edges, it fixes all vertices of the polygon. Therefore no σ other than the identity is in the subgroup fixing all the edges and so by part (b) the group acts faithfully.

(d) The map from σ to β_σ is a bijection if G acts faithfully.

(e) The edges are taken to edges; two adjacent edges are taken to two adjacent edges, but then the images of the remaining vertices and edges are determined. (Induct on the number of vertices if you want to prove this formally.)

(f) Two adjacent edges may be taken in two ways to each of n pairs of adjacent edges.

21. We know σ can be written in at least one way as a product of cycles by Theorem 2.9. Now suppose $\gamma_1, \gamma_2, \ldots, \gamma_n$ are cyclic permutations defined on disjoint sets, and $\sigma = \gamma_1\gamma_2\cdots\gamma_n$. Let i be in γ_j. Then i is not in γ_k for $k \neq j$, so $\gamma_k(i) = i$. Therefore $\gamma_j(i) = \gamma_1\gamma_2\cdots\gamma_n(i) = \sigma(i)$. Thus $\gamma_j(i) = \sigma(i)$ for every i in γ_j. Furthermore, if γ_j has r elements, then $(i, \sigma(i), \sigma^2(i), \ldots, \sigma^{r-1}(i))$ is a cycle of σ because $\sigma^r(i) = \iota$. Thus γ_j is a cycle of σ. Therefore the only way to write σ as a product of disjoint cycles is the way shown in Theorem 2.9.

23. Two rotations commute. However, if the (nontrivial) cyclic permutations permute different sets (neither of which is a subset of the other), then commutation is impossible. [If γ_1 and γ_2 are nontrivial cycles with i in γ_1 and γ_2 but $\gamma_2(i)$ not in γ_1, then $\gamma_1\gamma_2(i) = \gamma_2(i)$, while $\gamma_2\gamma_1(i) = \gamma_2(\gamma_1(i))$ and $\gamma_1(i)$ is not i.]

25. Note that each row of $M(\sigma)$ has exactly one nonzero entry. Thus each entry of $M(\sigma)M(\tau)$ must be either 0 or 1. Since $(M(\sigma)M(t))_{ij} = \sum_{k=1}^n M(\sigma)_{ik}M(t)_{kj}$, this entry will be 1 if and only if there is a k such that $M(\sigma)_{ik} = 1$ and $M(t)_{kj} = 1$; that is, if $t(j) = k$ and $\sigma(k) = i$. Further, the entry will be 0 if and only if there is no such k. Thus $M(\sigma)m(\tau)$ will be 1 in the (i, j) position exactly when $\sigma\tau(j) = \sigma(k) = i$, and will be zero otherwise, so $M(\sigma)M(t) = M(\sigma\tau)$.

27. If $\det M(\sigma) = 1$ and $\det M(\tau) = 1$, then $\det M(\sigma\tau) = \det M(\sigma)M(\tau) = 1 \cdot 1 = 1$. If τ is a permutation of sign -1, then $\tau\sigma$ also has sign -1 (by the product property for determinants) for each σ in the subgroup of permutations of sign 1. Further, if ω has sign -1, then $t^{-1}\omega$ has sign 1 (since the sign of t^{-1} must be -1) so ω is in the left coset generated by τ. The order of the subgroup is $n!/2$.

29. Choose a vertex, say vertex i. It can be sent to any of the six vertices. Then one line emanating from vertex i, say edge $\{i, k\}$, must be sent to one of the four lines emanating from vertex j, say edge $\{j, h\}$. But once we have chosen this line, we have determined where the two adjacent faces of edge $\{i, k\}$ must go and therefore where all vertices of the octahedron must go. Thus the group has $6 \cdot 4 = 24$ permutations.

30. Let $s'(m, k)$ be the number of permutations with exactly k cycles (including cycles of one element). Then either m is in a cycle by itself (there are $s'(m - 1, k - 1)$ such permutations) or it follows one of the other $m - 1$ elements in a cycle (since putting it at the beginning of a cycle is the same as putting at the end). Since there are $(m - 1)s'(m - 1, k)$ such permutations, $s'(m, k) = s'(m - 1, k - 1) + m - 1s'(m - 1, k)$. Using the similar formula from Theorem 3.7 of Chapter 2, you can prove (by induction, for example) that $s'(m, k) = |s(m, k)|$.

31. Yes, but you cannot appeal to Theorem 1.3 to prove the H is a subgroup; you must prove this directly or from Exercise 26 of Section 1 of this chapter.

Section 8.3, pages 578–580

1. As in Example 3.1, all six edges are fixed by the identity and two are fixed by ρ^2. Each flip fixes exactly two edges, so the number of orbits is $\frac{1}{8}(6+2+4\cdot2) = 2$.

3. Since we must use both colors, we cannot use all red beads or all blue beads. We apply the CFB theeorem to lists of six $R's$ and $B's$ under the action of the rotation group. All $2^6 - 2$ lists are fixed by the identity, none are fixed by ρ or ρ^5 (since each would equal the next element in a list fixed by ρ), two lists (alternating R and B) are fixed by ρ^2 or ρ^4, and six (put any distinct colors you want in the first three places) are fixed by ρ^3. Thus there are $\frac{1}{6}(2^6 - 2 + 2 + 2 + 6) = \frac{1}{6} \cdot 72 = 12$ ways to place the beads.

5. (a) There are $\binom{4}{2} = 6$ ways to place the beads and these are all fixed by the identity. For each edge we can rotate the tetrahedron 180 degrees around an axis through the center of the edge, and we have two ways of placing the beads that will be fixed by such a rotation. However, each of these axes of rotation passes through the center of two edges and thus there are $3\cdot2$ arrangements fixed by these motions. All other motions are rotations around an axis through the center of a face, moving three vertices, and thus fixing none of the arrangements. Therefore there are $\frac{1}{12}(6+6) = 1$ equivalence classes of arrangements.

(b) All $\binom{6}{2} = 15$ ways of coloring the edges are fixed by the identity. No ways of placing two reds and four blues is fixed by any element of order 3. A rotation through the bisectors of two sides breaks into three 2-cycles, any of which could be red. Thus each such rotation fixes three configurations. Since there are three such rotations, there are nine fixed configurations. Thus by the CFB theorem, we have two orbits. The two colors could be on either adjacent edges or opposite edges.

7. All paint jobs are fixed by the identity; there are $3 \cdot 6 \cdot 2$ such paint jobs. None are fixed by ρ or ρ^3 (since each color should equal the adjacent color) and none are fixed by ρ^2 (since each color should equal the opposite one). For each flip about a side bisector, we may paint the sides touching the axis with different colors in $3 \cdot 2$ ways, but a flip about an axis between vertices will fix no paint jobs. Thus we have $\frac{1}{8}(3 \cdot 6 \cdot 2 + 2 \cdot 3 \cdot 2) = 6$ equivalence classes of paint jobs.

9. $\frac{1}{6}(Z_1^6 + 2Z_6 + 2Z_3^2 + Z_2^3)$ **11.** $\frac{1}{12}(Z_1^4 + 8Z_3^1 Z_1^1 + 3Z_2^2)$.

13. $\frac{1}{24}(Z_1^6 + 6Z_4 Z_1^2 + 3Z_2^2 Z_1^2 + 8Z_3^2 + 6Z_2^3)$

15. Substitute $R^i + B^i + G^i$ for Z_i; then 1 for R, B, and G in the cycle index of Exercise 9, giving $\frac{1}{6}(3^6 + 2 \cdot 3 + 2 \cdot 9 + 27) = 130$.

17. In $\frac{1}{12}[(R+B)^6 + 4(R^2+B^2)^3 + 3(R^2+B^2)(R+B)^2 + 2(R^3+B^3)^2 + 2(R^6+B^6)]$ the coefficient of $R^2 B^4$ is $\frac{1}{12}(\binom{6}{4} + 4\binom{3}{1} + 3 \cdot 2 + 3) = 3$.

19. $\frac{1}{24}((R+B+G)^8 + 6(R^4+B^4+G^4)^2 + 9(R^2+B^2+G^2)^4 + 8(R^3+B^3+G^3)^2)$; $\frac{1}{24}(\binom{8}{4,2,2} + 9\binom{4}{2,1,1}) = 22$.

21. $\frac{1}{24}((x+y)^6 + 2(x^6+y^6) + 2(x^3+y^3)^2 + 4(x^2+y^2)^3 + 12(x+y)^2(x^2+y^2)^2 + 3(x+y)^4(x^2+y^2))$

23. Suppose inductively that $E(T_1 \times T_2 \times \cdots \times T_{n-1}) = E(T_1)E(T_2) \cdots E(T_{n-1})$. Now consider an n-tuple $(x_1, x_2, \ldots, x_n)$ as an ordered pair

$$((x_1, x_2, \ldots, x_{n-1}), x_n)$$

and apply Theorem 3.2. Thus

$$E[(T_1 \times T_2 \times \cdots \times T_{n-1}) \times T_n] = E(T_1)E(T_2) \ldots E(T_n),$$

and since $(T_1 \times T_2 \times \cdots \times T_{n-1}) \times T_n$ and $T_1 \times T_2 \times \cdots \times T_n$ will have the same enumerators, the principle of mathematical induction completes the proof.

INDEX

ISBN 0-12-110830-9

90182 >

9 780121 108304